Carbohydrate Chemistry

Fundamentals and Applications

Carbohydrate Chemistry

Fundamentals and Applications

Raimo Alén

University of Jyväskylä, Finland

NEW JERSEY • LONDON • SINGAPORE • BEIJING • SHANGHAI • HONG KONG • TAIPEI • CHENNAI • TOKYO

Published by

World Scientific Publishing Co. Pte. Ltd.
5 Toh Tuck Link, Singapore 596224
USA office: 27 Warren Street, Suite 401-402, Hackensack, NJ 07601
UK office: 57 Shelton Street, Covent Garden, London WC2H 9HE

Library of Congress Cataloging-in-Publication Data
Names: Alén, Raimo, 1951– author.
Title: Carbohydrate chemistry : fundamentals and applications / by Raimo Alén
(University of Jyväskylä, Finland).
Description: New Jersey : World Scientific Publishing Company, 2018. |
Includes bibliographical references and index.
Identifiers: LCCN 2018002052| ISBN 9789813223639 (hardcover : alk. paper) |
ISBN 9789813223646 (softcover : alk. paper)
Subjects: LCSH: Carbohydrates. | Carbohydrates--History. | Chemistry,
Organic. | Chemistry.
Classification: LCC QD321 .A44 2018 | DDC 547/.78--dc23
LC record available at https://lccn.loc.gov/2018002052

British Library Cataloguing-in-Publication Data
A catalogue record for this book is available from the British Library.

For any available supplementary material, please visit
http://www.worldscientific.com/worldscibooks/10.1142/10536#t=suppl

Typeset by Stallion Press
Email: enquiries@stallionpress.com

Printed in Singapore

Preface

Carbohydrates, which exist widely in nature, have always been important substances for humans. Carbohydrate chemistry has also been an integral part of organic chemistry for almost two centuries. However, from its early times dramatic changes have occurred in the field of carbohydrate chemistry, especially due to the realization that carbohydrates are also heavily involved in many biologically active molecules and that petroleum based industrial chemistry is gradually entering a new "green chemistry" era based on the more effective utilization of carbohydrates-rich renewable resources. In its broadest sense, and according to the traditional conception, carbohydrate chemistry is a comprehensive discipline, ranging from fundamental studies to practical applications. There are a great number of specialized books available that concern the detailed reactions of carbohydrates or address specific topics within this expanded field. On the other hand, there is undoubtedly a clear need for a condensed textbook that essentially covers the basic principles of carbohydrate chemistry as well as its versatile applications without delving too deeply into theoretical details. This book is intended to meet this demand. It can be primarily used as an advanced textbook for a wide range of readers in many disciplines. Its utility not only extends to students and teachers, but also researchers who work on production and technical planning, or everyone who needs a reliable resource on carbohydrates.

For this book, based on the original Finnish manuscript and several many-sided lectures given by the author over the past 30 years, 12 chapters were selected. It is hoped that the book will help readers to understand basic carbohydrate chemistry and to be able to realize the huge industrial potential of various carbohydrates-containing resources. The purpose of the first two chapters (1 and 2) is to give some background data and fundamental information, especially on the historical perspective within the utilization of carbohydrates and the general development of carbohydrate chemistry. The next five chapters (3 to 7) are devoted to the fundamental chemistry and chemical phenomena of carbohydrates including their relevant nomenclature. Chapter 8 deals with the biosynthesis of carbohydrates, and in this chapter, special emphasis is given to the biosynthesis of cellulose and hemicelluloses. In the two next chapters (9 and 10), some typical examples of carbohydrates and their derivatives as well as natural carbohydrate residues-containing substance groups are given. Finally, the book ends with chapters 11 and 12, which deal with common reactions of carbohydrates and provide readers with insight into various chemical, biochemical, and thermochemical conversion methods of lignocellulosic biomass for producing value-added chemicals, biofuels, and other products.

Raimo Alén
Laboratory of Applied Chemistry
Department of Chemistry
University of Jyväskylä
Jyväskylä, Finland

Contents

1. Introduction

Carbohydrates are naturally present in plants (70–75% of dry matter) and animals (5–10% of dry matter). They are mainly formed in green plants in a multitude of biosynthetic pathways from the products of photosynthetic reactions; their structures range from various monomers to giant polymer molecules. Carbohydrates serve as basic energy sources (sugars) or reserve fuels (starch and glycogen), but they also form supporting structures in plants and certain animals (cellulose and chitin). Carbohydrates and their derivatives are also found as structural members in numerous biologically important substances, such as *nucleic acids* (DNA and RNA).

The name "carbohydrate" was initially connected with compounds that could be seen as various hydrates of carbon and whose elemental composition followed the experimentally found formula $C_n(H_2O)_m$ ($n > 2$). Common compounds with a sweet taste, such as xylose $C_5(H_2O)_5$, glucose $C_6(H_2O)_6$, and sucrose $C_{12}(H_2O)_{11}$, were typically counted as carbohydrates. Nowadays, we know many carbohydrates — for example, deoxyribose $C_5H_{10}O_4$ — that do not follow this definition. On the other hand, enormous numbers of compounds exist that could be considered carbohydrates based on their elemental composition, but their chemical behavior is completely different. This shows that the traditional definition of carbohydrates is too narrow as the only criterion; the name is also misleading because they do not contain water molecules in their structures. In a broader context, these compounds can instead be seen as polyhydroxy compounds ($R\text{-}(OH)_n$) that either contain, or whose acid hydrolysis products contain, an aldehyde group (R-CHO, in *aldoses*) or a ketone group (R-CO-R′, in *ketoses*).

In practice, carbohydrates are mostly present as various ring structures where the functional hydroxyl and carbonyl groups participate in the formation of these cyclic structures. Among such structures are aldehyde-based *hemiacetals* (cyclic hemiacetals) and their derivatives, *acetals,* and the corresponding ketone-based *hemiketals* (cyclic hemiketals), and their derivatives, *ketals*. In aldoses, R is H and in ketoses, R is generally CH_2OH; the so-called "aglycone" part R' is normally an alkyl group — the carbohydrate part is then called a "glycone":

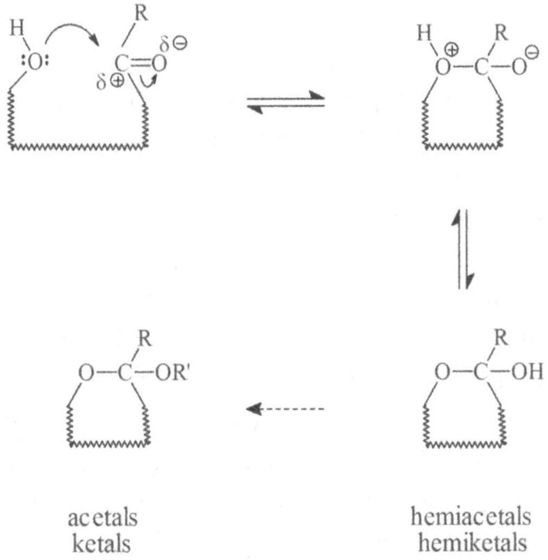

acetals
ketals

hemiacetals
hemiketals

It is rather difficult to give a simple characterization of the carbohydrate group of compounds because the traditional definition mentioned above only applies to "basic" carbohydrates but not to their many derivatives. Carbohydrates found in nature often contain other functional groups, for example, in compounds, such as *sugar phosphates*, *sugar sulfates*, *amino sugars*, and *thiosugars*. In addition, there are cases, such as well-known *sugar alcohols* (e.g., xylitol, mannitol, and glucitol or sorbitol), where the carbonyl group has been reduced to the corresponding alcohol. The formed *polyhydroxy-alkanes* have properties similar to carbohydrates, and they are commonly considered as carbohydrate derivatives. Similarly, *polyhydroxycycloalkanes* (e.g., inositols) resemble carbohydrates and are thus seen to belong to them.

There are also numerous *pectins* in nature, such as *polyuronides*, consisting of oxidized monosaccharide units, *uronic acids*. Because of the importance of carbohydrates in a variety of biochemical reactions, they are structural constituents, besides in the nucleic acids mentioned above, also in *glycoproteins* and *glycolipids* (so-called "glycoconjugates"). These conjugate molecules are no longer deemed pure carbohydrate derivatives; their classification among the compound groups of organic chemistry is primarily based on the parts of the molecule other than the carbohydrate component.

Carbohydrates are divided into three subgroups on the basis of their structure: (i) *monosaccharides*, (ii) *oligosaccharides*, and (iii) *polysaccharides* (as a group generally called "glycans"). These prefixes come from Greek, where "mono" means "one", "oligo" "a few", and "poly" "many". Monosaccharides are simple sugars, like ribose, xylose, glucose, and mannose, which cannot be hydrolyzed into smaller "carbohydrate units". Oligosaccharides comprise a few (two to nine) monosaccharide units linked together with *glycosidic bonds* formed by splitting water molecules. They are divided according to their number of monosaccharide units into *disaccharides*, such as lactose, maltose, sucrose, and cellobiose, *trisaccharides*, such as raffinose, and *tetrasaccharides*, such as stachyose, and so on. Except disaccharides, only a few oligosaccharides are found in nature. On the contrary, an abundance of natural polysaccharides are found: cellulose, starch, and hemicelluloses, such as xylans and glucomannans, contain monosaccharide units (their number can vary from ten to thousands) linked together with glycosidic bonds. In acid hydrolysis, the oligo- and polysaccharides can yield the same or dissimilar monosaccharides. If all these monomeric building blocks are identical, we speak about *homopolysaccharides*, such as cellulose and starch; otherwise we are dealing with *heteropolysaccharides*, such as hemicelluloses.

It is also possible to classify carbohydrates according to their solubility in water. Besides readily water-soluble carbohydrates (sugars), there are polymeric carbohydrates that only swell in water and form colloidal solutions (e.g., starches). Cellulose and other related polysaccharides are insoluble in water. Sugars are sweet and often exist as crystals, although they can also form syrups that are difficult to crystallize. Because of their many intermolecular hydrogen bonds, sugar crystals have generally a higher density and strength than the crystals of other organic compounds. This is also why sugars do not markedly dissolve in solvents, such as diethyl ether, chloroform, or toluene, which lack hydroxyl groups needed to form hydrogen bonds.

Carbohydrate structures often contain numerous *asymmetric carbon atoms* (called "chiral centers") whose four substituents are all different. Because of these chiral centers, such carbohydrates are often optically active and rotate the plane of plane-polarized (linearly polarized) light either to the right ((+)-form) or to the left ((−)-form). This optical activity arises from the three-dimensional structure of the compound and leads to the central importance of stereochemistry, a branch of organic chemistry that examines, for example, the spatial structure of carbohydrates. It can be claimed that a complete understanding of carbohydrate chemistry is not possible without a profound knowledge of the stereochemistry of either the simple molecules or the macromolecules examined.

In the context of certain carbohydrates, the term "reducing sugars" is used; this refers to the ability of their free aldehyde group (being present as hemiacetal) to reduce, for example, Fehling's solution. This property has been traditionally used for non-specific detection of sugars. In certain cases, such as sucrose which belongs to disaccharides and consists of two monosaccharide moieties, glucose and fructose, the hydroxyl group in the hemiacetal structure of glucose (an anomeric hydroxyl group) and the corresponding hydroxyl group in fructose's hemiketal structure form a mutual glycosidic bond. In this case, the above reduction process is no longer possible; hence, this carbohydrate belongs to "non-reducing sugars". The reducing end group in polysaccharides is of great importance in the carbohydrate degradation reactions that reduce the total cooking yield in alkaline pulping (cf., Chapt. "11.2. Polysaccharides").

Many small-molecule carbohydrates (the most common mono- and disaccharides) are widely used, among others, in the food and pharmaceutical industries. Similarly, starch and starch syrups are traditional commodities in the food industry. Starch is also employed, after modification (oxidation or heating with a catalyst), as a surface sizing agent and a strength additive of chemical pulps (containing mainly cellulose) in the paper industry. In addition to chemical pulps, smaller amounts of so-called "dissolving pulps" are produced; the goal in their cooking and bleaching processes is to maximize the cellulose content of the product. These pulps are suitable for the production of a variety of cellulose derivatives and, for example, viscose fibers. In addition to cellulose and starch, there are other important polysaccharides (such as hemicelluloses, pectins, and "biogums"), which will become more important as renewable resources for new processes. In general, carbohydrates have a wide spectrum of applications, and several of their products already replace a part of similar chemicals manufactured in the petrochemical industry. On the other hand, carbohydrates form a unique polyfunctional group of materials that can be converted with synthetic chemistry techniques to respond to the growing needs of the future commercial life and chemical industry. It should be emphasized that cellulose is the most common organic polymer on the planet; thus, its structural part, glucose, is the most common organic molecule in nature.

Carbohydrate chemistry is today an established branch of organic chemistry. The main reason for this special position is the hydrophilic nature of the carbohydrates, which causes them to be generally insoluble in organic solvents so that their handling often requires "uncommon conditions". In addition, sugars in particular are often seen as compounds that decompose easily and have a strong tendency to exist as mixed syrups that are difficult to investigate. The separation, purification, and structural study of unknown natural polysaccharides require special procedures that do not necessarily interest chemists outside this discipline. Another basic reason for the unique position of carbohydrate chemistry is the multiplicity of reactions that take place when starting from given initial compounds in the form of varying ring structures and the enhanced need for stereochemical evaluation in examining such reactions. Another special feature of carbohydrate chemistry is also the "unique system" of naming carbohydrates (about 3% of all organic compounds) that deviates in several aspects from the one used in other fields of organic chemistry. Clearly, our accumulated knowledge of carbohydrate chemistry is of fundamental importance, not only in the fields of organic and technical chemistry, but also for the understanding of biochemical processes and related areas, such as pharmacology.

2. Historical Background of Carbohydrate Utilization and Chemistry

2.1. The Era Before the 1800s

The instinctive or conscious use of carbohydrates must have begun in the early days of this planet's civilization. Foods containing carbohydrates were processed in many ways quite early; for example, in ancient Babylon and Egypt, local people made beer and wine by fermenting grain starch or grape sugar (glucose). Clay tablets found in Sumer (now Southern Iraq) dated about 4000 B.C. contain mentions of a cereal-based fermented drink named "sikaru". Moreover, besides starch, the use of sucrose and cellulose (see papermaking) for many purposes was known in early Far Eastern and Near Eastern cultures, from where this knowledge gradually spread to Europe. Developing uses for these "key material groups" until the 1800s will be discussed below; during the 1800s, carbohydrate chemistry gradually became a distinct branch of the emerging chemical knowledge.

Sucrose. The isolation of common sugar (sucrose or formerly known as saccharose) from the juice of sugarcane became an important milestone in the history of carbohydrates. The sugarcane plant probably originated from Northeastern India or the South Pacific; there is evidence from these areas (New Guinea ca. 10,000 B.C. and India ca. 6000 B.C.) of simple methods of making raw sugar. While it is still possible to see some wild sugarcane growing in India, the cultivation of this plant spread along with sugar use from India to China in 1800–1700 B.C. In the early days, sugar was separated for use as a yellowish syrup concentrate, which crystallized, on standing, into a brown mass of impure sucrose. In Sanskrit, the old cultural language in India, the word for sugar initially meant "brown sand", which well describes the appearance of crushed raw sugar. On the other hand, the word "saccharose" is derived from the Sanskrit word "sarkhara", which means "sweet", from which follows the Latin word "saccharum". Numerous Chinese texts from the 4th century B.C. describe in detail the process of concentrating the juice of sugarcane.

Arab traders brought the art of sugarcane cultivation to Mediterranean countries in ca. 1800 B.C. and developed methods for refining sugar mainly into products used to make early types of candy. Such candy was produced in large quantities, especially in Egypt in ca. 1000 B.C., where it was also utilized medicinally. The high nutritional value of sucrose was gradually recognized in many developing cultures. Pure, crystalline sucrose was finally described in India ca. 300 A.D., and its use extended along the caravan routes over North Africa into Spain. Importation of sucrose into Europe as an expensive sweetener increased during the 1300s and 1400s; many raw sugar refineries had already been established by the end of the 1400s. Since sugarcane could only be grown in tropical and subtropical climates, a search for alternate sugar-producing plants in temperate climates began in Europe. This led to selective breeding of the sugar beet in the late 1700s. Thereafter, the production of beet sugar as a substitute for cane sugar increased rapidly, especially while the Continental System (i.e., the British Navy blockade of Europe) established during the Napoleonic wars (1805–14) was in force. This meant the creation in Europe of a new and important branch of industry to utilize sources of carbohydrates.

Cellulose and paper. Many materials containing cellulose have long fulfilled human needs for dresses and housing. A widespread use of cellulose fiber as the main component in writing materials (such materials are collectively called "paper") became quite an early and significant application for cellulose fiber. The physical form of paper developed in China about 2000 years ago has been said to perhaps be humanity's most important invention. Paper has been made in many ways through the ages, and all the ancient methods are still being applied. On the other hand, machine production of paper began in the early 1800s, and its development has been so dramatic that the current products only remotely resemble the first paper materials. A further reason for this is the development of methods for producing large quantities of light and lignin-free chemical pulps from wood during the late 1800s. The rapid development of papermaking in the 1900s is among the decisive factors in the rise of our culture during that century — a rise that gave birth to the current information society.

According to the traditional view, paper comes from a filament product obtained by taking plant fibers that have been chemically or mechanically separated from each other. In the oldest method of making paper in China, a fixed amount of fiber suspension was poured into a mold partially submerged in water. The mold consisted of a wooden frame with a coarsely woven cloth stretched across. The water drained through the cloth, leaving a thin and uniform fiber web on top which could be separated and dried. Into this basic fiber material can be mixed a wide range of additives (nowadays, mineral fillers and pigments, functional chemicals, and process chemicals) for various purposes. Although the papermaking process has come a long way from its early days, and there have been many inventions that have affected the quality of paper products, paper making is still in essence a "rather simple process". In spite of this somewhat narrow view, it can be stated that the modern manufacture of paper embraces a wide range of technologies and fields of science, essential areas typically involve physical (i.e., surface and colloidal), organic, and inorganic chemistry, and even microbiology.

A short description of the development of coated papers offers a good example of the early origins of papermaking chemistry and paper converting: pigment-coated papers typically consist of a base paper covered by a mixture containing at least a binder and an inert pigment. An early Chinese document dated in 450 A.D. reported starch sizing and gypsum surface treatment to improve its properties. During the Tang Dynasty between 618 and 907 A.D., paper was coated with white mineral powders and wax, which filled the cavities between fibers to increase water repellency and smoothness for fine calligraphy. In addition, by the 8th century, Arab cultures modified paper with talc, gypsum, or chalk, which could also be mixed rice starch to coat the paper for increased whitness. During the Middle Ages (during the years 500–1500), paper was coated with white pigments to provide an appropriate surface for metal-point drawings. After this period, there are a lot of examples of applications using differently coated papers for a variety of purposes. Besides, the novel use of many chemicals, the 19th and 20th centuries also fostered the development of new coating application techniques. Modern coatings are composed not only of pigments and binders (i.e., a typical composition of 90–94% pigment and 6–10% binder), but also additives and water. The most important pigments include clay, calcium carbonate, and silicates, whereas the principal binders are either hydrophilic water-soluble colloids (e.g., starch and protein) or resin (or latexes) and resin emulsions in an aqueous medium (e.g., styrene-butadiene, polyacrylate, polyvinyl acetate, and polyvinyl alcohol resins). During the second half of the 1900s, large-scale papermaking became one of the key technologies with products that can be varied over a wide range.

Paper and its often multilayered stiffer variety, cardboard, are today among our most important and versatile products that serve us in all aspects of our lives. The distinction between paper and cardboard has been rather vague, especially in the past when products were named according to their purported use rather than their physical properties. Already in the 1800s, it was possible, in the chemical industry, to produce nearly pure cellulose for the use in the manufacture of cellulose derivatives with different properties. For example, in 1833, Henri Braconnot (1780–1855) was the first to describe commercially significant cellulose derivative, cellulose nitrate or nitrocellulose ("smokeless powder"), followed by Hilaire de Chardonnet (1839–1924) in 1899 with a product called "artificial silk" (rayon fiber — cf., Chapt. "9.4.2. Cellulose esters"), which caused a sensation at the Exposition Universelle of 1900 in Paris.

Cotton consists of nearly pure cellulose; it forms a protective capsule of fibers 10 cm to 40 cm long around the seeds of plants of the genus *Malvaceae* (the most important plant being *Gossypium hirsutum*). It is thought that their green seed pods, which contain an ample amount of sugar, were first used as food. There is archeological evidence that cotton was used in India to make cloth and yarn in ca. 3000 B.C., although this use may have started in India and elsewhere as early as

6000 B.C. By 1500 B.C., a center of cotton manufacture had arisen in India, from where the knowledge spread to Persia (Iran), China, and Japan. According to the writings of the Greek historian Herodotus (485–425 B.C.) and Roman Gaius Pliny the Elder or Gaius Plinius Secundus (23–79 A.D.), people of their time were familiar with cotton fibers. Christopher Columbus (1451–1506) noticed during his first journey to the Americas in 1492 that the native peoples of the New World were familiar with cotton. Later during the conquest of the Americas, cotton clothing was noted to be in common use in Mexico and Peru. The long time perspective of this use is evident in the fact that mummies found in Peruvian caves dated to 6000–5000 B.C. were covered with cotton strips; cloth samples from caves in Peruvian Andes have recently been found to be ca. 12,000 years old. About 8000-year-old remains of cotton seed capsules have been found in Mexico.

Flax, an annual fiber plant (*Linum usitatissimum*) was grown 4000 to 5000 years ago in Asia Minor (and in ancient Egypt), whence the Indo-Europeans brought the species to Europe. It is likely that flax was first grown for the useful oil that can be pressed from flax seeds, but gradually the art of preparing flax fibers into materials for cloth was learned. The oldest linen shrouds found in Egypt contained excerpts from the Book of the Dead, probably intended as guides for living beyond the grave; these are dated to the Sixth Dynasty of ancient Egypt (2345–2181 B.C.). There are numerous other plant-based varieties of fiber, such as manila, esparto, hemp, jute, and sisal, but their history will not be discussed in detail here.

Organized societies have used several different types of both disappearing (e.g., sand and leaves) and permanent surfaces (e.g., stone, clay tablets, bone objects, leather, fabrics, and walls of caves, as well as surfaces of tree bark, wood, and metals or their alloys) for writing. The most common, however, were papyrus and parchment.

Papyrus was made from veneers of the stem of a reed-type plant (a marsh grass) flourishing in the Nile River valley and delta in ancient times, and it was widely used to make rugs, boats, and sandals. This reed, *Cyperus papyrus*, belongs to the reed family (*Cyperaceae*); the ancient Egyptian word "pa-per-ah" means "pharaoh's own". Thin veneers, as wide as possible, were first soaked and then pounded into flat "basic sheets" that were then joined with either the plant's own "glue" or wheat starch. The final product was an even 20 m long (maximum width about 30 cm) strip that was strong and flexible enough to be easily rolled into a "book". The writing surface was treated with pumice to prevent the ink from spreading and to cause it to adsorb properly on the surface. Papyrus rolls have been found since the First Dynasty of ancient Egypt (3100–2900 B.C.); the most recent ones date from ca. 1000 A.D. Papyrus survives well in dry climates, like the desert, but it absorbs moisture in humid climates and becomes moldy. Especially, papyrus rolls that have not been exposed to oxygen or mold or those partially charred in fires have survived to the present time and tens of thousands of them can be seen in museums. An impressive, well preserved collection is in the library of Alexandria.

Structurally closer to present plywood than paper, papyrus was initially mainly produced in Egypt. From there, it spread — in spite of its "secret" method of preparation — as the preferred writing material of antiquity around the Mediterranean and nearby regions to the rest of Europe and Asia Minor. The world's first "real" plywood products were probably also made in Egypt ca. 3000 B.C. by attaching hand-sawn small pieces of veneer with albumin glue onto metal and ivory in the furniture and jewelry of pharaohs and dignitaries. In the excavations of the burial site of Tutankhamon (reign in the 1300s B.C.), one recovered crown was made of cedar wood covered with thin layers of ivory and ebony. The first definite mention of the preparation and use of veneer was found from the time of classical Rome in the writings of Gaius Pliny the Elder, where he describes the decorated table, which Egypt's Queen Cleopatra (69–30 B.C.) sent to Rome's first Caesar, Gaius Julius Caesar (102–44 B.C.).

While papyrus as such cannot fully be included within the concept of paper, the word "paper", for example, and its numerous varieties in European languages (starting from the Greek "papyrus" and Latin "papyrus") derive from its original name. During the period of Roman Caesars, the consumption of papyrus rolls was so large that their use had to be regulated. The production of papyrus decreased in the 800s and nearly ended in the 900s. The main reason for this was probably the introduction of a substitute, parchment, and gradually also actual "modern" paper. Moreover, the papyrus plant disappeared from large areas of Egypt at the same time. The Roman name for papyrus rolls was "charta" and the Greek one "byblos", from which came, for example, the English names "chart" and "bible". Although the latter initially meant books in general, it evolved into only meaning the "book of books", the Holy Bible.

From tree bark (in Latin "liber", from which comes, e.g., the word "library") prepared "bast fiber paper" has material properties similar to the papyrus of antiquity. Such paper was called "huun" in the Mayan culture and on it was printed, for example, the multicolored 45-page calendar dated from 900–1100 and is known as "Codex dresdenis". However, still older, from the 400s dated samples of ancient American paper have been found in Teotihuacan, the capital of Mexico in those days. The Aztecs developed the production of bast fiber paper from the bark of the fig tree into a substantial industry; in their culture, this writing material was known as "amatl", which corresponds to the amate paper of the current Otomi Indians curremtly living in Southern Mexico. Furthermore, in the South Pacific similar paper now known as "tapa paper" has been used since early times.

Parchment is a more expensive writing material than papyrus, and it is prepared from the skin of goats, sheep, or calves. It received its name from the city of Pergamon (currently in Turkey close to Bergama in the Mediterranean coastal region), where it is said to have already been made in 170 B.C. According to the historian Marcus Terentius Varro (116–27 B.C.), the Egyptians refused to sell papyrus to the King of Pergamon, Eumenes II (197–159 B.C.), who desired to build a library to rival the antique world's mightiest library in Alexandria — the city was founded by Alexander the Great (356–323 B.C.) in 331 B.C. Demetrius of Phalerum or Demetrius Phalereus (ca. 350–280 B.C.), a student of Aristotle (ca. 384–322 B.C.), founded this library in 295 B.C., and Kallimakhos (ca. 305–242 B.C.) served as its well-known librarian. It contained the essence of Hellenic knowledge in about 700,000 book rolls.

While historian Herodotos indicated that the skin of domestic animals was already used to write down text in ancient Greece, the project of the new library in Pergamon quickly started major production of parchment. Parchment (in Greek "pergamene") originally meant especially the material based on calf skin, which was later also known as "vellum" (in Old French, "vélin" and in Latin, "velum" means "cloth" or "sail" and in Latin, "vitulinum" means "made from calf"). The ability to write on both sides of it and to wipe off the written marks immediately, if needed, enhanced the success of parchment. However, we do not have a consensus about the size of this new library.

Parchment started clearly to replace papyrus as writing material in the 4th century when "codexes" resembling current books started appearing, mostly from monasteries. One reason for this preference was that the Christian Church authorities shunned "pagan" papyrus rolls. Even today in the Jewish tradition, "kosher", or from permitted animals made parchment, is the only material on which the Torah and other holy texts can be written. Even though the Middle Ages (500–1500) are considered the age of parchment in Europe, the use of this material can be seen to end in the 1400s along with the arrival of the art of printing and the need for paper. Many important government documents were, however, still written on parchment long afterwards honoring this tradition. The pergament or pergamine paper (parchment paper or bakery paper) nowadays is a modern product created by running sheets of chemical pulp through a bath of sulfuric acid. This treatment renders it resistant to grease and moisture, and the product is useful as wrapping paper (cf., "butter paper" with similar properties made by extended grinding).

Alexandria became the largest center of learning during this era. Its pride was the Museion, the Temple of Muses that was created as a counterweight for the Greek Hellenic civilization and was next to the finest library in ancient times described above. On the other hand, in Alexandria and elsewhere, there were several smaller libraries; thus, all the knowledge of those times was not concentrated in one large library complex. The so-called "old learning", known as "Egyptian art" or "occult art" ("khemeia", the origin of "chemistry"), prevailed in Egypt. This art involved in the beginning much chemical knowledge relating to the embalming of the dead, but it gradually included other Egyptian knowledge of chemistry, such as glass making, dyeing, and particularly metallurgy.

This chemical knowledge also became the basis for early alchemy, a word that was based on the Arabic word "al-khimiya" for the khemeia knowledge ("al" is the Arabic definite article). Alchemy originally meant the system of beliefs based on ancient secret knowledge about reaching eternal youth by means of the "philosophers" stone" or "elixir of life" (in Arabic, "al-iksir" means "dry, powdery substance") and also efforts to produce noble metals, primarily gold, from base metals by stroking them with that mystical miraculous stone. Generally, alchemy became a line of thought that existed outside of the Church and advocated the refinement of one's soul. Examples of this are current words, such as "alcohol", "alkali", "algebra", and "algorithm", all of Arabic origin.

The foundations of the current natural sciences were gradually born in the main library in Alexandria. Not only old literature was collected there; new material was produced, and many grand works of antique times were codified into their present form. According to the prevailing view, part of the library was destroyed in a fire when Alexander the Great stormed the city in 48–47 B.C. On the other hand, according to claims made during the Middle Ages to justify the Crusades (mainly in the 1090s and 1140s and at the end of the 13th century), Caliph Omar (died in 644) had ordered the library to be burned in 642 during the Arab conquest of Alexandria. It is said that the library was destroyed several times during later rebellions, but it is likely that Theophilus Alexandrius, Pope of Alexandria and Patriarch of all Africa destroyed this library and many other libraries in 391. This act was based on the edict by Caesar Theodosius I the Great (ca. 346–395) that designated Christianity as the only religion in the Roman Empire; this required, among other actions, the destruction of all pagan places of worship. The eradication of paganism continued and one of its notable victims was the female philosopher Hypatia (ca. 370–415); a zealous, Christian mob accused her of causing religious turmoil and cruelly assassinated her on the order of Patriarch Cyrillos. Hypatia was perhaps the most profound expert in the literature and science (philosophy, astronomy, and mathematics) of her time; she worked together with her father, the mathematician Theon Alexandricus (ca. 335–405), known as the last member of the Museion, around 380.

Ptolemaios III (284–222 B.C.) issued an edict that mandated any written material in the possession of merchants and travelers arriving in the city had to be handed over to scribes to be copied, which rapidly enlarged Alexandria's library collections. The originnal documents were placed in the library and, unfairly, well executed copies were returned to the original owners. It is clear that only a fraction of all this collected material has survived through the centuries to the present day — perhaps only one percent.

In Alexandria on the Mediterranean coast, a new library, the Bibliotheca Alexandrina, was opened in 2002 with a goal equally ambitious as that of its predecessor. Its objective is to acquire and preserve "all the world's book knowledge" in digital form. The initial event was the passionate goal of Suzanne Mubarak, the wife of the former President of Egypt, Hosni Mubarak. Brewster Kahle in San Francisco established its realistic foundation as "Electronic Alexandria", an enormous Internet-based archive of data. Their common objective was to preserve all recorded knowledge between San Francisco and Alexandria. The guiding principle was to provide "in the spirit of the old library" a worldwide right, as free as possible, to read and use the acquired material of knowledge and culture. In addition, a future goal was to give everyone the possibility to study the material in their own native language with the help of a translator.

In the Bible, the use of vinegar and tar (cf., the building of Noah's pinewood ark) is mentioned. Later, Caius Pliny the Elder collected in ca. 77 C.E. a tome "Naturalis Historia", which cited, for example, the testing of dyes and the use of tar water in the embalming of the body, the use of tar to treat roofs and ships, and the use of evaporated tar (pitch) for sealing barrels. Generally, the history of tar use extends from the time of tarred pole buildings (tar production in coal kilns), wood charring to make charcoal, and the subsequent development of charring ovens took place, respectively, in 4000–1800 B.C., 1100–500 B.C., and 1600–1800 A.D.

Perhaps the earliest, most systematic, organochemical observations of "tar chemistry" were made in Mesopotamia in ca. 3600 B.C., when it was found that the cooking of savory food droplets of fragrant liquid condensed (i.e., distilled) onto the underside of the pot lid could be wiped in linen fabric and thence isolated by pressing. Although the background processes were not investigated, the practices learned included, besides distillation, the extraction of aromatic oils, suitable raw materials for dyes and incenses and for fuels for oil lamps, from solid materials, such as mollusk shells and plant seeds. The first chemist whose name appears in cuneiform writings was Tapputi-Belat-ekalli (1250–1200 B.C., the last part of the name means "highly esteemed master"), who was a specialist in making perfumes and developing technology. These basic skills were known still earlier because a distillation vessel with a volume of about 40 liters was found in deposits more than 2000 years older.

The golden era of the Arabic culture (ca. 750–1260) can also be seen as an important period for the development of chemistry, including carbohydrate chemistry. The foundations for systematic scientific work were established. After the death of Prophet Muhammad (ca. 570–632), the Arabs had created with their conquests a wide empire and had absorbed progressive knowledge and customs from the nations they dominated. In this multiculture, along with poetry and history writing, sciences, such as mathematics, astronomy, and chemistry were further developed mainly based on the accomplishments of the Greeks. Simultaneously, Persian born scientist and medic Abu Bakr Muhammad ibn Zakariya al-Razi (865–925), among others, in his "Book of Secrets" ("Kitab al-Aznar") advanced many practical chemisty skills. Perfumes, medicines, acids (acetic as well as nitric, sulfuric, and hydrochloric acids and aqua regia), soda, potash, and from mineral oil ("black nafti") distilled light fraction ("light nafti") were among its central products. Others were, for example, soaps, dyes, writing inks, glass, and glazes. In addition, the textile industry was a visible part of these developments.

The Mongols finally stopped the development of the Arabic culture in 1258, when the grandson of the Genghis Khan (1161–1227), Hulagu Khan (Hülegü, 1218–65) conquered Baghdad and killed most of its population including thousands of scholars and scientists. The largest libraries in Baghdad were also destroyed and their books were thrown into the Tigris River. This destruction was reinforced in the middle of the 1300s when the Black Plague raged in the city.

The actual art of papermaking apparently did not arise accidentally. According to the documents of the Chinese ruler of the time, Emperor Ho-Ti of the Han Dynasty (206 B.C.–220 A.D.), Ts'ai Lun, a court officer and scholar who entered the service in 75, introduced in 105 to the public the art of making use of a variety of raw materials to create fiber products for different purposes. However, it is also said that paper materials dated from 73–49 B.C. have been found in China. Under Ho-Ti's systematic leadership, modern papermaking started and paper material evolved into a high-quality product; several mentions exist about its use during the following centuries. The Chinese used phloem fibers from the bark of the paper mulberry tree (silk tree, *Broussonetia papyrifera*) and the white mulberry tree (*Morus alba*) and fiber from Chinese grass (rami, *Boehmeria nivea*), and many other fiber materials, such as old rags and fish nets, which all were usually treated with potash and refined (i.e., macerated) until each filament was completely separate.

In China papermaking evolved into a significant home industry. Old Chinese paper found an unbelievably array of uses and was clearly of better quality than Egyptian papyrus. As a novelty, toilet paper was produced for the first time for the Emperor's court in 589, paper money used by the banks for receipts to the merchants ("flying money" that flew with the wind) in 807, playing cards in 969, casings for fireworks, banknotes for general use in ca. 1000, lamps, and even harnesses. In addition, the Chinese, since the 800s, began printing houses using wooden slabs that produced major printing works, such as the "Holy Books of the Buddha" that appeared in 972. The Chinese started using the fibers of rattan palm (*Calamus rotang*) as raw material in papermaking in the 900s, which became the final improvement of the method. Only after this did paper finally replace the traditional use of bamboo strips in making "bound books".

The Chinese tried for a long time to keep their art of papermaking secret to secure their monopoly. However, this art spread gradually first to Korea in the 300s or 400s and then to Japan in ca. 610. The Japanese used the phloem layers of the mulberry tree and local mitsumata (*Edgeworthia papyrifera*) and gambi (*Wickstroemia canhescens*) bushes as raw materials. They also learned to make new products, such as silk paper, which became famous due to its excellent strength. On the other hand, the Arabs, who had created a powerful army during the Turkestan war, captured in the Battle of Talas (nowadays in Kyrgyzstan) in 751 Chinese prisoners of war who knew the art of papermaking and printing. With the prisoners' labor, the city of Samarkand in Uzbekistan in Central Asia (southeast of Lake Aral) became an important center of these new arts. Although it is likely that paper was produced there already somewhat earlier, the Arabs were quick to realize the advantages of paper materials made from rags, flax, and hemp over traditional papyrus and parchment.

Thereafter, these papermaking skills spread expeditiously into other centers of the Arab empire (Baghdad 793, Damascus and Cairo 900, Fèz 1100, and Córdoba), thus migrating from Asia first into North Africa (Egypt and Morocco) and then to Spain. Caliph Harun ar-Rašid

(786–809), also known for his sagas, largely caused this development, as he understood the value of this knowledge obtained as war booty and decreed that papermaking was to be a state monopoly. This substantially enhanced the already powerful Arabic culture in its respect for the written word (ca. 750–1260). The Arabs further developed the techniques of rag grinding and paper sizing (see starch). They made colored papers for various purposes, for example, blue for sorrow, yellow for riches, and red for happiness and nobility.

The art of papermaking thus arrived in Europe, on one hand, from Morocco with Moor merchants and, on the other hand, with the Venetian explorer Marco Polo (1254–1324), who probably spent 17 years in the service of the Mongol emperor Kublai Khan (1215–94) in Peking. Polo, a member of an esteemed line of merchants and aristocrats, started his journey in 1271 along the Silk Road through the enormous Mongol empire with a retinue including his father and uncle, who had travelled in the East earlier. It is reported that Polo returned to Venice in 1295, but he was taken prisoner in 1298 when Venice and Genoa were at war. He probably prepared with the help of a fellow prisoner (Rustichcllo da Pisa) an accurate and extensive travelogue known as the book "Il Millione". It contains, for example, information on strange Asiatic plants and animals, the use of coal, the blue ultramarine dye obtained from lapis lazuli, porcelain, and the astonishing paper money even unknown at that time in Europe. However, there is surprisingly no mention of the Great Wall of China.

Although Arabian paper, among others, was in common use in many countries for a while, there was a clear trend to start independent production. The first "paper mills" were established, for example, in Spain (Xativa, Valencia, in 1139), France (1189), Italy (Fabriano, 1220), Germany (1390), Switzerland (1432), Austria (1463), Poland (1473), England (1490), Sweden (1565), Denmark (1573), Russia (1576), Holland (1586), Finland (1667), North America (1690), and Norway (1695), in a relatively rapid succession.

Starch. Food items containing starch have always been a part of the human diet. In the context of making other products, starch has been used as mentioned above (starting at least in ca. 3000 B.C.) as a sizing agent for papyrus, and according to the writings of Caius Pliny the Elder, there existed at his time well-preserved documents — about 200 years old (from 130 B.C.) — that were sized (i.e., to prevent the penetration of the dye into the writing material) with starch. In addition, he described the use of starch to bleach clothing and powder the hair. According to Chinese data from the year 312, starch from rice, wheat, and barley was used to size paper and to give the product other desirable properties (e.g., to prevent its cracking when folded). Between 700 and 1300, paper sized with starch was commonly still coated with a starch paste to give it more thickness, stiffness, and strength.

Besides these important areas of application, the Arabian teacher and pharmacologist Abu Mansur recommended the use of modified starch ("artificial honey") hydrolyzed with saliva (contains amylase enzyme that degrades starch) to treat wounds. Moreover, starch has traditionally been used (known in Egypt already for over 5000 years) to stiffen fabrics, which assists in their dyeing, especially with yellow, red, or blue dyes. Starch was also used in various cosmetic products. Starch became a familiar commercial product only in the 1350s, although, for example, the Roman politician Marcus Porcius Cato the Elder (234–149 B.C.) arranged for a detailed description of a new large-scale method for starch production in ca. 184 B.C.

2.2. The Time After the 1700s

Observations of the phenomena relating to plane-polarized light. Ordinary unipolarized natural light ("rain" of photons) can best be explained as electromagnetic vibrations that take place in all directions orthogonal to the propagation direction of the light. When light passes through a suitable filter, it becomes monochromatic; i.e., it contains light of only one wavelength, which can still vibrate in any direction. There are also crystalline materials that pass through only light with a given direction of vibration and they can be used to produce plane-polarized light. Étienne-Louis Malus (1775–1812) first observed this polarization effect in 1808; it can be easily produced either with a Nicol's prism made of crystalline calcium carbonate (discovered in 1828 by William Nicol 1770–1851) or, much later, with Polaroid material consisting of organic crystals imbedded in plastic (discovered in 1935 by Edwin Herbert Land 1909–91).

Polarized light will pass without absorption through a second piece of polarizing material only if the optical axes of the two pieces are parallel; its progress is totally stopped if the axes are at right angles to one another. In 1815, Jean-Baptiste Biot (1774–1862) made the observation that certain compounds were able to rotate the vibration plane of plane-polarized light. From 1815–1835, studies under his direction showed that many organic materials, among them the sugars known at that time, had this ability. Additionally, in 1846, Augustin Pierre Dubrunfaut (1797–1881) discovered the mutarotation caused by glucose when observing the changes in the optical activity of a "fresh" glucose solution as a function of time.

Early steps in the structural studies of carbohydrates. The long history of organic chemistry (i.e., the chemistry of carbon compounds) started with steps where modern organic chemistry gradually established itself apart from the numerous observations in general and inorganic chemistry. This involved the development of the concept of atoms (in Greek, "atomos" means "what cannot be divided"), which was based on early ideas presented by philosopher Leucippus in the 5th century B.C., and which philosopher Democritos (ca. 460–370 B.C.), who presented advanced views on the indivisibility of atoms, reinforced. Finally, it was completed in 1913, when Niels Bohr (1885–1962) first presented the modern model of atoms (the 1922 Nobel Prize in Physics). Regrettably, the "primitive" view of Aristotle that all matter was based on combinations of earth, water, air, and fire together with the devine substance "aether", displaced the earlier elegant theories for nearly two millennia.

The first synthetic preparation of an organic compound, urea, by Friedrich Wöhler (1800–82) in 1828 starting from an inorganic compound, ammonium cyanate, can be seen as an important milestone in organic chemistry. This synthesis also proved wrong the widely accepted "vitalism theory" about a special "force of life" in organic compounds, although it took many years before the absence of such a force was finally accepted. As the corresponding starting point of carbohydrate chemistry, Louis Pasteur (1822–95) observed in 1848 the different response to plane-polarized light of aqueous solutions of the ammonium sodium salt of tartaric acid with two different crystal structures. Further studies of this phenomenon led in 1858 (the final theory was verified in 1865), with the help of the theory of tetravalent carbon atoms (Friedrich August Kekulė 1829–96), to general definitions of the relationship of the structure of carbon compounds with their optical activity in 1874 by Jabobus Henricus van't Hoff (1852–1911, the 1901 Nobel Prize in Chemistry) and Joseph Achille Le Bel (1847–1930).

According to these theories, the asymmetry of the molecule causes this phenomenon (see an asymmetric carbon atom in Chapt. "3.4. Stereoisomerism"). However, already in 1824 Augustin Jean Fresnel (1788–1827) had expressed thoughts regarding the connection between the "spiral form" of a molecule and its optical activity. As the theories developed were applied to compounds, such as optically active sugars that were more complex than the basic examples already studied, the research activity of such compounds was invigorated. Still, during the 1880s, there were only a few monosaccharides (D-glucose, D-fructose, D-mannose, D-galactose, L-arabinose, and L-sorbose) and disaccharides (sucrose and lactose), whose chemical structures had been only partially clarified.

Emil Hermann Fischer (1852–1919, the 1902 Nobel Prize in Chemistry) continued Heinrich Kiliani's (1855–1945) work (Kiliani, e.g., suggested that fructose was a 2-hexanone) to clarify the structure of monosaccharides. He proposed in 1891 after a multiyear brilliant series of studies the stereochemical structures of D-glucose, D-mannose, D-arabinose, and D-fructose, which finally directed the general chemical interest into carbohydrates as well. During the following years, Fischer still clarified experimentally the structures of all missing D-aldohexoses and D-aldopentoses and made a stereochemical comparison, among others, between the three-dimensional configurations of the carbon atoms C_2 (C-2) and C_3 (C-3) of the forms of three tartaric acids (the L-(+) and D-(−) forms and the optically inactive *meso* form) with the configurations of the C_4 (C-4) and C_5 (C-5) atoms of the aldohexoses and the C_3 (C-3) and C_4 (C-4) atoms of the aldopentoses. By the time of Fischer's death in 1919, 14 of the possible 16 aldohexoses had been synthesized, and W.C. Austin and F.L. Humoller prepared the two remaining ones (L-allose and L-altrose) as crystalline compounds in 1934.

Fischer based his structural studies on the assumption that in his formulas, the hydroxyl group attached to last asymmetric carbon atom, C_5 in D-glucose (D-aldohexose) and C_4 in D-aldopentoses, pointed to the right.

This choice of orientation had only a 50% probability of being correct. The correctness of his choice became clear in 1951, when the three-dimensional structure of the rubidium salt of L-(+)-tartaric acid (i.e., its absolute configuration) was resolved with X-ray diffraction technique by Johannes Martin Bijvoet (1892–1980). The stereoisomer of this compound, D-(−)-tartaric acid, with a different spatial structure, was synthesized in a traditional way in 1917 in the context of so-called "relative configuration study" starting from D-(+)-glyceraldehyde without changing the expected orientation of its hydroxyl group at carbon atom C_2 in the series of reactions (Fig. 2.1). With this information, it was easy to confirm the current structures of D-aldoses and L-aldoses. If Fischer's initial assumption had been incorrect, all stereochemical structures in the literature until 1951 would have been mirror images of the actual structures.

Deepening knowledge of carbohydrates. The development of carbohydrate chemistry after the fundamental work on structural concepts passed many milestones, such as the naming of the dominant ring structures (see forming of acetals and hemiacetals) in 1909 (Claude Silbert Hudson 1881–1952), questions about their size in 1920–30 (Walter Norman Haworth 1883–1950, the 1937 Nobel Prize in Chemistry and Edward L. Hirst 1898–1975), and their stereochemical form (conformational isomerism), in addition to numerous structural studies.

As an example of the general evolution of subsequent "later times", carbohydrate chemistry is the important question about the bond between the structural components D-glucose and D-fructose of sucrose, a non-reducing sugar. This study actually started in 1827, when William Prout (1785–1850) determined together with his contemporaries (Justus von Liebig 1803–73, Eugene-Melchoir Peligot 1811–90, Jöns Jacob Berzelius 1779–1848, and Dubrunfaut) that the molecular formula of sucrose was $C_{12}H_{22}O_{11}$. Thereafter, several chemical structures were proposed for it (in 1883 by Bernhard Christian Toller 1841–1918, in 1893 by Fischer, in 1914 by Tollens, in 1916 by Haworth, and in 1916 by Hudson), until finally, Hudson elucidated the correct ring structures of the structural components with his classic methylation experiments. Subsequently, in 1953, Raymond Urgel Lemieux (1920–2000) showed via X-ray diffraction technique and synthesizing sucrose that the compound in fact

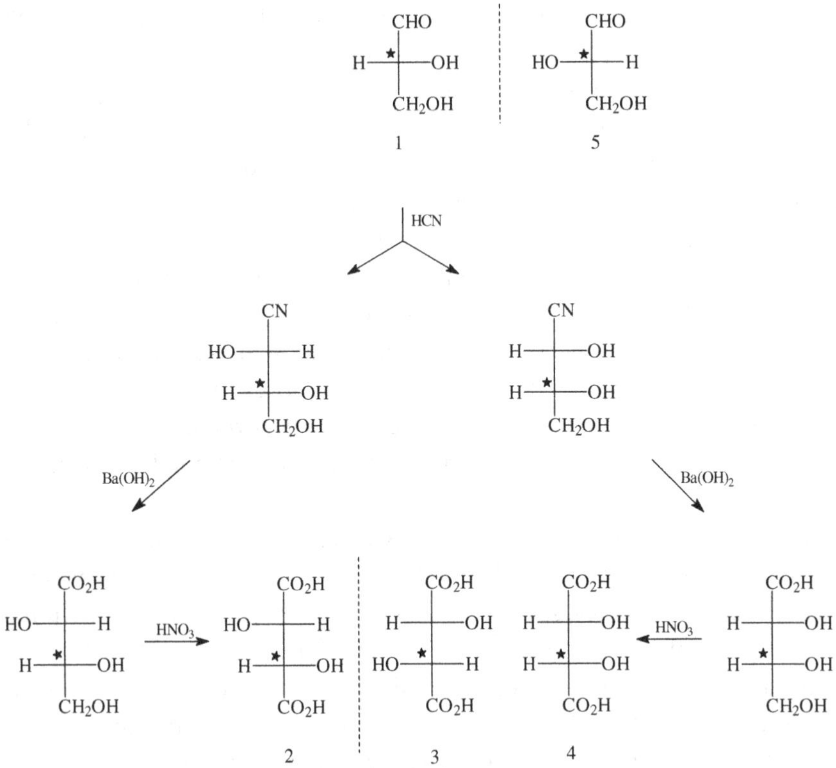

Fig. 2.1. The synthesis of D-(−)-tartaric acid (2) and the optically inactive *meso*-tartaric acid (4) from D-(+)-glyceraldehyde (1) without changing the configuration of the carbon atom C_2 (marked with *) in the initial compound. D-(−)-Tartaric acid is the mirror image of L-(+)-tartaric acid (3), which is not produced in the reaction. If this L-(+) form is desired (and along with it the *meso* form), one has to start with the mirror image of D-(+)-glyceraldehyde, L-(−)-glyceraldehyde (5).

consisted of the components β-D-fructofuranose and α-D-glucopyranose (Fig. 2.2), which were linked together with a (2↔1)-glycosidic bond. Lemieux studied carbohydrate chemistry from the viewpoint of organic chemistry and biology, and his contributions to modern carbohydrate chemistry (especially NMR studies and the synthesis of oligosaccharides) have been substantial. Among the many other researchers and writers, who have had a decisive impact on modern carbohydrate chemistry,

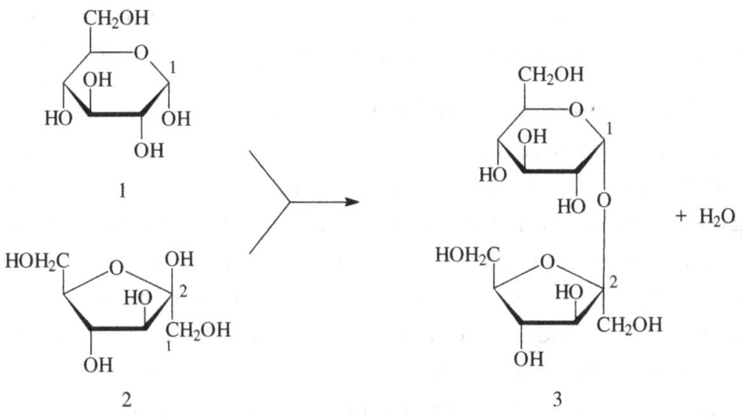

Fig. 2.2. The chemical structure (3) of sucrose (β-D-fructofuranosyl-(2↔1)-α-D-glucopyranoside). It can be thought of arising through the formation of a glycosidic bond between β-D-fructofuranose (1) and α-D-glucopyranose (2).

Gerald Oliver Aspinall, James N. BeMiller, Peter Collins, Derek Horton, James Colquhoun Irvine, Horace S. Isbell, John F. Kennedy, William Ward Pigman, Roy L. Whistler, and Melville Lawrence Wolfrom can be particularly mentioned.

Current status of carbohydrate chemistry. Our knowledge of the structures, properties, and chemical reactions of both natural and synthetic carbohydrates (and of other classes of substances that contain carbohydrates) has increased at an almost incredible rate since the middle 1900s. This of course went along with the fast progress of general organic chemistry, as well as that of the methods of research and analysis. This rapid evolution would not have been possible without a fruitful interaction between basic research and applied science that directly benefits human society. Furthermore, new methods in spectrometry (e.g., NMR) and chromatography (e.g., GC and LC), which became popular during the second half of the 1900s, had a decisive impact on the structural research of carbohydrates. In addition, the explosive development of information technology has substantially accelerated the handling of complex and advanced analytical data, and also dramatically improved our ability to evaluate and perceive the spatial structure of macromolecules. In this way, the basic characteristics of modern science, deeply probing theory, and refined instrumental technique, meet in carbohydrate chemistry.

Pulp production and papermaking. As an example of industrial carbohydrate chemistry, herein briefly outlines the main stages of chemical pulp production, especially in the 1800s, when the decisive steps in technology were taken. At that time, starch refining also gradually developed into a substantial industry; its worldwide production is currently about 55 million tons. In the West, actual paper was primarily made, in order to obtain a uniform product, from pulped flax rags until 1843–44. During those days and unknown to each other, Friedrich Gottlob Keller (1816–95 in Saxony, Germany) and Charles Fenerty (1821–92 in Nova Scotia, Canada) produced raw material for papermaking from pulped wood into which rag pulp was added to increase its strength. Then, in the early 1850s, Heinrich Voelter (1817–87) and Johan Matthäus Voith (1803–74, who bought the patent from Keller in 1846) in Heidenheim, Germany presented the principles of a grinder suitable for large-scale production of groundwood pulp. After entering commercial production, the grinder won in 1854 a gold medal in the First General German Industrial Exhibition in Munich. It also evoked great interest at the Paris Exposition in the following year. Consequently, it was possible to quickly develop industrial production of pulp from wood material by this groundwood process in Scandinavia and Germany in the 1860s, when the first pulp mill was founded in the city of Poix in Belgium in 1861. At the Exposition Universelle of 1867 in Paris, it was possible to exhibit a modern pulp mill that even showed techniques for sorting, reject control, and drainage.

In the manufacture of traditional rag paper (or cotton paper) made from cotton linterns or normally cotton cloths (rags), the collected and sorted rag material was reduced by cutting it into smaller pieces and removed buttons, pins, and other foreign matter. This material was then treated with lye solution, which removed fats and part of dyes. Thereafter, the rags were allowed to "rot" in the retting or fermentation process at a warm temperature (the process was followed by observing the formation of mold), which caused the fiber strands to swell and become suitably brittle to enable the grinding of the material into a watery pulp. This grinding was first done by hand, but later, stampers driven by water wheels, windmills, or horses were used. The ground watery pulp was drained to form paper sheets. The dried sheets were still dipped into a solution of animal glue or starch to improve their properties, and finally they were hung on lines to dry.

The actual "machinization" of papermaking can be said to have started in the 1670s in Holland, where a grinder called the "Hollander beater" or the "Hollander" was developed to replace stamping machines, which had previously been used for the disintegration of rags and beating pulp. This machine clearly expedited the production of the pulp and improved its quality. The reign of the rag pulp ended along with the rapidly increasing demand for paper in the 1870s, when pulp makers learned to produce mechanical pulp from wood on an industrial scale (see above). On the other hand, in 1789, along with the rapidly increasing number of newspapers during the French Revolution, a shortage of raw materials caused the price of rags to rise, rendering papermaking unprofitable. The development work of Keller described above followed the trends of the times. It was primarily based on the sharp observation of René-Antoine Ferchault de Réaumur (1683–1757) in 1719 that wasps (*Vespa vulgaris*) made very fine, grey, and paper-like texture for the walls of their nests from fibers of ordinary trees with their saliva and mastication. Inspired by this same observation, it took Bavarian Jacob Christian Shäffer (1718–90) until 1765 to describe a non-commercial process for producing fiber material suitable for papermaking from sawdust and other woodworking residues.

The production could be "theoretically increased" only after the final principle of a paper machine had finally been invented (Nicholas-Louis Robert 1761–1828) in France (the patent was awarded in 1799), preceded by a close cooperation between Robert and Didot Saint-Lèger, whom Robert hired to work in his factory in Essonnes — close to Paris — in 1793. This was the first hand-operated paper machine (the typical width of French wall paper was 64 cm, length 260 cm, and the machine speed 9 m/min), and it only contained a belt of wire cloth but no dryer section, which meant that the paper sheets still had to be lifted off to be hung over a rope or wooden rod to air dry.

The paper machine based on Robert's patent remained on an experimental level. The next step was taken by the English banker brothers Henry Fourdrinier (1766–1854) and Sealy Fourdrinier together with

John Gamble (the brother-in-law of Didot Saint-Léger). After acquiring the patent in question (see above) in 1803, they built in their machine workshop with Bryan Donkin (under John Hall) a modified paper machine — one that could be seen as an improved version of Robert's machine (width 120 cm and, e.g., including a wet press). With this machine, it was possible to produce continuously paper, and it was granted a new patent in 1807. Regarding production, a machine of this type was still rather small, and it soon proved uneconomical. However, the name of the brothers has survived in papermaking until this day; a traditional longitudinal wire papermaking machine is still called "a Fourdrinier machine" (or, more commonly, the forming section of a paper machine is called "a fourdrinier"). In spite of its lack of commercial success, the first continuously operating paper machine initiated a rapid technical development that has continued until this day; the largest width of the web is now over 11 m and the driving speeds, especially in the production of LWC ("light-weight-coated") and newsprint paper, are at best about 2000 m/min, being continuously improved. The pilot machines used in product development can currently be driven at about 2400 m/min. The time interval from the invention of the paper machine to its wide industrial use was about 30 years with no significant differences emerging between paper and cardboard machines. However, the methods of papermaking are continuously developing; one innovation being investigated is so-called "foam forming" which reduces, besides the consumption of water and energy, the fiber material.

The introduction of surface sizing was another important breakthrough in papermaking from plant fibers. It enabled the production of a smooth surface on which the printing ink behaved properly. Papers made in the West were surface sized, before the introduction of large-scale starch sizing of machine papers in early 1800s, with animal glues (gelatin sizing) originated by the Italians by boiling bones and skins (the Arabs used traditional starch sizing). However, the gelatin sizing method was an expensive and tedious process. Hence, in 1807, Moritz Friedrich Illig (1777–1845) in Germany published a sizing method based on rosin and alum. By the mid-1800s, this technique was in worldwide use; it substantially improved the quality of printing and wrapping paper. Additionally, the Italians were the first to introduce the use of watermarks as a warranty of the high quality of the product. The first watermark found in handmade paper (the letter "F" after the Italian mill Fabriano) originated in 1271.

After the end of the 1700s, handmade paper could no longer satisfy the needs of Western society. As the production and consumption of machine-made paper increased, a shortage of raw material, until then collected rags and other textile fabric, became a major problem. The steadily increasing demand for paper in Europe was closely connected with the societal evolution taking place; Johannes Gutenberg (ca. 1398–1468) initiated this cultural shift with his mechanical, movable-type printing invention in ca. 1438. This technique started the Printing Revolution and is widely regarded as the most important invention of the second millennium. However, the use of movable type and printing press was already known early in China, where movable wooden type was used for ideograms. Together with many other Gutenberg's contributions to printing, it was possible to create a practical system that allowed the mass production of printed books. This progress (i.e., books were quickly printed in a smaller page size with a smaller font) was economically viable for printers and readers alike, thus making books with reduced prices accessible to more readers. Along with the art of printing based on movable type, centuries-old vocal tradition of the troubadours gradually became printed literature. The Reformation further influenced this trend, which struck the Catholic world and liberated the spread of information. In 1455, the new technique was used to print on paper, for example, the first important work published by Johann Fust (ca. 1400–1466) and Peter Schöffer (ca. 1425–1503), the Bible in Latin, of which about 50 copies are still in existence. Printing an opus like this on parchment would have required the skin of some 300 sheep.

The delignification (i.e., chemical removal of lignin) from wood, the obvious and most common source of raw fiber material, proved surprisingly difficult and time consuming. In Europe in 1800, paper pulp was first made by cooking straw (its chemical defibration is easier than that of wood) in an alkaline aqueous solution in an open vessel. Only in the early 1850s was a more efficient method (A. Miller) of using a closed vessel under pressure introduced, enabling the use of higher temperatures. According to a patent (by Charles Watt and Hugh Burgess) granted in 1854, in addition to straw and other plant fiber pulps, wood raw material could also be cooked in still stronger solution of sodium hydroxide at still a higher temperature (150°C). In this case, the losses of the cooking chemical were replaced with soda (Na_2CO_3), whence the process was called "soda pulping" (the first factory was founded in the United States in 1854). Their patent also included chlorine-based bleaching of the product. The principles of "general chlorine bleaching" were already established nearly 100 years earlier in 1744, when Carl Wilhelm Scheele (1742–86) noticed the bleaching effect of the gaseous matter (containing chlorine and oxygen) he had prepared. Soon thereafter, Claude Louis Berthollet (1748–1822) used mixed chlorine gas to bleach rags of poor quality for use as raw material for white paper. This also enabled the use of colored rags as such. It took another 30 years before Humphry Davy (1778–1829) realized that chlorine was an element, for which he proposed the name "chloride" (in Greek "light green").

Asahel Knowlton Eaton (1822–1906) patented in 1870 a method for replacing the sodium losses in alkaline cooking with sodium sulfate ("sulfate cooking process" or also known as "kraft cooking process" or "kraft process"); it was named according to the high strength of the resulting product (in German, "Kraft" means "strength"). Finally, based on the same idea, Carl Ferdinand Dahl presented in 1884 a practical process of employing wood as raw material, which can be seen as the decisive impetus towards the modern kraft industry. This led quickly to large-scale production; the first factory was founded in Sweden in 1885. The burning and causticizing of the cooking chemicals (the sodium carbonate formed in the burning is converted to "caustic soda", sodium hydroxide, with calcium hydroxide) regenerates the active cooking chemicals into a mixture of sodium hydroxide and sodium sulfide formed from sodium sulfate in the reduction zone of the furnace. This method did not contain a completely new idea because it had been observed already in the early 1800s that the addition of sulfur and sodium sulfide clearly accelerated the delignification of straw under alkaline conditions.

From 1837–42, Anselme Payen (1795–1871), an open-minded experimenter, laid the foundations of the other main line of development, acid cooking. He prepared rather pure cellulose by treating wood and other plant materials with acidified ammonia (after this treatment followed washing with water, ethanol, and ether) and concluded that cellulose formed a uniform basic component of plant cells (the component of a cell, in Latin, "cellula" means "small chamber"). For the technical realization, Benjamin C. Tilghman patented in 1867 a method, where the defibration of wood took place in a pressurized system of aqueous calcium hydrogen sulfite and sulfur dioxide ("sulfite cooking"). This process was not immediately realized on a large scale. When the world's first sulfite pulp factory started operations in Sweden in 1874, Carl Daniel Ekman (1845–1904) conducted pivotal research that realized the practical execution of large-scale processes. Alexandre Mitscherlich (1836–1918) also devised a similar large-scale production independently of Ekman. Before the actual acid calcium-based sulfite cooking, Peter Claussen had in 1851 proposed the possibility of making fiber suitable for papermaking from wood with the help of sodium hydroxide and sulfur dioxide (this formed the precursor for "neutral sulfite cooking"). Several so-called "high-yield cooking" methods were proposed later in the 1880s, but their economy and products were not found to be very competitive in the following decades.

During their evolution of about 150 years, the kraft and sulfite processes have evolved in many ways, giving birth to several process solutions and modifications (cf., Chapt. "12.4. Chemical Delignification"). In addition to the progress in the cooking phase, major development has taken place, especially in the bleaching of the product, where the trend nowadays is toward a strong reduction in the use of chlorine and its compounds to protect the environment. Many countries no longer use elemental chlorine in the bleaching and the use of chemicals containing oxygen (oxygen, hydrogen peroxide, and ozone) is rapidly growing. As a fundamental way of chemical defibration, the kraft process has become the dominant cooking method since the 1950s with a share of over 90% of the total chemical pulp production, currently about 130 million tons per year. This chemical pulp production is about 70% of the total pulp production, most of the remainder being based on mechanical pulping. In addition, increasing amounts (about 145 million tons) of recycled fiber, mostly recycled paper and cardboard, are used annually. The world's annual production of paper and cardboard (including large amounts of additives and fillers) totals about 400 million tons, and the demand is expected to grow to nearly 500 million tons by 2020.

3. Isomerism

3.1. General

A general molecular formula (e.g., $C_xH_yO_zN_n$) expresses the kind and number of the constituent atoms of a compound, but it insufficiently represents the structure of the compound in question. A molecular formula can thus correspond to several compounds (*isomers*) that normally have different chemical and physical properties. Isomers can be defined as chemical compounds with identical molecular formulas (i.e., contain the same number of atoms of each element) that differ from one another in the arrangements of their atoms. This phenomenon is called "*isomerism*" (in Greek, "isos" means "equal" and "meros" "part"), and it is divided into two main types (Fig. 3.1): (i) *constitutional isomerism* or structural isomerism and (ii) *stereoisomerism* or space isomerism. Upon examining certain isomers (such as aldoses with the same number of carbon atoms), one does not necessarily find differences based on constitutional isomerism, and finding the actual differences requires detailed comparison of the stereoisomeric properties of the structures.

Constitutional isomerism can be divided into three subgroups: (i) *functional group isomerism* ("function isomerism"), (ii) *chain isomerism* ("skeletal isomerism"), and (iii) *position isomerism* ("regioisomerism"), which are discussed in the next chapters with the help of illustrative examples. The general name "structural isomerism" is traditionally used for "constitutional isomerism". However, since the structure of the compounds can be thought to cause all isomerism, the use of the former term is not recommended.

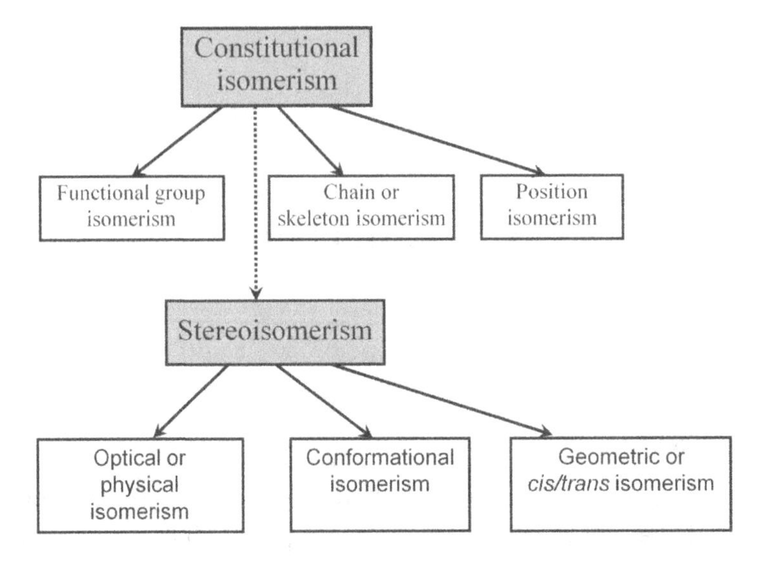

Fig. 3.1. The main types of isomerism and their subtypes.

The branch of organic chemistry that examines the three-dimensional structures of molecules, *stereochemistry*, has gained importance when striving to understand the physical and chemical properties of various compounds. In carbohydrate chemistry, it is also essential to know the stereochemical structure of the compounds. Stereoisomerism can be seen to generally represent the form of isomerism where compounds with the same chemical structure (i.e., the order of attachment of the atoms involved and the location of the bonds between them) differ from each other only in the spatial direction of their atoms or atom groups. This isomerism is divided into (i) *optical isomerism* ("physical isomerism"), (ii) *conformational isomerism*, and (iii) *geometric isomerism* ("*cis*/*trans* isomerism"). As the first two types are characteristic of carbohydrates, they will be emphasized in the following discussion.

3.2. Constitutional Isomerism

Constitutional isomers generally differ from one another only in the order of attachment of their atoms and the location of their bonds. In the functional group isomerism, the isomers have the same molecular formula, but their functional groups are different. The following compounds are examples of such isomers:

$C_4H_{10}O$

$CH_3CH_2{-}O{-}CH_2CH_3$ — diethyl ether (ether)

$CH_3CH_2CH_2CH_2OH$ — 1-butanol (alcohol)

C_4H_8O

$CH_3CH_2CH_2C(=O)H$ — butanal (aldehyde)

$CH_3CH_2C(=O)CH_3$ — 2-butanone (ketone)

$C_4H_8O_2$

$CH_3CH_2CH_2C(=O)OH$ — butanoic acid (carboxylic acid)

$CH_3CH_2C(=O)OCH_3$ — methyl propionate (ester)

CHON

$HO{-}C{\equiv}N$ — cyanic acid

$H{-}N{=}C{=}O$ — isocyanic acid

Chain isomers have the same molecular formula, but the skeleton (usually carbon skeleton) differs by having branches or otherwise. The following compounds (C_5H_{12}) are examples:

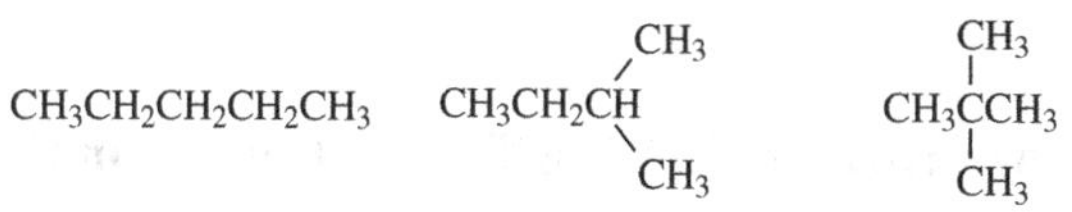

$CH_3CH_2CH_2CH_2CH_3$ — *n*-pentane

$CH_3CH_2CH(CH_3)CH_3$ — 2-methylbutane (isopentane)

$CH_3C(CH_3)_2CH_3$ — 2,2-dimethylpropane (neopentane)

The number of chain isomers increases very rapidly with the increase in the number of carbon atoms in the compound. Theoretically, for 6, 7, 8, 15, and 20 carbon atoms in an aliphatic hydrocarbon, the numbers of possible chain isomers are 5, 9, 18, 4347, and 366,319, respectively.

In position isomers, a functional group can be located at different points on the carbon chain, or it can be attached to a different point in the ring structure of an aromatic derivative, in which case the isomers have identical carbon chains (or aromatic ring structures). This type of isomerism is found, for example, in some alkanes, alcohols, ketones, and aromatics. The following compounds are examples of typical position isomers:

C_4H_8

$CH_3CH_2CH{=}CH_2$ — 1-butene

$CH_3CH{=}CHCH_3$ — 2-butene (*cis* or *trans* form)

$C_4H_{10}O$

$CH_3CH_2CH_2CH_2OH$ — 1-butanol

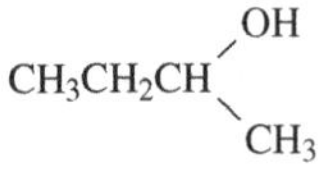

$CH_3CH_2CH(OH)CH_3$ — 2-butanol

C_4H_{10}

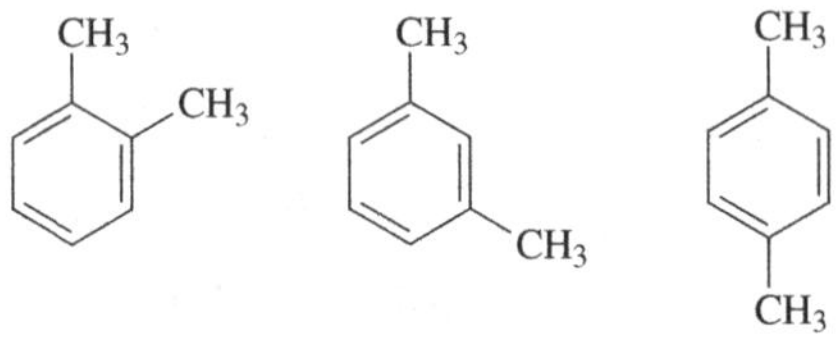

1,2-, 1,3- and 1,4-dimethylbenzene

o-, *m*- and *p*-dimethylbenzene

3.3. Basic Factors Influencing the Three-dimensional Compound Structures

The direction of bonds. The basis of the three-dimensional structure of a molecule is the orientation in space of the bonds between its atoms. Table 3.1 contains the ground-state electron configurations of the elements that typically occur in organic compounds. When atoms form covalent bonds with each other, their *atom orbitals* (expressed with orbital quantum numbers 0, 1,... or letters *s*, *p*, *d*,...) partially overlap. This forms common

Table 3.1. Electron configurations of elements typically found in organic compounds

Element	Symbol	Electron configuration
Hydrogen	H	$1s^1$
Carbon	C	$1s^2\ 2s^2\ 2p_x^1\ 2p_y^1$
Nitrogen	N	$1s^2\ 2s^2\ 2p_x^1\ 2p_y^1\ 2p_z^1$
Oxygen	O	$1s^2\ 2s^2\ 2p_x^2\ 2p_y^1\ 2p_z^1$
Phosphorus	P	$1s^2\ 2s^2\ 2p^6\ 3s^2\ 3p_x^1\ 3p_y^1\ 3p_z^1$
Sulfur	S	$1s^2\ 2s^2\ 2p^6\ 3s^2\ 3p_x^2\ 3p_y^1\ 3p_z^1$

molecule orbitals where, similarly to atomic orbitals, a maximum of two electrons can occupy each molecule orbital. Only the valence electrons occupying the outermost electron shell (expressed with the principal quantum numbers 1, 2, 3,... or letters *K, L, M*,...) participate in chemical reactions.

However, *hybridization* (i.e., explaining the actual orientation of bonds), where the atom orbitals of individual atoms, usually *s* and some or all of the occupied orbitals p_x, p_y, and p_z, "harmonize" with each other creating *hybrid orbitals* with different energies and shapes. Linus Carl Pauling (1901–94, the 1954 Nobel Prize in Chemistry) developed this theory to explain the structure of simple molecules, such as methane. These hybrid orbitals are, especially in the case of carbon atoms, similar to each other in their geometry and energy but differ from the original atom orbitals. This way, four identical hybrid orbitals are formed in the *sp^3 hybridization* ($1s^2 2s^2 2p_x^1 2p_y^1 \rightarrow 1s^2 2s^1 2p_x^1\ 2p_y^1 2p_z^1$) forming four identical (i.e., having the same energy) sp^3 hybrid orbitals with one electron in each, which enables the atom to bond neighboring atoms with four separate σ bonds (see the tetravalent carbon atom). Correspondingly, in the *sp^2 hybridization*, the one 2*s* orbital and two 2*p* orbitals of the outer *L* shell form three identical sp^2 hybrid orbitals (able to form three identical σ bonds), and in the *sp hybridization*, one 2*s* orbital and one 2*p* orbital in the outermost shell form two identical *sp* orbitals (able to form two separate

Table 3.2. The angles between the axes of hybrid orbitals in different cases of hybridized carbon atoms

				Example	
Hybridization	Number of ligands	Geometry	Internuclear angle	Component	Internuclear angle
sp^3	4	tetrahedral	109.5°	CH_4	∠H-C-H 109.5°
sp^2	3	planar (trigonal)	120.0°	$H_2C{=}CH_2$	∠H-C-H 116.7° ∠H-C-C 121.6°
sp^1	2	linear	180.0°	HC≡CH	∠H-C-C 180.0°

σ bonds). In the first case, one *p* orbital and in the second one two *p* orbitals remain unchanged, and they can participate in the formation of either a double bond (one σ bond and one π bond) or a triple bond (one σ bond and two π bonds). Table 3.2 shows the theoretical bond angles in different bond types of hybridized carbon atoms.

The electron configuration of nitrogen is $1s^2 2s^2 2p_x^1 2p_y^1 2p_z^1$ and that of oxygen $1s^2 2s^2 2p_x^2 2p_y^1 2p_z^1$ (Table 3.1). This is why in their hybridization, in the case of nitrogen (five *L* electrons), three hybrid orbitals form with one electron in each and one with two electrons ("the fourth ligand"), and in the case of oxygen (six *L* electrons), two hybrid orbitals form with one electron and two with a pair of electrons ("the third and fourth ligands"). In both cases, all four hybrid orbital axes tend to orient, similarly to sp^3 hybridized carbon, in the manner of a tetrahedron (the free electron pair forms one ligand), even though the bond angles deviate somewhat from the ideal value of 109.5° because of the different values of the ligands. Thus, for example, in ammonia (NH_3), water (H_2O), and dimethyl ether (H_3COCH_3):

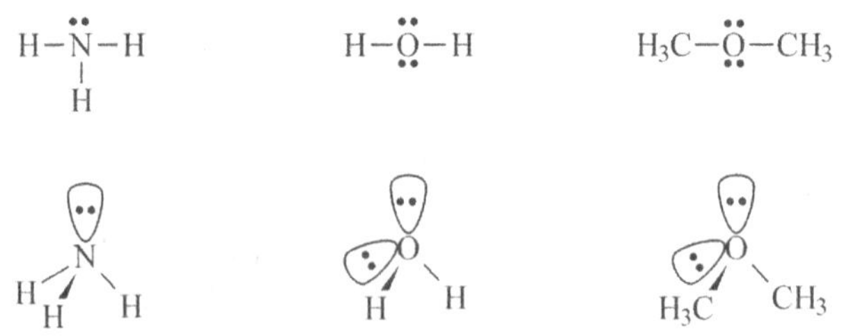

where the angles ∠ H-N-H, ∠ H-O-H, and ∠ C-O-C are, respectively, 107.3°, 104.5°, and about 110°, and the corresponding shapes are "trigonal pyramidal" (NH_3, $NH_4^{\oplus}$ is more clearly tetrahedral) or "angular" (H_2O and H_3COCH_3). When oxygen or nitrogen form double bonds with a carbon atom (sp^2 hybridization), the angle ∠ C-C-O(N) is generally close to 120°. The character and orientation of bonds formed by elements with a more complex electron configuration, and in phosphorus (five electrons in the outermost *M* shell) and sulfur (six electrons in the *M* shell) that are found in certain carbohydrate derivatives, depends strongly on the compound in question.

The length of bonds. X-ray diffraction technique reveals that the length of the bonds between atoms varies between certain limits. Examples of lengths of bonds frequently found in organic compounds are shown in Table 3.3, although the majority of these bonds do not normally occur

Table 3.3. Examples of bond lengths between atoms

Type of bond	Length of bond/nm	Type of bond	Length of bond/nm
C-H	0.109	C-N	0.147
		C=N	0.130
C-C	0.154	C≡N	0.115
C=C	0.134		
C≡C	0.120	C-S	0.182
C-C=	0.150	C=S	0.170
C-C≡	0.146	C-P	0.184
=C-C=	0.148	C-Si	0.187
C=C=	0.131		
		O-O	0.147
C-O	0.143	N-N	0.145
C=O	0.123		
		C-F	0.135
O-H	0.096	C-Cl	0.178
N-H	0.101	C-Br	0.193
S-H	0.135	C-I	0.214

in carbohydrates. It is assumed that these bonds are "isolated" (except some C-C bonds) so that the effects of a resonance phenomenon caused by double or triple bonds possibly conjugated with the atom are not taken into account. An illuminating example of this is the uniform distribution of the C-C bonds (0.139 nm) in a benzene ring, which indicates an even distribution (delocalization) of the π electrons over the bond types "C-C" and "C=C" between the atoms in the ring. The values in the table also show that the bonds between carbon atoms vary in the interval 0.120 nm to 0.154 nm and those containing oxygen between 0.096 nm and 0.147 nm. In addition, for each type of bond its possible "degree of unsaturation" (single $\rightarrow$ double $\rightarrow$ triple) generally shortens the distances between the atoms. Tradiotionally, bond lengths have also been expressed in Ångström units (1 nm = 10^{-9} m = 10 Å), after the Swedish physicist Anders Jonas Ångström (1814–74), but this obsolete unit is no longer included in the SI system of units. Ångström was a professor at Uppsala and did pioneering work in spectral analysis determining, among others, the wavelengths of the Fraunhofer (Joseph von Fraunhofer 1787–1826) lines in the spectrum of the sun.

Knowledge about the bond lengths between the atoms or the angles between its bonds determined by their electron configurations is not in many cases be sufficient to define the exact stereochemical structure of the molecule. Atoms or atom groups bonded together by single bonds are able to rotate about their axis of bonding enabling them to form alternative spatial structures with different degrees of stability. This bond-centered rotation is significantly influenced by interactions with chemically unbonded atoms, both inside the molecule and those in neighboring molecules. These interactions can be either repulsive or attractive, and their magnitude depends primarily on the size, polarity, and distance. The interactions are especially important in the cases where the "rotation degree of freedom" is either sterically limited or totally hindered. A typical case is the bonds between the ring atoms in monosaccharides (cf., Chapt. "6.3. Naming of Conformations").

Two atoms (A and B), connected by a maximum of three bonds, can be either (i) *adjacent* (A-B), (ii) *geminal* (A-C-B), or (iii) *vicinal* (A-C-C-B). In the first of these cases, the spatial structure is determined by, besides the sizes of A and B, only the length of the bond A-B. If there is a third atom

C between the atoms in question (case ii), in addition to the bond lengths A-C and C-B, the magnitude of the angle ∠ A-C-B (α) must be taken into account. Instead, in the accurate determination of the three-dimensional structure of a molecule or molecular part formed by vicinally placed atoms, it is necessary at least to know, in addition to the bond lengths A-C, C-C, and C-B and the bond angles ∠ A-C-C (α_1) and ∠ C-C-B (α_2), the existing *dihedral angle* or torsion angle (ϕ) that expresses the angle between the planes that contain the atoms ACC and CCB:

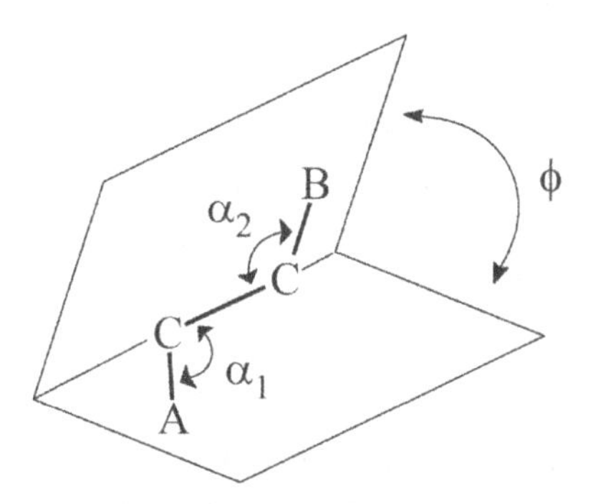

Hydrogen bond. The hydrogen bond is the most significant interaction (a unique dipole-dipole interaction) between atoms that are not chemically bonded to each other. Therein, a hydrogen atom, covalently bonded to an electronegative atom, is in an electrostatic interaction with another electronegative atom that contains a free electron pair (Fig. 3.2). Particularly strong hydrogen bonds (length 0.27–0.34 nm) arise between electronegative atoms with small atomic radii (primarily F, O, and N) (Table 3.4). Linus Carl Pauling (see above), who worked as a professor at

(a) (b)

Fig. 3.2. The formation of intramolecular (a) and intermolecular (b) hydrogen bonds through hydroxyl groups.

Table 3.4. Examples of the relative electronegativity values (Pauling scale) and the bond lengths of elements

Element	Electronegativity	Radius of bond/nm	Element	Electronegativity	Radius of bond/nm
F	4.0	0.071	B	2.0	0.082
O	3.5	0.073	Cu	1.9	0.138
Cl	3.0	0.099	Si	1.8	0.111
N	3.0	0.075	Fe	1.8	0.125
Br	2.8	0.114	Al	1.5	0.118
I	2.5	0.133	Mn	1.5	0.139
C	2.5	0.077	Mg	1.2	0.130
S	2.5	0.102	Ca	1.0	0.174
P	2.1	0.106	Na	0.9	0.154
H	2.1	0.037	K	0.8	0.196

the California Institute of Technology, devised the concept of the relative electronegativity values for elements. On this scale, the highest value, 4.0, is given to fluorine (F) and the lowest one, 0.7, to cesium (Cs), so that the difference of their electronegativity is 3.3. It can generally be observed that if the difference in electronegativity of the atoms forming a bond is <0.6, we speak of a "nonpolar covalent bond" (e.g., C-C, C-H, C-N, C-S, and S-H), while in the range of 0.6–1.9, the bonds are called "polar covalent", and at >1.9 the bonds are called "ionic" (e.g., Na-Cl).

The hydrogen bonds in molecules are usually much weaker (5–35 kJ/mol) than the covalent bonds (300–800 kJ/mol). However, attraction through hydrogen bonds has a major effect on the physical properties, such as boiling and melting points, solubility, and crystallinity of monomeric compounds. In the same way, the formation of hydrogen bonds determines the most stable stereochemical structures (e.g., cellulose and proteins; especially DNA that carries genetic information). For example, the adhesion of the fibers containing carbohydrates onto one another in papermaking is extensive due to hydrogen bonds. It is also easy to understand why monosaccharides and disaccharides (sugars) that contain many hydroxyl groups are easily soluble in water and partly in ethanol (in both cases hydrogen bonds are formed between the sugar molecules and those of the solvent),

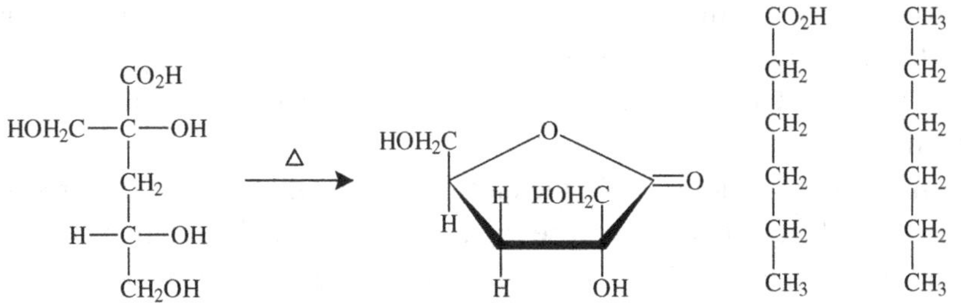

Fig. 3.3. The chemical structures of α-glucoisosaccharinic acid (1), α-glucoisosaccharino-1,4-lactone (2) (boiling point 205°C at a pressure of 0.5 mmHg), hexanoic acid (3) (boiling point 205°C at a pressure of 760 mmHg), and hexane (4) (boiling point 69°C at a pressure of 760 mmHg).

but not in toluene or ether. This "truth" was expressed long ago in Latin: "similia similibus solvuntur" or "like dissolves like". In addition, sugar crystals are among the strongest crystals formed by organic compounds because of the formation of hydrogen bonds.

Figure 3.3 shows an example of the influence of intermolecular hydrogen bonds on the boiling points of selected compounds with six carbon atoms. In cellulose cooking, the substantial content of glucomannan (cf., Chapt. "9.4.5. Other polysaccharides") in the raw wood material is converted into α-glucoisosaccharinic acid among its main reaction products (cf., Chapt. "11.2. Polysaccharides") in an alkaline environment. This can easily form hydrogen bonds so that the distillation of this compound without its disintegration (in free form this compound exists as a 1,4-lactone) is possible only under a very low pressure. In hexanoic acid, the possibility of forming hydrogen bonds is more limited, and it is totally absent in hexane. One can reason that it takes additional thermal energy first to unravel the intermolecular hydrogen bonds before the actual evaporation and distillation of the molecule. On the other hand, differences exist already in the heats of vaporization of the compounds shown as examples.

Molecular models. Molecular structure, or its influence on the stability and reactivity of the molecule under different conditions, can be examined with the help of different molecular models or using advanced and illustrative computer programs (computer modeling and graphics). Two main types of molecular models, each with its own characteristics, have traditionally been used.

In "space filling models", the electron shells of atoms are pictured as spheres of different colors whose radii correspond to the relative sizes of the electron shells. Figure 3.4 (model a) shows the chemical structure of methane, where the merging of the electron shells of the atoms is depicted by partially imbedding the spheres into each other. The use of this "old-fashioned model" is somewhat laborious, and this mode of presentation is not very useful for examining the relationships between atoms or the angles formed by their bonds. Therefore, more illustrative models are so-called "skeletal models" ("ball and stick models"), where atoms are represented with spheres of different colors (and possibly size) that are joined with rods of different lengths depicting chemical bonds, also taking into account likely double or triple bonds (models b and c in Fig. 3.4). In addition, there are models comparable to the above that aim at accurately illustrating both the angles and lengths of the bonds and strive for a realistic presentation of the structure of the bonds. With the help of skeletal models, one can easily examine, for example, phenomena related to

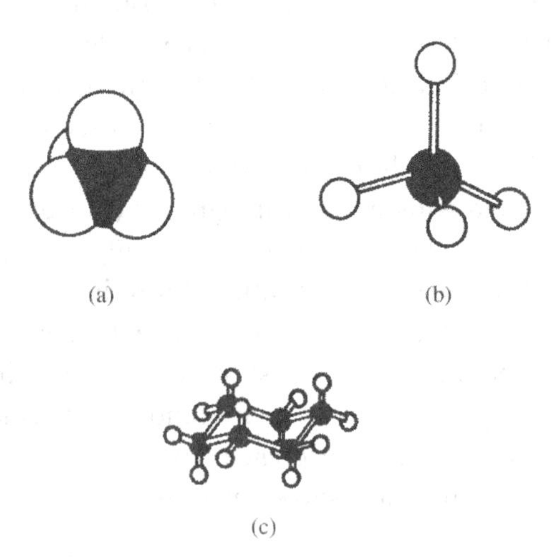

Fig. 3.4. Examples of molecular models about the chemical structure of methane (a) and (b) and cyclohexane (c).

the rotational motion about a single bond and configurations (the orientation of substituents) of a carbon atom, especially in the ring structures of carbohydrates (cf., Chapts. "3.6. Conformational Isomerism" and "6.3. Naming of Conformations"). However, it should be kept in mind that in reality the atoms and the bonds between them are in a continuing oscillatory motion, whose frequencies can be used to spectrometrically investigate the chemical structure and shape of the molecule.

3.4. Stereoisomerism

The observation of optical isomerism and related definitions. With the help of a polarimeter, one can examine the property of some organic compounds (i.e., optically active substances) to rotate the plane of linearly or plane-polarized or linear-polarized light (cf., Chapts. "1. Introduction" and "2.2. The Time After the 1700s") (Fig. 3.5). In this method, monochromatic, plane-polarized light is passed through a sample cuvette that contains the organic compound examined, either in a liquid form or dissolved in a suitable solvent. An analysator will pass through only light with a plane of oscillation that is parallel to its optical axis. If the analysator is rotated in order to maximize the intensity of light passing through, it is possible to determine the angle of rotation caused by the sample. On the other hand, turning the analysator ±90° from the maximal position will cause the light

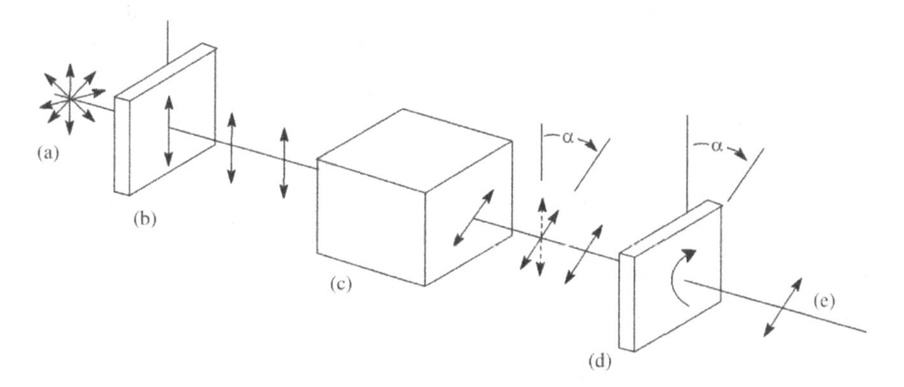

Fig. 3.5. Main parts of a polarimeter. Monochromatic light source (a), polarizing prism (b), sample cuvette (c), analysator prism (d), and detector system (e). The specific rotation of the compound [α] can be calculated from the observed angle of rotation α.

to vanish totally, which is easier to observe visually. As it is difficult to detect small changes in light intensity, simple conventional polarimeters have optical systems that divide the field of view into two halves of equal size. A comparison between these halves (according to the principle: in the beginning "extremes", then "equalities") already yields rather accurate values of the angles. Nowadays, a quick and accurate determination of the rotation angle is made automatically using photoelectric detection.

An *optically active* compound will rotate the plane of plane-polarized light either to the right (clockwise) or to the left (counter-clockwise). A *right-rotating compound* (right-handed compound) is denoted by a prefix (+)-before its name, while a *left-rotating compound* (left-handed compound) receives a prefix (−)-, or by letters *d-* (dextrorotatory, in Latin, "dexter" means "right") and correspondingly *l-* (laevorotatory, in Latin, "laevus" means "left"). If the compound investigated instead has no influence on plane-polarized light, it is *optically inactive.* Equal amounts of (+)- and (−)-isomers form an optically inactive "mixture" (cf., Chapt. "3.5. Racemic Modifications") that is denoted with a prefix (±)- or alternatively *dl-*. It should be emphasized that prefixes *d-*, *l-*, and, *dl-*, common in "old" literature, are being systematically replaced with the corresponding prefixes (+)-, (−)-, and (±)-.

The angle of rotation α depends primarily on the type and number of the molecules met by the beam of light, since every optically active molecule contributes to the rotation of its plane of oscillation. The angle of rotation thus depends on the concentration of the solution (or on the density of the pure liquid) and on the length of the path of the light ray through the sample. As the specific rotation can vary markedly using different solvents (which must be optically inactive) or light with different wavelength (most frequently the D line of sodium at 589 nm), the conditions have to be accurately determined. In addition, the temperature has a small effect on the specific rotation. The *specific rotation* [α] is defined for solutions (Eq. 3-1) or undiluted liquids (Eq. 3-2) as follows:

$$[\alpha]_{\lambda}^{t} = (100 \bullet \alpha)/(l \bullet c), \tag{3-1}$$

$$[\alpha]_{\lambda}^{t} = \alpha / (l \bullet d), \tag{3-2}$$

where λ is the wavelength of the light used, t the temperature, α the measured angle of rotation in degrees, l the length of the sample cuvette in dm, and c the concentration of the sample solution in g/100 cm^3 at temperature t.

The dimension of the specific rotation is thus (deg)cm^2/g (not often misleadingly used "degrees"), although the dimension is normally not included. Properly expressed, for instance, $[\alpha]_D^{25} = +50.5$ (c 2.5, EtOH) means that the specific rotation of the compound over the distance of 1 dm at the temperature of 25°C using monochromatic (wavelength 589 nm) plane-polarized light is 50.5 (deg)cm^2/g (to the right) in a solution containing 2.5 g of the compound in 100 cm^3 of ethanol. The specific rotation of a compound is thus defined precisely as, for example, its melting or boiling point or its solubility. The (+)- and (−)-isomers of a compound, (enantiomers, see the following subparagraph "The structure of a compound and its optical activity") have the same physical properties except their reaction to plane-polarized light. Such isomers can, however, have differences in their "*chiral properties*", which include various biological and physiological (such as taste and smell) effects and in certain cases, differences in their reaction rates.

For example, (+)-limonene (*R* form, cf., Chapt. "5.1. Fundamental Definitions") tastes like an orange, whereas *S*-(−)-limonene tastes like a lemon. Another such well-known difference is between *R*-(−)-carvone (tastes like green mint) and *S*-(+)-carvone (tastes like caraway). The most unfortunate example is the use of a medicine known as racemate, (±)-thalidomide, in the 1960s. Pure (+)-thalidomide is an efficient medicine against nausea, but its (−)-form causes serious fetal deformities. Substantial amounts of this drug were sold, for example, against nausea in pregnancy with horrible results before its production and sale were prohibited. It should be noted that even the use of the corresponding (+)-form would not have been risk free because it can be gradually racemized (cf., Chapt. "3.5. Racemic Modifications") in the organism to the (−)-form. On the other hand, many synthetic drugs are used as racemates; for example, the *S*-(+)-enantiomer of ibuprofen (dexibuprofen) is physiologically active, and the *R*-(−)-enantiomer relieves pain only as far as it is converted into dexibuprofen in the organism. These differences in properties have not been reported between enantiomers in enantiomer pairs formed by carbohydrates.

Besides specific rotation the *molecular rotation* $[M]$ (or ϕ), defined as follows:

$$[M]_{\lambda}^{t} = \left([\alpha]_{\lambda}^{t} \cdot M\right)/100, \qquad (3\text{-}3)$$

where λ is the wavelength of the light used, t the temperature, $[\alpha]$ the specific rotation of the compound, and M its molar mass.

According to Eq. 3-3, differences in the rotational properties can be explained as direct consequences of differences in the molar mass; this opens the possibility to investigate the relationships between the chemical structure and the optical rotation of similar compounds, such as different esters of an optically active carboxylic acid (see chiral properties).

The structure of a compound and its optical activity. If a figure, such as the structure of a molecule, contains *similarity* or *symmetry*, so-called "symmetry examination" will find either a midpoint, axis, or plane such that the parts of the figure are symmetrical with respect to it. This can be illuminated in practice, for example, with capital letters (Fig. 3.6), and similar symmetry can be shown to also exist in certain compounds (Fig. 3.7). If it is possible to place a *plane of symmetry (plane of mirror)*

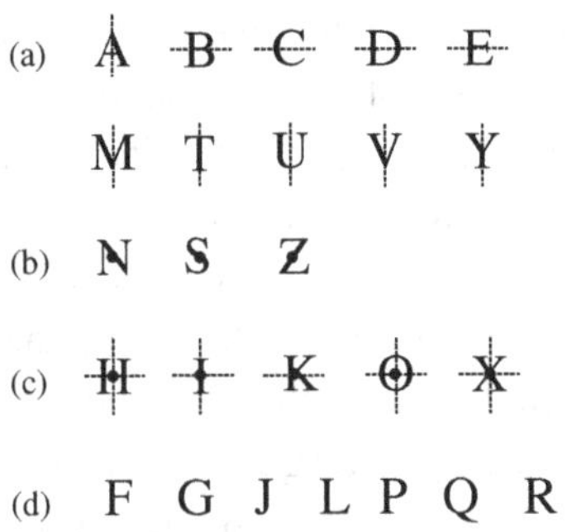

(a) A B C D E

M T U V Y

(b) N S Z

(c) H I K O X

(d) F G J L P Q R

Fig. 3.6. Capital letters of the alphabet can be classified according to their type of symmetry in the following groups: (a) letters that can be divided into two symmetrical halves with an axis so that a rotation of 180° about such an axis will lead to a letter superimposable with the original one, (b) letters that can be made superimposable with the original one by rotating them for 180° about a midpoint (vertical axis) in the horizontal plane, (c) letters with both symmetries described above, and (d) letters with no symmetries.

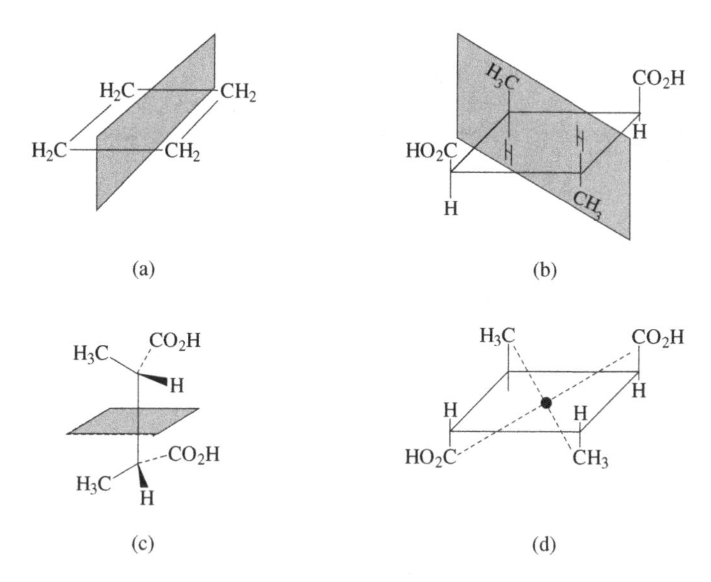

Fig. 3.7. Examples of plane of symmetry (a→c) and midpoint of symmetry (d) present in molecules.

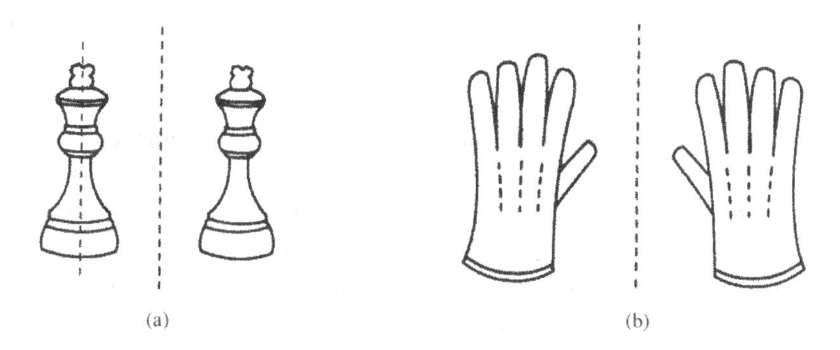

Fig. 3.8. A symmetrical object (a) is superimposable with its mirror image, but an unsymmetrical object and its mirror image (b) are non-superimposable.

in a three-dimensional object (such as a molecule), the parts separated by such a plane are superimposable mirror images. Similarly, an object that contains a plane of symmetry is superimposable with its mirror image (Fig. 3.8a). A mirror image that deviates from the original structure is obtained in the cases where the object examined is non-superimposable.

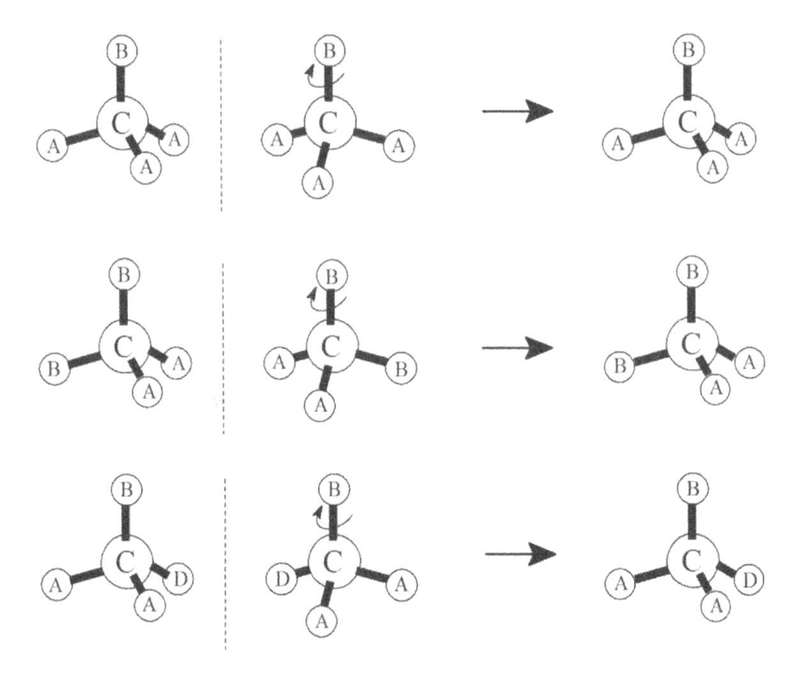

Fig. 3.9. These molecular models (CA_3B, CA_2B_2, and CA_2BD; A ≠ B ≠ D) can be made superimposable with their mirror images by suitably rotating the marked C-B bonds, and as symmetrical models they possess planes of symmetry.

In this way, for example, a left-hand glove and its mirror image, a right-hand glove, are non-superimposable when placed on top of one another (Fig. 3.8b). This property arising from asymmetry is generally called "*chirality*" (in Greek, "cheir" means "hand"), and it is the reason for the optical activity seen in unsymmetrical molecules (*chiral molecules*).

As mentioned before, most molecules are not chiral but *achiral* or symmetrical (i.e., contain symmetry elements), such that they are optically inactive and their mirror images are superimposable. The tetrahedral molecule models shown in Fig. 3.9 (CA_3B, CA_2B_2, and CA_2BD) and their mirror images can in each case be easily made superimposable by rotating the marked C-B bonds about their axes for a suitable amount. If the substituents of the atom differ from one another (model CABDE), the mirror image of the chiral compound will deviate from the original structure, and the structures can be made superimposable only by a rearrangement involving the breaking of two bonds (Fig. 3.10). The existing

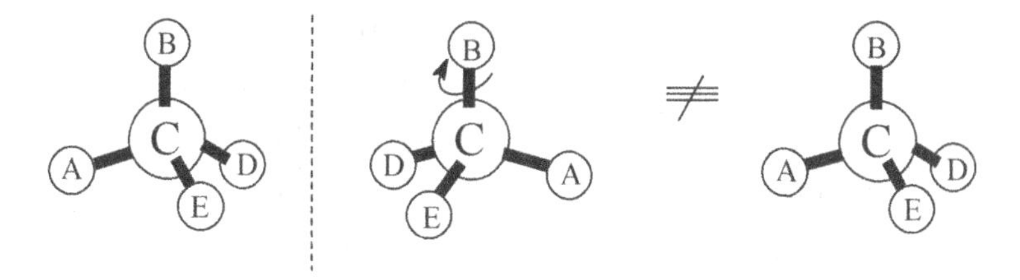

Fig. 3.10. When four different substituents (A ≠ B ≠ D ≠ E) bond with the atom C, two stereoisomeric forms are produced that are mirror images of one another. They cannot be superimposable without a rearrangement involving two bonds.

stereoisomers are *enantiomers* or *mirror isomers* (in Greek, "enantio" means "opposite" and "meros" "part") of each other, and they are also corresponding (+)- and (−)-isomers forming a *pair of enantiomers* (this type of stereoisomerism is called "enantiomorphism"). Enantiomers, hence, have the same physical properties except for their opposite response to plane-polarized light, which is why they are also *optical isomers* (the phenomenon is called "optical isomerism").

The molecular asymmetry described above arises from an *asymmetric* carbon atom (such carbon atoms are indicated with an asterisk * when necessary), with four different substituents (atoms or atom groups) bonded to them, which then forms the *chiral center* or *chirality center* (i.e., the center of asymmetry). The structural inequality of these enantiomers that arises from this center of chirality is also easily identified from molecular models; one or several asymmetrical carbon atoms cause almost without exception, among others, the optical activity of carbohydrates. (cf., Chapt. "1. Introduction"). On the other hand, it should be noted that even though an atom contains asymmetric carbon atoms, it does not necessarily lead to optical activity (cf., *meso* forms, Chapt. "4.2. The Total Number of Stereoisomers"). In these cases, internal symmetry exists between parts of the molecule, causing the chiral centers to "cancel" each other's effect.

Even though optical activity was first observed in compounds containing asymmetric carbon atoms, it also arises from centers of asymmetry formed by other atoms, such as silicon, nitrogen, phosphorus, and sulfur. Furthermore, optical activity arising from the asymmetry of the molecule is found in certain organic compounds that form rigid structures, either

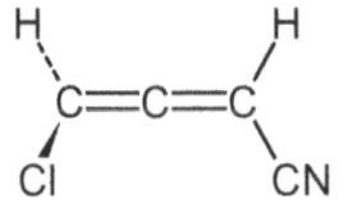

Fig. 3.11. Examples of optically active compounds with no asymmetric carbon atoms in their structures. 1-Chloro-3-cyanoallene (1) and 2,2′-dinitro-6′-carboxy-6-methylbiphenyl (2).

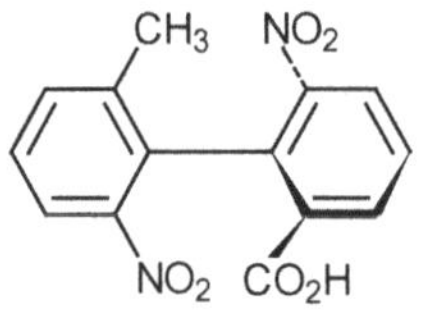

due to saturated bonds (e.g., allene derivatives containing an axis of chirality) or caused by hindered free rotation about a bond (e.g., certain so-called "atropisomers", such as biphenyl derivatives) (Fig. 3.11).

3.5. Racemic Modifications

The crystallization of enantiomers. When a solution contains two compounds that form a pair of enantiomers, they can under suitable conditions crystallize in two main modes depending on the type of the enantiomer pair: (i) the individual crystals consist of equal amount of both enantiomers or (ii) the individual crystals each consist of one enantiomer only (Fig. 3.12). The first case (ordered achiral crystals) is called "*racemate*" or "*racemic form*" or a "*racemic compound*" (true racemate) (in Latin, "racemus" means "bunch of grapes"). The melting point of the racemate deviates from that of its pure enantiomers, and it can be either higher or lower than the latter (Fig. 3.13). The other case (enantiomeric chiral crystals) is called "*racemic conglomerate*" (conglomerate) or a "*racemic mixture*". This forms a true eutectic mixture with a melting point that is always lower than that of the pure enantiomers. Generally, an eutectic mixture contains certain proportions of different components with a melting point lower than that of either of its components, and also lower than that of any other mixture of the same components. Upon crystallization, however, a racemate is formed with a probability of over 90%, since nowadays, fewer than 300 cases (e.g., in 1972 only 124) of enantiomers crystallizing as a racemic conglomerate are known.

The relationships between the melting point of a solid mixture and its enantiomeric composition have been widely investigated with thermal

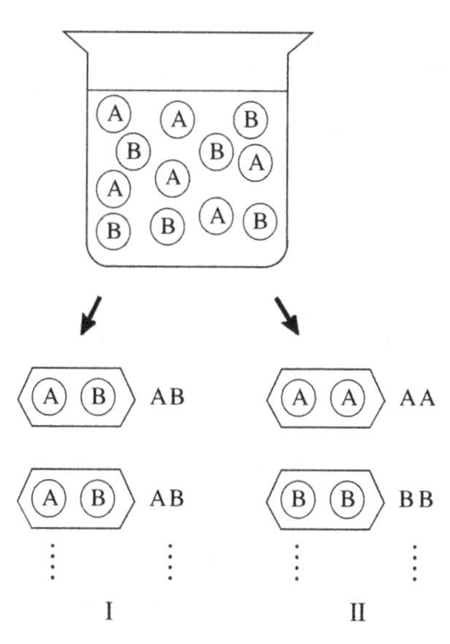

Fig. 3.12. Pair of enantiomers can crystallize from a solution in two different modes: (I) individual crystals contain equal amounts of each enantiomer (racemate), or (II) individual crystals consist of only one or the other pure enantiomer (racemic conglomerate).

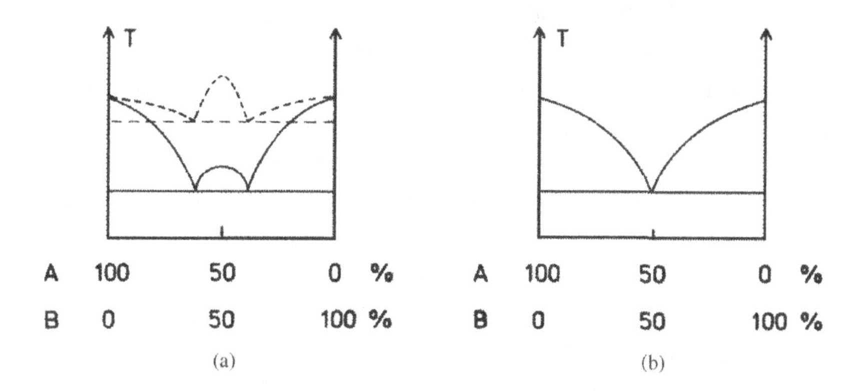

Fig. 3.13. Binary melting point diagrams of a racemate (a) and a racemic conglomerate (b) formed by two enantiomers (A and B) compared to the melting points of their enantiomers (T is the temperature). In the first case, the melting point of the racemate can be higher or lower than that of the enantiomers, and in the second case, it is always lower.

analysis, which reveals the actual nature of the racemic modification. These studies have shown that in extremely rare cases (third mode of crystallization), when the enantiomers are nearly identical in their shape, they can easily replace each other in the crystals that form. The melting

Table 3.5. General physical properties of the stereoisomers of tartaric acid

Form*	Melting point/°C	Density (20°C) /kg/dm^3	Specific rotation $[\alpha]_D$ (20°C)	Solubility (20°C) /g/100 gH_2O
(+)-	170	1.760	+12	139
(−)-	170	1.760	−12	139
(±)-	206	1.788	0	21
meso	140	1.666	0	125

*For chemical formulae, see Chapt. "4.2 Total number of stereoisomers".

point of the crystals with varying enantiomeric composition (a so-called "racemic solid solution" or "pseudoracemate") is in all cases nearly the same as (usually slightly higher than) the melting point of the pure enantiomers. Also in this case, it is not possible to separate the pure enantiomers from each other only by means of crystallization. A "quasiracemate" is a mixture of two similar but distinct compounds, i.e., one left-handed and the other right-handed.

As an example, Table 3.5 contains physical properties of the different stereoisomers of tartaric acid (cf., Chapt. "4.2. Total Number of Stereoisomers"). These (+)- and (−)-forms constitute a pair of enantiomers; therefore, they have the same physical properties except their opposite responses to plane-polarized light. Instead of this, the general physical properties of a *meso* form and a racemic (±)-forms differ, and these forms are not optically active either. Compared to other forms, the higher melting point and density of the racemic form and, on the other hand, its lower solubility in water, can be explained with the ability of the (+)- and (−)-enantiomers to position themselves alternately into the crystal lattice in a more solid structure than that formed by each individual compound. Racemic tartaric acid is thus among the cases with a melting point that is higher than that of its enantiomers.

Racemization. The synthetic preparation in the laboratory of chemically pure enantiomers that are found in nature is often difficult. Consequently, reactions — especially S_N1 reaction, in an S_N2 reaction, the configuration always becomes the opposite and so-called "*inversion*" takes place — is usually formed a racemate, which is optically inactive. Thus, the reduction

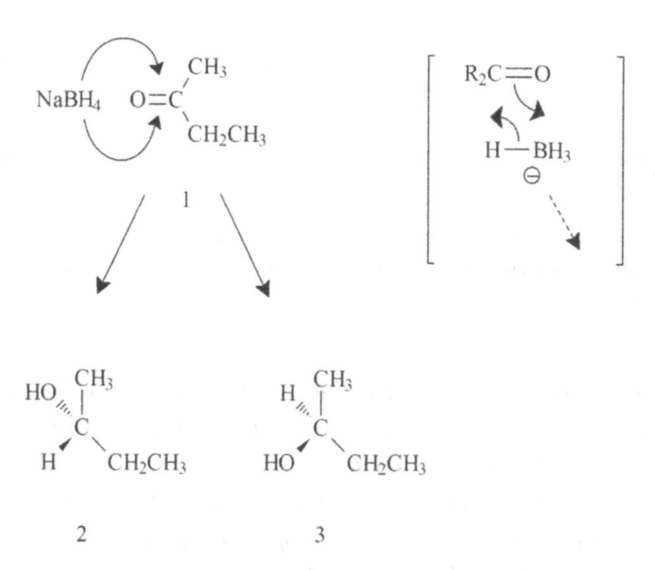

Fig. 3.14. The reduction of butan-2-one with sodium borohydride ($NaBH_4$) leads to racemic (±)-butan-2-ol such that a racemate consisting of equal amounts of enantiomers (2) and (3) is formed.

of butan-2-one with sodium borohydride leads to a racemic (±)-form because of the two possible directions of approach (Fig. 3.14). It can be generally stated that a synthesis in the laboratory of chiral compounds starting from achiral source materials will always lead to a racemic modification, so that optically inactive initial substances lead to an optically inactive product. Instead of this, many biological reactions that are catalyzed by chiral enzymes, chiral or achiral initial source materials will lead to chiral products.

A racemic modification can also be formed in some cases as a product of *racemization*, where a pure enantiomer or an optically active mixture of enantiomers gradually changes to a racemic modification under the influence of heat, light, or certain solvents (i.e., loses optical activity with time). For example, (+)-tartaric acid with a small amount of water is converted upon heating into (±)-tartaric acid. It is believed that in racemization, absorption of energy, and molecular oscillations cause the momentary appearance of planar intermediate products that have equal probabilities of forming tetrahedral enantiomers (Fig. 3.15).

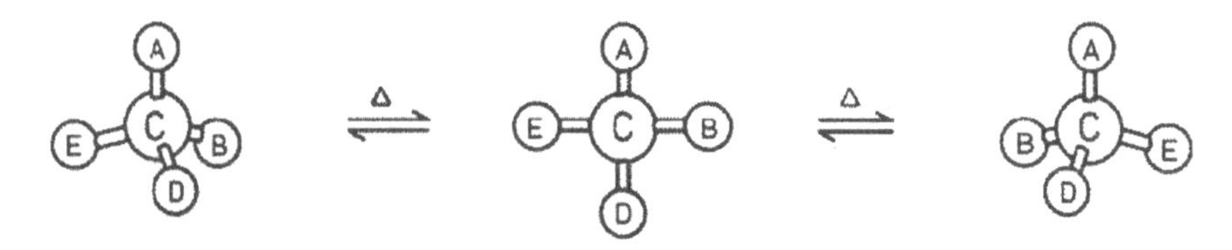

Fig. 3.15. A pure enantiomer can be racemized, for example, by absorbing heat, which causes the formation of a planar intermediate structure that leads to the equilibrium formation of the enantiomers.

The purity of racemic modifications. Mixtures of enantiomers where one component is dominant and whose specific rotation depends on their composition are frequently encountered. If $[\alpha]_{max}$ (or $[\alpha]_{100\%}$) is the specific rotation of the pure enantiomer under the same conditions and no optically active impurities are present, then the *optical purity OP* of the mixture (specific rotation $[\alpha]$) can be calculated from the following equation:

$$OP = 100 \bullet |[\alpha]| \, / \, |[\alpha]_{maks}| \, \% \qquad (3\text{-}4)$$

Assuming that the rotation of the plane of polarized light caused by each molecule of the enantiomer is an additive property, the following equation is valid (*EP* is the *enantiomeric purity* or *enantiomeric excess*, *EE*):

$$EP = 100 \bullet ([A] - [B])/([A] + [B]) \, \%, \qquad (3\text{-}5)$$

where $[A]$ and $[B]$ are the numbers of moles (or molecules) of the enantiomers in question ([A] > [B]).

If in this case *OP* is known for this mixture of enantiomers, by making the usual assumption that $OP = EP$, the *enantiomer ratio* or *enantiomeric composition* is as follows:

$$[A]/[B] = (100 + OP)/(100 - OP) \qquad (3\text{-}6)$$

Finally, the *enantiomer concentrations* ($A\%$ and $B\%$) can be calculated from the following equations:

$$A\% = 100 \bullet [A]/([A] + [B]) \, \% \qquad (3\text{-}7)$$

$$B\% = 100 \bullet [B]/([A] + [B]) \, \% \qquad (3\text{-}8)$$

It is, however, possible that the optical purity of the mixture differs from its enantiomeric purity, and the enantiomer ratio (in this example $[A]/[B]$) cannot always be determined only on the basis of polarimetric measurements. This case is encountered, for example, when the enantiomers form strong hydrogen bonds. In addition, specific rotation ($[\alpha]_{max}$) of the pure enantiomer is not always known, or it is too small to measure. In this case, the enantiomeric purity has to be measured by other techniques.

The concentration of each enantiomer in a solution can in principle be determined if (i) the enantiomers can be quantitatively converted with a suitable chiral reagent to products that no longer are mirror images of one another (i.e., they have different physical properties, see the following subchapter "The resolution of enantiomers") and their concentrations can be separately determined, either spectroscopically (NMR) or chromatographically (GC or HPLC), or (ii) directly chromatographically (HPLC or GC) or using capillary electrophoresis (CE), where the separation in direct methods is based on the use of a chiral stationary phase. In addition, in certain cases, it is possible to determine the concentrations of the enantiomers directly with NMR using a suitable chiral solvent (that attaches to the enantiomer with a hydrogen bond) or a chiral transition reagent (that forms a complex with the enantiomer). The analysis of enantiomers with these methods directly or with the help of a derivative (often requiring a large excess of the enantiomer) is not a simple task and it depends on the case under investigation. The general trend is toward replacing the historically employed concept "optical purity" (i.e., based on polarimetric measurements only) with the term "enantiomeric purity".

The resolution of enantiomers. As two enantiomers that form a pair have the same general physical properties, they cannot be separated from one another with ordinary chemical or physical methods. Therefore, the separation usually requires chiral reagents or chiral catalysts. This separation of racemic variants is called "*resolution*" ("chiral resolution" or "resolving"), and methods used for this purpose are briefly discussed below.

(1) *Mechanical resolution.* Racemic modifications can spontaneously crystallize as pure enantiomer crystals only in the case of a racemic conglomerate (see subchapter "The crystallization of enantiomers" above).

This possibility is very rare and, therefore, unsuitable for general use. However, it has major historical significance, especially in the chemistry of carbohydrates (cf., Chapt. "2.2. The Time After the 1700s"), since in 1848 Louis Pasteur at age 26 prepared crystals of ammonium sodium tartrate by slowly evaporating an aqueous solution of this pair of enantiomers. He noticed that the crystals formed were mirror images of one another (i.e., they had differently oriented hemihedral surfaces), allowing them to be separated using a magnifying glass and tweezers. Since Pasteur worked in Paris that summer and not in the torrid Mediterranean climate, he was able to achieve this spontaneous mode of crystallization that only takes place below 27°C. Otherwise, the beginning of carbohydrate chemistry could have been substantially delayed. A stroke of luck was also that this particular salt crystallizes as a racemic conglomerate and not as a racemate.

Nowadays, only a few technical applications, such as the preparation of given enantiomers of glutamic acid, α-methyl-L-dopa (α-methyl-3′,4′-dihydroxy-L-phenylalanine) and chloroamphenicol, involve their crystallization in a pure form. In these cases, only one enantiomer can be crystallized under controlled conditions from a supersaturated solution of the racemic modification using seed crystals of the pure enantiomer in question (while the other enantiomer remains in the solution). It is also possible (e.g., in the resolution of α-methyl-L-dopa) to pass the supersaturated solution through two "crystallization chambers", each containing pure seed crystals of one enantiomer such that the L-(+)- and D-(−)-forms (cf., Chapt. "5.1. Fundamental Definitions") can be separately crystallized. Of these enantiomers, only the L-(−)-form is physiologically effective; as the precursor of dopamine that forms from it in the organism, it is used to treat Parkinson's disease.

(2) *Resolution based on diastereoisomers.* When the enantiomers of a racemic modifycation react with the same achiral compound (e.g., acids that constitute a pair of enantiomers create esters with an optically inactive alcohol), two enantiomeric products are formed. If the same enantiomers react with a chiral compound (e.g., acids that constitute a pair of enantiomers create esters with an optically active alcohol), two *diastereoisomers* or *diastereomers* are formed. These compounds are stereoisomers that

no longer are mirror images of one another and, consequently, are not enantiomers. Diastereoisomers have different physical and often also chemical properties such that they can be separated from each other with ordinary laboratory procedures, such as crystallization, chromatography, or distillation. Besides the requirement of an efficient resolution, it is important for the general success of the resolution that the reagent is a suitable chemical; it needs to be easily available, to be easy to regenerate, to react readily and completely with the enantiomers comprising the racemic modification, and does not, for example, itself racemize during the resolution.

The synthesis of diastereoisomeric compounds offers the best preparative opportunity for separating the enantiomers comprising a racemic modification. This is traditionally done by preparing a salt from racemic acid enantiomers (or basic enantiomers) and a chiral base (or acid) (Fig. 3.16). The reagents most frequently used are natural or synthetic compounds, among them bases, such as amphetamine, brucine, ephedrine,

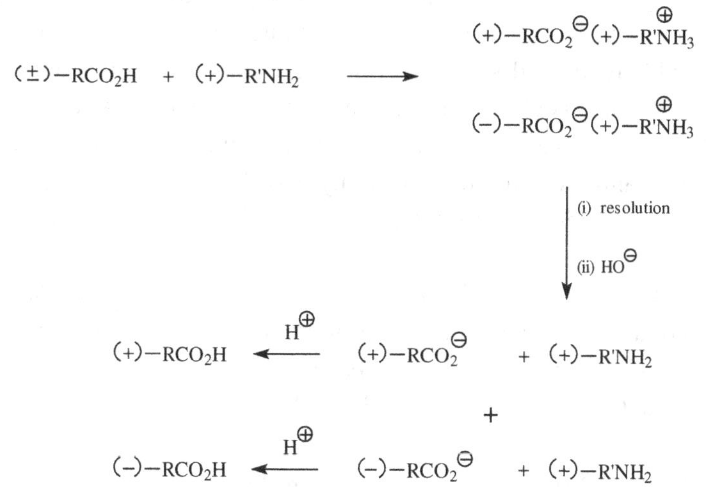

$$(\pm)\text{–}RCO_2H + (+)\text{–}R'NH_2 \longrightarrow \begin{matrix} (+)\text{–}RCO_2^{\ominus}(+)\text{–}R'\overset{\oplus}{N}H_3 \\ (-)\text{–}RCO_2^{\ominus}(+)\text{–}R'\overset{\oplus}{N}H_3 \end{matrix}$$

$$\downarrow \text{(i) resolution; (ii) } HO^{\ominus}$$

$$(+)\text{–}RCO_2H \xleftarrow{H^{\oplus}} (+)\text{–}RCO_2^{\ominus} + (+)\text{–}R'NH_2$$

$$+$$

$$(-)\text{–}RCO_2H \xleftarrow{H^{\oplus}} (-)\text{–}RCO_2^{\ominus} + (+)\text{–}R'NH_2$$

Fig. 3.16. As an example of resolution, racemic acid ((±)–RCO_2H) is allowed to react with chiral base ((+)–$R'NH_2$) and the diastereoisomeric salts formed are first separated from each other with physical methods. Then, the acid enantiomers are liberated as pure compounds ((+)–RCO_2H and (−)–RCO_2H). The amine base ((+)–$R'NH_2$) liberated earlier in alkaline hydrolysis can be separated from the carboxylate salts, for example, by extracting it into ether that does not dissolve carboxylate salts.

quinidine, quinine, cinchonadine, cinchonine, strychnine, 2-amino-1-butanol, dehydroabietylamine, and 1-phenylethylamine, as well as acids, such as camphoric acid, malic acid, pyroglutaminic acid, and tartaric acid. For example, the D-(−)- and L-(+)-forms (corresponding *R* and *S* forms) of racemic (±)-lactic acid can be separated from each other with *S*-1-phenylethylamine, which first forms the corresponding diastereoisomeric *R*,*S*- and *S*,*S*-salts.

Thus far, only the resolution of enantiomers contained in racemic acid and basic modifications has been discussed. In separating, for example, enantiomeric alcohols, difficulties may arise in certain cases. One option, then, is to turn the alcohol into an acid derivative ("half ester") by allowing it to react with a divalent acid, after which the resolution can be accomplished with a chiral base (that will react with the free acid group). In this way, an (±)-alkyl alcohol can react with the anhydride of succinic or phthalic acid, forming mono-(±)-alkyl phthalate ester in the latter case. Similarly, racemic esters can be directly converted to, for example, diastereoisomeric urethanes ($RR'NCO_2R''$). These techniques have nowadays been mostly replaced by chromatographic separations. Various diastereomeric compounds, such as acetals and ketals, can also be produced from racemic aldehydes and ketones. Aldoses, ketoses, and their derivatives (such as aldonic acids) also contain many reactive functional groups (i.e., the anomeric carbon atom and anomeric hydroxyl group in aldoses and ketoses and the often primary ω-hydroxyl group, as well as the carboxyl group in aldonic acids) that can be used in the resolution.

(3) *Chromatographic and capillary electrophoretic resolution.* As mentioned above, the analytical (and often also preparative) separation of enantiomers is in certain cases possible with liquid and gas chromatography (HPLC and GC) by using a solid (or liquid) chiral stationary phase (so-called "selector"). To obtain a good separation, the difficulty often lies in the choice of the selector; the chiral phases can be classified into several main groups that differ in their mechanism of function. Generally, the adsorption of the enantiomers onto the surface of this phase varies and causes the repeated formation of momentary "diastereomeric complexes". The "more weakly bonded" enantiomer is gradually being eluted away from column faster than the other enantiomer.

This method of separation has been put into practice by using, among others, lactose, starch, cellulose, cellulose derivatives, and optically active quartz as selectors (adsorbents) in traditional liquid chromatography, but in modern HPLC, there are many different special selectors available for different purposes. They are prepared, for example, by attaching suitable chiral compounds (such as proteins) on the surface of a silica phase, or they are spiral polymers, cyclodextrines, or cyclic ethers attached to a carrier material or optically active ion exchange resins. Selectors frequently employed in CE are cyclodextrines and their derivatives. These compounds consist of cyclic oligosaccharides containing ring structures with cavities whose size depends on the starting carbohydrate. In addition, polysaccharides and cyclic ethers are frequently used as selectors. The chiral selectors usually have their best separation efficiency at a certain optimum concentration.

Another possibility for chiral resolution is the use of a chiral mobile phase and a normal "ordinary" stationary phase. In this case, the separation requires different ratios of distribution between the mobile and stationary phases for the two enantiomers; this can be influenced to a certain extent by adjusting the conditions of the HPLC run. The simplest way is to add the chiral compound directly to the eluant and, in the CE method, to the electrolyte. In addition to this "direct method", it is possible to use an "indirect method", where the diastereoisomers are prepared with the help of a chiral reagent before the separation. This indirect procedure is particularly suitable in GC, although chiral stationary phases for both capillary and packed GC columns are already available. Further, in certain cases, the resolution of racemic modifications has been accomplished with paper chromatography by using plain paper (cellulose) or paper impregnated with another chiral auxiliary reagent. In this way, for example, the enantiomers of (±)-aminophenylethanoic acid ($C_6H_5CH(NH_2)CO_2H$) can be separated from each other using paper impregnated with (±)-camphor-10-sulfonic acid.

(4) *Biological resolution.* Optically active compounds are common in nature because living organisms generally produce only one enantiomer of each enantiomer pair. Thus, for example, in the yeast fermentation of starch, (−)-2-methyl-1-butanal is formed, and under anaerobic stress of a

muscle (+)-lactic acid is produced, while (−)-malic acid is found in fruit juices and (−)-quinine in the bark of cinchona trees. Because of this, one possibility for the isolation of racemic modifications, or for obtaining one of the enantiomers in a pure state, is to treat the modification in a cultivation environment with a microorganism (such as specific bacteria, fungi, and algae) that can consume only one of the enantiomers present. With this method one can, for example, produce pure *R*-(+)-nicotine (β-pyridyl-α-methylpyrrolidine) by incubating *R,S*-(±)-nicotine with bacterium *Pseudomonas putida*, which oxidizes the *S*-(−)-form but not the *R*-(+) form. In the same way, upon feeding, for example, racemic (±)-mevalonic acid (3,5-dihydroxy-3-methyl-pentanoic acid) to rats, one enantiomer is metabolized while the other one is excreted in urine.

The drawbacks of this method are the possible toxicity of the enantiomers or the possibility that they are not metabolized at all, making it difficult to find a suitable biological system. On the other hand, this method can be seen as a special application of the separation of racemic modifications based on the formation of diastereoisomers because the reactions in living organisms are controlled by chiral enzymes (protein catalysts). Here, the ability of an organism to metabolize a material (a substrate) depends on the presence of an enzyme that is able to adsorb the substrate for the required chemical transformation. The formation of an enzyme/substrate combination thus represents the integration of the enantiomer into a chiral enzyme, where the enantiomer that easiest forms this stereoisomeric complex will be metabolized.

3.6. Conformational Isomerism

Conformational isomerism, a type of stereoisomerism, is traditionally seen to occur among open-chained organic compounds with single covalent bonds, to a large extent C-C bonds. Single bonds enable in principle "free rotation" of about the bond axis so that the same molecule can momentarily change into several spatially "unequal" positions without the breaking or forming of bonds. It can be generally stated that the *conformations* or *conformers* (conformational isomers) of a molecule signify its different spatial states (i.e., the present spatial placement of the atoms comprising the molecule) that appear as the atoms or atom groups rotate with respect to one another within the limits set by the bonds. It follows that chemical properties of the conformations are the same and they do not, in spite of their different energy levels, represent isolable individual compounds.

In the context of a confirmation study of the simplest alkane, ethane, it was noticed in the 1930s (by the research groups of American Kenneth Sanborn Pitzer 1914–97 and Japanese San-Ichiro Mizushima 1899–1983) from its spectroscopical and thermodynamical properties that this molecule exists in a given stable form and that the rotation about its C-C bond is not completely "free" but contains certain "thresholds". As a consequence of this bond-centered rotation, it is possible for ethane to form an infinite number of different conformations. Figure 3.17 presents two ways of expressing (sawhorse projection and Newman projection according to Melvin Spencer Newman 1908–93 in 1952) the corresponding "extreme conformations": *staggered* and *eclipsed conformations.*

In the Newman projection, the ethane molecule is viewed along the direction of the C-C bond; therefore, it is easy to visualize the distances between the hydrogen atoms bonded to the carbon atoms that are

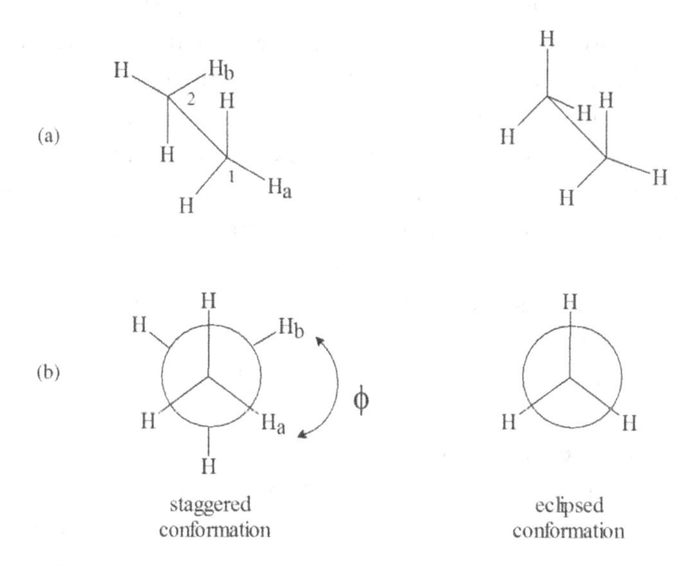

Fig. 3.17. The staggered and eclipsed conformations of ethane depicted as (a) sawhorse projection and (b) Newman projection. The angle between the planes that contain the $H_aC_1C_2$ and $C_1C_2H_b$ atoms is called the "dihedral angle" (ϕ).

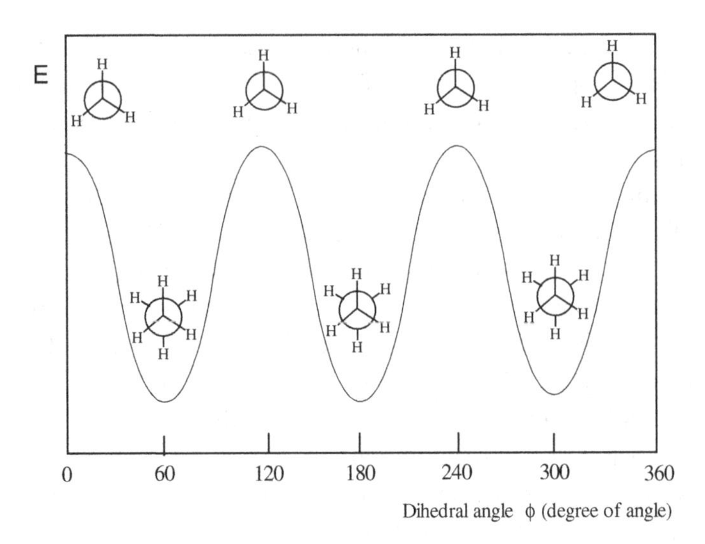

Fig. 3.18. Relative change in the torsion strain energy of an ethane molecule as a function of dihedral angle φ (see Fig. 3.17).

projected onto planes at right angles to the bond axis, as the dihedral angle φ changes. In the staggered conformation (φ is 60°, 180° or 300°, or ±60° or ±180°), the hydrogen atoms bonded to the C_1 and C_2 atoms are as far apart as possible, so that their interaction is at minimum, and this conformation corresponds to a minimum of the torsion strain energy (torsional energy) (Fig. 3.18). Instead, in the eclipsed conformation (φ is 120°, 240° or 360°, or ±0° or ±120°), H_a and H_b are, like other corresponding pairs of hydrogen atoms, as close to each other as possible, and this conformation represents the maximum of torsion strain energy. Between these two extreme conformations, there naturally are an infinite number of different "intermediate conformations" ($0° < φ < 60°$), with torsion strain energies being intermediate between the minimum and maximum energy values mentioned.

The difference in torsion strain energy between the staggered and eclipsed confirmations is relatively low at 25°C (about 12.2 kJ/mole); this is why the rotation of the CH_3 groups about the C-C bond is quite fast at room temperature (the interval between extreme conformations is about 10^{-12} s). In practice, this difference is, however, appreciably higher than, for example, the value of the factor RT in the equation of state for gases per mole, about 2.4 kJ/mole at 25°C. The minimum energy of torsional oscillations in ethane at this temperature is about 1.7 kJ/mole or about 17% of the energy threshold. Thus, under these circumstances, only a very small fraction of ethane molecules momentarily possess sufficient rotational energy to exceed the threshold. In order to further illustrate this point, one can assume that individual photographs with extremely short exposure times could be taken of an ethane molecule. Then, the photographs would show with a very high probability a frozen staggered conformation (or some "energetically favorable" oblique conformations). The photo with a much longer exposure time would look like a spinning top. Thus, it cannot be stated that even ethane is in a free state of unhindered rotation about its C-C bond. On the other hand, from the practical point of view, there is no great need in organic chemistry to know the energy levels of the rotational vibration of the various conformations.

In alkanes, more complex that ethane, rotation about the axis of a bond (C(a,b,c)-C(A,B,C) (the letters denote substituents that can be the same or different) can lead to complicated rotation energy profiles that depend on several dihedral angles. Thus, it is possible, for example, to observe — depending on the bond in question — several energy thresholds for the formation of different conformations so that the resulting energy profile may repeat itself only after each full rotation (360°). An example of this is *n*-butane and the rotation about its C_2-C_3 bond with three-fold barriers (Fig. 3.19).

In cycloalkanes, rotation about the C-C bonds is inhibited; the interactions of the atoms or atom groups chemically substituted onto the carbon atoms cause only spatial twisting of the ring structure into an energetically most favored ring conformation (cf., Chapt. "6.3. Naming of Conformations"). This phenomenon is especially significant in carbohydrate chemistry because the common monosaccharides consist of five- and six-atomic rings. Research relating to this subject started essentially with Norwegian Odd Hassel's (1897–1981, worked as professor of physical chemistry at University of Oslo) observations of cyclohexane and its derivatives. Hassel and British chemist Derek Harold Richard Barton (1918–98) later developed (they shared the 1969 Nobel Prize in Chemistry) *conformational analysis*, which covers the subject and enables, for example, predictions of the spatial structure of organic compounds (e.g., their conformation vs. configuration).

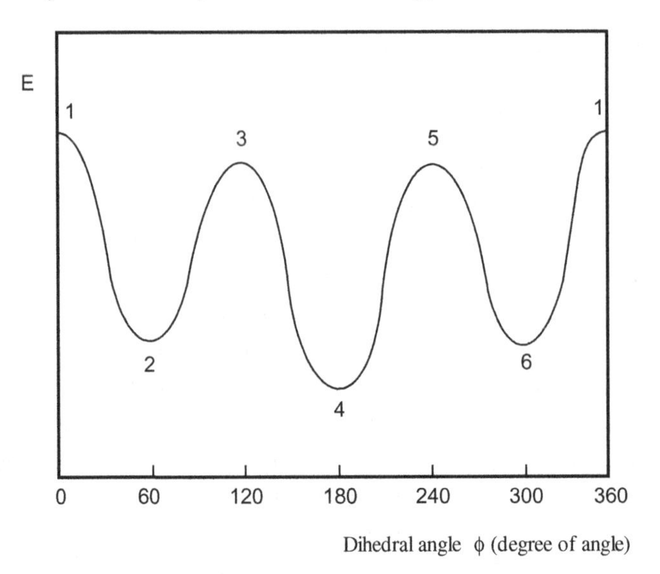

Fig. 3.19. Relative change in the torsion strain energy of the conformations arising in the rotation of a butane molecule about its C_2-C_3 bond as a function of an increasing dihedral angle (ϕ) from 0° to 360° (for clarity, the hydrogen atoms have been omitted). About 70% of the momentarily appearing conformations are of type (4) (*anti* conformation), while the remaining ones are mainly of types (2) and (6). The energy threshold between conformations (1) (*cis* conformation) and (4) is about 19 kJ/mole, between conformations (1) and (2 or 6) about 15 kJ/mole and between (1) and (3 or 5) about 4 kJ/mole.

3.7. Geometric Isomerism

The C=C bond is rigid; thus, a rotation about it is not possible. It follows that atoms or atom groups substituted on a carbon atom that has a double bond can in certain cases (i.e., when the two substituents are not identical) be located in two different ways with respect to the double bond. This stereoisomeric phenomenon is called "*geometric isomerism*" or "*cis/trans*

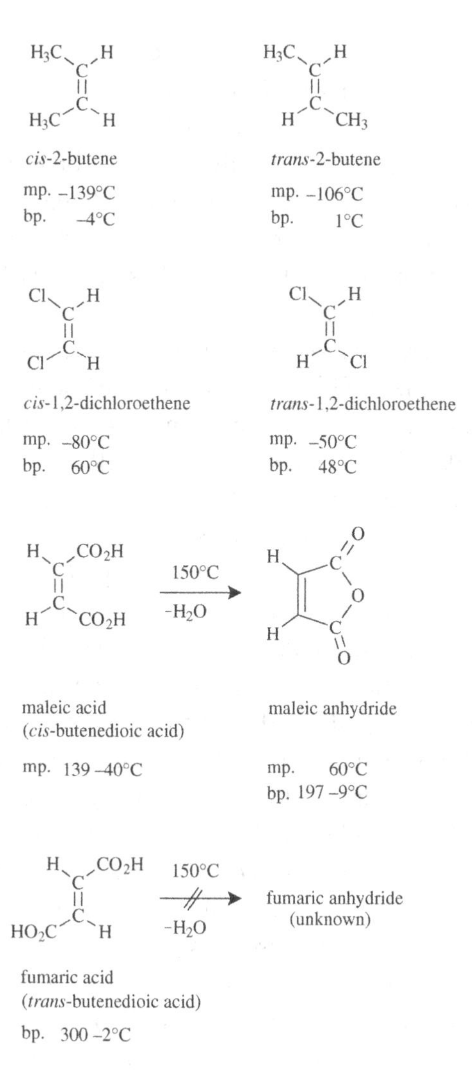

Fig. 3.20. Examples of *cis* and *trans* isomerism. It should be noted that upon heating of maleic acid its anhydride is formed, but in the case of fumaric acid, the corresponding anhydride is not formed.

isomerism"; if similar substituents are located on the same side with respect to the double bond, we speak of *cis isomers,* and in the reverse case of *trans isomers* (in Latin, "cis" means "on this side" and "trans" "on the other side", "across"). Examples of simple *cis* and *trans* isomers are shown in Fig. 3.20. It can generally be stated that *cis* and *trans* isomers are individual compounds that differ from each other in their chemical and physical properties. Geometric isomerism is, however, rare in the chemistry of carbohydrates because double bonds occur relatively seldom. On the other hand, compounds that only possess geometric isomerism are not optically active. If, after all, there are double bonds in carbohydrates, they are likely to have also one or several chiral centers (i.e., one or several asymmetric carbon atoms). This way, these carbohydrate derivatives can also have optical isomerism.

Cis isomers generally have lower melting points and higher boiling points than *trans* isomers. This phenomenon is explained by *cis* isomers usually being polar (having large dipole moments), while *trans* isomers usually are nonpolar (having dipole moments close to zero) molecules. This follows from the fact that, for example, in normal 2-butene (Fig. 3.20), the double bond causes the carbon atoms of the methyl groups (C_1 and C_4 atoms) to have a small positive charge ($\delta^{\oplus}$) and, correspondingly, the C_2 and C_3 atoms to have a small negative charge ($\delta^{\ominus}$). In 1,2-dichloroethene (Fig. 3.20), due to the difference in electronegativity between chlorine and carbon atoms, the chlorine atoms have a small negative charge ($\delta^{\ominus}$), while the carbon atoms on both sides of the double bond have a small positive charge ($\delta^{\oplus}$). This causes the charge vectors of *trans* isomers to cancel each other (the dipole moment μ of *trans*-1,2-dichloroethene is 0 D), while that does not happen in the *cis* forms (μ of the *cis*-1,2-dichloroethene is 1.85 D). Thus, the *cis* isomers have weak dipole-dipole interactions and dispersion forces whose unraveling requires extra energy compared to the *trans* isomers (this is seen in the corresponding boiling points). In addition, it is easy to understand that the crystal lattices of the solid *cis* and *trans* isomers are built from different basic units, which causes differences (due to different packing and stability of the crystal lattices), especially in the melting points and solubilities of these isomers. The unit of dipole moment, 1 D (debye), is 10^{-18} esu·cm (esu is the electrostatic unit) = $3.336 \cdot 10^{-30}$ Cm (coulomb-meter, the SI unit), received its name after Dutch physicist Peter Joseph William Debye (1884–1966). He received the 1936 Nobel Prize in Chemistry for, among others, his studies of the dipole moments of molecules.

Information about the dipole moments of compounds can be utilized in clarifying the stereochemical structure of isolated isomers. Other resources for this are the isomer's melting and boiling points, spectroscopical (IR and NMR) data, and their chemical reactivity. On the other hand, the general physical properties of *cis* and *trans* isomers are often quite similar, and it is not always possible to separate them from one another by fractional distillation or crystallization. For this reason, their separation often requires chromatographic techniques. A special case is presented by the non-volatile maleic and fumaric acids (*cis*- and *trans*-butenedioic acids) that form under normal conditions by heating malic acid. These isomers can be separated by distillation because only maleic acid is capable of forming a volatile anhydride.

Geometric isomerism is not limited to compounds whose structural rigidity only derives from double bonds. The ring structure of cyclical compounds also forms a "reference structure" relative to which different substituents can be located in different ways (either above or below the ring). For example, dimethylcyclohexane can exist altogether in nine stereoisomeric forms (Fig. 3.21), as one 1,1 isomer, six 1,2 and 1,3 isomers, and two 1,4 isomers. These stereoisomers are separate compounds and they have, at least in principle, individual physical properties. This type of geometric isomerism has a certain significance in carbohydrate chemistry.

It is not always possible to unequivocally express the isomerism of compounds containing double bonds in the terms of the *cis/trans* system. If three or four mutually different atoms or atom groups are attached to the carbon atoms having a double bond ($A \neq B \neq D \neq E$ or either $A \neq B = D \neq E$ or $A \neq B = E \neq D$),

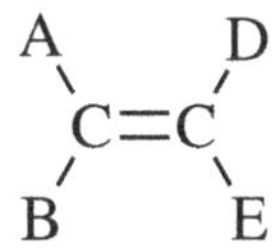

```
A     D
 \   /
  C=C
 /   \
B     E
```

the corresponding stereoisomers are named using the prefixes *E* (in German, "entgegen" means "opposite") and *Z* (in German, "zusammen" means "together"). The use of these prefixes is based on the relative priority of the substituents attached to the carbon atoms ($A > B$ and $D > E$), so that if the substituents with the highest priority (A and D) are situated on the same side of the double bond, we have a *Z* isomer, and in the opposite

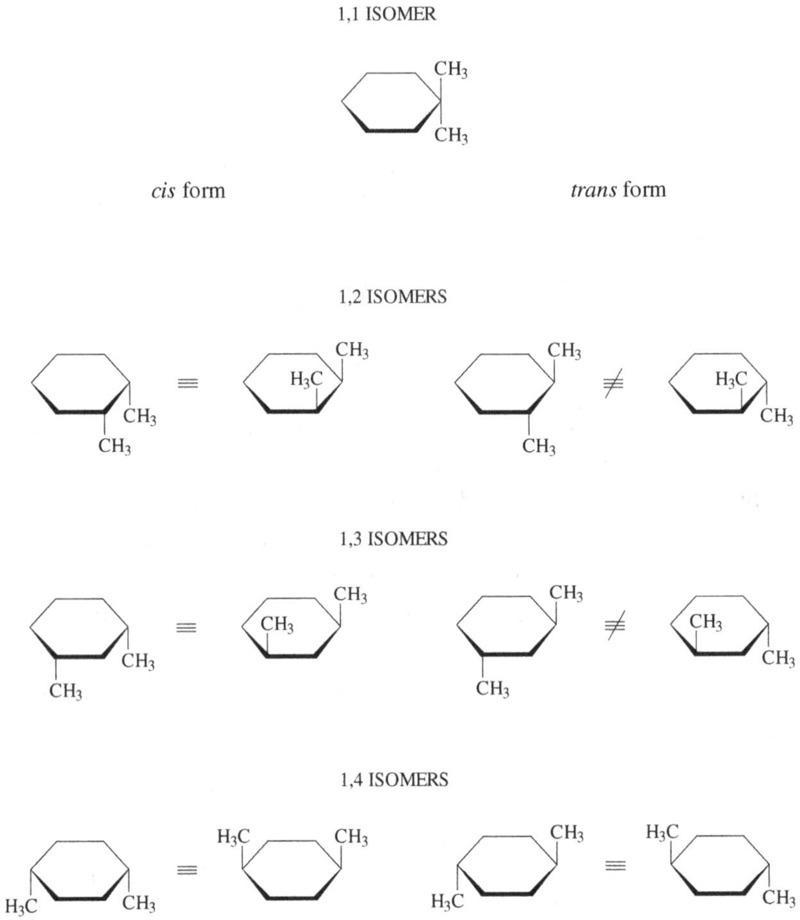

1,1 ISOMER

cis form

trans form

1,2 ISOMERS

1,3 ISOMERS

1,4 ISOMERS

Fig. 3.21. By presenting the ring structure of dimethylcyclohexane as a plane, it is easy to elucidate the spatial structure of the possible stereoisomers. In four cases (1,2-*trans*- and 1,3-*trans*-dimethylcyclohexanes), the compounds do not exhibit internal symmetry (i.e., there is no plane or midpoint of symmetry), and they are optically active isomers.

case, an *E* isomer. This system can also be applied to the *cis* and *trans* isomers, if needed. Three researchers share the praise for creating the order of priority, the "Cahn-Ingold-Prelog system" (CIP system), developed for the naming of the absolute configuration of an asymmetric carbon atom: British Robert Sidney Cahn (1899–1981), British Christopher Kelk Ingold (1893–1970), and Croatian-Swiss Vladimir Prelog (1906–98, the 1975 Nobel Prize in Chemistry) in 1956 (cf., Chapt. "5.2. Relative and Absolute Configuration"). The priority rules of the CIP system (the *sequence rules*) can be summarized in the order of priority as follows:

(1) The four substituents of a tetrahedral chiral center are ranked in order of decreasing atomic number (Z) of the atoms directly bonded to the chirality center. The group having the atom of higher Z receives higher priority. Therefore, there is, for example, an order I > Br > Cl > S > F > O > N > C > H (or, e.g., $-SH > -OH > -NH_2 > -CH_3$). Correspondingly, isotopes of the same chemical element are listed in order of decreasing atomic mass; ${}^3_1H$ (T) > ${}^2_1H$ (D) > ${}^1_1H$ (H). However, the priority for a lone pair of electrons is the lowest.

(2) If two or more atoms bonded to the chiral center have the same Z, the second atoms are used to rank the substituents. However, if the second atoms are also the same, the third atoms are used, etc.; at the earliest difference, the group containing the atom of higher Z receives higher priority. Thus, the priority of the groups $-CH_2Cl > -CH_2OH > -CH_3$ is based on the order of Cl > O > H. Correspondingly, the priority of the following alkyl groups (note also the priority ranking $-SCH_3 > -OCH_3 > -N(CH_3)_2 > -CF_3 > -CH_2OH$) is:

$$-C(CH_3)_2-CH_3 \;>\; -CH(CH_3)-CH_3 \;>\; -CH_2-CH_3$$

(3) Multiple bonds are counted as the corresponding number of single bonds: a double-bonded atom is counted twice and a triple-bonded atom three times. In this case, the equivalents are, for example, as follows:

$$-C(=O)OH \;>\; -C(=O)NH_2 \;>\; -C(=O)H \;>\; -C{\equiv}N$$

$$\parallel \qquad \parallel \qquad \parallel \qquad \parallel$$

$$-C(O)(O)-O \;>\; -C(O)(N)-O \;>\; -C(O)(H)-O \;>\; -C(N)(N)-N$$

According to this rule, it is possible to obtain, for example, the priority order (note also the priority ranking -SO_3H > -NO_2): -CO_2CH_3 > -CO_2H > -$CONH_2$ > -$COCH_3$ > -CHO > -CH(CH_3)OH > -C_6H_5.

(4) In rare cases, the substituents may differ from each other only according to their three-dimensional structures (i.e., they are stereoisomers); subsequently, the priority order is as follows: *Z* isomer > *E* isomer (*cis* isomer > *trans* isomer) and *R* configuration > *S* configuration (cf., Chapt. "5.2. Relative and Absolute Configuration").

Figure 3.22 shows some examples of *E* and *Z* isomers having double bonds C=C, C=N, and N=N.

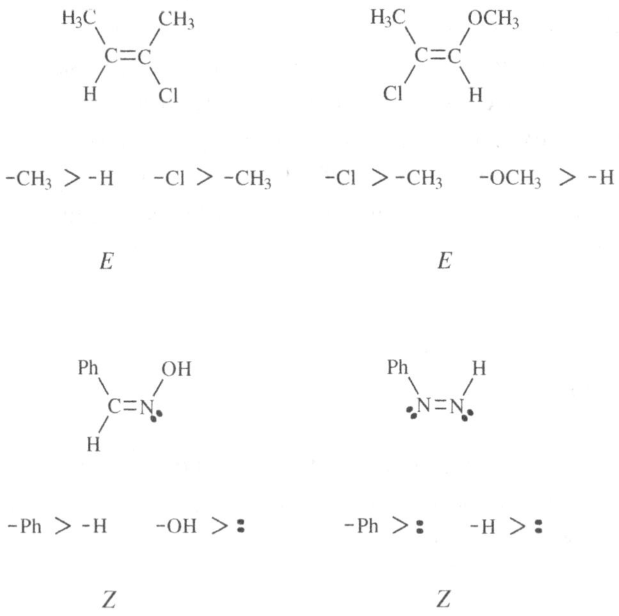

H_3C(H)C=C(CH_3)(Cl) H_3C(Cl)C=C(OCH_3)(H)

-CH_3 > -H -Cl > -CH_3 -Cl > -CH_3 -OCH_3 > -H

E *E*

Ph(H)C=N(OH) Ph–N=N–H

-Ph > -H -OH > : -Ph > : -H > :

Z *Z*

Fig. 3.22. Examples of *E* and *Z* isomers. Ph is a phenyl group.

4. Representation of Open-chain Chiral Molecules as Planar Formulas

4.1. The Fischer Projection

Three-dimensional molecular models are well-suited for the structural study of compounds with small molecules, but their utility diminishes as one attempts to illustrate molecules with many chiral centers. In addition, it is naturally essential to be able to express the desired stereostructures unequivocally and quickly also as planar formulas. The currently favored system in carbohydrate chemistry to illustrate open-chain monosaccharides that contain asymmetrical carbon atoms was born during the 1800s as a result of the research of especially Emil Fischer (cf., Chapt. "2.2. The Time After the 1700s"). Figure 4.1 shows *Fischer projections* for both optical isomers of the simplest aldose, glyceraldehyde (2,3-dihydroxypropanal), which form a pair of enantiomers.

Fischer-projections are meaningful only in the cases where molecules contain at least one asymmetric carbon atom. In drawing the formulas, one should generally consider the following (see Fig. 4.1):

(1) The main carbon chain is drawn vertically so that the carbon atom with the lowest order number is on the top of the chain.
(2) Horizontal bond lines that start from an asymmetric carbon atom depict bonds that are directed above the plane of writing (toward the viewer).

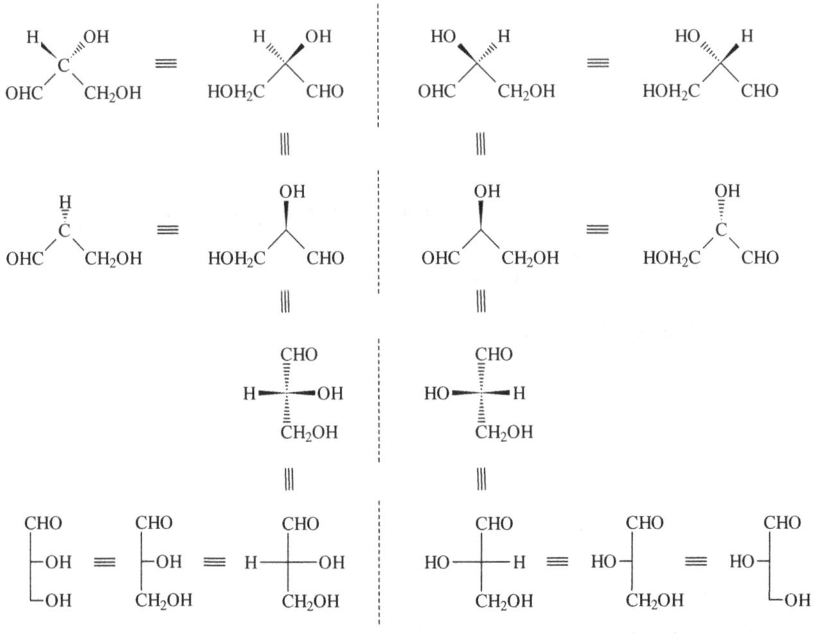

Fig. 4.1. Fischer projections of D-(+)- (1) and L-(−)-glyceraldehyde (2).

(3) Vertical bond lines that start from an asymmetric carbon atom depict correspondingly bonds that are directed below the plane of writing (away from the viewer).

If for some reason it is not desired to indicate, or it is not known, which way the bond from an asymmetric carbon atom is directed, the situation can be expressed as follows (the asymmetric carbon atom can also be marked with an asterisk *):

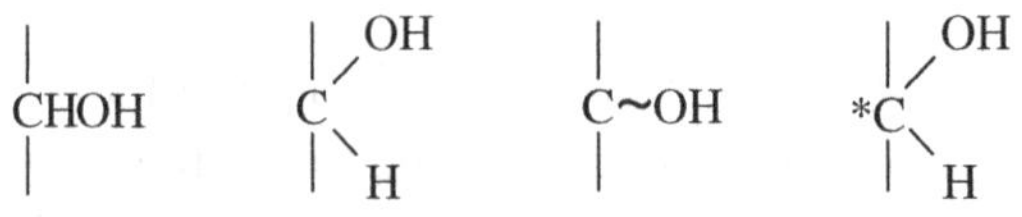

CHOH   C(OH)(H)   C~OH   *C(OH)(H)

Using the Fischer projections, one should particularly cosider the orientation of the bonds with respect to the plane (Fig. 4.2). If the formula is rotated for 180° in the plane, one obtains a structure that is

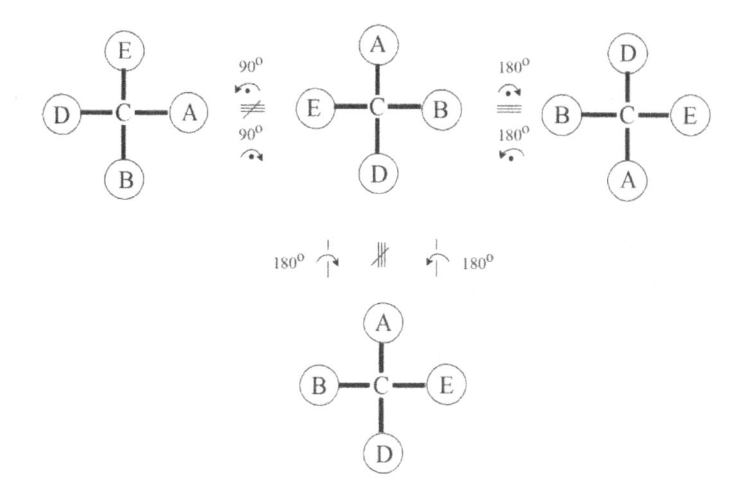

Fig. 4.2. If the Fischer projection is rotated for 90° in the plane, or rotated for 180° about the axis A-C-D, a mirror image of the original compound is formed. Instead, only a rotation for 180° in the plane leads to a situation where the assumed directions of both the horizontal and vertical bonds remain unchanged.

stereochemically superimposable with the original one. A corresponding rotation for 90° leads instead to a mirror image (enantiomer) of the compound because, according to the definition, the assumed orientation of both the vertical and horizontal bonds with respect to the plane changes. Thus, a rotation of 180° about the axis of the carbon chain leads to the mirror image of the original compound.

The utility of the Fischer projection becomes evident in illustrating the stereochemical structure of molecules that contain several asymmetric carbon atoms. Figure 4.3 shows the possible stereochemical alternatives (1→4) with the generally used manners of presentation. It is necessary here to perform, in the sawhorse projections on the top of the figure, rotations of 180° about carbon atom C_2 in order to obtain "correct" directions of the bonds. However, it is not possible to change these stereoisomers to one another with similar rotations because it would require the breaking of bonds connecting the groups to the asymmetric carbon atoms C_2 and C_3. The Newman projections are quite useful in these cases as well, and with their help, it is possible to elucidate the conformation alternatives that arise from the rotation about the bond between carbon atoms C_2 and C_3.

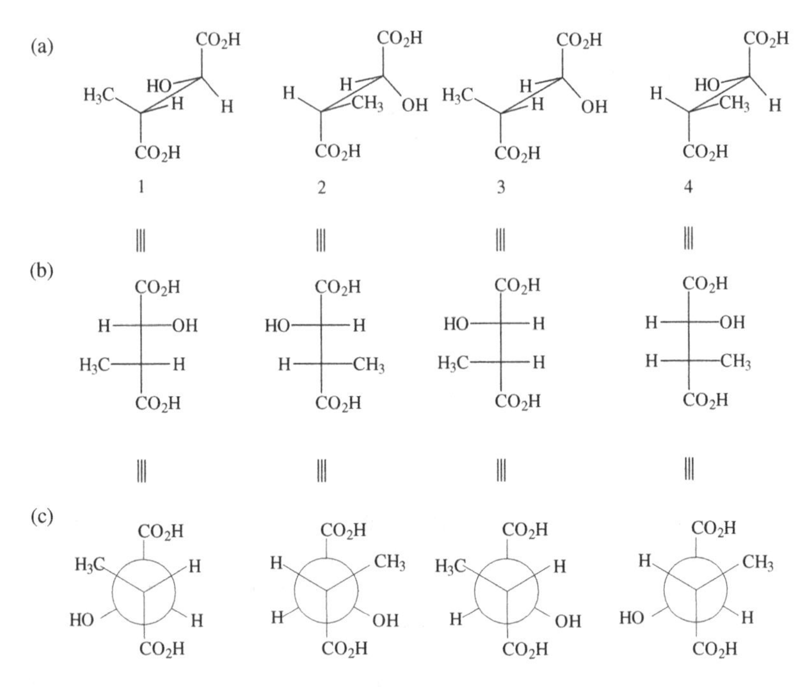

Fig. 4.3. The possible stereoisomers of 2-hydroxy-3-methylbutanedioic acid depicted as (a) sawhorse, (b) Fischer, and (c) Newman projections.

In Fig. 4.3 compounds 1 and 2 as well as 3 and 4 are mirror images of one another; they form two pairs of enantiomers (1–2 and 3–4). In this example, compounds 1–3, 1–4, 2–3, and 2–4 are instead stereoisomers of one another (cf., Chapt. “3.5. Racemic Modifications”), and they can be seen as “pairs of diastereoisomers”. Although diastereoisomers cannot form a pair of enantiomers (or, in analogy, two enantiomers cannot be diastereoisomers of one another), two diastereoisomers can naturally each have their own enantiomers.

The Fischer projections are most popular in carbohydrate chemistry and, for example, the naming of monosaccharides has traditionally been based on their use. Other practices have, however, also emerged in the presentation of open-chain monosaccharides as planar formulas (Fig. 4.4). In this figure, the Fischer projections have been converted into corresponding *Masamune projections*, where the zigzag-formed main chain is

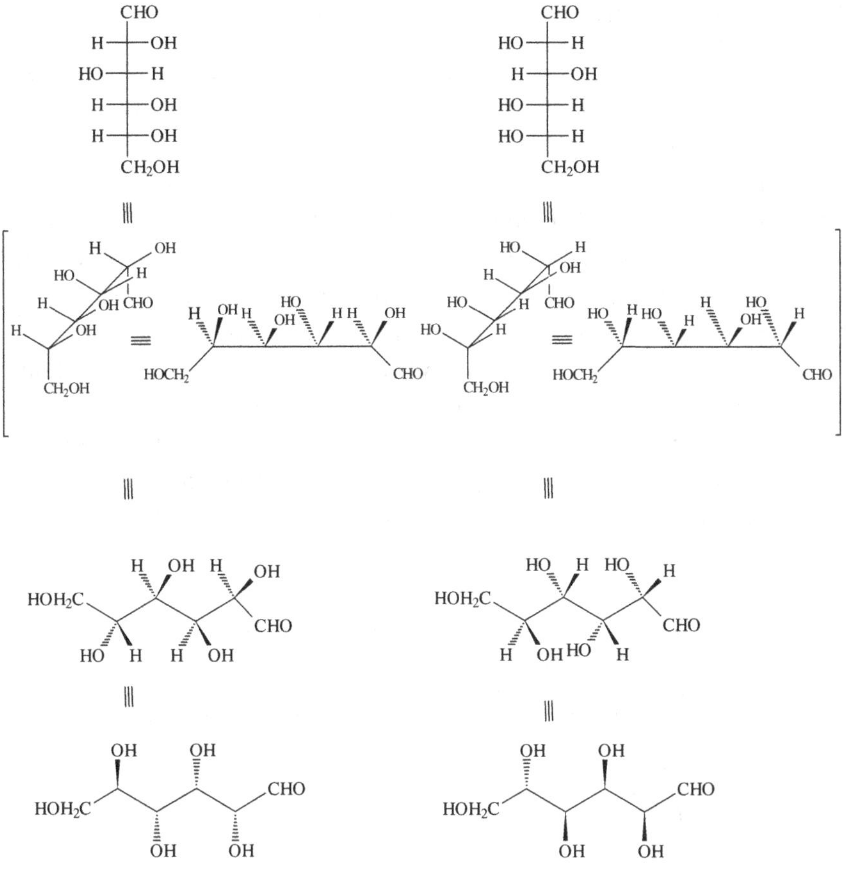

1

2

Fig. 4.4. The conversion of the Fischer projections of D-(+)-glucose (1) and L-(−)-glucose (2) (above) to zigzag projections (Masamune projections) (below). For explanations, see the text.

drawn horizontally and the bonds of the alternating hydroxyl groups (the hydrogen atoms are not shown) are directed either above or below the plane of the carbon chain. In the "intermediate stages" of conversion shown in parentheses, the bonds between carbon atoms C_2 and C_3, C_3 and C_4, and C_4 and C_5 are consecutively rotated for 180° in order to arrive at the rather illustrative Masamune projections.

4.2. The Total Number of Stereoisomers

Compounds that have several (number n) asymmetric carbon atoms (e.g., aldoses and aldonic acids, Table 4.1), have, according to the rule of Le Bel-van't Hoff in 1874, 2^n stereoisomeric forms and $2^{(n-1)}$ pairs of enantiomers. If $n \geq 2$, diastereoisomers can occur according to the above discussion. In alditols, the aldehyde group of aldoses (or the carboxyl group of aldonic acids) is replaced by a hydroxymethyl group; this is why the terminal groups of their carbon chains are identical and the potential for forming different stereoisomers is reduced (e.g., n = 4 — see also aldaric acids, Table 4.2). When n is *even* in alditols, the total number of chiral and achiral compounds are, respectively, $2^{(n-1)}$ and $2^{(n-2)/2}$ (the total number of stereoisomers is then $2^{(n/2-1)}(2^{n/2}+1)$)). If the number n is *odd*, then the corresponding total number of chiral and achiral compounds are, respectively, $2^{(n-1)}-2^{(n-1)/2}$ and $2^{(n-1)/2}$, while the total number of stereoisomers is $2^{(n-1)}$.

Table 4.1. Total number of stereoisomers in aldoses $HOH_2C(CHOH)_nCHO$ (or the corresponding aldonic acids $HOH_2C(CHOH)_nCO_2H$)

Aldose	n	Number of stereoisomers	Number of enantiomer pairs
Triose	1	2	1
Tetrose	2	4	2
Pentose	3	8	4
Hexose	4	16	8
Heptose	5	32	16

Table 4.2. Total number of stereoisomers in alditols $HOH_2C(CHOH)_nCH_2OH$ (or the corresponding aldaric acids $HO_2C(CHOH)_nCO_2H$)

Alditol	n	Number of chiral compounds	Number of achiral compounds	Number of stereoisomers
Tetritol	2	2	1	3
Pentitol	3	2	2	4
Hexitol	4	8	2	10
Heptitol	5	12	4	16

Tables 4.1 and 4.2 show that aldoses have clearly more stereoisomers than alditols with the same number of asymmetric carbon atoms; in addition, symmetrical (achiral) and thus optically inactive stereoisomer are found in the latter. For further illustration this matter, tartaric acid (aldaric acid) (see Table 3.5 in Chapt. "3.5. Racemic Modifications") with two asymmetric carbon atoms (C_2 and C_3) is also discussed here. Due to identical terminal groups (-CO_2H) of its carbon chain, only three stereoisomeric forms can be found (Fig. 4.5). As compound 3 and its assumed mirror image compound 4 can easily be seen as superimposable (by a 180° rotation in the plane), they represent stereochemically the same compound. In them, the spatial location of the same groups (-H, -OH, and -CO_2H) bonded to both asymmetric carbon atoms creates a symmetry between the "top and bottom" halves of the molecules so that a plane of symmetry can be placed between carbon atoms C_2 and C_3. These stereoisomers that are superimposable with their mirror image are generally called "*meso forms*" (in Greek, "mesos" means "center" or "midpoint") (cf., Chapt. "3.5. Racemic Modifications").

Although *meso* forms contain chiral centers, the influences of those centers on plane-polarized light cancel one another, and the compounds are optically inactive. Among similar compounds are certain ketoses, where the carbonyl carbon is located symmetrically at the center of the carbon chain and the groups bonded to it are mirror images of one another. The total number of stereoisomers of these symmetric ketoses can thus be calculated in the same way as for alditols (with even n). For all other ketoses, the total number of stereoisomers can instead be calculated in the same manner as for aldoses, with the number of isomers for a given

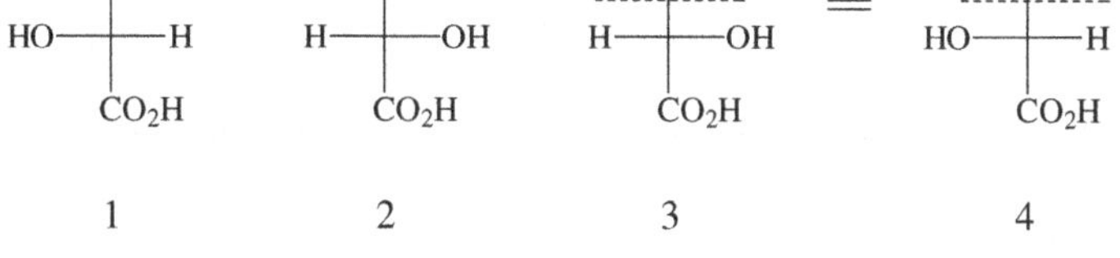

1	2	3		4
CO_2H	CO_2H	CO_2H		CO_2H
H–C–OH	HO–C–H	H–C–OH	≡	HO–C–H
HO–C–H	H–C–OH	H–C–OH		HO–C–H
CO_2H	CO_2H	CO_2H		CO_2H

Fig. 4.5. Stereochemical structures of (+)- (1, L or 2*R*,3*R* form), (−)- (2, D or 2*S*,3*S* form), and *meso* forms (3, 2*R*,3*S* form) tartaric acid.

number of carbon atoms being always lower than that for the corresponding aldoses due to the smaller number of asymmetric carbon atoms. Among the ketoses important in nature the carbon atom C_2 is usually the carbonyl carbon.

Certain molecules can contain carbon atoms to which are bonded, in addition to two chiral substituents (R) consisting of the same atoms or atom groups, two achiral groups (A and B):

B A A B
C C
R R R R

In the simplest case, R contains one chiral center; it is then possible to form four stereoisomers. Of the four stereoisomers of 2,3,4-trihydroxyglutaric acid in Fig. 4.6 (see Table 4.2), 1 and 2 form a pair of enantiomers, where carbon atoms C_2 and C_4 are stereochemically identical with one another (in compound 1 2*S* and 4*S* configurations and in compound 2 2*R* and 4*R* configurations) (*R*,*S* system, cf., Chapt. "5.2. Relative and Absolute Configuration"). For achiral diastereoisomers 3 and 4, it is possible to place planes of symmetry within the molecule because their carbon atoms C_2 and C_4 are mutually enantiomeric (both compounds have a 2*R* and 4*S* configuration). These compounds can be made superimposable only by exchanging the positions of the achiral groups (-H and -OH) bonded to the

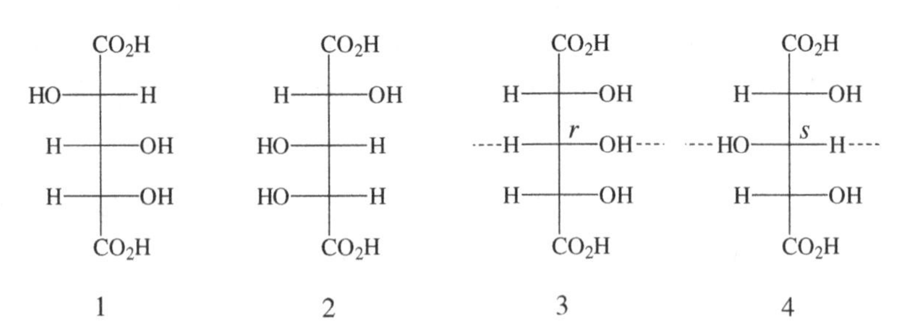

Fig. 4.6. Stereoisomers of 2,3,4-trihydroxyglutaric acid. Compounds 1 and 2 form a pair of enantiomers, and the *meso* forms 3 and 4 can be made superimposable by exchanging the positions of the achiral substituents (-H and -OH) of the pseudochiral carbon atom C_3 (in either *r* or *s* configuration).

so-called "*pseudochiral*" or "*pseudoasymmetric*" carbon atom C_3 (in Greek, "pseudo" means in compound words "apparent" or "quasi"). The absolute configuration of the pseudochiral carbon in these *meso* forms is expressed with the letters *r* and *s*, which are derived in analogy with the priority rules of the CIP system (i.e., *R* configuration > *S* configuration) (cf., Chapt. "3.7. Geometric Isomerism").

5. Configuration

5.1. Fundamental Definitions

The spatial organization of the atoms in a given stereoisomer is called the "*configuration*" of the compound in question. Thus, it is possible to distinguish between stereoisomers on the basis of their different configurations. The concept of configuration in organic chemistry is, however, often associated with only a certain chiral center, such as an asymmetric carbon atom (the configuration of a carbon atom), so that the possibly several centers of asymmetry will be examined separately.

Along with the progress of carbohydrate chemistry (cf., Chapt. "2.2. The Time After the 1700s"), a need emerged to unequivocally express their spatial structure (configuration). At that time, there were no techniques yet available to determine the exact three-dimensional structure of the compounds or even of their parts. In order to clarify the matter, (+)-glyceraldehyde (Fig. 5.1, see also Fig. 4.1) was chosen, on the suggestion of Ukrainian-American Martin André Rosanoff (1874–1951) in 1906, as the "official fundamental compound", which was deemed to have the *D form,* while the (–)-stereoisomer had the *L form* (in Latin, "dexter" means "right" and "laevus" "left"). This choice was based on Fischer's earlier designation (+)-glyceraldehyde as a reference compound resulted from its being the simplest optically active aldose (aldotriose) and also from the fact that as a reactive compound, it was easy to convert into many other compounds. It must be emphasized that the practical choice that led to the existing usage was made, following Fischer, completely haphazardly with the probability of 50% that it represented the actual structure (i.e., that it showed the actual orientation of the hydroxyl group

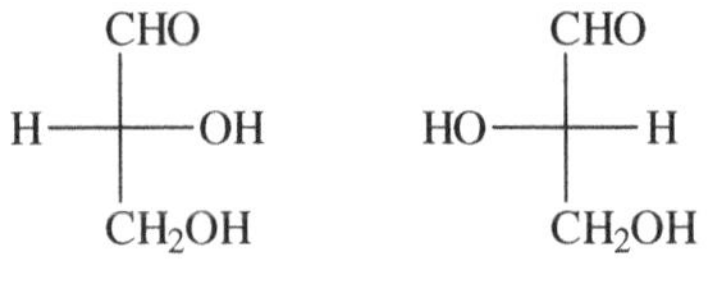

CHO CHO

H—OH HO—H

CH_2OH CH_2OH

1 2

Fig. 5.1. Fischer projections for D-(+)-glyceraldehyde (1) and L-(−)-glyceraldehyde.

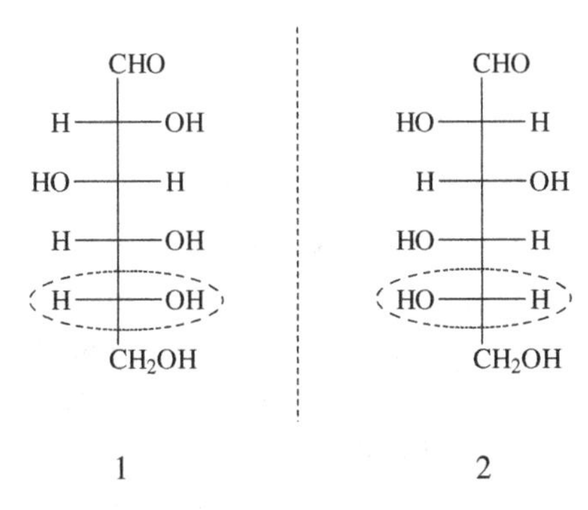

Fig. 5.2. Fischer projections for D-glucose (1) and L-glucose (2). Note that these compounds are mirror images of one another and constitute a pair of enantiomers.

in the Fischer projection). Hence, it was not possible to know, for example, which alternative structural configuration the (+)-stereoisomer in fact had. The "validity" of this fortunate choice was finally verified with the X-ray diffraction technique in the early 1950s.

According to the general definition, a carbohydrate is classified as a D form if the hydroxyl group attached to its last asymmetric carbon atom (i.e., the asymmetric carbon atom with the highest number) drawn as a Fischer projection points to the right (Fig. 5.2). Accordingly, in an L form, this hydroxyl group points to the left. In this case, we speak of *D* and *L series carbohydrates* (within so-called "aldose series" and "ketose series"). It should be noted that the prefixes D- and L-only refer to the configuration and do not indicate the optical activity of the compounds (which is expressed with symbols (+)- and (−)-). As the optical activity of compounds was traditionally expressed in carbohydrate chemistry with prefixes *d*- (corresponds to prefix (+)-) and *l*- (corresponds to prefix (−)-), those prefixes were gradually abandoned because of the confusion. As the D and L forms of a given carbohydrate are mirror images of one another (i.e., the configurations of all corresponding asymmetric carbon atoms in them are mirror images of one another), such forms constitute a pair of enantiomers. It follows that it is not sufficient to change only the spatial structure of the last asymmetric carbon atom of an aldose (with ≥2 asymmetric carbon atoms) to the opposite configuration in order to change it from a D form to an L form.

For aldoses (cf., Chapt. "1. Introduction") with at most four asymmetric carbon atoms (in total at most six carbon atoms), their trivial names (with an ending *-ose*) are generally used. Figure 5.3 presents these names and the corresponding configurations (for clarity, the hydrogen atoms

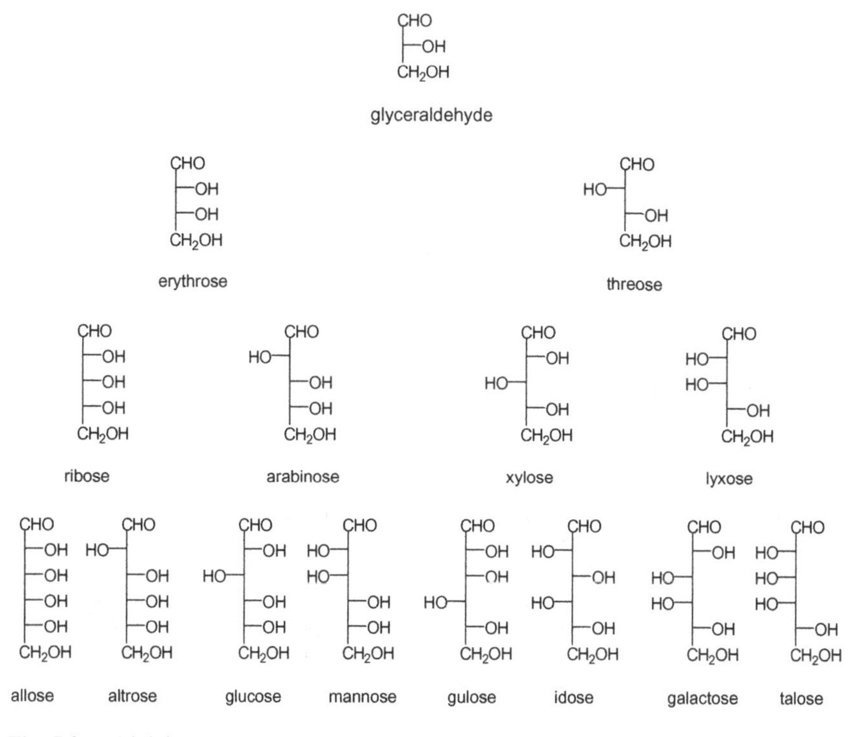

Fig. 5.3. Trivial names and configurations of the D-aldoses (with one to four asymmetric carbon atoms) represented by the Fischer projections.

attached to their asymmetric carbon atoms are not shown, while the hydroxyl groups are) for all D series aldoses; the configurations of the corresponding L-aldoses can easily be found as the mirror images. This also illustrates the simplicity and utility of the Fischer projections in perceiving different stereoisomers of aldoses.

In ketoses, the number of asymmetric carbon atoms relative to the total number of carbon atoms is less than in aldoses due to the ketone group within the carbon chain (see Table 4.1), so that the number of possible stereoisomers relative to the total number of carbon atoms is also less than in aldoses. Figure 5.4 shows the trivial names of the D-ketoses with up to six carbon atoms (the endings being typically *-ulose*) and the corresponding configurations; only the most common 2-keto derivatives in nature have been included. In this case, the corresponding L-ketoses can also be easily derived from the D-ketoses presented.

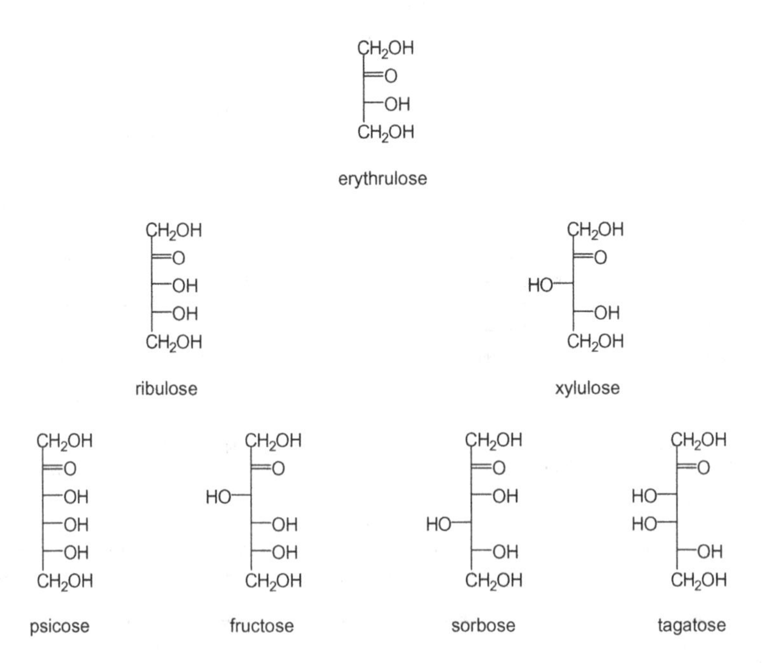

Fig. 5.4. Trivial names and configurations of the D-2-ketoses (with one to three asymmetric carbon atoms) represented by the Fischer projections.

Table 5.1. Epimers among aldopentoses and aldohexoses (see Fig. 5.3 and the text)

Epimer	Aldopentoses			Aldohexoses		
C-2	Ribose	↔	Arabinose	Allose	↔	Altrose
	Xylose	↔	Lyxose	Glucose	↔	Mannose
				Gulose	↔	Idose
				Galactose	↔	Talose
C-3	Ribose	↔	Xylose	Allose	↔	Glucose
	Arabinose	↔	Lyxose	Altrose	↔	Mannose
				Gulose	↔	Galactose
				Idose	↔	Talose
C-4	D-Ribose	↔	L-Lyxose	Allose	↔	Gulose
	D-Arabinose	↔	L-Xylose	Altrose	↔	Idose
				Glucose	↔	Galactose
				Mannose	↔	Talose
C-5		—		D-Allose	↔	L-Talose
				D-Altrose	↔	L-Galactose
				D-Glucose	↔	L-Idose
				D-Mannose	↔	L-Gulose

Monosaccharides (or carbohydrates in general) that differ from one another only in the configuration of one asymmetric carbon atom are called "*epimers*". For example, D-glucose and D-mannose are C_2 epimers, while D-glucose and D-galactose are C_4 epimers (Table 5.1). If *epimerization* is related to the asymmetric carbon atom with the highest number (e.g., the C_4 carbon in aldopentoses and C_5 carbon in aldohexoses), the corresponding epimers always belong to different series (D- and L-aldoses). This is why the C_2 and C_3 epimers in aldopentoses and also the C_2, C_3, and C_4 epimers in aldohexoses always belong to the same series (either D- or L-aldoses).

In addition to trivial names, aldoses and ketoses can be systematically named by adding in front of the so-called "stem name" a prefix that indicates the configuration (cf., Chapts. "7.1. Aldoses" and "7.2. Ketoses"). Figure 5.5 presents the stem names of aldoses and ketoses with at most six carbon atoms. For example, the systematic names of D-glucose and L-xylose are D-*gluco*-hexose and L-*xylo*-pentose, respectively.

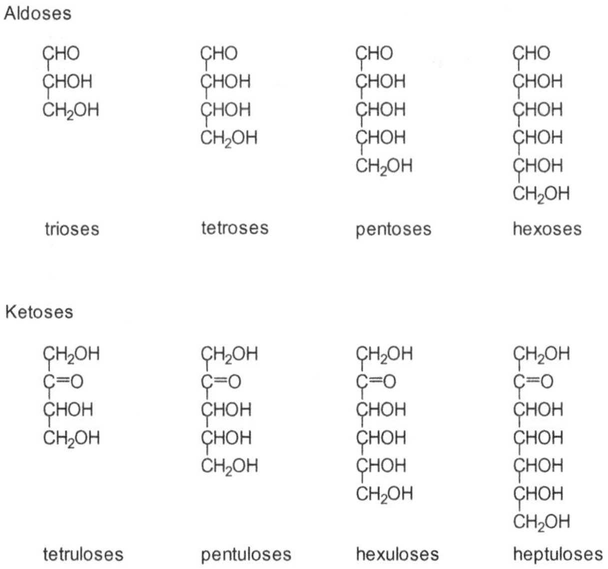

Aldoses

CHO	CHO	CHO	CHO
CHOH	CHOH	CHOH	CHOH
CH_2OH	CHOH	CHOH	CHOH
	CH_2OH	CHOH	CHOH
		CH_2OH	CHOH
			CH_2OH
trioses	tetroses	pentoses	hexoses

Ketoses

CH_2OH	CH_2OH	CH_2OH	CH_2OH
C=O	C=O	C=O	C=O
CHOH	CHOH	CHOH	CHOH
CH_2OH	CHOH	CHOH	CHOH
	CH_2OH	CHOH	CHOH
		CH_2OH	CHOH
			CH_2OH
tetruloses	pentuloses	hexuloses	heptuloses

Fig. 5.5. Stem names of aldoses and ketoses containing fewer than five asymmetric carbon atoms (only possible 2-ketoses are included).

5.2. Relative and Absolute Configuration

According to the traditional study of configuration, several chemically pure compounds were prepared from D-glyceraldehyde without ever breaking the chemical bonds between the initial asymmetric carbon atom and its substituents. The *relative configuration* of a compound could thus be determined in all cases where the configurational connection between the product compound and the initial substance, in this case, D-glyceraldehyde (or its derivative obtained in a controlled process), was accurately known. As an illuminating example, Fig. 5.6 presents the preparation of D-(−)-lactic acid (*R* form), which is stable in the organism, from D-(+)-glyceraldehyde in a multistage reaction, where the configuration of the asymmetric carbon atom (C_2) remains constant in all stages. It is similarly possible to arrive at the relative configuration of L-(+)-lactic acid (*S* form) that is converted in the organism to pyruvic acid (2-oxopropanoic acid). It should be reiterated that the configurational prefix of a compound does not indicate the nature of its optical activity (see D-(+)- and L-(−)-glyceraldehyde or D-(−)- and L-(+)-lactic acid).

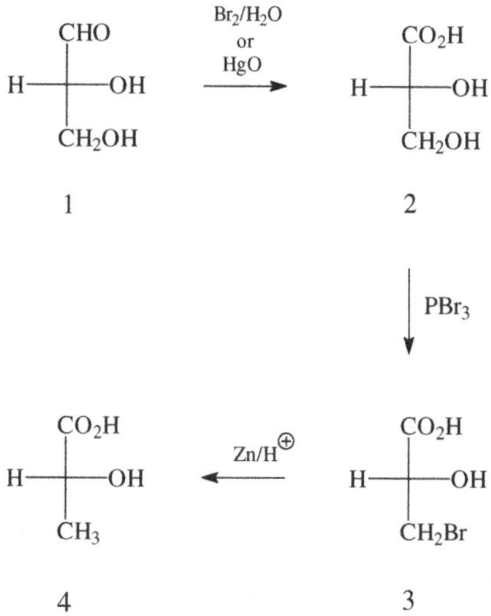

CHO / H—C—OH / CH_2OH (1) $\xrightarrow{Br_2/H_2O \text{ or } HgO}$ CO_2H / H—C—OH / CH_2OH (2)

(2) $\xrightarrow{PBr_3}$ CO_2H / H—C—OH / CH_2Br (3)

(3) $\xrightarrow{Zn/H^{\oplus}}$ CO_2H / H—C—OH / CH_3 (4)

Fig. 5.6. Configurational connection between glyceraldehyde and lactic acid. D-(+)-Glyceraldehyde (1), D-(−)-glyceric acid (2), D-(−)-3-bromo-2-hydroxypropanoic acid (3), and D-(−)-lactic acid (4).

The term "relative configuration" is now generally used in relative examinations where it is sufficient to know whether the reaction product has the same or the opposite configuration compared to the initial compound. Therefore, it is not always necessary to know the *absolute configuration* of a compound. However, the concept of relative configuration automatically evolved into the concept of absolute configuration when, as mentioned above, a stereochemical link was finally found between the three-dimensional structure of D-(−)-tartaric acid (2*S*,3*S* form) (which was deduced from the exactly defined three-dimensional structure of L-(+)-tartaric acid (2*R*,3*R* form) and the its predecessor from the D-(+)-glyceraldehyde-synthesized, identical structure (see Fig. 2.1 in Chapt. "2.2. The Time After the 1700s").

As a rule, the absolute configuration of a center of asymmetry in organic chemistry is reported using the *R*,*S* or CIP system (cf., Chapt. "3.7. "Geometric Isomerism"); however, except for special cases (e.g., some carbohydrate derivatives), its general use is rather rare. In this system, for example, the exact spatial location of substituents bonded to a carbon atom is expressed with prefixes *R*- (in Latin, "rectus" means "right") and

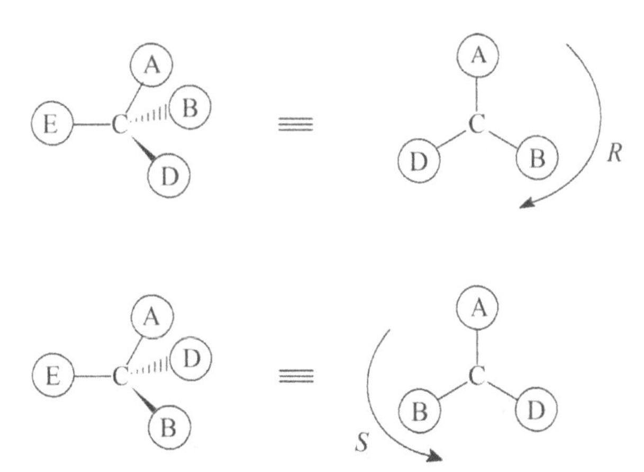

Fig. 5.7. Use of prefixes *R*- and *S*- according to the CIP system in case of C(ABDE) (the order of substituent priority is A > B > D > E, see the text).

S- (in Latin, "sinister" means "left"). Here, the groups bonded to the asymmetric carbon atom are placed in an order of priority (such as A > B > D > E) according to the priority rules presented in detail in Chapt. "3.7. Geometric Isomerism". Once the relative priorities of these four substituents have been determined, the asymmetric carbon atom is viewed from the side opposite the lowest priority group (E). In practice, this means that only groups A, B, and D are visible — in addition to the asymmetric carbon atom (Fig. 5.7). If now the order of priority (A→B→D) decreases clockwise, the configuration is an *R* form. In the opposite case, one uses the prefix *S*-. Here, it should be again emphasized that prefixes *R*- and *S*- indicate only the configuration and not the optical activity of the compound.

As a practical example of configurational study with the CIP system, Fig. 5.8 shows the naming of the asymmetric carbon atoms C_2 and C_3 in the stereoisomers of tartaric acid (see Table 3.5 and Fig. 4.5). It should be mentioned that 2*R*,3*R*-tartaric acid and 2*S*,3*S*-tartaric acid are optically active and form a pair of enantiomers, whereas 2*R*,3*S*-tartaric acid is an optically inactive *meso* form. In the latter case, the chiral centers C_2 and C_3 are mirror images of one another (i.e., they have intramolecular symmetry) such that their influences on plane-polarized light cancel one

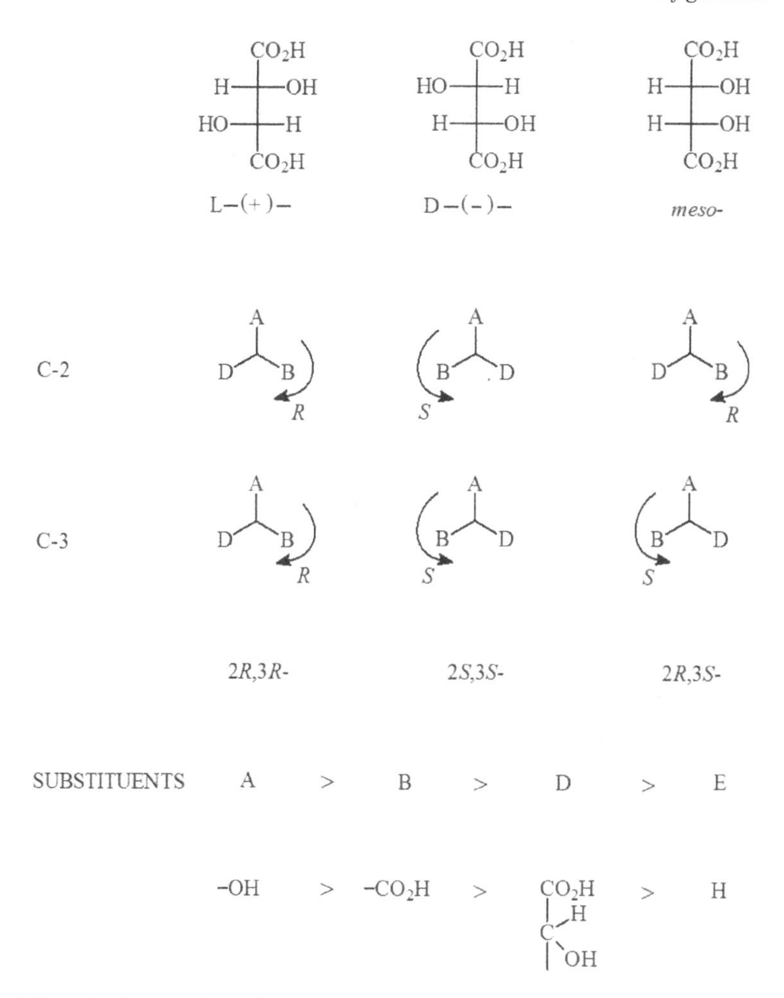

Fig. 5.8. Absolute configurations at the asymmetric carbon atoms C_2 and C_3 of tartaric acid.

another. If needed, the CIP system offers an opportunity to define the absolute configurations of the asymmetric carbon atoms in all aldoses and ketoses (Table 5.2). Since the D and L forms are mirror images of one another, the *R* configurations of the D forms appear as *S* configurations of the corresponding L forms (and naturally, vice versa, the *S* configurations will appear as *R* configurations).

Table 5.2. Absolute configurations at the chiral centers (asymmetric carbon atoms) in D-aldoses and D-2-ketoses with less than six carbon atoms (see Figs. 5.3 and 5.4)

Monosaccharide	C-2	C-3	C-4	C-5
Aldoses				
Glyceraldehyde	*R*	—	—	—
Erythrose	*R*	*R*	—	—
Threose	*S*	*R*	—	—
Ribose	*R*	*R*	*R*	—
Arabinose	*S*	*R*	*R*	—
Xylose	*R*	*S*	*R*	—
Lyxose	*S*	*S*	*R*	—
Allose	*R*	*R*	*R*	*R*
Altrose	*S*	*R*	*R*	*R*
Glucose	*R*	*S*	*R*	*R*
Mannose	*S*	*S*	*R*	*R*
Gulose	*R*	*R*	*S*	*R*
Idose	*S*	*R*	*S*	*R*
Galactose	*R*	*S*	*S*	*R*
Talose	*S*	*S*	*S*	*R*
Ketoses				
Erythrulose	—	*R*	—	—
Ribulose	—	*R*	*R*	—
Xylulose	—	*S*	*R*	—
Psicose	—	*R*	*R*	*R*
Fructose	—	*S*	*R*	*R*
Sorbose	—	*R*	*S*	*R*
Tagatose	—	*S*	*S*	*R*

6. Cyclic Forms of Monosaccharides

6.1. Mutarotation

During the early days of carbohydrate chemistry (cf., Chapt. "2.2. The Time After the 1700s"), Fischer solved the fundamental structure of both aldopentoses and aldohexoses, which is the prevailing opinion. However, conflicting results that could not be explained, as open-chain structures were emerging. The biggest problem was observations that aldoses did not participate in all reactions typical for aldehydes or that some of those reactions were markedly slower than expected. On the other hand, in some common reactions, such as the acetylation of glucose or forming its methyl glucoside (see reactions later), two isomeric pentaacetates were formed in the first case, while two isomeric products were also obtained in the second case. As the formation of these reaction products could not be explained starting from the then known glucose structure that contained an aldehyde group, several additional studies and conclusions drawn from them gradually led Hudson, Haworth, and Hirst, among others, through many intermediate phases, to the discovery of cyclic monosaccharide structures.

This complex issue was also related to the interesting, and in those days unexplained observation that the specific rotation of the two modifications of crystalline D-glucose (α form $[\alpha]_D^{20}$ + 112.2 and β form $[\alpha]_D^{20}$ + 18.7) changed in a water solution. With both these compounds, the end result under the same conditions gradually became a constant +52.7; Dubrunfaut observed this phenomenon in 1846. In addition, two crystalline forms were known for, among others, lactose (α form $[\alpha]_D^{20}$ + 92.6 and β form $[\alpha]_D^{20}$ + 34.2), whose specific rotations changed similarly (observed in 1856) to a constant value of $[\alpha]_D^{20}$ + 52.3. Instead, several cases were known (e.g., sucrose and certain methyl glucosides) that did not exhibit two crystalline forms or similar changes in their specific rotation. This phenomenon was primarily seen to relate to so-called "reducing sugars" and "non-reducing sugars" (cf., Chapt. 1. "Introduction").

According to general practice (as finally proposed by Hudson in 1909), the α form was defined for the D series sugars as compounds that, of the crystalline forms, rotated the plane of plane-polarized light "more to the right" and the β form accordingly defined as the compound that rotated it "more to the left" (the opposite definition was applied to the L series sugars). The prefixes α- and β- were thus directly connected with the optical activity of the compound. Besides these α and β forms, it was incorrectly thought that the same compound could also exist in a γ form, which, for example, in the case of glucose was a "compound" with a specific rotation of +52.7.

We know now that upon dissolving the *tautomeric forms* separately in water, the end result is always (often in a few minutes) the same state of equilibrium that contains these tautomeric forms in a certain ratio and will have a specific rotation determined by this ratio of equilibrium. This *tautomerism* can be exactly defined as a phenomenon of equilibrium between two isomeric compounds. The term "tautomerism" is, however, generally used for this kind of equilibrium in carbohydrate chemistry, even though the tautomeric forms of a monosaccharide (cyclic forms) are not necessarily in equilibrium directly with each other but have a certain open-chain intermediate compound (oxo or carbonyl compound) between them. As, in reality, several tautomeric equilibrium states may exist, primarily between acyclic oxo compounds and various cyclic molecular structures (i.e., a mixture in equilibrium), this type of tautomerism has been traditionally called "*oxo-cyclo-tautomerism*".

In practice, a common monosaccharide (e.g., an aldopentose or an aldohexose) will usually exist in neutral or mildly acid or basic aqueous solutions as six tautomeric forms. Of these forms, two are open-chain (aldehyde forms in aldoses and ketones in ketoses and their corresponding hydrate forms), and four are heterocyclic ring forms: two six-atom *pyranoid forms* and two five-atom *furanoid forms,* thus including all thermodynamically stable cyclic hemiacetals of aldoses and the cyclic hemiketals of ketoses (see aldose in Fig. 6.1 and Chapt. "1. Introduction"). The terms "pyranoid" and "furanoid" are, respectively, based on the structurally analogous organic compounds pyrane and furane (Fig. 6.2), leading to the terms "*pyranoses*" and "*furanoses*" in the context of

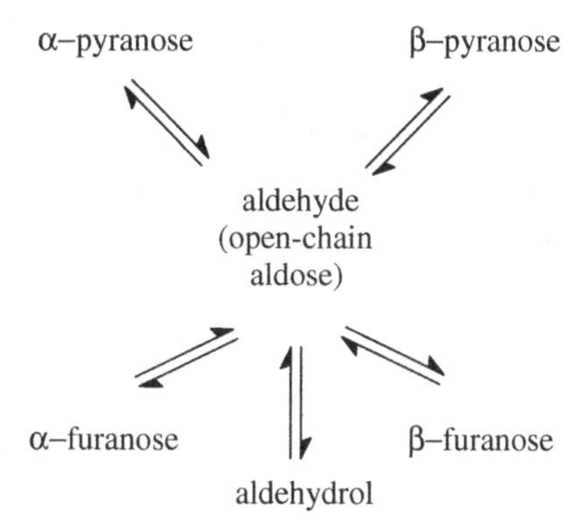

Fig. 6.1. An open-chain aldose (aldehyde) exists in aqueous solution generally as an equilibrium mixture of two pyranoid six-atom ring structures (pyranoses) and two five-atom ring structures (furanoses) together with an acyclic hydrate, aldehydro-form (aldehydrol or *gem*-diol). At equilibrium, the relative concentrations of these forms are based on their thermodynamic stabilities and clearly, cyclic forms are heavily favored over acyclic forms and pyranose rings over furanose forms. For the proper use of prefixes α- and β-, see the text and Fig. 6.3.

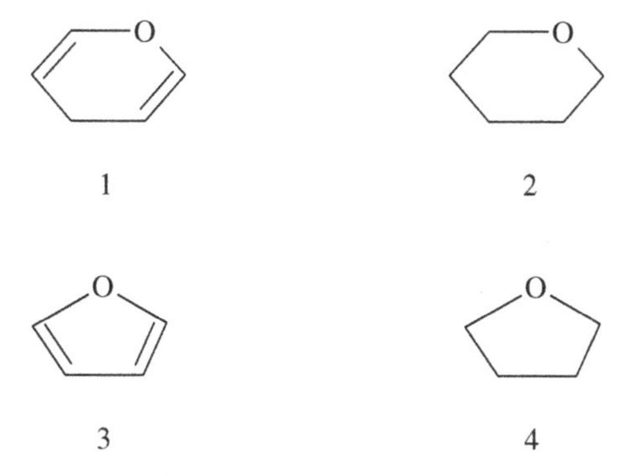

Fig. 6.2. Chemical structures of pyran (γ-pyran) (1), tetrahydropyran (2), furan (3), and tetrahydrofuran (4).

monosaccharides (e.g., glucopyranoses and fructofuranoses). The formed five- and six-atom heterocyclic rings are, as mentioned before, thermodynamically more stable than possible three-atom (*oxiroses*) or four-atom (*oxetoses*) rings, and also seven-atom structures (*septanoses*) formed in certain cases in extremely small concentrations.

The settling down of the equilibrium can be followed by observing the change in the specific rotation of the solution; this spontaneous change is called "*mutarotation*". Table 6.1 contains examples of the specific rotation of the equilibrium states for some monosaccharides common in nature. A new center of asymmetry is formed in the monosaccharide structure during the process of cyclization. This *anomeric carbon atom* (in aldoses carbon atom C_1 and in 2-ketoses carbon atom C_2) also acquires a new hydroxyl group, which is called either a "*hemiacetal hydroxyl group*" (in aldoses) or a "*hemiketal hydroxyl group*" (in ketones), but in both cases also generally an "*anomeric*" or a "*glycosidic hydroxyl group*". An anomeric hydroxyl group is more reactive (see preparation of glycosides) than the other hydroxyl groups previously existing in the monosaccharide molecule.

The hydroxyl group attached to the anomeric carbon atom can be located either above or below the "plane" of the formed ring. These diastereoisomers (i.e., they are not mirror images of one another) can easily

Table 6.1. Examples of specific rotations $[\alpha]_D$ of some monosaccharides in aqueous solution at equilibrium (Guthrie 1974)

Monosaccharide	$[\alpha]_D$
D-Ribose	−24
L-Arabinose	+105
D-Xylose	+19
D-Glucose	+53
D-Mannose	+14
D-Galactose	+80
D-Fructose	−94
L-Sorbose	−43
D-Tagatose	−5

transform to one another, but they are individual compounds with their own physical properties. According to a general definition, carbohydrate isomers that only differ from one another in the spatial orientation of their anomeric hydroxyl group are called "*anomers*". Following a practice developed in 1932 by Karl Johann Freudenberg (1886–1983) that is now in common use, anomers are divided into α and β forms (*α and β anomers*), after the Hudson system, but the illustrative foundation is primarily the Haworth projections of the compounds (Fig. 6.3, cf., Chapt. "6.2. Planar Formulas"). Here, for example, if the glycosidic hydroxyl group in D series aldoses is placed above the ring plane, it corresponds to a β form, while the opposite position corresponds to an α form. It should be noted that for the L series aldoses (which are mirror images of the D aldoses), these prefix definitions are the opposite. This is why, for example, the enantiomer of α-D-glucopyranose is α-L-glucopyranose (and not β-L-glucopyranose). The α and β forms of ketoses are named, in analogy with aldoses, according to the orientation of the hemiketal hydroxyl group with respect to the ring plane.

Consequently, according to the traditional study of configuration, the α form is considered to be a compound where the formed anomeric carbon atom is located in the Fischer projection (cyclic formulas, cf., Chapt. "6.2. Planar Formulas") on the same side of the vertically drawn carbon chain as the hydroxyl group (which can participate in the formation of the ring) attached to the highest numbered asymmetric carbon atom that defines the belonging of the monosaccharide to either the D or L series. According to the CIP system, the α form corresponds to a situation where the configurations of the anomeric carbon atom (present in an either *R* or *S* configuration) and that of the highest-numbered carbon atom (also present in an either *R* or *S* configuration) of the monosaccharide are opposite (either *R,S* or *S,R*). If the configurations are the same (either *S,S* or *R,R*), we have a β form. Table 6.2 presents the absolute configurations of both the anomeric (C_1) and the highest numbered asymmetric carbon atoms (C_5) in the α and β anomers of certain aldopyranoses and the specific rotations of the compounds in question. Based on this information, it is easy to conclude the "correctness" of the original Hudson definitions. However, it is not known that some cases do not conform to his definitions. Therefore, the original, parallel view that the prefixes α and β

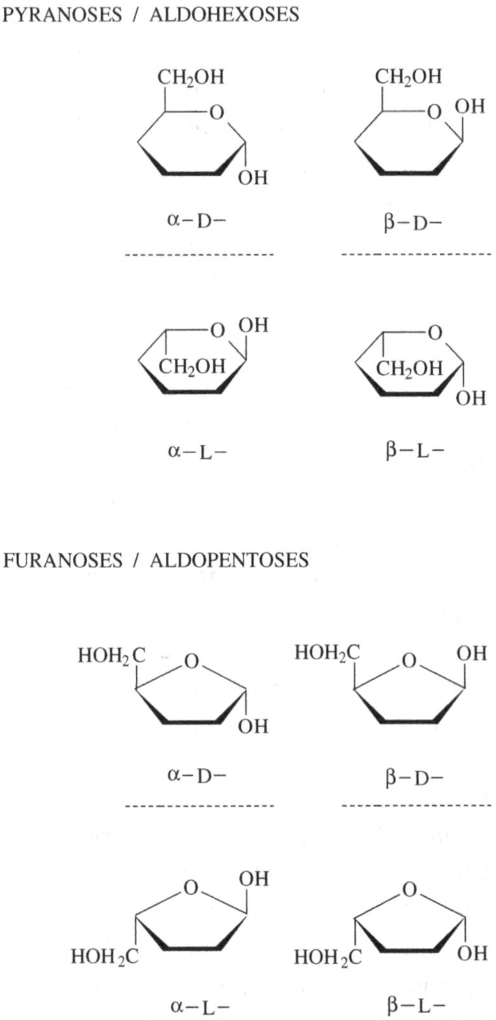

PYRANOSES / ALDOHEXOSES

CH_2OH O OH α−D−

CH_2OH O OH β−D−

O OH CH_2OH α−L−

O CH_2OH OH β−L−

FURANOSES / ALDOPENTOSES

HOH_2C O OH α−D−

HOH_2C O OH β−D−

O OH HOH_2C α−L−

O HOH_2C OH β−L−

Fig. 6.3. Examples of the names for the α and β forms of aldopyranoses and aldofuranoses in D and L series. Since D-aldoses and L-aldoses are mirror images of one another, the prefixes α-D- and α-L- and, on the other hand, the prefixes β-D- and β-L- indicate the opposite orientation of the anomeric hydroxyl group with respect to the ring plane. For further clarity, in each case, the hydroxymethyl groups attached on the last asymmetric carbon atoms are also shown.

Table 6.2. Examples of absolute configurations of carbon atoms C_1 and C_5 in the α and β anomers of aldopyranoses as well as of the corresponding specific rotations $[\alpha]_D$

Aldopyranose	C-1	C-5	$[\alpha]_D$
α-L-Arabinose	*R*	*S*	+40
β-L-Arabinose	*S*	*S*	+191
α-D-Glucose	*S*	*R*	+112
β-D-Glucose	*R*	*R*	+18
α-D-Mannose	*S*	*R*	+29
β-D-Mannose	*R*	*R*	+17
α-D-Galactose	*S*	*R*	+151
β-D-Galactose	*R*	*R*	+53

are firmly connected with the specific rotation of the compound is no longer seen as absolutely correct.

The chemical structures of the tautomeric forms of D-glucose in aqueous solution and the percentage compositions of these forms for aldopentoses, aldohexoses, and 2-ketohexoxes at equilibrium are shown in Fig. 6.4 and Table 6.3, respectively. Concentration changes in the tautomeric forms during mutarotation can be monitored by NMR spectrometry (^{1}H-NMR and ^{13}C-NMR) or polarography. It is also possible to determine the corresponding equilibrium constants and, especially clarify the effect of conditions on equilibrium. However, it is apparent that the tree-dimensional structure of the pure tautomerric form can be best obtained by X-ray diffraction techniques.

The pyranoid form is almost without exception more stable than its furanoid counterpart; the primary reason is that a six-atom ring is generally more stable than a five-atom one. In certain cases (D-ribose, D-altrose, D-idose, D-talose, D-psicose, and D-fructose), the furanoid form is also found in substantial amounts. Simply examining the mutual positions of the "large substituents" -OH, -CH_2OH, and -$CH(OH)CH_2OH$ at the adjacent ring carbon atoms of these monosaccharides and their possible steric interaction (cf., Chapt. "6.2. Planar Formulas") offers one explanation.

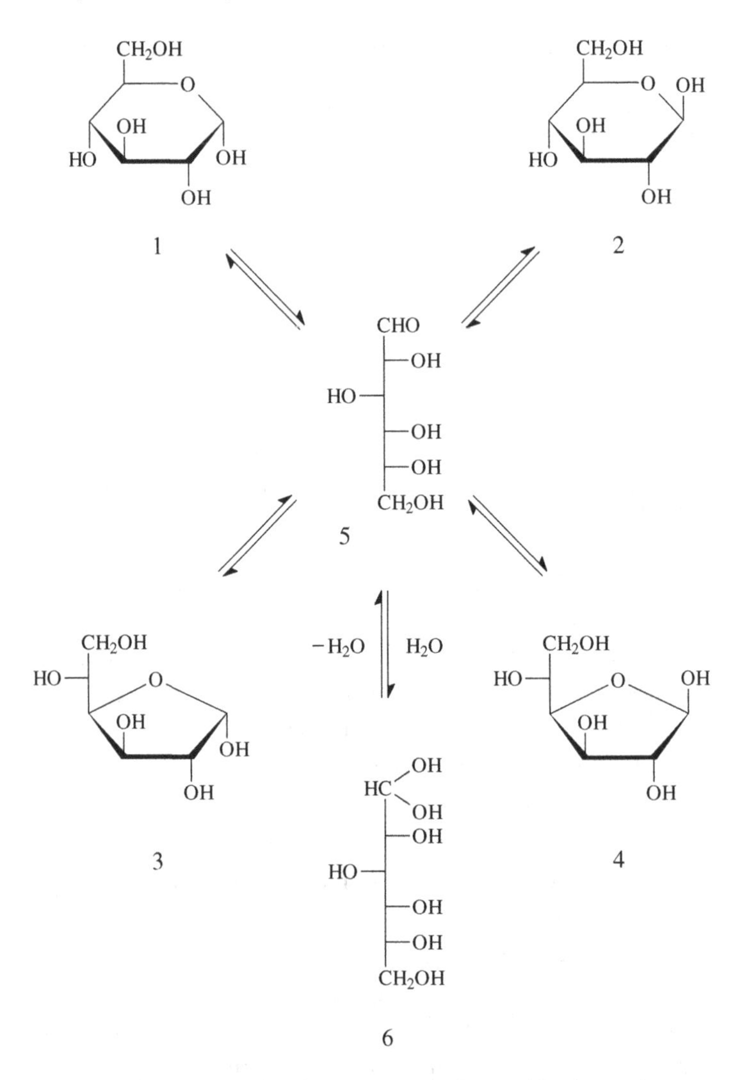

Fig. 6.4. Chemical structures of various tautomeric forms of D-glucose in aqueous solution. α-D-Glucopyranose (1), β-D-glucopyranose (2), α-D-glucofuranose (3), β-D-glucofuranose (4), open-chain D-glucose (aldose) (5), and acyclic hydrate (aldehydrol) (6).

A similar analysis can lead to the main reason for the percentage ratio of the α and β anomers of the monosaccharide structure. It is also easy to understand the high abundance of, for example, the pyranoid form of the β anomer in D-glucose (or D-xylose), because the substituents -OH

Table 6.3. The percentage compositions of aldopentoses, aldohexoses, and 2-ketohexoses in aqueous solution at equilibrium*

	Pyranose			Furanose		
Monosaccharide	α	β	Total	α	β	Total
Aldopentoses						
D-Ribose	21	59	80	6	14	20
D-Arabinose	61	35	96	2	2	4
D-Xylose	36	63	99	<1	<1	1
D-Lyxose	70	28	98	1	<1	2
Aldohexoses						
D-Allose	15	76	91	4	5	9
D-Altrose	28	42	70	18	12	30
D-Glucose	38	62	100	+	+	+
D-Mannose	65	34	99	<1	<1	1
D-Gulose	21	78	99	+	<1	1
D-Idose	39	36	75	11	14	25
D-Galactose	29	64	93	3	4	7
D-Talose	42	29	71	16	13	29
2-Ketohexoses						
D-Psicose	22	24	46	39	15	54
D-Fructose	3	65	68	7	25	32
L-Sorbose	93	2	95	4	1	5
D-Tagatose	75	17	92	2	6	8

*Adopted primarily from Lehman (1971), Collins & Ferrier (1998), El Khadem (1998), and Rao *et al.* (1998). Data are based on aqueus-solution determinations mainly performed at slightly higher temperatures than room temperature. The proportion of acyclic forms is generally <0.05 % and <0.5 % for aldoses and 2-ketoses, respectively. Entries <1 and + indicate the average contents of 0.5 % and 0.1 %, respectively.

and -CH_2OH at the adjacent ring carbon atoms are all in a *trans* position with one another in these structures. Because of the prevailing equilibrium mixtures, it is not easy to isolate these tautomeric forms in a pure state; it is primaryly limited to pyranoid forms (see D-glucose). In all cases, very little of the open-chain aldehyde is present under the equilibrium conditions, and the hydrated form is clearly still rarer. However, in ketoses the open-chain forms may be present in slightly larger quantities than in aldoses.

While the stability of tautomeric forms is primary influenced by the actual spatial structure of the ring (cf., Chapt. "6.3. Naming of Conformations"), it is reinforced by the interaction between the hydroxyl groups attached to the ring and water molecules, which depends on the structure. For example, in the thermodynamically favored conformation of the pyranoid monosaccharide, the distances between certain hydroxyl groups from one another are generally optimal for the monosaccharide to form a stable overall structure with water molecules. Further illustrative examples are D-glucose and D-xylose, which behave in a very similar manner in aqueous solutions.

In addition to the spatial structure, the pH of the solution and the temperature, among others, also influence the mutual equilibrium between the tautomeric forms. Because of the amphoteric nature of water, mutarotation of monosaccharides takes place in a solution in pure water, but it is catalyzed already by small amounts of acid or base. On the other hand, as the pH of the solution becomes increasingly acid or basic, the tautomeric equilibriums increase and, due to a momentary increase of the open-chain aldehyde, various further reactions, such as enolization or dehydration become more probable. Increasing the temperature of the solution accelerates the mutarotation of the monosaccharides and generally increases the proportion of the thermodynamically less stable furanoid and open-chain aldehyde forms. The speed rate of mutarotation increases on the average by a factor of about 2.5 for each 10°C rise in temperature but, as the temperature approaches 100°C, the general stability of tautomeric forms decreases substantially.

Mutarotation is markedly influenced by the use of solvents other than water (e.g., dimethyl sulfoxide) or their various aqueous mixtures. It is also possible that when a "typical" solvent (such as *N,N*-dimethylformamide or pyridine) is employed, the mutarotation is very slight without an acid or basic catalyst. Likewise, the other components of an aqueous solution ("impurities") affect the stability of the monosaccharides and thus, their mutarotation. For example, certain metal cations can have an effect on the equilibrium of the tautomeric forms if the spatial structure of the monosaccharides (i.e., the spatial orientation of the substituents) favors the formation of a metal complex, especially if the anomeric hydroxyl group can participate in the formation of the complex. In such cases, the equilibrium shifts towards the tautomeric form that is able to form the metal-monosaccharide complex. However, in certain cases, the metal ions can also form coordination bonds with water molecules more strongly than with hydroxyl groups of monosaccharides, leading to the weakening of the stabilizing effect of water on monosaccharides.

6.2. Planar Formulas

Open-chain structural formulas are illustrative for examining the different configurations of monosaccharides, and such formulas are used as the foundation for naming compounds. However, as the monosaccharides tend to cyclize and be present in nature mainly as ring structures, it is also meaningful to be able to represent the corresponding three-dimensional ring structures unambiguously as planar formulas.

The ring structures of monosaccharides were illustrated in the early days of carbohydrate chemistry in the same way as open-chain forms, as Fischer projections (traditionally also known as "*Fischer-Tollens projections*"). Later, the use of *Haworth projections* became popular in the illustration of monosaccharides (and generally of parts of oligosaccharides and polysaccharides), and such formulas are still today the most common way of representing carbohydrate structures. Here, the pyranoid ring of a monosaccharide is drawn as planar, and it is examined from a direction such that the ring oxygen is located at the upper corner of the ring (in furanoid rings at the middle of the top) (Fig. 6.5). The substituents are drawn, according to the configuration being presented, either above or below the plane of the ring; as a rule, the hydroxyl groups that are located in the open-chain Fischer projection on the right side of the vertical carbon chain and are attached to the ring carbons after cyclization, are placed below the ring plane. In order to enhance the impression of three-dimensional depth, the C-C bonds in the front of the ring are generally shown in bolder print, and usually the hydrogen atoms are not shown for clarity.

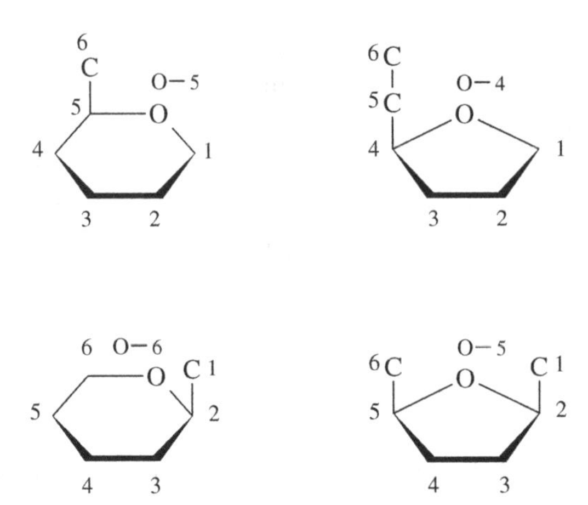

Fig. 6.5. Numbering of the pyranoid and furanoid ring atoms of aldohexoses and 2-keto-hexoses. Aldopyranose and aldofuranose (above) and 2-ketopyranose and 2-ketofuran-ose (below).

In organic chemistry, especially in the cases where the carbohydrate component only forms a part of a macromolecule, or in the cases (such as protecting reactions of hydroxyl groups) where, as a result of a reaction of two hydroxyl groups with one molecule, an additional ring (or, as a result of a reaction by four hydroxyl groups, two rings) have been attached to the basic monosaccharide ring, the *Mills formulas* are increasingly used. In these formulas, the planar ring is seen from above; the bonds of the substituents located above the plane are shown bolder, and those of the substituents below the plane are shown dashed. Also here, the hydrogen atoms of the ring are almost without exception not shown.

In Fig. 6.6 are examples of different ring structures of the monosaccharides discussed. The use of prefixes α- and β- should be noted, in particular with the D and L series monosaccharides (cf., Chapt. "6.1. Mutarotation") and the illustration of the possible mirror images of the structures (Fig. 6.7). In some cases, it is also necessary to represent, as an exception to the basic definitions, for example, the Haworth projections with the ring oxygen not occupying the place specified by the definition (Fig. 6.8). In these instances, one has to be quite careful with the spatial

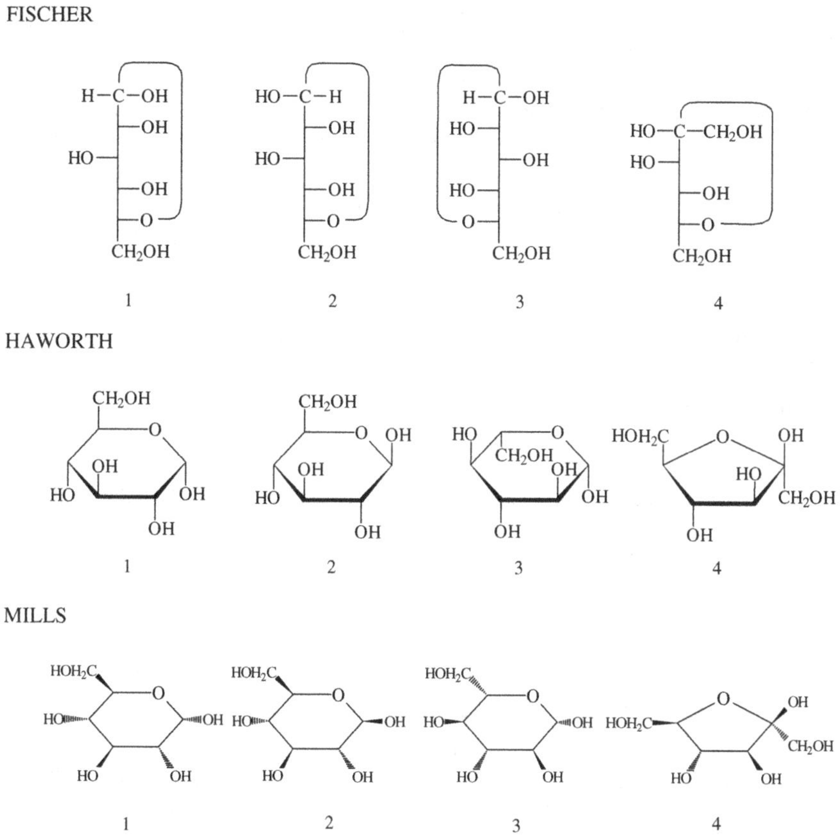

FISCHER

HAWORTH

MILLS

Fig. 6.6. Examples of Fischer projections, Haworth projections, and Mills formulas. α-D-Glucopyranose (1), β-D-glucopyranose (2), β-L-glucopyranose (3), and β-D-fructofuranose (4).

orientation of the substituents; the examination may not always be successful without the help of molecular models.

In Figs. 5.3 and 5.3 in Chapt. "5.1. Fundamental Definitions" the structures (configurations) of open-chain D-aldoses (one to four asymmetric carbon atoms) and D-2-ketoses (one to three asymmetric carbon atoms) were presented as Fischer projections together with their corresponding trivial names. Figs. 6.9 and 6.10 now contain the furanoid and pyranoid forms as Haworth projections of all D-aldoses (or α,β aldoses

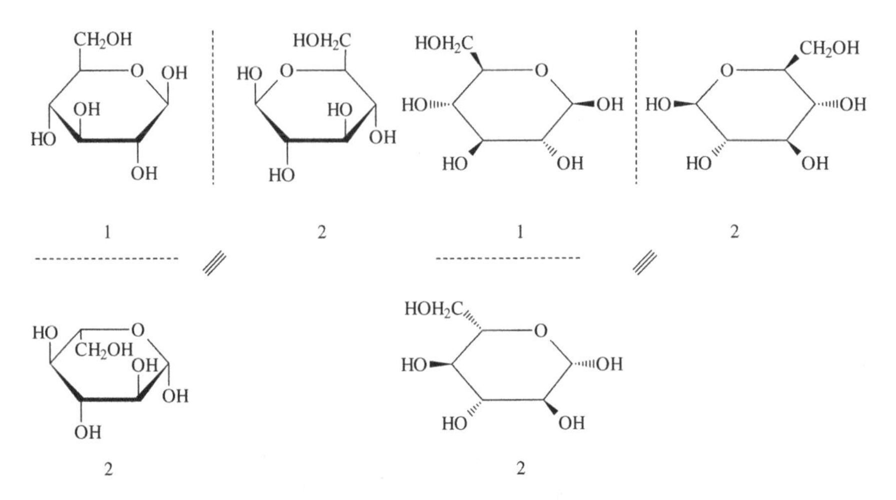

Fig. 6.7. Mirror images of the enatiomer pair of β-D-glucopyranose (1) and β-L-glucopyranose (2).

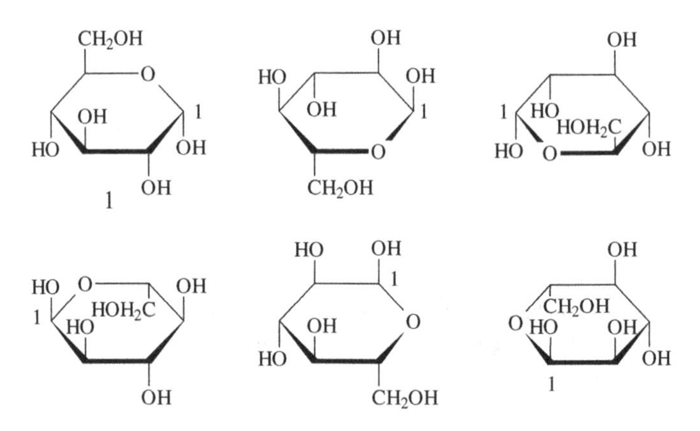

Fig. 6.8. Examples of the Haworth projections of α-D-glucopyranose represented in different ways. Only the formula (1) corresponds to the case drawn according to the "correct definition". The anomeric hydroxyl group C_1 is separately numbered in all the alternatives.

with two to four asymmetric carbon atoms in addition to the anomeric carbon atom) and α-D-2-ketoses with two or three asymmetric carbon atoms in addition to the anomeric carbon atom), together with their corresponding trivial names. If it is not desired to indicate the anomeric form of a monosaccharide (i.e., the α,β form), the following expressions are used (see also Figs. 6.9a and 6.9b):

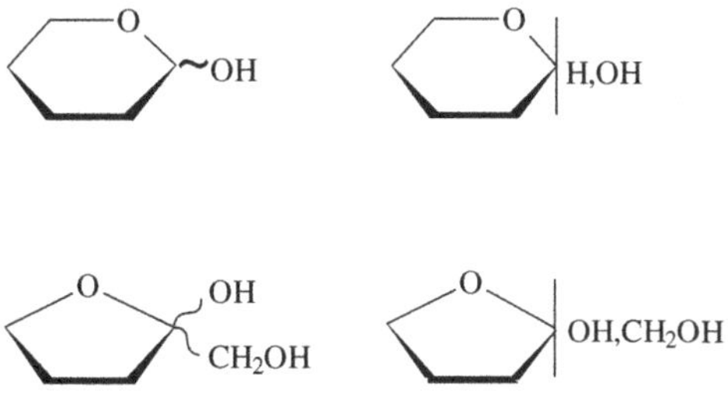

O ~OH   O H,OH

O OH CH$_2$OH   O OH,CH$_2$OH

6.3. Naming of Conformations

In Chapt. "3.6. Conformational Isomerism", the phenomenon of stereoisomerism is generally discussed in the light of open-chain organic compounds. There, the conformations or conformers (Haworth introduced the term "conformation" to describe the shape of a molecule) are defined as the different spatial states form as the atoms or atom groups of a molecule rotate with respect to one another within the limits imposed by the bonds. However, conformation isomerism is also important in the context of cyclic structures, where the rotation about the ring atoms is inhibited and the interactions between the atoms or atom groups bonded to the ring atoms only cause the ring structure to be distorted to the energetically most favorable ring conformation in each situation. Thus, this phenomenon has a rather central position in the study of monosaccharides that primarily occur as pyranoid six-atom ring structures or, less stable than those, as furanoid five-atom ring structures (cf., Chapt. "6.1. Mutarotation").

In addition to carbohydrates, six-atom ring structures of other compounds are common elsewhere in nature, which is why their conformations have been investigated in detail, especially with the help of the "basic compound", cyclohexane. Observations of cycloalkanes (begun by Hermann Sachse in the 1890s and continued by Ernst Mohr in the 1910s) can analogously be applied to the ring structures of heterocyclic carbohydrates where an oxygen atom is one of the ring atoms. In cycloalkanes, the formation of the ring generates in the structure a measurable tension

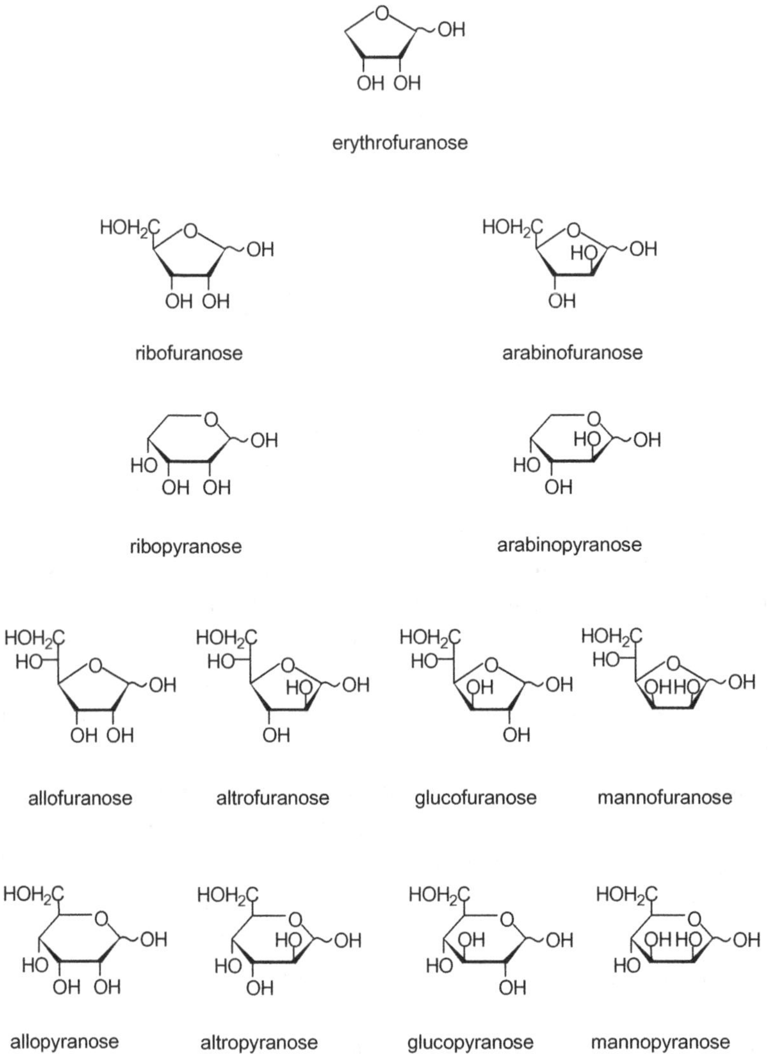

erythrofuranose

ribofuranose arabinofuranose

ribopyranose arabinopyranose

allofuranose altrofuranose glucofuranose mannofuranose

allopyranose altropyranose glucopyranose mannopyranose

Fig. 6.9a. Haworth projections for the furanoid and pyranoid forms of D-aldoses (with two to four asymmetric carbon atoms in addition to the anomeric carbon atom) and the corresponding trivial names (see also Fig. 6.9b).

(heat of combustion/CH_2 group) that varies with the size of the ring. This is why the strain that reflects the stability of the structure can be examined generally, for example, as a function of the number of the CH_2 groups in

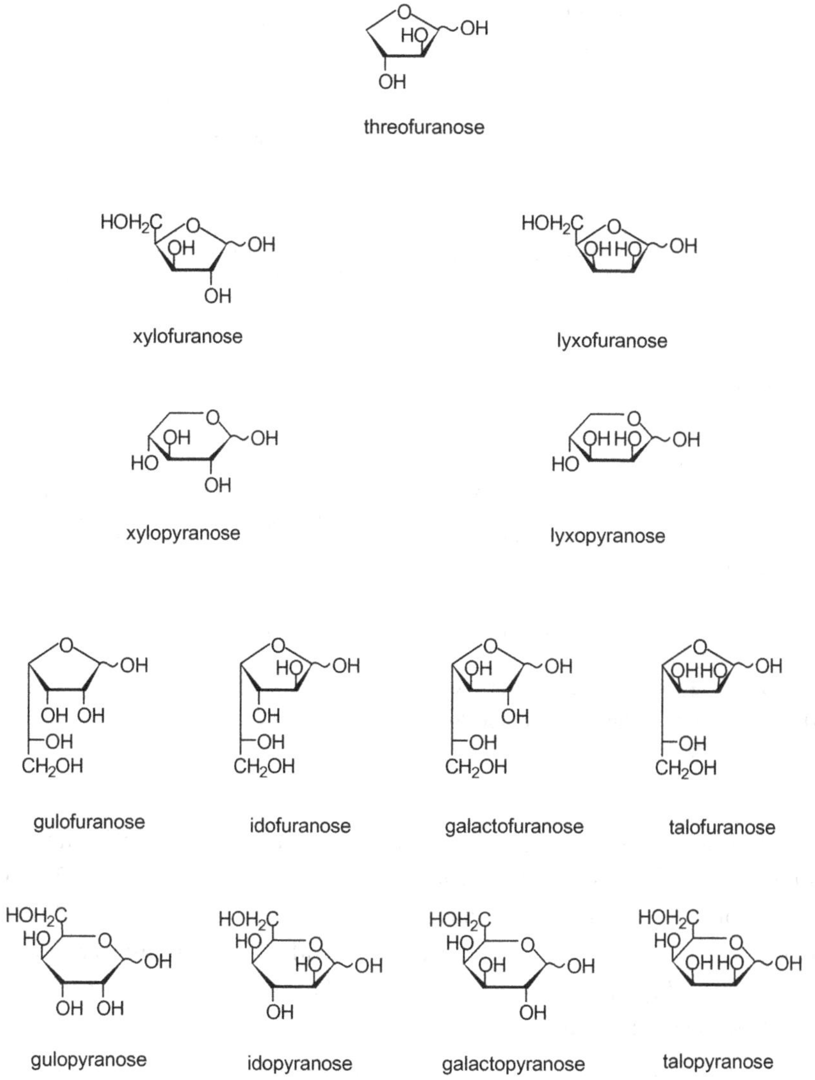

threofuranose

xylofuranose

lyxofuranose

xylopyranose

lyxopyranose

gulofuranose

idofuranose

galactofuranose

talofuranose

gulopyranose

idopyranose

galactopyranose

talopyranose

Fig. 6.9b. Haworth projections for the furanoid and pyranoid forms of D-aldoses (with two to four asymmetric carbon atoms in addition to the anomeric carbon atom) and the corresponding trivial names (see also Fig. 6.9a).

the molecule (total strain/CH_2 group, with the value for the most stable structure, cyclohexane, set at 0 kJ/mole) with the following results: for three CH_2 groups in the ring, each 39 kJ/mole (for cyclopropane a total of

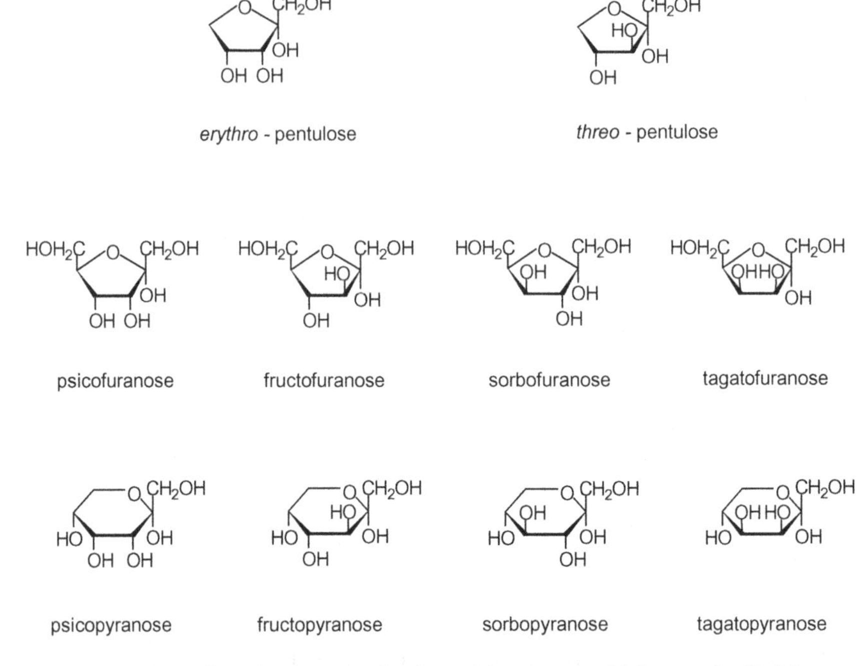

Fig. 6.10. Haworth projections for the furanoid and pyranoid forms of α-D-2-ketoses (with two or three asymmetric carbon atoms in addition to the anomeric carbon atom) and the corresponding trivial names.

116 kJ/mole), for four CH_2 groups, each 27 kJ/mole (for cyclobutane 109 kJ/mole), for five CH_2 groups, each 5 kJ/mole (for cyclopentane 25 kJ/mole), for seven CH_2 groups, each 4 kJ/mole (for cycloheptane 27 kJ/mole), and for eight to ten CH_2 groups a total of 40–54 kJ/mole. From these energy data, it is easy to understand that the most stable structures are, besides six-rings, those with five or seven atoms.

Six-atom rings. If cyclohexane existed as a planar and regular hexagon with a carbon atom at each corner, every angle ∠ C-C-C would be 120°. This angle would substantially deviate from the ideal angle, 109.5°, between the tetrahedrically oriented substituents in an ideal carbon atom (cf., Chapt. "3.3. Basic Factors Influencing the Three-dimensional Compound Structures") and would induce in the ring tensional strain that would reduce its stability. In addition, in a planar cyclohexane structure, the hydrogen atoms attached to adjacent carbon atoms would be placed in an unfavorable conformation opposite to one another (cf., Chapt. "3.6. Conformational Isomerism"), which would create a "total instability strain" that could be estimated to be at least about 50 kJ/mole.

Because of these factors, cyclohexane tends to "pucker", resulting in two spatially rather different fundamental conformations (called after their basic forms either "*chair conformation*" or "*boat conformation*", see the text below), where in both all ∠ C-C-C angles are close to the ideal value of 109.5°. In practice, it is not possible to distinguish between these conformations because they change (in a complex reversible equilibrium reaction) at room temperature through *half-chair* and *skew-boat conformations* to one another relatively easily (with an energy threshold of about 46 kJ). The chair conformation (accurately characterized by Hassel in 1943) is, however, more stable than the boat conformation, and its share in equilibrium at room temperature is about 99% (Fig. 6.11). The smaller abundance of the boat conformation can primarily be explained by the unfavorable interactions between the hydrogen atoms attached to the adjacent carbon atoms (C_2 and C_3 and also C_5 and C_6). These hydrogen atoms are situated opposite one another; moreover, two hydrogen atoms attached to carbon atoms C_1 and C_4 are closest to one another (so-called "flagpole" hydrogens), causing steric hindrance due to their closeness (distance 0.183 nm) with a repulsion energy of about 12 kJ.

As can be seen in Fig. 6.11, the conformation of cyclohexane contains both axially ("a" bonds) and equatorially ("e" bonds) oriented bonds. In some cases, it may be difficult to ascertain the orientation; it can be generally stated that if an "average" plane is imagined to pass through the carbon atoms of cyclohexane, the axial bonds will be parallel to an axis placed at right angles through this plane (three above the plane and three below it), whereas the equatorial bonds are almost parallel to the plane. In *ring inversion* ($^4C_1 \leftrightharpoons {}^1C_4$, frequency at room temperature 10^6 times/s), all axial bonds become equatorial and vice versa.

Monosaccharides exist in aqueous solutions mainly in the tautomeric forms required by this equilibrium (principally pyranoid structures), while alternative conformations of all these forms arise. The main chains of

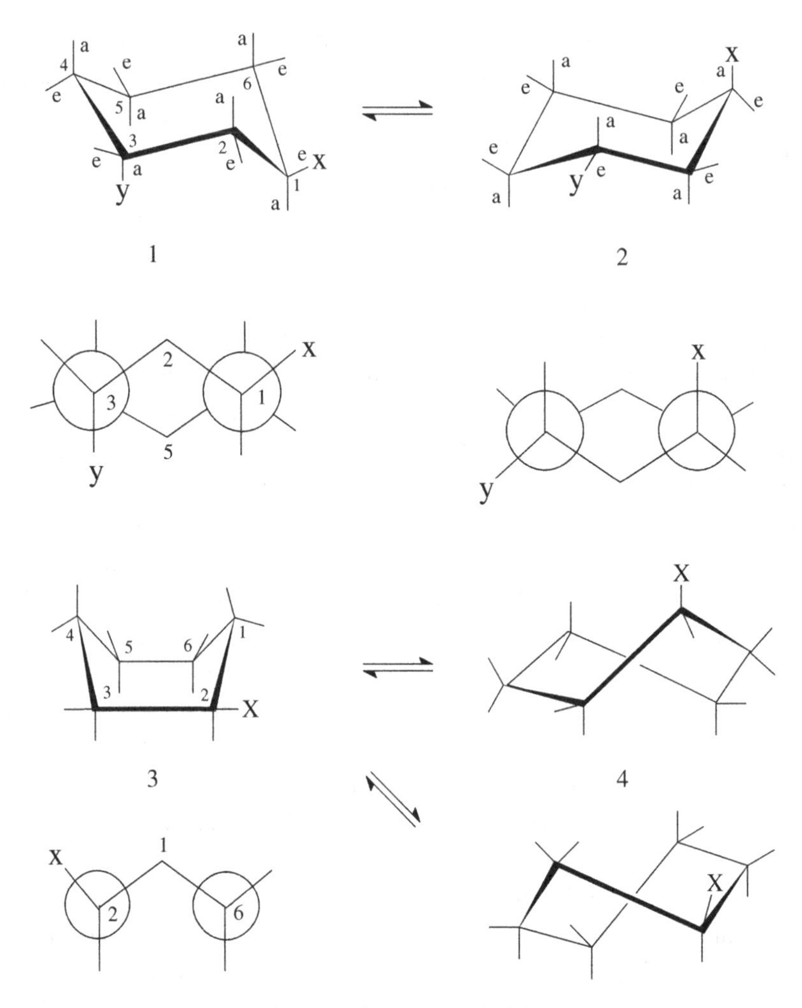

Fig. 6.11. In the chair conformations $^4C_1$ (1) and $^1C_4$ (2) of cyclohexane (see the text below), the substituents attached to carbon atoms C_2 and C_3, as well as those attached to carbon atoms C_5 and C_6 exist in the staggered conformation, while in the boat conformation (3) these substituents exist correspondingly in the eclipsed conformation. When the ring goes from conformation $^4C_1$ to conformation $^1C_4$ by flipping (ring inversion), all axial (a) bonds become equatorial (e) and vice versa. The intermediate skew-boat conformation (4) is also shown and besides all these conformations, half-chair conformation is possible.

solid oligo- and polysaccharides form almost without exception pyranoid monosaccharide units of only one conformation; in this case, for example, furanoid monosaccharide units are found only in small amounts, particularly in side groups of carbohydrate chains or as structural parts of other natural substances, such as DNA and RNA.

The configurative differences in the structure of the pyranoid ring carbon atoms in monosaccharides are highly important because of the stability of these ring structures and, consequently, their actual presence in macromolecules. The most stable alternative is typically the structure where, due to the interactions between the substituents, the majority of so-called "large substituents" (-OH and -CH_2OH) are equatorially oriented. On the contrary, the majority of D-aldopyranoses have been found to exist in conformation $^4C_1$, which corresponds to conformation $^1C_4$ in L-aldopyranoses (which are mirror images of the former). Table 6.4 presents a summary of the spatial orientation of the substituents bonded to the

Table 6.4. Orientations of substituents -OH and -CH_2OH in the most stable structures of aldopentoses and aldohexoses. It is usually in D series $^4C_1$ conformation corresponding to $^1C_4$ conformation in L series ("a" refers to axial and "e" to equatorial)

	Substituent						
Aldopyranose	HO-1(α)	HO-1(β)	HO-2	HO-3	HO-4	HOH_2C-5	Conformation*
Aldopentoses							
Ribose	a	e	e	a	e	—	$^4C_1$
Arabinose	e	a	e	e	a	—	$^1C_4$**
	a	e	a	a	e	—	$^4C_1$
Xylose	a	e	e	e	e	—	$^4C_1$
Lyxose	a	e	a	e	e	—	$^4C_1$***
	e	a	a	a	a	—	$^1C_4$
Aldohexoses							
Allose	a	e	e	a	e	e	$^4C_1$
Altrose	a	e	a	a	e	e	$^4C_1$***
	e	a	e	e	a	a	$^1C_4$
Glucose	a	e	e	e	e	e	$^4C_1$
Mannose	a	e	a	e	e	e	$^4C_1$
Gulose	a	e	e	a	a	e	$^4C_1$
Idose	e	a	e	e	e	a	$^1C_4$**
	a	e	a	a	a	e	$^4C_1$

(Continued)

Table 6.4. (*Continued*)

Aldopyranose	Substituent HO-1(α)	HO-1(β)	HO-2	HO-3	HO-4	HOH_2C-5	Conformation*
Galactose	a	e	e	e	a	e	$^4C_1$
Talose	a	e	a	e	a	e	$^4C_1$***
	e	a	e	a	e	a	$^1C_4$

*The conformations only for the D-series of aldoses are shown. All the equatorial subtituents in the $^4C_1$ (D-aldoses) (or $^1C_4$ (L-aldoses)) become axial in the $^1C_4$ (D-aldoses) conformation (or $^4C_1$ (L-aldoses)) and *vice versa*.

**The conformation $^1C_4$ represents the preferred conformation.

***The conformation $^4C_1$ and $^1C_4$ are almost equally favored conformations.

pyranoidic ring structures of aldopentoses and aldohexoses in their most stable conformations according to general opinion, and Fig. 6.12 shows a few illuminating examples. It should be noted that the most stable conformation of an individual monosaccharide may differ from that of a certain derivative of the same monosaccharide.

In reality, the angle ∠ C-C-C is about 111° in the chair conformation of cyclohexane with a dihedral angle (cf., Chapt. "3.3. Basic Factors Influencing the Three-dimensional Compound Structures") of about 56° (the ideal values being 109.5° and 60°, respectively). As, for example, in D-glucopyranose, the angle ∠ C-O-C is about 111°, a conformational view based on cyclohexane (or generally cycloalkanes) is also well-suited for pyranoses. However, a detailed study of conformation isomers is seldom necessary, and in many practical applications, it is more illustrative and suitable to use, for example, Haworth projections.

Naming practice. R.E. Reeves devised the naming of monosaccharide conformations in the 1940s, which R.D. Guthrie, H.S. Isbell, and R.S. Tipson gradually sharpened in some parts in the late 1950s. Today, the following main principles are applied:

(1) The basic form of conformation is expressed for pyranoid monosaccharides with letters C ("chair conformation"), B ("boat conformation"), S ("skew-boat conformation" or "twist-boat conformation"),

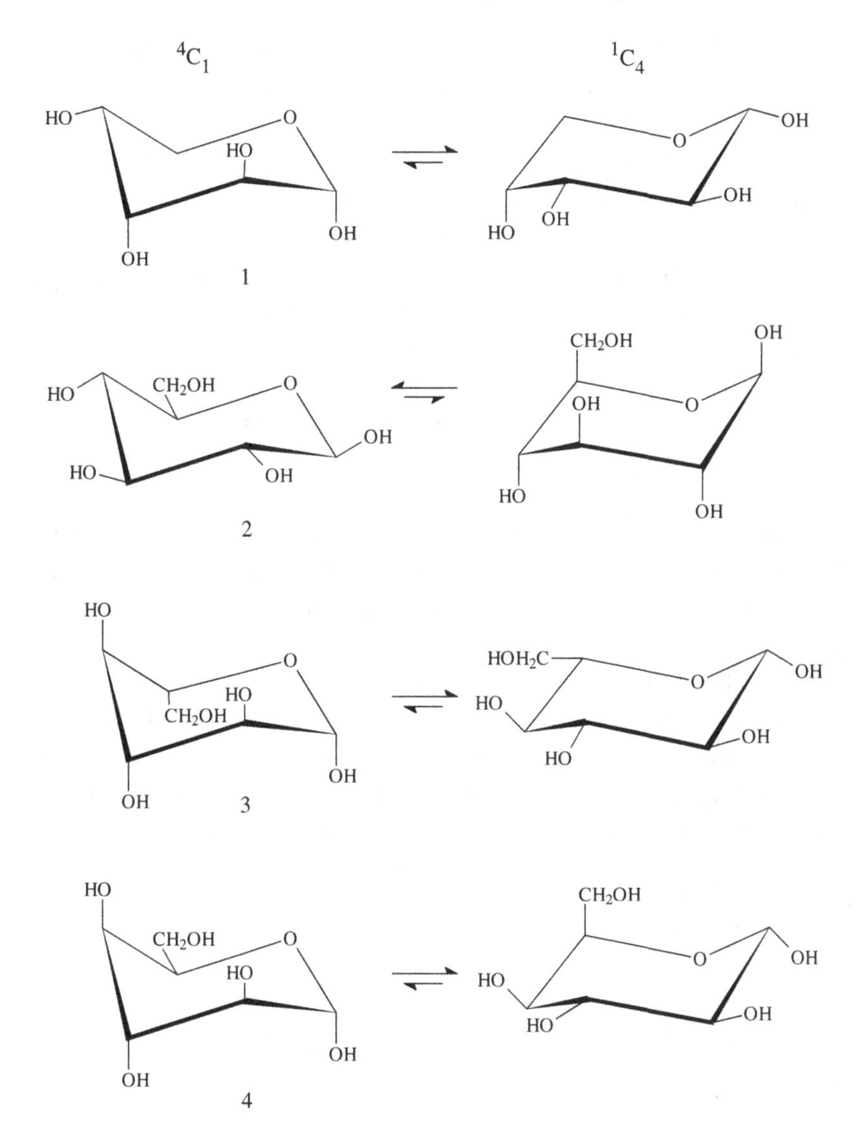

Fig. 6.12. Examples of the aldopyranose conformations $^4C_1$ and $^1C_4$ and their mutual stabilities including the orientations of substituents -OH and -CH_2OH. α-D-Arabinopyranose (1) ($^4C_1$; with three axial substituents and $^1C_4$; with one axial substituent), β-D-glucopyranose (2) ($^4C_1$; with no axial substituents and $^1C_4$; with five axial substituents), β-L-glucopyranose (3) ($^4C_1$; with five axial substituents and $^1C_4$; with no axial substituents), and α-D-idopyranose (4) ($^4C_1$; with four axial substituents and $^1C_4$; with one axial substituent).

H ("half-chair conformation"), and E ("envelope conformation") and for furanoid monosaccharides (see the text below) with letters E and T ("twist conformation").

(2) The ring atoms of pyranoid and furanoid monosaccharides are numbered as a rule according to Fig. 6.5.

(3) In pyranoid structures, a reference plane containing four atoms (not necessarily all adjacent) (in rare, only in certain derivatives occurring E conformations five atoms), where the lowest numbered carbon atom (the anomeric carbon atom in monosaccharides) is not included in the reference plane. In the H conformation, the four-atom reference plane instead includes four adjacent atoms (in aldopyranoses often C_1, C_2, C_3, and C_4).

Two reference planes can be found for each S conformation; the choice of the plane thus is not unambiguous as it is with C and B conformations. Here, the reference plane contains three adjacent atoms of the pyranoid form and one non-adjacent atom; of the alternatives one should select the alternative where the lowest-numbered carbon atom or the one following it is left outside the plane. In the E conformation of the furanoid structures, the reference plane includes four adjacent atoms, and in the corresponding T conformation, it includes three adjacent ring atoms, while the atoms outside the plane are placed on its opposite sides.

Based on the rules above the reference planes in the common conformations $^4C_1$ and $^1C_4$ of pyranoid structures are as follows:

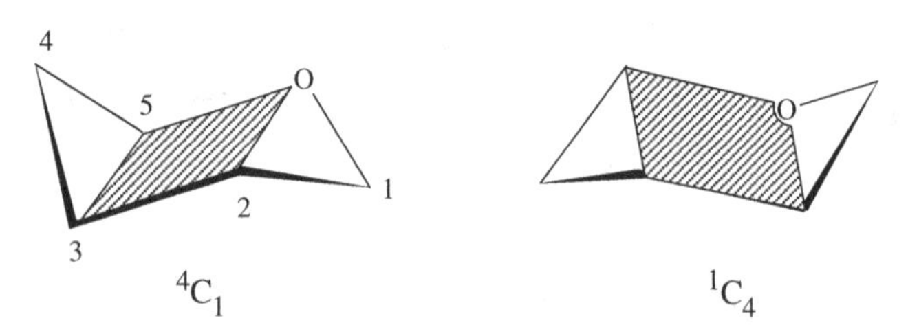

(4) The number of the atom situated above the reference plane (or the numbers of the atoms so situated) are placed as superscripts before the letter indicating the conformation and, correspondingly, the number of the atom situated below the reference plane (or the numbers of

the atoms so situated) are placed as subscripts after the letter. It should be noted that heterocyclic oxygen is designated with the letter "O" and, when appearing together with a number, is placed last. Conformations can thus be expressed, for example, as $^4C_1$ (old symbol C1), $^1C_4$ (old symbol 1C), $^{1,4}B$, $B_{3,O}$, $^1S_3$, $^OS_2$, ^OE, and E_2.

Altogether, six structural variants exist for B conformations (Fig. 6.13); they form three pairs of mirror images, where the direction of the atoms is indicated with the numbering. Similarly, S conformations have six alternative structures (Fig. 6.14). As the reference plane can be placed in two ways in all cases, two different usages can be found in the literature (e.g., $^OS_2 \equiv {}^3S_5$ and $^2S_O \equiv {}^5S_3$). The following potential equilibrium exists between the B and S conformations: $\leftrightharpoons {}^{1,4}B \leftrightharpoons {}^1S_5 \leftrightharpoons B_{2,5} \leftrightharpoons {}^OS_2$ ($^3S_5$) $\leftrightharpoons {}^{3,O}B \leftrightharpoons {}^3S_1 \leftrightharpoons B_{1,4} \leftrightharpoons {}^5S_1 \leftrightharpoons {}^{2,5}B \leftrightharpoons {}^2S_O$ ($^5S_3$) $\leftrightharpoons B_{3,O} \leftrightharpoons {}^1S_3 \leftrightharpoons {}^{1,4}B \leftrightharpoons$. In addition, the following equilibria are present: $\leftrightharpoons {}^4C_1 \leftrightharpoons {}^4H_5 \leftrightharpoons {}^1S_5 \leftrightharpoons {}^1H_2 \leftrightharpoons {}^1C_4 \leftrightharpoons$, $\leftrightharpoons {}^4C_1 \leftrightharpoons {}^OH_5 \leftrightharpoons {}^OS_2 \leftrightharpoons {}^3H_2 \leftrightharpoons {}^1C_4 \leftrightharpoons$, $\leftrightharpoons {}^4C_1 \leftrightharpoons {}^OH_1 \leftrightharpoons {}^3S_1 \leftrightharpoons {}^3H_4 \leftrightharpoons {}^1C_4 \leftrightharpoons$, $\leftrightharpoons {}^4C_1 \leftrightharpoons {}^2H_1 \leftrightharpoons {}^5S_1 \leftrightharpoons {}^5H_4 \leftrightharpoons {}^1C_4 \leftrightharpoons$, $\leftrightharpoons {}^4C_1 \leftrightharpoons {}^2H_3 \leftrightharpoons {}^2S_O \leftrightharpoons$

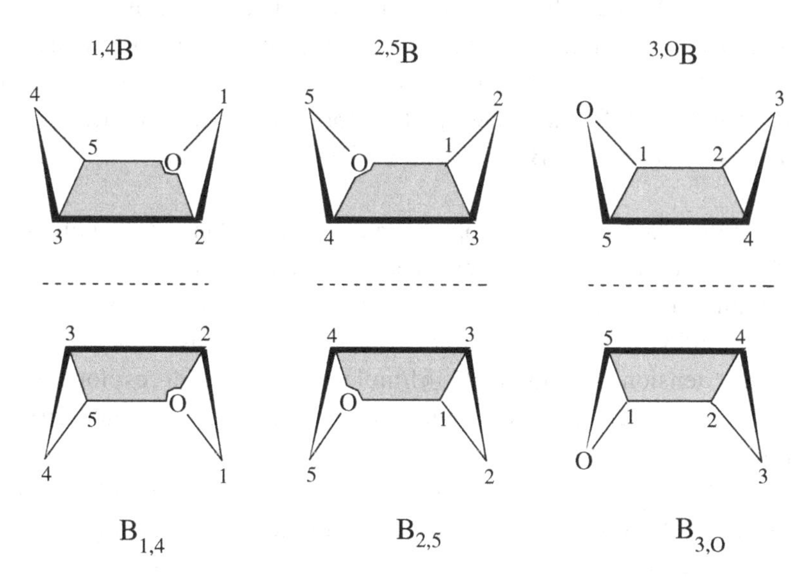

Fig. 6.13. Schematic representation of the possible B conformations of the pyranoid ring and the corresponding reference planes. $^{1,4}B$ and $B_{1,4}$, $^{2,5}B$ and $B_{2,5}$, and $^{3,O}B$ and $B_{3,O}$, are pairs of mirror images in which the atom numbering is always opposite.

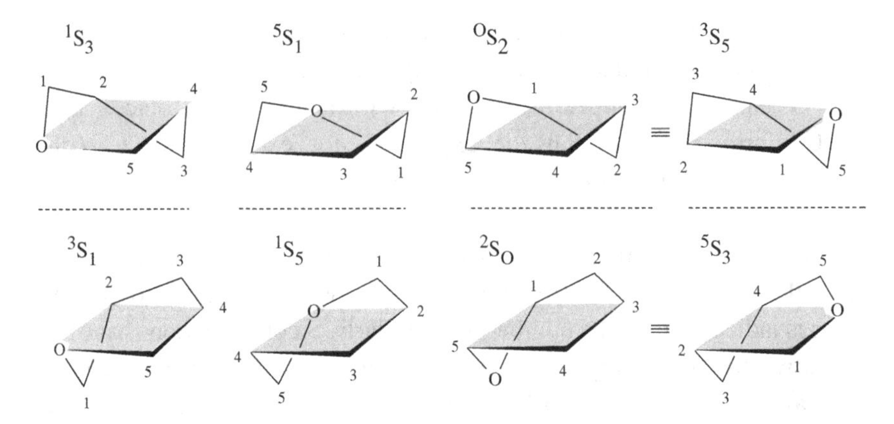

Fig. 6.14. Schematic representation of the possible S conformations of the pyranoid ring and the corresponding reference planes. $^{1}S_{3}$ and $^{3}S_{1}$, $^{5}S_{1}$ and $^{1}S_{5}$, and $^{O}S_{2}$ ($^{3}S_{5}$) and $^{2}S_{O}$ ($^{5}S_{3}$) are pairs of mirror images in which the numbering order of carbon atoms are always opposite.

$^{5}H_{O} \leftrightharpoons {}^{1}C_{4} \leftrightharpoons$, and $\leftrightharpoons {}^{4}C_{1} \leftrightharpoons {}^{4}H_{3} \leftrightharpoons {}^{1}S_{3} \leftrightharpoons {}^{1}H_{O} \leftrightharpoons {}^{1}C_{4} \leftrightharpoons$. It is thus possible to construct a detailed chart of equilibria that contains all conformational inter-actions of pyranoid ring structures. As a summary, Fig. 6.15 contains general formulas of the different conformations of cyclohexane, their mutual potential differences, and the transformations between them: $\leftrightharpoons C \leftrightharpoons H \leftrightharpoons S \leftrightharpoons B \leftrightharpoons \ldots \leftrightharpoons B \leftrightharpoons S \leftrightharpoons H \leftrightharpoons C \leftrightharpoons$.

Other ring structures. If cyclopentane exists as a planar pentagon with a carbon atom at each corner, every angle ∠ C-C-C is close to optimal (108°), but the hydrogen atoms bonded to adjacent carbon atoms would be in an unfavorable conformation to each other. This leads to "total instability tension" of about 42 kJ/mole, which causes cyclopentane to "pucker" into either an envelope conformation (E conformation) or a skewed one (T conformation). These conformations can be seen to represent, in terms of their bond angles and the location of their hydrogen atoms, the best "compromise structures", and their potential energies are also nearly equal — the planar form is about 20 kJ less stable. Hence, they easily and quickly transform to one another at room temperature. Every furanoid structure can form ten E and ten T conformations; the

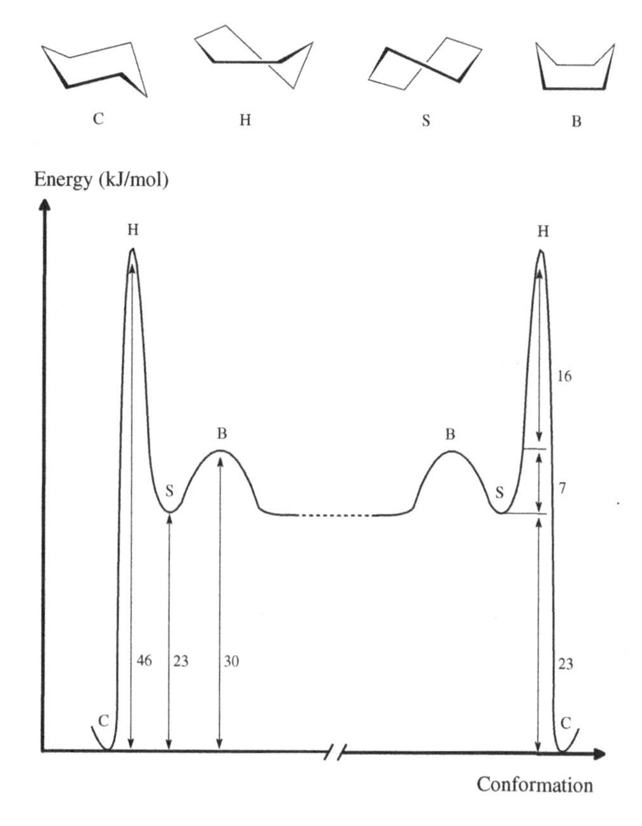

Fig. 6.15. Conformations of cyclohexane, their mutual potential energy differences, and the transformations between them (⇌ C ⇌ H ⇌ S ⇌ B ⇌ ... ⇌ B ⇌ S ⇌ H ⇌ C ⇌). C refers to chair, H to half-chair, S to skew-boat, and B to boat conformations.

continuing transformation between these conformations is often called "*pseudorotation*". Figure 6.16 shows examples of the conformations found in β-D-xylofuranose.

The ring structures of seven-atom septanoses also show different conformations analogous to those found in pyranoidic rings (chair, half-chair, boat, and skew-boat). In planar cyclobutane, the angle ∠ C-C-C is 90°, which is why rare oxetoses can be thought to have quick equilibrium states between various non-planar "folder" conformations that change to one another in motions that resemble, for example, the motion of butterfly wings.

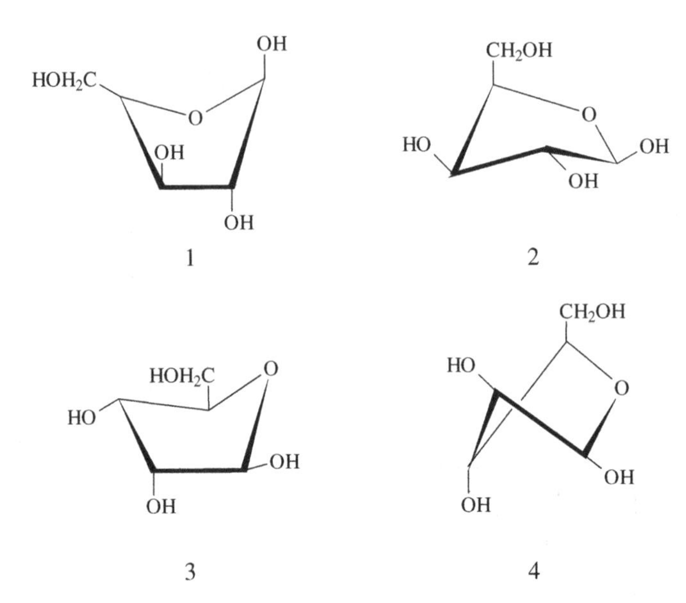

Fig. 6.16. Examples of three E conformations (1→3) and one T conformation (4) of β-D-xylofuranose.

7. Naming of Monosaccharides

Until about the 1940s, carbohydrates had been generally named as proposed by individual researchers, and they were either accepted by the scientific community or failed to gain such popularity. Only thereafter, the "official circles" started to pay more attention to standardizing the nomenclature of carbohydrates; this development led through many intermediate phases to the publication of comprehensive rules and recommendations in the late 1990s. Currently, a more specific practice of naming has been published as a joint operation of the "International Union of Pure and Applied Chemistry" (IUPAC) and the "International Union of Biochemistry and Molecular Biology" (IUBMB). It is clear that, along with the continuing development of the field, many details of carbohydrate naming will be further clarified and complemented.

7.1. Aldoses

Chapt. "1. Introduction" provides general definitions of carbohydrates and also, among others, of aldoses and ketoses; Edward Frankland Armstrong 1878–1945 introduced these terms. Aldoses can be given, depending on the case, either *trivial names* or *systematic names*. In the latter, both D and L series aldoses with at most four asymmetric carbon atoms (a total of at most six carbon atoms) are primarily assigned trivial names with an *-ose* ending (cf., Chapt. "5.1. Fundamental Definitions"): erythrose and threose with four carbon atoms; arabinose, lyxose, ribose, and xylose with five carbon atoms; and allose, altrose, galactose, glucose, gulose, idose, mannose, and talose with six carbon atoms. However, the simplest aldose with three carbon atoms is almost without exception called "glyceraldehyde" (2,3-dihydroxypropanal) following the general nomenclature of organic chemistry. The systematic name of an aldose is normally formed from a *stem name* and a *configurational prefix*. The stem names for the aldoses mentioned are tetrose (with four carbon atoms), pentose (with five carbon atoms), and hexose (with six carbon atoms).

Table 7.1 presents the trivial names and the corresponding systematic names together with their abbreviations used in expressing, for example, the ring forms of monosaccharides and the structures of large oligo- and polysaccharide molecules based on such forms. The configurational prefixes can easily be derived from the spatial structures of the asymmetric carbon atoms of the aldoses presented as Fischer projections

Table 7.1. Trivial and systematic names of aldoses (with two to four asymmetric carbon atoms) together with their common abbreviations

Trivial name	Systematic name	Abbreviation
Erythrose	*erythro*-Tetrose	Eryth*
Threose	*threo*-Tetrose	Thro*
Arabinose	*arabino*-Pentose	Ara
Xylose	*xylo*-Pentose	Xyl
Lyxose	*lyxo*-Pentose	Lyx
Ribose	*ribo*-Pentose	Rib
Allose	*allo*-Hexose	All
Altrose	*altro*-Hexose	Alt
Galactose	*galacto*-Hexose	Gal
Glucose	*gluco*-Hexose	Glc
Gulose	*gulo*-Hexose	Gul
Idose	*ido*-Hexose	Ido
Mannose	*manno*-Hexose	Man
Talose	*talo*-Hexose	Tal

* Usage not established.

(see Fig. 5.3). If needed, the aldose series (e.g., L-glucose or L-*gluco*-hexose) can also be indicated. For glyceraldehyde, the configurational prefixes used are D-*glycero*- and L-*glycero*-.

Systematic names are always used when the number of asymmetric carbon atoms in an aldose is >4 (the total number of carbon atoms ≥7). In these cases, two or more prefixes have to be added to the basic name to indicate the conformation. The general principle is that the carbon chain presented according to the Fischer projection is divided, starting from the lowest-numbered asymmetric carbon atom, into sequences of four asymmetric carbon atoms (n sequences), the last one perhaps having fewer (m) such carbon atoms. Then, the total number of asymmetric carbon atoms in the aldose is 4n + m (n = 1, 2, … and m = 0–3). Now the systematic name of the monosaccharide is built by adding the configurational prefixes denoting the structures of the sequences onto the stem name starting with the sequence located farthest from the aldehyde group (carbon atom C_1) (Fig. 7.1). The most common stem names are heptose (with seven carbon atoms), octose (with eight carbon atoms), nonose (with nine carbon atoms), and decose (with ten carbon atoms), although practically such aldoses are quite rare. This method of naming is easy to understand, while the need for trivial names would involve a minimum (i.e., using also the D,L system) of 16, 32, 64, and 128 names for heptoses, octoses, nonoses, and decoses, respectively, forming a set that is difficult to manage.

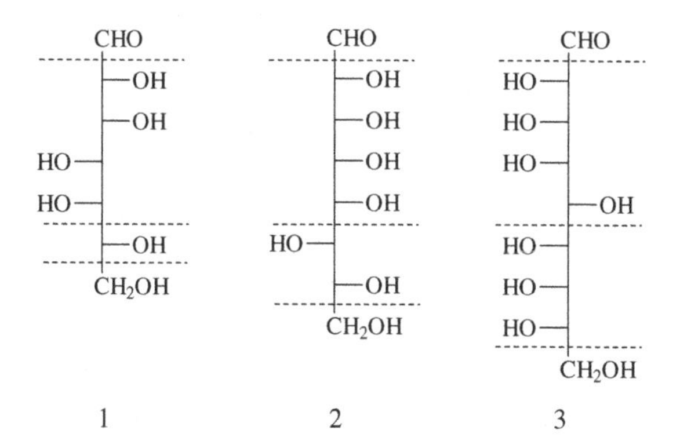

Fig. 7.1. Examples of aldoses containing five, six, or seven asymmetric carbon atoms. D-*glycero*-L-*manno*-Heptose (1), D-*threo*-D-*allo*-octose (2), and L-*ribo*-D-*talo*-nonose (3).

The most common derivatives of monosaccharides will be discussed later in the context of groups of compounds; a few examples are presented here to illustrate their relation to the general naming of aldoses.

Deoxy sugars are generally monosaccharides where one or several hydroxyl groups in their structure are replaced with hydrogen atoms:

$$>CHOH \longrightarrow >CH_2$$

$$-CH_2OH \longrightarrow -CH_3$$

Here, the hydroxymethylene group becomes a methylene group and, respectively, the hydroxymethyl group of the ω-carbon becomes a methyl group. Deoxy sugars are primarily named with systematic names; exceptions are some relatively common *ω-deoxy derivatives* (6-deoxy derivatives) (Fig. 7.2). These monosaccharides are thus not classified as, *C*-methyl derivatives of aldopentoses. Similarly, according to the established usage, the 2-deoxy-D-*erythro*-pentose is called "2-deoxy-ribose" and the 2-deoxy-D-*arabino*-hexose is called "2-deoxy-glucose".

It is naturally possible that more achiral deoxy carbon atoms are present in a monosaccharide. In these cases, such deoxy carbon atoms are not taken into account in forming the sequences dividing the carbon chain of the aldose as described above (Fig. 7.3).

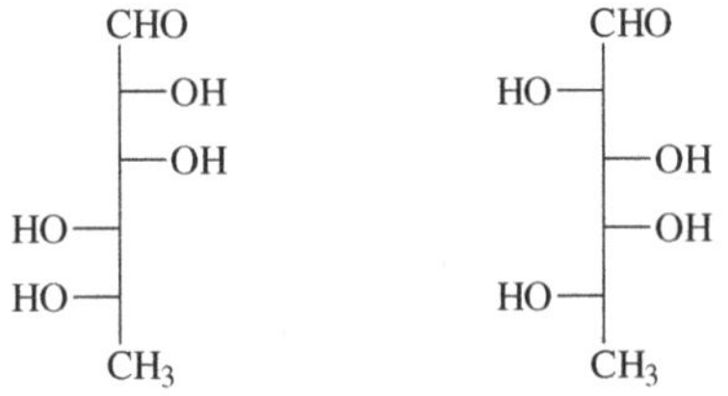

Fig. 7.2. Examples of common ω-deoxy sugars. L-Rhamnose or 6-deoxy-L-mannose or 6-deoxy-L-*manno*-hexose (Rha) (1) and L-fucose or 6-deoxy-L-galactose or 6-deoxy-L-*galacto*-hexose (Fuc) (2).

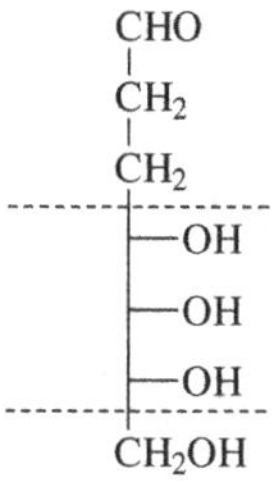

1

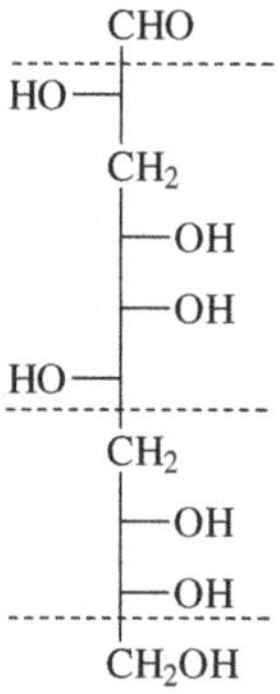

2

Fig. 7.3. Example of aldoses containing several deoxy carbon atoms. 2,3-Dideoxy-D-*ribo*-heptose (1) and 3,7-dideoxy-D-*erythro*-L-*galacto*-decose (2).

One or several of the hydrogen atoms of a hydroxyl group in an aldose, or those directly bonded to a chain carbon atom, may be substituted by an organic group (R):

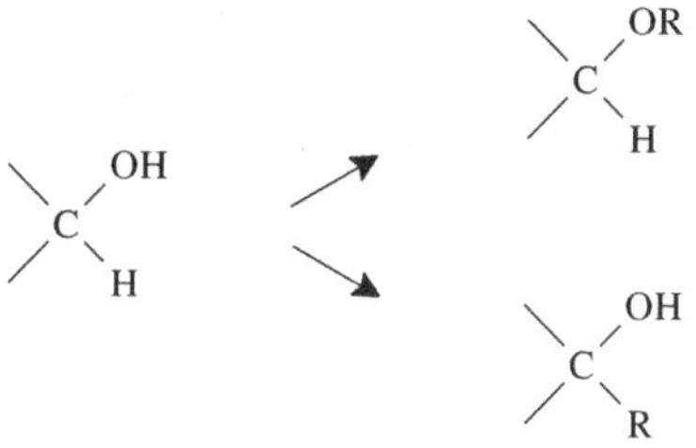

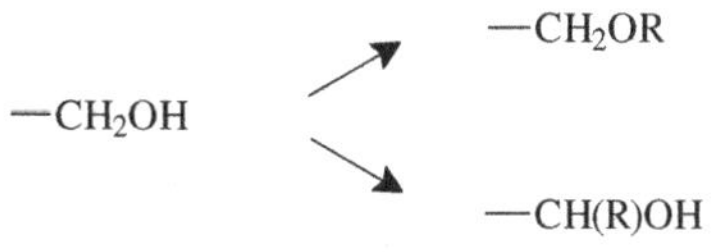

$$\mathrm{C(OH)(H)} \rightarrow \mathrm{C(OR)(H)} \quad \text{or} \quad \mathrm{C(OH)(R)}$$

$$-\mathrm{CH_2OH} \rightarrow -\mathrm{CH_2OR} \quad \text{or} \quad -\mathrm{CH(R)OH}$$

Here, the substituting groups do not influence the configuration of the aldose. The presence of such groups is expressed, depending on the case, with markings placed in front of the name of the aldose, such as

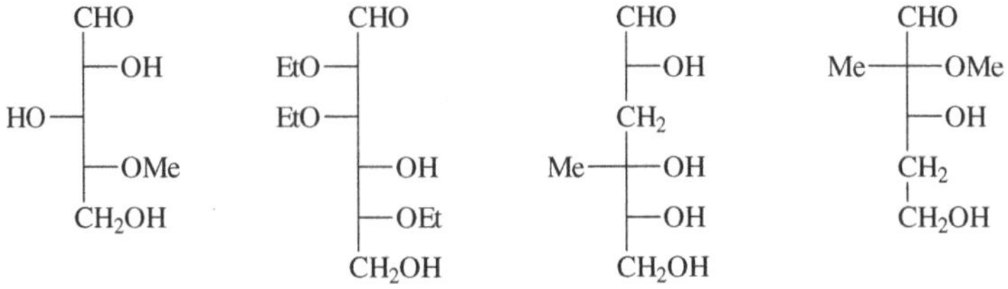

CHO | CHO | CHO | CHO
OH | EtO | OH | Me OMe
HO | EtO | CH_2 | OH
OMe | OH | Me OH | CH_2
CH_2OH | OEt | OH | CH_2OH
 | CH_2OH | CH_2OH |

1 2 3 4

Fig. 7.4. Examples of aldoses containing organic groups. 4-*O*-Methyl-D-xylose (1), 2,3,5-tri-*O*-ethyl-D-mannose (2), 3-deoxy-4-*C*-methyl-D-*ribo*-hexose (3), and 4-deoxy-2,2-di-*C*,*O*-methyl-D-*erythro*-pentose (4).

carbon number-*O*-R or carbon number-*C*-R, or "suitable combinations" of these. Established abbreviations Ac- for acetyl-, Et- for ethyl-, Me- for methyl-, Ph- for phenyl-, and Ts- for tosyl- (*p*-toluenesulfonyl-) can be used if required. Fig. 7.4 shows a few examples of the possible aldose derivatives. When several different groups are attached to a monosaccharide, they are expressed in front of the compound name in alphabetical order (e.g., 5-*O*-ethyl-2-*O*-methyl-D-mannose or 5-*O*-Et-2-*O*-Me-D-mannose). The number of the deoxy carbon (or the numbers of such carbons) is placed before other prefixes of the stem name. We do not, however, have today a flawless set of rules covering all cases; this is why one should strive for expressions that are as simple and unequivocal as possible.

7.2. Ketoses

Depending on the case, ketoses can be named like aldoses either using trivial names ending with *-ulose* or systematic names (Table 7.2) (see Fig. 5.4). Here, the systematic name is also formed from the stem name with an added prefix that indicates the configuration. In practice, nearly without exception, only ketoses where carbon atom C_2 belongs to the characteristic ketone group (*2-ketoses*) are present in nature. This is also why it is not necessary in these cases to express the number of the carbonyl carbon. It should be noted that, due to the existence of the ketone group within the carbon chain, the number of the asymmetric carbon

Table 7.2. Trivial and systematic names of 2-ketoses (with one to three asymmetric carbon atoms) together with their common abbreviations

Trivial name	Systematic name	Abbreviation
Erythrulose*	*glycero*-Tetrulose	Eul**
Xylulose*	*threo*-Pentulose	Xul
Ribulose*	*erythro*-Pentulose	Rul
Fructose	*arabino*-Hexulose	Fru
Psicose	*ribo*-Hexulose	Psi
Sorbose	*xylo*-Hexulose	Sor
Tagatose	*lyxo*-Hexulose	Tag

*Usage not recommended.
**Usage not established.

atoms relative to the total number of carbon atoms is less than in the aldoses (see Table 4.1). In addition three-carbon-atom glycerone, typically known by its trivial name dihydroxyacetone following the organic nomenclature, or by its systematic name 1,3-dihydroxy-2-propanone, is not considered a ketone due to its lack of an asymmetric carbon atom. All trivial names of ketoses (erythrulose, ribulose, and xylulose) are not fully correct because of the "scarcity" of the asymmetric carbon atoms present, so that the use of the respective systematic names is recommended. Similarly, in the literature one occasionally sees the corresponding trivial names threulose, lyxulose, and adonose for the occasionally present ketoses named above; the latter names are not recommended.

Ketoses with larger molecules than those shown in Table 7.2, such as heptuloses (e.g., seduheptulose or D-*altro*-heptulose), octuloses, and nonuloses, are found in small amounts in nature. For them, a systematic name is formed analogously with the corresponding aldoses if there are ≥5 asymmetric carbon atoms (Fig. 7.5).

The naming of ketoses is often more difficult than that of aldoses because one cannot immediately decide which of the terminal carbons of equal value (C_1 or C_ω) is taken as the first atom of the carbon chain in the Fischer projection (see also alditols, dialdoses, and aldaric acids). Here, one has to name these two alternatives separately (the transformation

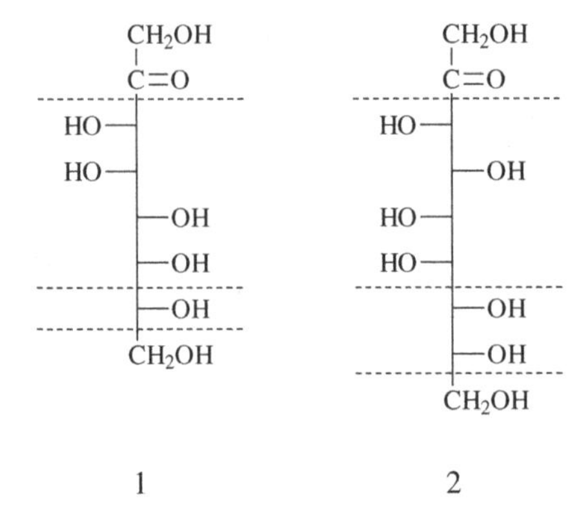

Fig. 7.5. Examples of ketoses with five or six asymmetric carbon atoms. D-*glycero*-D-*manno*-Octulose or D-*glycero*-D-*manno*-2-octulose (1) and D-*erythro*-L-*gluco*-nonulose or D-*erythro*-L-*gluco*-2-nonulose (2).

between them takes place by rotating the projection formulas in the plane for 180°), and to choose between the alternatives according to precisely defined rules; this way the properly drawn projection formula is also obtained. The prefix or prefixes of the stem name are chosen under the following rules presented here in their order of priority:

(1) The alphabetical order of the configurational prefix first dictates the order, so that, for example, *allo-* > *altro-* or *galacto-* > *gluco-*; therefore, the order *allo-* > *altro-* > *arabino-* > *erythro-* > *galacto-* > *gluco-* > *glycero-* > *gulo-* > *ido-* > *lyxo-* > *manno-* > *ribo-* > *talo-* > *threo-* > *xylo-* is followed. If there are several prefixes in the name of a ketose indicating its configuration, they are examined one pair at a time starting with the first prefixes among them.
(2) Then, the alphabetical order of the prefixes D- and L-; D- > L- govers the order. Similarly, although applicable only to ring structures, the order of priority of the prefixes showing the orientation of the anomeric hydroxyl group, is α- > β-. Thus, the order is, for example, β-D- > α-L- and α-L- > β-L-.
(3) Subsequently, the numbering of the ketone group (e.g., 2-ketose > 5-ketose) determines the order.

(4) Next, when only one type of substituent is present (once or several times) in a ketose, it should be given the lowest possible numbering (e.g., 3-*O*-Me- > 6-*O*-Me- or 2,3-di-*O*-Me- > 4,5-di-*O*-Me-). For example, in the case of alditol or dialdose, the naming with the lowest possible numbering is chosen (e.g., 2,3,5-tri-*O*-Me- > 2,4,5-tri-*O*-Me-).

(5) Lastly, the first mentioned (of the alphabetically presented) substituent with the lower numbering sets the order (e.g., 3-*O*-Et-5-*O*-Me- > 4-*O*-Et-2-*O*-Me-).

In most cases, the choice of the name of a ketose can be made according the first two rules; Fig. 7.6 presents a few illustrative examples. Ketoses can also have deoxy derivatives that are named following the rules presented above (Fig. 7.7). Ketoses can also possess intramolecular symmetries depending on the position of the ketone group; this leads to *meso* forms (cf., Chapt. "4.2. The Total Number of Stereoisomers"), such as *meso-erythro*-pentulose or *meso*-D-*erythro*-pentulose. The use of a prefix, such as 3,3-di-*C*,*O*-Me- for the structure H_3CO-C_3-CH_3 or H_3C-C_3-OCH_3, should be noted.

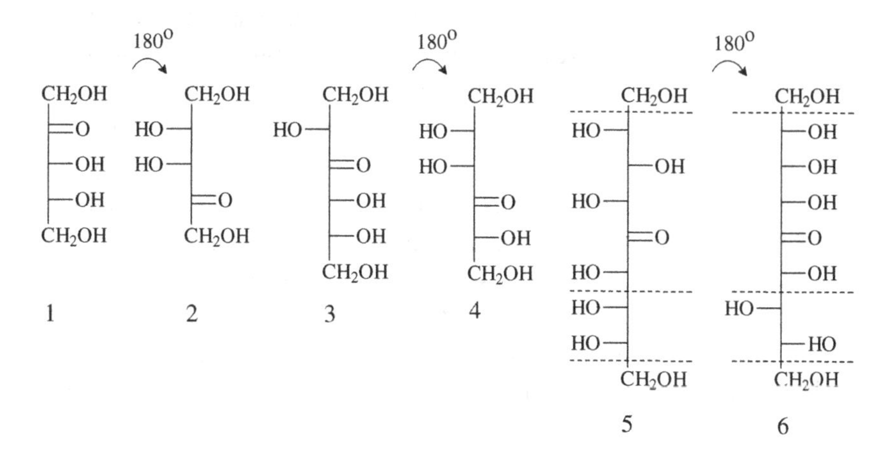

Fig. 7.6. Examples of the names of ketoses. D-*erythro*-Pentulose or D-*erythro*-2-pentulose (1) > L-*erythro*-4-pentulose (2) (D- > L-), D-*arabino*-3-hexulose (3) > D-*lyxo*-4-hexulose (4) (*arabino*- > *lyxo*-), and L-*erythro*-L-*gluco*-5-nonulose (5) > D-*threo*-D-*allo*-5-nonulose (6) (*erythro*- > *threo*-).

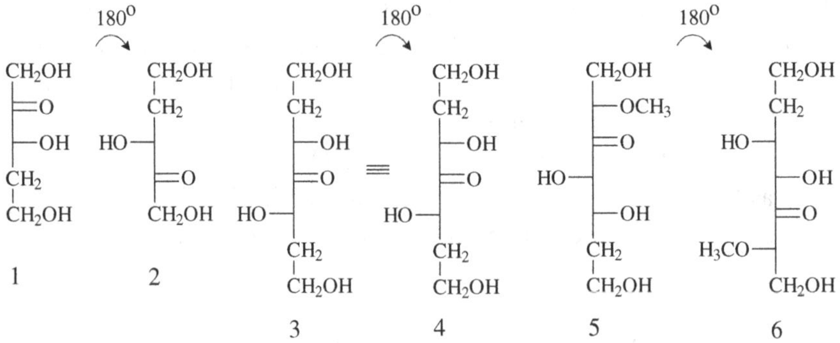

Fig. 7.7. Examples of the names of deoxy ketoses. 4-Deoxy-D-*glycero*-pentulose or 4-deoxy-D-*glycero*-2-pentulose (1) > 2-deoxy-L-*glycero*-4-pentulose (2) (D- > L-), 2,6-dideoxy-L-*threo*-4-heptulose (3) ≡ 2,6-dideoxy-L-*threo*-4-heptulose (4) (with respect to naming, formulas are identical), and 6-deoxy-2-*O*-methyl-D-*xylo*-3-heptulose (5) > 2-deoxy-6-*O*-methyl-L-*xylo*-5-heptulose (6) (D- > L-).

7.3. Cyclic Forms

The naming principles for monosaccharides have been presented earlier in Chapts. "6.1. Mutarotation" and "6.2. Planar Formulas". The structures and corresponding trivial names of both the pyranoid and furanoid forms of, among others, D-aldoses (with two to four asymmetric carbon atoms in addition to the anomeric one, Fig. 6.9) and α-D-2-ketoses (with two or three asymmetric carbon atoms in addition to the anomeric one, Fig. 6.10) were shown with the help of the Haworth projection. Here, it is only necessary to mention that those forms can also be unambiguously named with the abbreviations of monosaccharides presented in Chapts. "7.1. Aldoses" and "7.2. Ketoses". In this process, a postscript *p* (for a pyranoidic ring) or *f* (for a furanoidic ring) is added after the abbreviated name of the monosaccharide. Thus, for example, the name of β-D-glucopyranose can be abbreviated as β-D-Glc*p* and that of β-D-fructofuranose as β-D-Fru*f*.

8. Carbohydrate Biosynthesis

8.1. Photosynthesis

Living organisms require energy for three main tasks: (i) for performing mechanical work, (ii) for actively moving molecules and ions, and (iii) for synthesizing molecules. *Photosynthesis* (in Greek, "phos" means prefix "light-" and "synthesis" "put together") or *assimilation* is the most important and unique biochemical process in nature, where plants and certain other organisms produce, from carbon dioxide and water with the help of solar radiation energy (photons), oxygen and glucose for the needs of the Earth's biosphere according to the following general reaction:

$$6CO_2 + 6H_2O \xrightarrow{h\upsilon} 6O_2 + C_6H_{12}O_6$$

British Joseph Priestley (1733–1804) made the first preliminary observations relating to photosynthesis in 1776, when he reported that plants had the ability to improve air that had become poisonous from the breathing of animals and that was suitable for a kind of nutrition for plants. Then, the British scientist Jan Ingenhousz (1730–99, born in the Netherlands) presented in 1799 that photosynthesis was a reaction generally taking place in green plants, where they produce sugar, and as a side product, oxygen gas from carbon dioxide and water with the help of bright sunlight. Soon after these individual observations, the questions relating to thermal energy formed an aggregate that dominated the thoughts of German Julius Robert von Mayer (1814–78), which led him to the first rule of thermodynamics: that energy cannot be destroyed. As a result of this profound deliberation, he showed in 1845 that green plants continuously transmute the radiation energy of the Sun to chemical energy. However, it took until 1937 to finally prove (Robin Hill 1899–1991) that the oxygen liberated in photosynthesis originated from water and not from carbon dioxide.

In 1678, Olaus Borrichius (1626–90) liberated pure oxygen from saltpetre, about half a century after the basic recognition of oxygen (Michael Sendivogius 1556–1636) as the "elixir of life" in air. The discovery of oxygen as an element was then made, unbeknownst to one another, in 1772 by Carl Wilhelm Scheele (1742–86), in 1774 by Priestley, and in 1775 by Antoine Laurent Lavoisier (1743–94), who also gave it the international Latin name "oxygenium" (in Greek, "oxygen" means "acid forming"). This name was based on the erroneous thought of oxygen being present in all acids. The thinking in those days was still strongly burdened by German Johann Joachim Becher's (1635–82) "phlogiston theory" of chemistry, which attempted to explain the phenomenon of burning. According to this theory, burning matter liberates colorless, odorless, tasteless, and massless phlogiston (in Greek, "flogistos" means "incombustible"); it was assumed that air had a special ability to absorb a certain amount of this substance. However, experiments revealed contradictions with this theory and, for example, Lavoisier proved that burning actually required oxygen.

Scot Joseph Black (1728–99) first identified carbon dioxide in the 1750s and called it "fixed air". Belgian Jan Baptist van Helmont (1580–1644) realized that it was the same gas liberated, among others, in fermentation and also combustion and that was also produced by plants, calling it "wild spirit" (in Latin "spiritus sylvestre"). In 1773, Lavoisier clarified the nature of this gas and it finally received the name "carbon dioxide".

In photosynthesis the radiation energy of the Sun is thus transformed in biomolecules into chemical energy; the oxygen produced, instead of carbon dioxide, gradually became the other main constituent of the atmosphere along with nitrogen. This substantially reduced the acidity of air and water, which enabled the development of the enormous biodiversity of our planet and now secures the continuity of life here. Photosynthesis uses only about 1% of the average radiation energy (about 4.3×10^7 kJ/m^2 or totally about 2×10^{22} kJ) annually arriving at the outer layers of our atmosphere. In spite of the smallness of this fraction, photosynthesis is estimated to consume annually about 10^{12} tons of carbon dioxide; about one third of this is done by photosynthetic marine microorganisms.

Autotrophs are self-sufficient organisms that can use radiation energy in photosynthesis to produce organic compounds. Their opposites, *heterotrophs*, instead use as carbon sources in their reactions, among others, ready made carbohydrates, such as glucose. *Chemosynthesis* again is a process where certain microorganisms obtain the energy they need for their metabolism by splitting inorganic compounds of sulfur and nitrogen. These microorganisms (archaea or archaebacteria) are found in hot springs and the depths of oceans; their contribution to the total production of biomass is, however, vanishingly small. From the point of view of energy sources, *phototrophs* utilize the energy of sunlight, and *chemotrophs* produce energy by oxidizing the compounds formed by phototrophs.

Even though the spectrum of organisms capable of photosynthesis is wide, extending from the simple procaryotic cells (i.e., cells that do not contain a nucleus, e.g., bacteria) all the way to the largest organisms of the plant kingdom (giant Californian redwood, *Sequoia sempervirens*), typical similarities exist between all processes of photosynthesis. The catalyzing plant pigment of photosynthesis (in Greek, "kata" means "fully" and "lyein" "to liberate"; Jöns Jakob Berzelius introduced the term in 1836), *chlorophyll* (in Greek, "chloros" means "green" and "phyllon" "leaf") (in bacteria bacteriochlorophyll), is found in eukaryotic plant cells (i.e., cells that contain a nucleus), *chloroplasts* (in Greek, "plastes" means "the one who forms"), which belong to plant-specific organelles or plastids. Richard Martin Willstätter (1872–1942, the 1915 Nobel Prize in Chemistry) presented the empirical formula of chlorophyll. He also showed that chlorophyll is a mixture of two compounds, chlorophyll a and chlorophyll b.

The chloroplasts surrounded by a double membrane resemble in many ways mitochondria; they contain, regardless of the species, the typical and regular inner membrane system consisting of so-called "thylakoid membranes". These membranes are organized into paired folds that extend throughout the organelle. The paired folds or lamellae give rise to flattened sacks or discs, thylakoid vesicles (in Greek, "thylakos" means "sack"). The inner space of thylakoid vesicle is called "thylakoid lumen" or "thylakoid space", and the inner space of the chloroplast is called "stroma". Tens of thylakoid vesicles form several "sacks" that are attached together.

The chloroplasts contain, among others, both oxidizing and reducing enzymes. The typical lipid structures of their thylakoid membranes are impermeable to most ions and molecules, while the mitochondrial inner membranes pass several ions and molecules. "Light antennae" are attached on the bag-like thylakoid membranes and consist of about 300 pigment molecules (e.g., chlorophylls a and b and carotenoids). They are able to capture light quanta and to transmit their energy to the chlorophyll molecules (chlorophylls a) that are the centers of action in the photosynthetic reactions. The pigments and electron transport proteins form a coherent photosynthetic unit or pigment unit with a detailed structure that has been essentially elucidated only in the last few years. Besides chlorophylls and carotenoids, crucial parts of the process include various proteins and iron-containing cytochromes that structurally resemble chlorophyll, plastoquinones, and copper-containing plasto-cyanin ($Ca^{2\oplus}$ ions are also present). The electron transport molecules are asymmetrically placed on the membrane, and a part of them also transports hydrogen atoms from the stroma to the thylakoid lumen. A remarkable detail is that the central metal ($Mg^{2\oplus}$) of chlorophyll is attached with coordination bonds to, among others, certain amino acids in the surrounding protein. In prokaryotic cells, the photosynthetic reactions basically follow the same course in the membranes that fill the inner space of the cell.

Chemically, chlorophylls belong to porphin (I) derivatives or *porphyrins*, which characteristically contain a large number of conjugated double bonds that lend color to these compounds. Porphin can be synthesized by condensing pyrrole and formaldehyde leading to a molecule with four nitrogen-heterocyclic structures. The most important porphyrin derivatives include those found in nature: magnesium-containing chlorophyll a (II) (in chlorophyll b, the methyl group indicated by the arrow is replaced by a formyl group -CHO), and the iron-containing hemoglobin (III) present in blood (participates in absorption and transport of oxygen):

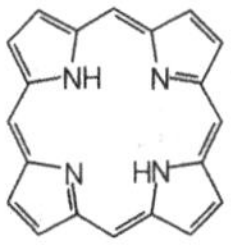

I 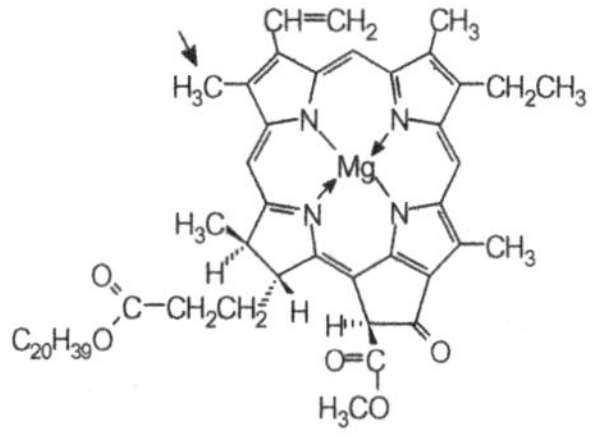 II 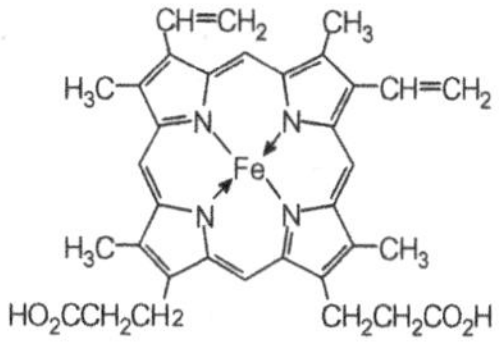 III

Carotenes ($C_{40}H_{56}$), closely related to terpenes, resemble tetraterpenes ($C_{40}H_{64}$) and are found, for example, in carrot (*Daucus carota*), many flowers, animal fats, egg yolk, mushrooms, algae, and, as mentioned, green leaves. Their chemical structure usually consists of eight interconnected isoprene units; the chromophoric unit generally contains eleven conjugated double bonds. *Carotenoids* is a general name for carotenes and "basic carotenes" can also contain functional oxygen-containing groups.

Vitamins A ($C_{20}H_{29}OH$) are produced in the organism from carotenoids, which are, therefore, known by their collective name "provitamins" (cf., Chapt. "10.3. Vitamins"). In this process, water is added to the molecule, and it splits into two parts. Out of about four hundred carotenoids found in nature, only about one in ten are known to metabolize into those vitamins in the liver. The most common and important carotenoid is β-carotene, after which the whole group has been named. Carotenoids are colored compounds, but they are easily oxidized, when exposed to air, into fully colorless products. The structural formulas of β-carotene (I) and vitamin A_1 (II) are as follows:

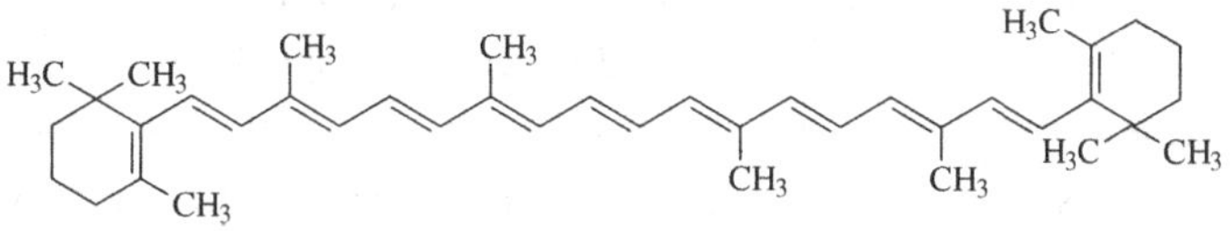

I

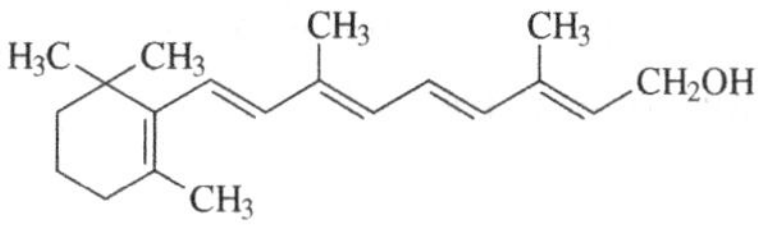

II

As carotenoids are able to absorb visible light at wavelengths different from those absorbed by chlorofylls a and b, photosynthesis utilizes the energy spectrum of sunlight (active range 400–700 nm) quite efficiently. Because of their aromatic nature, chlorophylls are excellent light-collecting compounds. Their structures contain π-electrons delocalized among the carbon atoms that are excited by radiation energy (i.e., raised to orbitals higher than the ground state), which increases the probability of their transfer to a suitable acceptor (i.e., an oxidation/reduction reaction takes place). The end result is a transduction of radiation energy into chemical energy. Again, carotenoids have many π-electrons capable of being excited. In addition, carotenoids as antioxidants are able to efficiently destroy very reactive and thus harmful oxygen compounds formed in photosynthesis.

Anabolism or tissue-building metabolism in cells is, in its physiological meaning, a synthetic process where structurally complex biomolecules of great variety (polysaccharides, proteins, nucleic acids, and lipids) are built from clearly simpler precursors (monosaccharides, amino acids, certain heterocyclic nitrogen compounds, and fatty acids). Its opposite is *catabolism* or metabolism that breaks up materials of the cell; this primarily happens when complex nutritional molecules (polysaccharides, lipids, and proteins) are decomposed into small-molecule products (e.g., lactic acid, ethanol, carbon dioxide, urea, and ammonia) normally under oxidizing conditions. Catabolic pathways are characteristically energy yielding, whereas anabolic pathways require energy. It should be pointed out that the basic product of photosynthesis, glucose, is the precursor in all biosyntheses of the material components of biomass.

Photosynthesis happens in one to three seconds depending on the wavelength (i.e., energy) of the light. Broadly speaking, chlorophylls utilize violet and blue, and then also orange and red light, while carotenoids absorb blue and green. The first reaction of photosynthesis that happens in thylakoid lumena is called a "*light reaction*" (or "*light reactions*"), because chlorophylls bind solar radiation energy. In this reaction, the protons and electrons of water are separated, while oxygen is liberated according to the following general formula:

$$H_2O + ADP + P_i + NADP^{\oplus} \xrightarrow{h\upsilon} O_2 + ATP + NADPH + H^{\oplus}$$

In this reaction, ADP is adenosine 5′-diphosphate (I), ATP adenosine 5′-triphosphate (II), $NADP^{\oplus}$ nicotinamide adenine dinucleotide phosphate (III), and NADPH reduced $NADP^{\oplus}$ (IV):

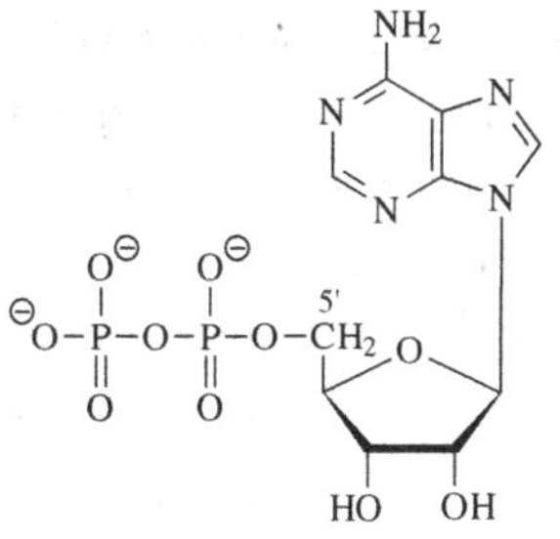

I

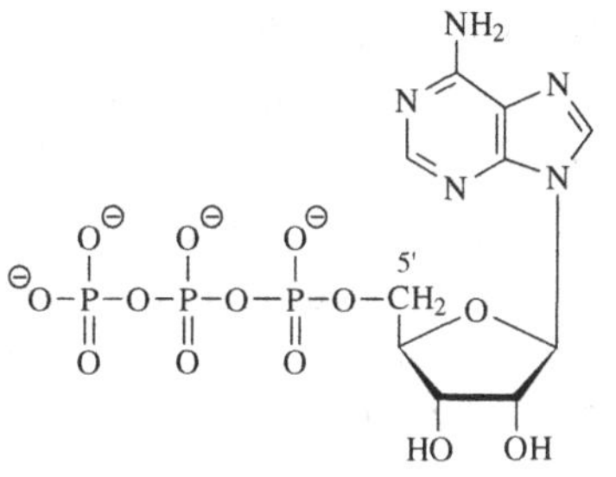

II

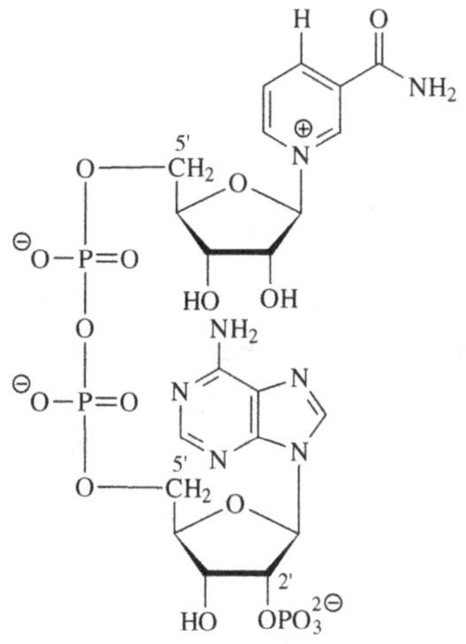

III

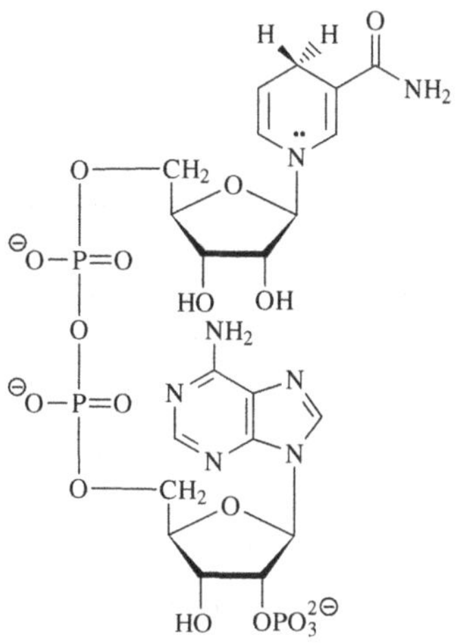

IV

ADP and ATP belong to so-called “high-energy phosphate compounds”, because the changes of the Gibbs free energy (ΔG) in their hydrolysis (liberating an “active phosphate group” P_i), ATP → ADP + P_i and ADP → AMP (monophosphate) + P_i, are strongly negative, in these cases, respectively, –30.5 kJ/mole and –35.7 kJ/mole. This is why the “reverse reaction”, the light reaction, does not happen spontaneously but is endergonic (i.e., a forced process), the reverse of spontaneous, exergonic reactions (which have a negative ΔG), such as typical katabolic reactions. Generally, exothermic reactions (the change in enthalpy ΔH is negative) liberate heat, forming an antithesis for endothermic reactions (with positive ΔH) that absorb heat. These phosphate compounds (ATP and ADP) are able to transfer and store (without spontaneously disintegrating) energy for brief periods, for example, for local enzymatic reactions in the cell. In summary, in light reactions, water decomposes into oxygen that leaves the system and hydrogen that binds to a hydrogen transfer molecule such as coenzyme NADP$^{\oplus}$, which is reduced to NADPH, while the energy is absorbed by an ATP molecule.

The second reaction stage of photosynthesis does not require light; it is called a "*dark reaction*" (or "*dark reactions*"), where carbon dioxide is gradually converted into glucose according to the following general equation:

$$CO_2 + ATP + NADP + H^{\oplus} \longrightarrow C_6H_{12}O_6 + ADP + P_i + NADP^{\oplus}$$

This reaction takes place in the space between the stacks of thylakoid vesicles in the chloroplasts; there, the "reducing power" stored in the ATP molecule (NADPH + H$^{\oplus}$) is used to convert the oxidized CO_2 into a more reduced compound, glucose. The conversion proceeds through the so-called "*Calvin cycle*" or "Calvin-Benson cycle" or "Calvin-Benson-Bassham cycle" (Melvin Calvin 1911–97, the 1961 Nobel Prize in Chemistry), where CO_2 first binds, catalyzed by so-called "RuBisCO enzyme" (ribulose-1,5-bisphosphate carboxylase/oxygenase), into ribulose-1,5-bisphosphate (RuBP). The formed product then disintegrates under the influence of two enzymes (phosphoglycerate kinase and glyceraldehyde-3-phosphate dehydrogenase) to two glyceraldehyde-3-phosphate molecules (G-3-P or GAP). These gradually form (the participating enzymes are triose phosphate isomerase, aldolase, fructose bisphosphatase, and phosphoglucoisomerase) glucose-6-phosphate (G-6-P), which is then converted into glucose under the influence of glucose-6-phosphatase. The main reaction chain from carbon dioxide to glucose thus proceeds as follows: CO_2 + RuBP →→ 2G-3-P →→ G-6-P → glucose. Several other enzymatic reactions occur with the final result that the starting material RuBP and some other monosaccharides and their derivatives (e.g., D-fructose-6-phosphate) are manufactured for the needs of the next cycle.

The slow operation of the RuBisCO enzyme is the most critical phase of the Calvin cycle; it is also the main reason for the small amount of radiative energy that can be utilized in photosynthesis. This forces the plants to synthesize, in spite of the large amount of energy required, major quantities of this very massive (550 kDa), key enzyme. The proteins of the RuBisCO enzyme comprise over one half of the soluble proteins in leaves and over 15% of the proteins in chlorophyll; they are probably the most common proteins in the entire biosphere. On the other hand, the RuBisCO enzyme is rather nonspecific in its function, a characteristic well reflected in its name. Therefore, it connects, instead of the carboxylase function, an oxygenase function into the Calvin cycle in certain cases when oxygen is present in enhanced quantities relative to CO_2. When this happens, part of the energy is spent on assimilation of nitrogen, while a part is wasted in so-called "photorespiration".

The speed of photosynthesis is influenced by many factors whose function has not always been fully elucidated. Among the crucial factors are the amount of light intercepted by the leaf, the structure of the leaf (general velocity of diffusion of CO_2), and conditions at the boundary on the surface of the leaf (the influence of wind on the boundary and thus, on the diffusion velocity of CO_2), the abundance of CO_2, temperature, availability of water, and access to nutrients, such as nitrogen and certain salts. In general, increasing amounts of light and a high abundance of CO_2 will increase the speed of photosynthesis (i.e., more carbon is passed through the Calvin cycle). Water (of which only <5% is used in actual photosynthesis) is a key factor since the processes take place in an aqueous environment and water tends to evaporate into the drier outside air. This explains why the plant actively controls the uptake of CO_2 and the loss of water by opening and closing the stomata on the underside of the leaves. Gases also exchange to some extent directly through the wax layer on the surface on the leaves, but this so-called "nocturnal leaf respiration" is minor, especially in species adapted to drought. A rise in temperature mostly increases the assimilation of carbon in the dark reactions; it also increases evaporation, closes the stomata, and slows down the fluid transport to the leaf. Moreover, it is not possible for the dark reactions of photosynthesis to consume the light reaction products formed during unfavorable dry and cold periods.

Photosynthesis generates in practice more AT.P and NADPH than the cell requires, so that a suitable mechanism to regulate the consumption of its primary carbohydrate product, glucose, is needed. The glucose produced can be stored in the leaves, or it can be transported (generally as sucrose) to other parts of the plant. Sucrose can then be either degraded back into glucose and spent in energy production or the biosynthesis of cellulose and other compounds. It can also be stored further as, for instance, starch. The gradual energy-generating (combining with ATP) *cellular respiration* is the reverse of the photosynthetic process; it proceeds according to the following general reaction:

$$C_6H_{12}O_6 + 6O_2 \longrightarrow 6CO_2 + 6H_2O$$

This reaction occurs in a multiphasic electron transport chain, where glucose is generally oxidized, while oxygen is reduced. The first separate main phase is *glycolysis* in the cytoplasm, where glucose or other carbohydrates that have been converted into glucose are oxidized into pyruvic acid. Thereafter, the *citric acid cycle* or Krebs cycle or TCA-cycle (which German Hans Adolf Krebs 1900–81, the 1953 Nobel Prize in Physiology or Medicine, discovered in the 1930s) forms in the mitochondria, in oxidizing decarboxylation of pyruvic acid, CO_2, and coenzyme acetyl-CoA (Ac-CoA). Then, in the citric acid cycle, Ac-CoA is converted back (liberating more CO_2) into coenzyme A (HS-CoA or CoASH), which will then react in the first phase of the cycle. The energy liberated in the oxidizing reactions retains its "energy of reduction", causing the reduction of the coenzymes $NAD^{\oplus}$ (nicotinamide adenine dinucleotide) and Q (ubiquinone) to the corresponding coenzymes NADH and QH_2 (ubiquinol). As an essential part of the process in the mitochondria, these reduced coenzymes are oxidized back in a complex respiratory electron transport chain with certain enzymes acting as electron transporters. The energy liberated is primarily spent in the transfer of the formed protons ($H^{\oplus}$) from the mitochondrial matrix to the intermembrane space, where a concentration gradient containing the free energy of the protons is set up on the exterior surface of the membrane. Finally, the protons will be channeled back through the interior membrane to the matrix, where this energy is utilized by ATP synthase, an enzyme that catalyzes the reaction ADP + P_i → ATP + H_2O. On its part, oxygen acts as an acceptor of the electrons in this chain of transfers, so that the general reaction becomes $1/2O_2 + 2H^{\oplus} + 2e^{\ominus} \rightarrow H_2O$. These reactions taking place after the citric acid cycle are generally called "*oxidative phosphorylation*". British Peter Dennis Mitchell (1920–92) discovered these reactions in the early1960s; for these observations, he was awarded the 1978 Nobel Prize in Chemistry.

Scientists examining the mechanisms of photosynthesis are inspired by the possibility of developing technological applications that could offer solutions to the nutrition and energy problems facing mankind. The objective of this research into the behavior of these molecules has been to find out how a plant transforms the energy of solar radiation into a stream of electrons; another goal is to see how the rate of production in useful plants could be increased by changing their protein structure. The trend is to broaden this research by including, besides common useful plants, such as spinach — plants living in nature under a maximum variety of conditions. Plant breeding and gene technology are becoming areas requiring stronger research into photosynthesis; the overall goal is to develop plants with better yield. The short-term objective, however, is for improving the efficiency of existing mechanisms of production, rather than finding "artificial" replacements for the process of photosynthesis. As a final comment, the modification and ulilization of photosynthesis is still at its beginning stages; the annual energy consumption of the human society is still clearly less than the amount transformed in photosynthesis.

8.2. Polysaccharides

We now have only partial knowledge of the biosynthesis of the numerous polysaccharides that form the cell walls in plants. The principal reason is the large number of enzymes that participate in the biosynthetic reactions; it has not yet been possible to isolate all those enzymes in pure form. However, the biosyntheses of sucrose, cellulose, and starch are already known in substantial detail. The following discussion will cover the main features of the biosynthetic formation of the most common polysaccharides.

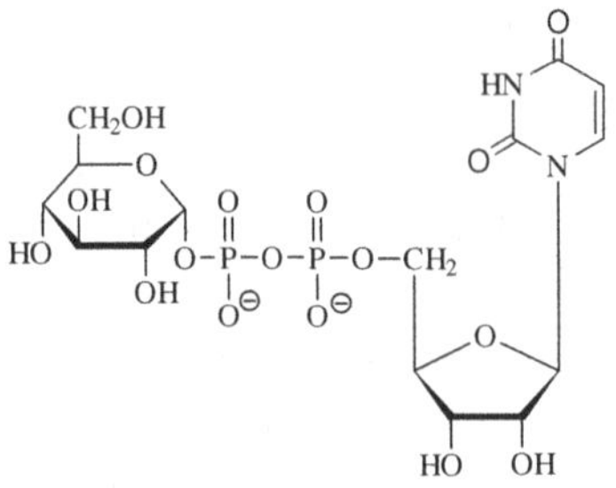

Fig. 8.1. Examples of nucleoside diphosphate sugars. Uridine diphosphate α-D-glucopyranose (UDP-D-glucose) (1), adenosine diphosphate α-D-glucopyranose (ADP-D-glucose) (2), and guanosine diphosphate α-D-mannopyranose (GDP-D-mannose) (3).

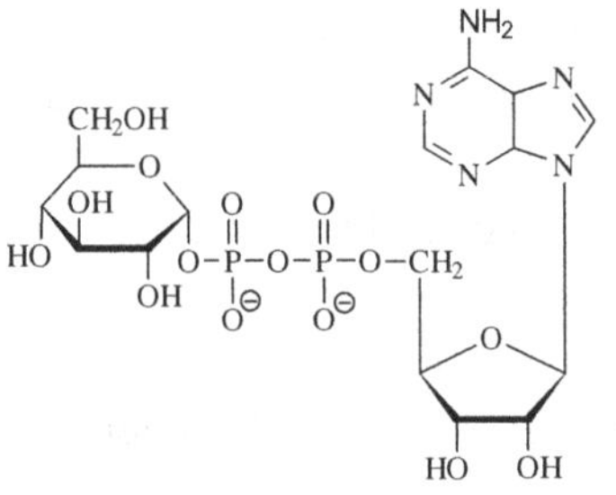

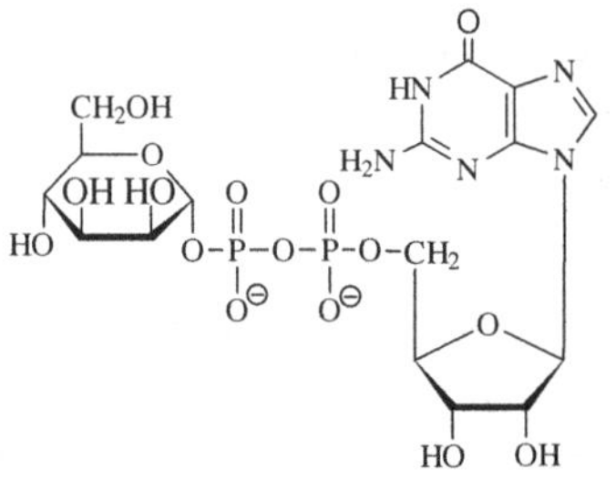

Uridine diphosphate α-D-glucopyranose (UDP-D-glucose, Fig. 8.1) is an important intermediate product in the biosynthesis of cellulose (cf., Chapt. "9.4.1. Cellulose and its derivatives"); it is formed from α-D-glucopyranose-1-phosphate and uridine triphosphate with the help of an enzyme, UDP-glucose pyrophosphorylase (Fig. 8.2). Argentinian Luis Federico Leloir (1906–87), who received the 1970 Nobel Prize in Chemistry for clarifying the metabolic pathway of lactose, pioneered the work on this and many similar compounds containing phosphorus. Cellulose ((1→4)-β-D-glucan) is synthesized from UDP-D-glucose, catalyzed by the cellulose synthase enzyme, in the cell membrane as follows:

$$\text{UDP-D-glucose} + ((1 \rightarrow 4)\text{-}\beta\text{-D-glucopyranosyl})_n \rightarrow$$
$$((1 \rightarrow 4)\text{-}\beta\text{-D-glucopyranosyl})_{n+1} + \text{UDP}$$

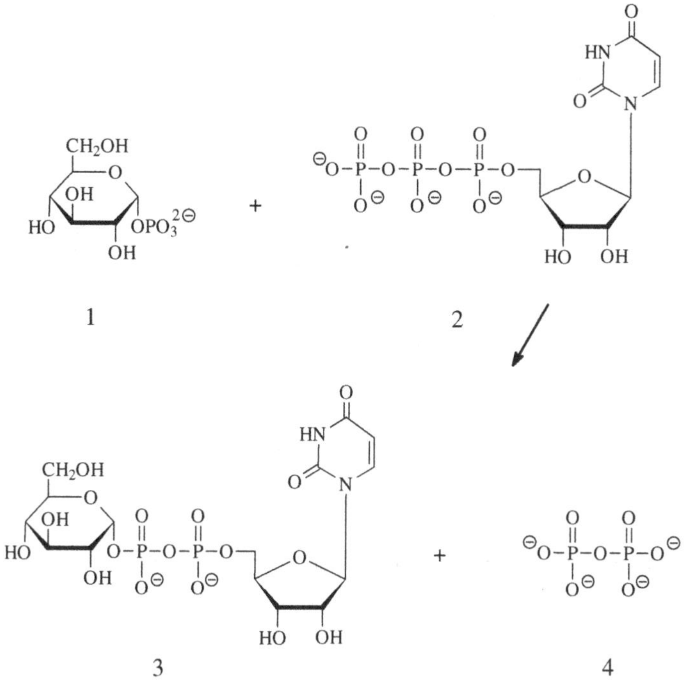

Fig. 8.2. Reaction between α-D-glucopyranose-1-phosphate (1) and uridine triphosphate (2) results in the formation of uridine diphosphate α-D-glucopyranose (UDP-D-glucose) (3) with the simultaneous release of pyrophosphate (4).

Cellulose synthase complexes are proteins that synthesize cellulose chains (growing microfibrils of cellulose) in a controlled way on the exterior surface of the cell wall (Fig. 8.3). Cytosol is the fluid and gel-like part of the cytoplasm; it fills slightly over half of the volume of the cell, and many important chemical reactions (such as glycolysis and protein synthesis) take place in it. The UDP-D-glucose that participates in the building of individual cellulose chains (where D-glucose connects via (1→4)-glycosidic bonds to the non-reducing end of the chain) can form sucrose or directly cytoplasm UDP-D-glucose: β-D-glucose → β-D-glucose-6-phosphate → α-D-glucose-6-phosphate → α-D-glucose-1-phosphate → UDP-α-D-glucose. If UDP-D-glucose connects with D-fructose, sucrose is formed with the liberation of UDP as follows: UDP-α-D-glucose + D-fructose-6-phosphate → sucrose-6′-phosphate → sucrose.

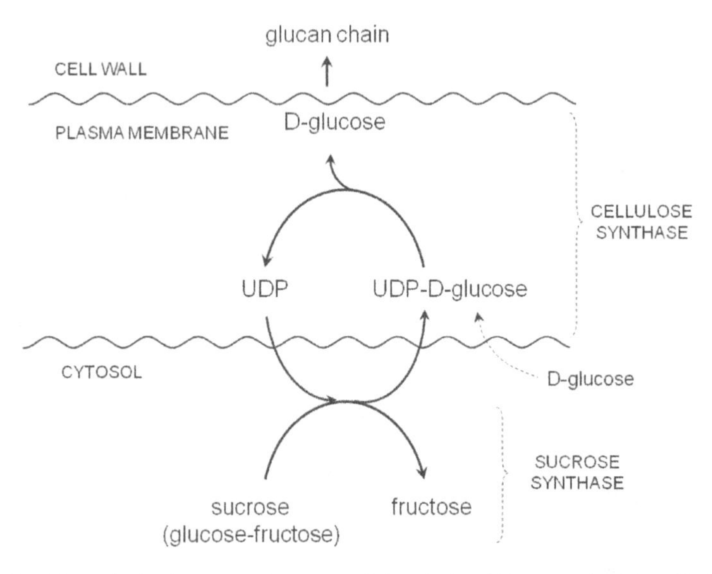

Fig. 8.3. Hypothetical conventional model of the biosynthesis of cellulose in plant cells.

The biosynthesis of starch ((1→4)-α-D-glucan, cf., Chapt. "9.4.5. Other polysaccharides"), and also its storage, takes place in plastids (chloroplasts); its central compounds are adenosine diphosphate α-D-glucose (ADP-D-glucose, Fig. 8.1) and, in the final phase of the formation of this intermediate product, the enzyme ADP-glucose pyrophosphorylase: D-fructose-6-phosphate → β-D-glucose-6-phosphate → α-D-glucose-6-phosphate → α-D-glucose-1-phosphate → ADP-α-D-glucose → (1→4)-α-D-glucan. ADP-D-glucose probably forms in the cytosol before its transport to the plastids, where either linear amylose (with the enzyme starch synthase), or branched amylopectin (with enzyme amylo-(1→4), (1→6)-transglycosylase), is synthesized. Among other glucans, for example, callose ((1→3)-β-D-glucan) is also found in small quantities in plants. It is formed from UDP-D-glucose with the enzyme (1→3)-β-D-glucose synthase, which is widely found in various plant cells.

Although UDP-D-glucose and, on the other hand, ADP-D-glucose, are the two most significant nucleoside diphosphate sugars in abundance, other nucleosides, such as uridine, a nitrogen-heterocyclic base formed from β-D-ribose with *N*-glycosidic bonds (cf., Chapt. "10.2. Nucleic

Acids") and also adenosine, may play significant parts in, among others, the biosynthesis of polysaccharides. These nucleosides include guanosine, cytidine, thymidine, and inosine; see guanosine diphosphate α-D-mannopyranose, GDP-D-mannose (Fig. 8.1). The corresponding NDP (nucleotide diphosphate) compounds and, for example, glycoproteins are formed in animals in the so-called "Golgi apparatus" or "Golgi complex"; in plant cells the corresponding cellular structure is called "dictyosome". The Golgi apparatus is an organelle found in most eukaryotic cells; Italian physician Camillo Golgi (1843–1926, the 1906 Nobel Prize in Physiology or Medicine) identified it in 1897 and named it after himself in 1898.

As structural components of hemicelluloses (cf., Chapt. "9.4.5. Other polysaccharides"), various monosaccharides and their derivatives that may be attached via glycosidic bonds to each other, are found. Figure 8.4 shows a simple scheme of the biosynthetic path from sucrose to (1→4)-β-D-xylan that is common in both hardwoods and non-woods. This also indicates the possibilities of forming various other basic

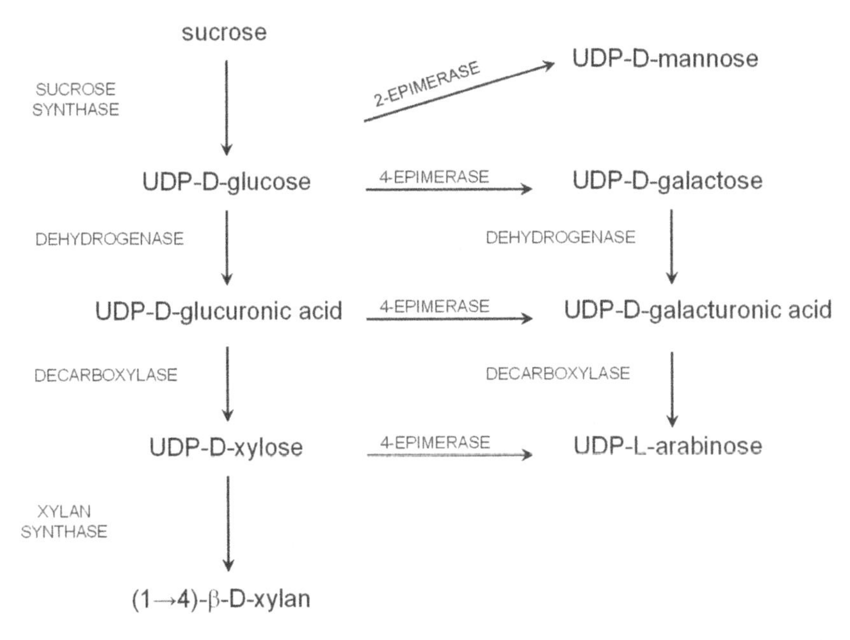

Fig. 8.4. Simplified representation of the biosynthetic formation of (1→4)-β-D-xylan and some hemicellulose precursors from sucrose.

monosaccharides and their derivatives. It should be emphasized that on the main chain of the hemicelluloses (e.g., xylan and glucomannan) various side groups are often attached, causing complications in their biosynthesis. Hemicelluloses thus form a heterogenic group of compounds; they are located in the space between cellulose microfibrils (and cellulose lamels) as unorganized polymers (polysaccharides) that form additional chemical bonds (phenyl glycolsides, benzyl ethers, and benzyl esters) with the other unorganized macrocomponent of wood, lignin (in Latin, "lignum" means "wood").

9. Natural Carbohydrates and Their Derivatives

This chapter primarily describes the most common natural carbohydrates and their derivatives. Our inquiry is largely based on the previous chapters, especially regarding the naming and classification of the basic compounds. In this context, some new examples will be presented of the naming process of compounds belonging to carbohydrate derivatives. It should be emphasized that among the carbohydrate derivatives, there are substantial numbers of compounds (cf., Chapt. "12. Utilization of Biomass") that are not necessarily found in nature.

We follow as much as possible, for example, in the key figures describing properties of compounds, the fundamental units of the SI (Système International d'Unités) system. An exception is pressure, whose SI unit is pascal (Pa) = 1 N/m^2. Here, the unit mmHg (760 mmHg = 1 atm = 101.3 kPa) is regularly used, primarily in order to follow the standard handbook information; it was not considered appropriate to convert these primary measurement results into another system of units. Regarding optically active compounds, the dimension of their specific rotation is (deg)cm^2/g (cf., Chapt. "3.4. Stereoisomerism"), following general usage that is left unexpressed in this review. Similarly, the values of specific rotation [α] have normally been measured in dilute aqueous solutions (1–5 g sugar/100 mL water) at 20–25°C, without a more accurate expression of the conditions. Moreover, it should be noted that the key figures reported in various literature sources, such as melting and boiling points of a compound (or its (+)- ja (–)-enantiomers), are not always congruent; certain choices between different literature sources are necessary.

9.1. Monosaccharides

Monosaccharides, either as such or as building blocks for oligo- and polysaccharides, are important because of their biological activity and their industrial applications. In many cases, these natural sugars are chemically various types of derivatives of monosaccharides.

Natural monosaccharides contain almost without exception three to nine carbon atoms, although most common are compounds with five and six carbon atoms. They are primarily aldoses (a carbonyl group at carbon atom C_1 — cf., Chapt. "7.1. Aldoses"), but 2-ketoses (a carbonyl group at carbon atom C_2 — cf., Chapt. "7.2. Ketoses") are also found. The monosaccharides are usually connected together with glycosidic bonds forming oligo- and polysaccharides (cf., Chapts. "9.3. Oligosaccharides and Their Derivatives", and "9.4. Polysaccharides and Their Derivatives"). In these cases, the monosaccharide components can be completely or in part liberated with acid or enzymatic hydrolysis. This is often the first stage in industrial processes of utilizing, especially through fermentation, biomass materials that contain carbohydrates (cf., Chapt. "12.2. Production of Chemicals"). According to their definition, sugars are soluble in water, and thus, their solubility in organic solvents is usually low. They change chemically, for example, at higher temperatures and under the influence of acids and bases (cf., Chapts. "11. Characteristic Reactions of Carbohydrates" and "12.2. Production of Chemicals"). The purification of monosaccharides and the known oligosaccharides is almost without exception based on crystallization, which does not require a substantial rise in temperature; it may be preceded by ion-exchange treatment or fractionation for the removal of impurities.

9.1.1. *Aldose monosaccharides*

Glyceraldehyde, glycerose, 2,3-dihydroxypropanal

D-Glyceraldehyde, D-(+)-glyceraldehyde, *R*-glyceraldehyde, 2*R*,3-dihydroxypropanal

CHO
H–C–OH
CH₂OH

Molar mass 90.08 ($C_3H_6O_3$).

D-Glyceraldehyde was synthesized for the first time in 1917 (Alfred Wohl 1863–1939 and Franz Momber) by deamination and hydrolysis of 3-amino-3-deoxy-D-glyceraldehyde dimethyl acetal. The preparation was also of historical significance because this compound was converted into D-(−)-tartaric acid (D-threaric acid), which represented a final experimental link between the classification and structures of D,L-carbohydrates (classified in 1906 by Martin André Rosanoff — cf., Chapt. "5.1. Fundamental Definitions"). (+)-Glyceraldehyde had indeed been considered the simplest sugar and, due to its reactivity, also a parent substance for making numerous stereoisomeric molecules. In such cases, the synthesized products contained the same configuration of the asymmetric carbon atom as no chemical bonds were broken between the asymmetric carbon atom and its substituents (cf., Chapt. "5.2. Relative and Absolute Configuraton"). The present production of D-glyceraldehyde originated in 1956 and is based on the oxidation of D-fructose with a limited proportion of lead tetraacetate, followed by the hydrolysis of the resulting derivative. Its specific rotation $[\alpha]_D$ in water (2 g/100 mL H_2O) is +8.7. Glyceraldehyde is readily polymerized in water; hence, the D,L form is considered by some to be more stable than its pure stereoisomers. D-Glyceraldehyde is found as a 3-phosphate in cell metabolism and similar biological processes.

D,L-Glyceraldehyde. Molar mass 90.08 ($C_3H_6O_3$), melting point 145°C, boiling point 140–50°C (0.8 mmHg), density D_{18} 1.455 kg/dm³ (18°C).

D-Erythrose, D-(−)-erythrose, D-*erythro*-tetrose, 2*R*,3*R*-2,3,4-trihydroxybutanal

CHO
—OH
—OH
CH_2OH

Molar mass 120.10 ($C_4H_8O_4$).

D-Erythrose (in Greek, "erythros" means "red" — originally the red material isolated from a certain alga) was first synthesized from D-arabinose in 1893 by the Wohl degradation (Alfred Wohl), but nowadays, many other methods (e.g., the Ruff degradation — Otto Ruff 1871–1939) (cf., Chapt. "11.1. Monosaccharides") are employed. Furthermore, it is obtained, for example, by oxidation of D-glucose with a limited proportion of lead tetraacetate. Its specific rotation $[\alpha]_D$ in water (1 g/100 mL H_2O)

at 22°C is –43.5. D-Erythrose is found in the muscles in extremely small concentrations as a 4-phosphate.

D-Apiose, 3-*C*-(hydroxymethyl)-*glycero*-tetrose, 2*R*-2,3,4-trihydroxy-3-(hydroxymethyl)butanal

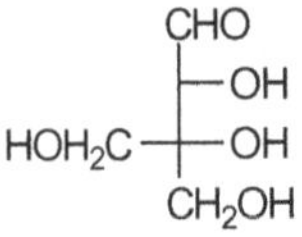

CHO
OH
HOH_2C — OH
CH_2OH

Molar mass 150.13 ($C_5H_{10}O_5$).

Arabinose, *arabino*-pentose, pectinose, pectin sugar

D-Arabinose, D-(–)-arabinose, D-*arabino*-pentose

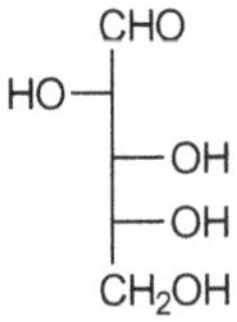

CHO
HO
OH
OH
CH_2OH

Molar mass 150.13 ($C_5H_{10}O_5$), melting point 159–60°C.

D-Arabinose (in Latin, "arabia" means "appearing in gum arabic") occurs in the furanoid form in nature as a constituent of the polysaccharide fraction of tubercle bacilli and in certain glycosides (aloins), such as barbaloin, isobarbaloin, nataloin, and homonataloin in plants of the genus *Aloe*. Because the asymmetric carbon atoms C_2, C_3, and C_4 of D-arabinose have the same configuration as the lower three asymmetric carbon atoms C_3, C_4, and C_5 of D-glucose, all methods (e.g., the Wohl degradation) for removing C_1 from D-glucose lead to D-arabinose. The water solution of this monosaccharide at 31°C contains 61% α-pyranoid, 35% β-pyranoid, 2% α-furanoid, 2% β-furanoid, and a small amount (<0.1%) of open-chain aldehyde forms. Its specific rotation $[\alpha]_D$ in water (1 g/100 mL H_2O) at 20°C is –104.5. D-Arabinose is a suitable starting compound for synthesizing products used in cancer therapy.

L-Arabinose, L-(+)-arabinose, L-*arabino*-pentose, L-pectinose

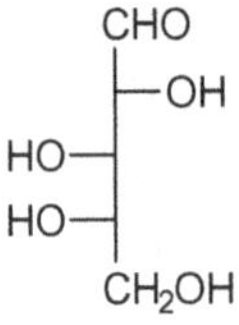

CHO
OH
HO
HO
CH_2OH

Molar mass 150.13 ($C_5H_{10}O_5$), melting point 160°C.

L-Arabinose occurs free in nature in the heartwood of many coniferous trees and is more widely distributed in the furanoid form as a structural unit in gums, softwood hemicelluloses (arabinoglucuronoxylan and arabinogalactan), pectin materials, and bacterial polysaccharides. It can be isolated from the mesquite gum of a plant (*Prosopis juliflora*) common in the southwestern United States. This gum consists of L-arabinose, D-galactose, and 4-*O*-methyl-D-glucuronic acid moieties. The isolation procedure comprises controlled hydrolysis without liberating the other constituents to any great extent, and the product is then partially purified by dialysis or various ion-exchange methods and finally crystallized from ethanol. Another possible source is cherry gum which also has, in addition, some D-xylose and D-mannose residues. Its specific rotation $[\alpha]_D$ in water (4 g/100 mL H_2O) at 20°C is +104.5.

D,L-Arabinose, D,L-*arabino*-pentose. Molar mass 150.13 ($C_5H_{10}O_5$), melting point 164°C.

Lyxose, *lyxo*-pentose

D-Lyxose, D-(–)-lyxose, D-*lyxo*-pentose

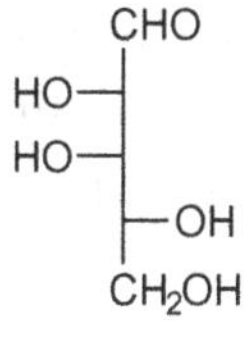

CHO
HO–
HO–
–OH
CH_2OH

Molar mass 150.13 ($C_5H_{10}O_5$), melting point 106–7°C.

D-Lyxose is mainly a synthetic monosaccharide that can be obtained from calcium D-galactonate by the Ruff degradation. It forms hygroscopic crystals with a sweet taste. The water solution of D-lyxose at 31°C contains 70% α-pyranoid, 28% β-pyranoid, 1% α-furanoid, <1% β-furanoid, and a small amount (<0.1%) of open-chain aldehyde forms. Its specific rotation $[\alpha]_D$ in water at 20°C is –14.0. D-Lyxose is also a suitable starting compound for synthesizing products used in cancer therapy.

L-Lyxose, L-(+)-lyxose, L-*lyxo*-pentose

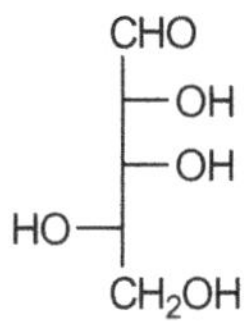

CHO
–OH
–OH
HO–
CH_2OH

Molar mass 150.13 ($C_5H_{10}O_5$), melting point 105°C.

L-Lyxose was first synthesized from calcium L-galactonate by the Ruff degradation in 1914 (Willem Alberda van Ekenstein 1858–1937 and Jan Johannes Blanksma 1875–1950). Another route is the treatment of L-*galacto*-heptulose with periodic acid to cause a selective C_2-C_3 cleavage. It is rare in nature, but has been identified among the hydrolysis products of the antibiotic curamycin. L-Lyxose is hygroscopic and its specific rotation $[\alpha]_D$ in water at 20°C is +13.5.

D,L-Lyxose, D,L-*lyxo*-pentose. Molar mass 150.13 ($C_5H_{10}O_5$), melting point 95°C.

Ribose, *ribo*-pentose

D-Ribose, D-(–)-ribose, D-*ribo*-pentose

CHO
├OH
├OH   Molar mass 150.13 ($C_5H_{10}O_5$), melting point 87°C.
├OH
CH_2OH

D-Ribose in the furanoid form is a carbohydrate component of nucleic acids which are found in all plant and animal cells. It is also a constituent of several coenzymes. D-Ribose can be prepared by stepwise acid hydrolysis or enzymatic hydrolysis of ribonucleic acid (RNA) or yeast nucleic acid. It is synthetically prepared, for example, from D-arabinose by alkaline isomerization or through pyridine-catalyzed epimerization of D-arabinonic acid (the product is D-ribonic acid) followed by a reduction. D-Ribose forms hygroscopic and flaky crystals and its water solution at 31°C contains 21.5% α-pyranoid, 58.5% β-pyranoid, 6.5% α-furanoid, 13.5% β-furanoid, and a small amount (<0.1%) of open-chain aldehyde forms. Certain solutions of this monosaccharide may contain a considerable proportion of the furanoid form. Its specific rotation $[\alpha]_D$ in water at 20–25°C is –23.7. D-Ribose is not fermentable by ordinary yeast.

Xylose, *xylo*-pentose, wood sugar

D-Xylose, D-(+)-xylose, D-*xylo*-pentose

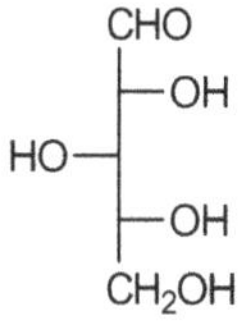

CHO
HO— —OH —OH
CH$_2$OH

Molar mass 150.13 ($C_5H_{10}O_5$), melting point 144–5°C density 1.525 kg/dm^3 (20°C).

D-Xylose (in Greek, "xylon" means "wood" — the name is based on its occurrence in nature) is found in the pyranoid form as a significant structural unit of many agricultural wastes (e.g., corncobs, cottonseed hulls, and various straws) and wood hemicelluloses, xylans. It can be isolated from these sources by acid hydrolysis followed by crystallization after chromatographic purification of the hydrolysate. D-Xylose crystallizes readily forming monoclinal needles or prisms whose water solution at 31°C contains 36% α-pyranoid, 63% β-pyranoid, <1% α- and β-furanoid, and a small amount (<0.02%) of open-chain aldehyde forms. Its specific rotation $[\alpha]_D$ in water at 20°C is +18.8. D-Xylose is not fermentable by ordinary yeast, but many bacteria and certain yeasts are able to ferment it, for example, with the formation of ethanol, lactic acid, and acetic acid. An industrial-scale catalytic reduction of D-xylose results in the formation of xylitol (cf., Chapt. "9.2.3. Diols and polyalcohols") well suited for a sweetening agent expecially in diabetic food. In addition, it is used to a small extent for tanning and dyeing, as well as a diagnostic in the malabsorption of nutrients in the human being.

L-Xylose, L-(–)-xylose, L-*xylo*-pentose. Molar mass 150.13 ($C_5H_{10}O_5$), melting point 144°C.

D,L-Xylose, D,L-*xylo*-pentose. Molar mass 150.13 ($C_5H_{10}O_5$), melting point 129–31°C.

Hamamelose, 2-*C*-(hydroxymethyl)-D-ribose, 2-*C*-(hydroxymethyl)-D-*ribo*-pentose

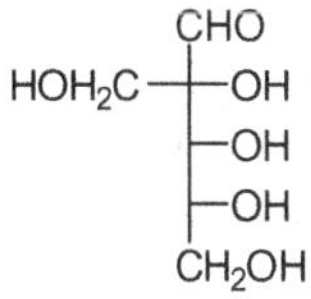

CHO
HOH$_2$C— —OH —OH —OH
CH$_2$OH

Molar mass 180.16 ($C_6H_{12}O_6$).

Allose, *allo*-hexose

D-Allose, D-(+)-allose, D-*allo*-hexose

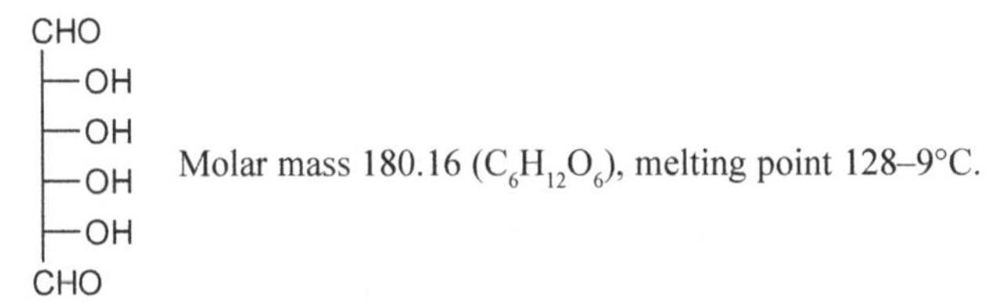

Molar mass 180.16 ($C_6H_{12}O_6$), melting point 128–9°C.

D-Allose (in Greek, "allo" means "other" — i.e., originally denotes the more stable form of D and L forms) is a rare monosaccharide occurring in extracts of the fresh-water alga *Ochromas malhamensis*. It is also found as a 6-*O*-cinnamyl glycoside in the leaves of the African shrub *Protea rubropilosa*. D-Allose can be synthesized from D-ribose by the Fischer-Kiliani synthesis. The water solution of D-allose at 31°C contains 15% α-pyranoid, 76% β-pyranoid, 4% α-furanoid, 5% β-furanoid, and a small amount (<0.01%) of open-chain aldehyde forms. Its specific rotation $[\alpha]_D$ in water at 20°C is +14.4. D-Allose is suitable for the care of chronic leukaemia.

D,L-Allose, D,L-*allo*-hexose. Molar mass 180.16 ($C_6H_{12}O_6$), melting point 181–3°C.

Galactose, *galacto*-hexose, cerebrose, brain sugar

D-Galactose, D-(+)-galactose, D-*galacto*-hexose

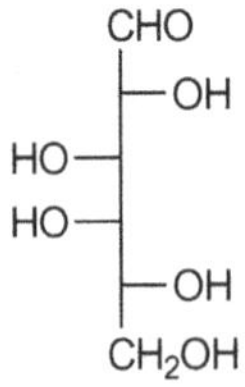

Molar mass 180.16 ($C_6H_{12}O_6$), melting point 167 (anhydrous) or 118–20 (monohydrate) °C.

D-Galactose (in Greek, "galactos" means "milk") is a common constituent of many natural oligo- and polysaccharides. Especially softwood hemicelluloses, galactoglucomannans, and galactans in the coniferous reaction wood (compression wood) contain D-galactose in the pyranoid form (cf., Chapt. "9.4.5. Other polysaccharides"). It is mainly prepared from lactose (milk sugar), various plant gums or coldwater red seaweed (purple laver) *Porphyra umbilicalis* by hydrolysis. Many of the cerebrosides and gangliosides, occurring in brain and spinal cord, are D-galactosides. It crystallizes as prisms whose water solution at 31°C contains 29% α-pyranoid, 64% β-pyranoid, 3% α-furanoid, 4% β-furanoid, and a small amount (<0.02%) of open-chain aldehyde forms. Its specific rotation $[\alpha]_D$ in water is +80.2.

L-Galactose, L-(–)-galactose, L-*galacto*-hexose. Molar mass 180.16 ($C_6H_{12}O_6$), melting point 163–5°C.

L-Galactose occurs to some extent as a constituent of certain natural oligo- and polysaccharides, although synthetic methods are most convenient for its preparation. In one reaction, D-galacturonic acid (I) is first reduced to L-galactonic acid (II) from which D-galactose (III) can be obtained by reduction:

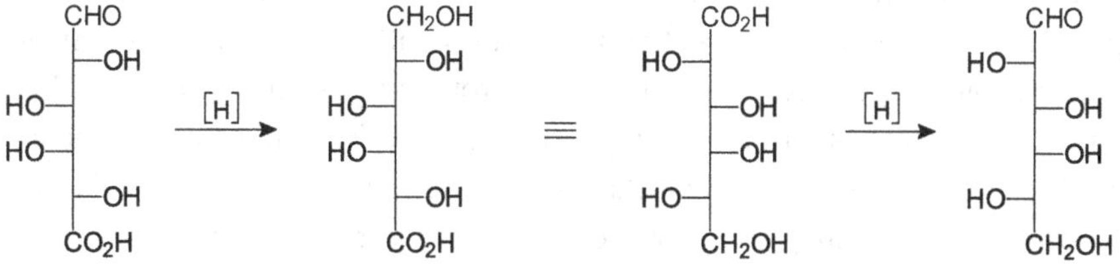

I —[H]→ II ≡ —[H]→ III

D,L-Galactose, D,L-*galacto*-hexose. Molar mass 180.16 ($C_6H_{12}O_6$), melting point 143–4°C.

Glucose, *gluco*-hexose, dextrose, grape sugar, corn sugar, blood sugar

D-Glucose, D-(+)-glucose, D-*gluco*-hexose

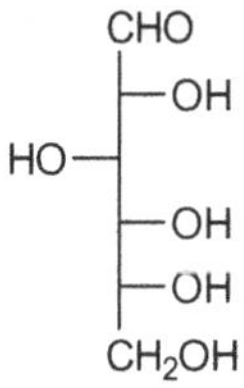

Molar mass 180.16 ($C_6H_{12}O_6$), melting point 146 (anhydrous) or 83–6 (monohydrate) °C, density 1.544 kg/dm^3.

D-Glucose (in Greek, "glykos" or "glykeros" means "sweet") is conventionally called "dextrose", since it rotates the plane of plane-polarized light to the right. It is, in the free or combined form, the most abundant organic compound. It occurs free in nature in grapes and other sweet fruits and berries and together with D-fructose (fruit sugar) in honey and syrup. In addition,

D-glucose is a major component of many polysaccharides (particularly cellulose, glycogen, and starch) and also in significant amounts in certain hemicelluloses (cf., Chapt. "9.4.5. Other polysaccharides"). It is manufactured in the pyranoid form on a large scale from potato starch (Europe) or corn starch (the United States) by acid hydrolysis, followed by several purification and crystallization steps. Below 50°C α-D-glucopyranose monohydrate forms a stable crystalline phase, and slightly above 50°C, the anhydrous form is obtained. At higher temperatures, β-D-glucopyranose forms the solid phase. "Hydrol" is the mother liquor remaining from the manufacture of D-glucose, and it corresponds to the "molasses" from the refining of cane sugar (cf., Chapt. "9.3.1. Disaccharides"). Hydrol carbohydrates contain typically about 65% of D-glucose and 35% of disaccharides and other oligosaccharides. The water solution of D-glucose contains 38% α-pyranoid, 62% β-pyranoid, <0.1% α- and β-furanoid, and small amounts of open-chain aldehyde and septanose forms. Its specific rotation $[\alpha]_D$ in water at 20°C is +52.7. D-Glucose is fermented by yeast. The use of this monosaccharide as a sweetening agent is less than that of sucrose, but it has certain value in industry and medicine, being the most important intravenous nutrient.

Insulin (in Latin, "insula" means "island") is a peptide hormone produced in the pancreas. It is provided within the body in a constant proportion to excess D-glucose in the blood; otherwise, it would be toxic. In type 1 diabetes, insulin is no longer produced internally, while in type 2 diabetes, patients are often insulin resistant and suffer from an insulin deficiency. In diabetes mellitus, the D-glucose content of blood rises as a consequence of a decrease in insulin secretion, and it (D-glucose), for example, appears in the urine in which it is not usually found. For this reason, patients need (always with type 1 diabetes) external insulin, which is most commonly injected subcutaneously.

Mannose, *manno*-hexose, seminose, carubinose

D-Mannose, D-(+)-mannose, D-*manno*-hexose

CHO
HO—
HO—
—OH
—OH
CH_2OH

Molar mass 180.16 ($C_6H_{12}O_6$), melting point 132 (decomposes) °C, density 1.539 kg/dm^3 (20°C).

D-Mannose ("mannose" and "mannitol" originated from "manna", which already the Bible records as food) occurs free only to some extent in nature, for example, in apples, peaches, and blood serum. In contrast, it is found in the pyranoid form as a structural unit especially in softwood hemicelluloses (galactoglucomannans) (cf., Chapt. "9.4.5. Other polysaccharides"). For preparative purposes, the important sources include the seeds of the tagua palm ("vegetable ivory" or "ivory nut", *Phytelephans macrocarpa*) and the Canary Island date palm (*Phoenix canariences*). Besides from the hydrolysate of these feedstocks, it can also be chromatographically separated (followed by crystallization) from the hydrolysates of konjac flour, which is commonly available from the corms of *Amorphophallus konjac*. In addition, D-mannose can be obtained from D-mannitol by oxidation. It crystallizes as needles whose water solution contains 65% α-pyranoid, 34% β-pyranoid, <1% α-furanoid, <1% β-furanoid, and a small amount of open-chain aldehyde forms. Its specific rotation $[\alpha]_D$ in water at 20°C is +14.5. D-Mannose is fermented by yeast. It is used in biochemical research.

D,L-Mannose, D,L-*manno*-hexose. Molar mass 180.16 ($C_6H_{12}O_6$), melting point 132–3°C.

Talose, *talo*-hexose

D-Talose, D-(+)-talose, D-*talo*-hexose

CHO
HO—
HO—
HO—
—OH
CH_2OH

Molar mass 180.16 ($C_6H_{12}O_6$), melting point 128–32 (anhydrous) or 89–90 (monohydrate) °C.

D-Talose occurs in free only to some extent in nature; it is obtained from hydrolysis of the antibiotic hygromycin and cocacitrin. It can be synthesized by reduction of the D-talono-1,4-lactone obtained by epimerization of D-galactonic acid. The water solution of D-talose at 22°C contains 42% α-pyranoid, 29% β-pyranoid, 16% α-furanoid, 13% β-furanoid, and a small amount (0.03%) of open-chain aldehyde forms. Its specific rotation $[\alpha]_D$ in water at 27°C is +19.7. D-Talose is fermented by yeast.

D-*glycero*-D-*galacto*-Heptose

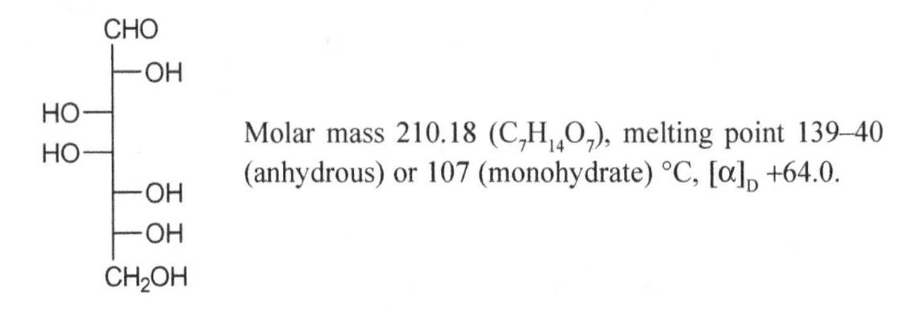

Molar mass 210.18 ($C_7H_{14}O_7$), melting point 139–40 (anhydrous) or 107 (monohydrate) °C, $[\alpha]_D$ +64.0.

D-*glycero*-D-*gluco*-Heptose

Molar mass 210.18 ($C_7H_{14}O_7$), melting point 156–7°C, $[\alpha]_D$ +46.2.

D-*glycero*-L-*gluco*-Heptose

Molar mass 210.18 ($C_7H_{14}O_7$).

D-*glycero*-D-*manno*-Heptose

Molar mass 210.18 ($C_7H_{14}O_7$).

L-*glycero*-D-*manno*-Heptose

```
       CHO
  HO—|
  HO—|
     |—OH
     |—OH
  HO—|
       CH2OH
```

Molar mass 210.18 ($C_7H_{14}O_7$).

D-*glycero*-L-*manno*-Heptose

```
       CHO
     |—OH
     |—OH
  HO—|
  HO—|
     |—OH
       CH2OH
```

Molar mass 210.18 ($C_7H_{14}O_7$).

L-*glycero*-D-*talo*-Heptose

```
       CHO
  HO—|
  HO—|
  HO—|
     |—OH
     |—OH
       CH2OH
```

Molar mass 210.18 ($C_7H_{14}O_7$).

The symbols for other aldohexoses are Alt (altrose), Gul (gulose), and Ido (idose).

9.1.2. *Ketose monosaccharides*

The compound **1,3-dihydroxy-2-propanone** or 1,3-dihydroxyacetone

```
CH2OH
 |
 C=O
 |
CH2OH
```

can be considered to represent the simplest ketose. However, it does not have any asymmetric carbon atoms typical for carbohydrates, and for this reason, it is often classified only as a ketoalcohol.

D-*erythro*-Pentulose, D-ribulose, D-riboketose, D-arabinulose, D-arabino-ketose, D-adonose

CH_2OH / =O / –OH / –OH / CH_2OH Molar mass 150.13 ($C_5H_{10}O_5$).

Phosphorylated D-*erythro*-pentulose is an intermediate in the metabolism of D-glucose and is an early product of plant photosynthesis. It can be synthesized from D-arabinose by isomerization.

D-*threo*-Pentulose, D-xylulose, D-xyloketose, D-lyxulose, D-lyxoketose, Xul (or Xylulo)

CH_2OH / =O / HO– / –OH / CH_2OH Molar mass 150.13 ($C_5H_{10}O_5$).

D-*threo*-Pentulose 5-phosphate occurs as an intermediate in the metabolism of D-xylose and is formed from D-*erythro*-pentulose 5-phosphate in the presence of epimerase. It is synthesized by the isomerization of D-xylose in hot pyridine or by the oxidation of D-arabinitol with *Acetobacter xylinum*.

L-*threo*-Pentulose, L-xylulose, L-xyloketose, L-lyxulose, L-lyxoketose

CH_2OH / =O / –OH / HO– / CH_2OH Molar mass 150.13 ($C_5H_{10}O_5$).

L-*threo*-Pentulose is found in high concentrations in the urine of patients with pentosuria, which has been characterized as an inborn error of carbohydrate metabolism. For this reason, it may lead to a false diagnosis of diabetes. L-*threo*-Pentulose can be synthesized by boiling L-xylose with pyridine.

L-*erythro*-Pentulose, L-ribulose, L-riboketose, L-arabinulose, L-arabino-ketose, Rul (or Ribulo)

CH₂OH / =O / HO— / HO— / CH₂OH

Molar mass 150.13 ($C_5H_{10}O_5$).

Fructose, *arabino*-hexulose, levulose, fruit sugar

D-Fructose, D-(−)-fructose, D-*arabino*-hexulose

CH₂OH / =OH / HO— / —OH / —OH / CH₂OH

Molar mass 180.16 ($C_6H_{12}O_6$), melting point 102–4°C.

D-Fructose is conventionally called "levulose" (first named by Edward Frankland Armstrong 1878–1945), since it rotates the plane of plane-polarized light to the left. It is found, usually accompanied by sucrose (like D-glucose it is also a structural unit of sucrose), in an uncombined form in honey and fruit juices. It has been reported that apples and tomatoes contain significant amounts of this monosaccharide. In addition, the plants of the family *Compositae* (e.g., the roots of dahlia and artichoke) utilize the polysaccharides of D-fructose (inulins) as reserve sources, from which it can be obtained as a relatively pure product by hydrolysis. Its industrial production is, however, mainly based on the inversion of sucrose by acids or invertase enzymes (i.e., the production of "invert sugar" or "inverted sugar"), thus resulting in an equimolar mixture of D-fructose and D-glucose. During this inversion process, the rotation of the plane of plane-polarized light is also simultaneously inverted from negative to positive; Jean-Baptiste Biot discovered this phenomenon in 1836. Because D-fructose is clearly sweeter and thus a more valuable sweetening agent than D-glucose, the latter monosaccharide is often converted into the former one by alkali or enzyme. The end product is purified by chromatography and crystallization. The water solution of

D-fructose at 30°C contains 3% α-pyranoid, 65% β-pyranoid, 7% α-furanoid, 25% β-furanoid, and a small amount (0.7%) of openchain ketose forms. Its specific rotation $[\alpha]_D$ in water at 20–24°C is –92.4. D-Fructose is fermented by yeast.

D-Psicose, D-(+)-psicose, D-*ribo*-hexulose, Psi

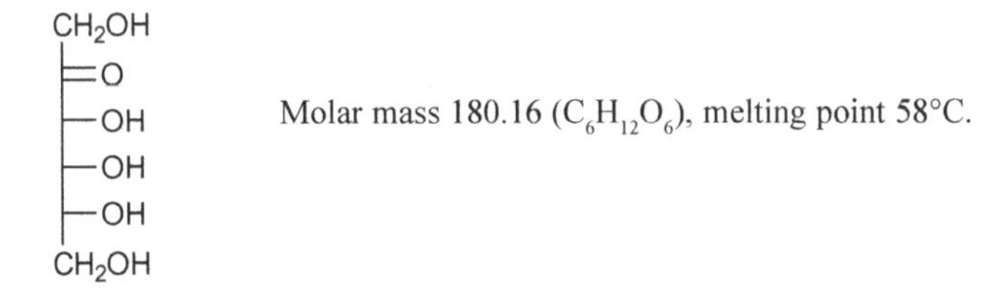

Molar mass 180.16 ($C_6H_{12}O_6$), melting point 58°C.

D-Psicose has been isolated from the antibiotic psicofuranine by hydrolysis. It also occurs in the plant *Itea*, from which it can be separated as the diisopropylidene acetal. In these plants, D-psicose is converted in light into allitol (in darkness, the reverse process takes place), which accumulates during photosynthesis. D-Psicose can be prepared from D-allose in anhydrous pyridine by isomerization or it can be obtained from D-fructose by isomerization in hot methanol to a mixture of D-psicose, D-glucose, and D-mannose, followed by the fermentation with yeast (D-psicose is not fermentable by yeast). The water solution of this monosaccharide at 27°C contains 22% α-pyranoid, 24% β-pyranoid, 39% α-furanoid, 15% β-furanoid, and a small amount (0.3%) of open-chain ketose forms. Its specific rotation $[\alpha]_D$ in water at 25°C is +4.7. D-Psicose is considered one of the most promising novel sweetening agents.

Sorbose, *xylo*-hexulose, sorbinose

L-Sorbose, L-(–)-sorbose, L-*xylo*-hexulose, L-sorbinose

CH_2OH
=O
HO–
–OH
HO–
CH_2OH

Molar mass 180.16 ($C_6H_{12}O_6$), melting point 165°C, density 1.612 kg/dm^3 (17°C).

L-Sorbose is formed by bacterial fermentation (*Acetobacter suboxydans* and *A. xylinum*) in the juice of mountain-ash berries (*Sorbus aucuparia*) which contains significant amounts, for example, D-glucitol (D-sorbitol). It is also obtained by hydrolysis from the pectin in edible passion fruit (*Passiflora edulis*) belonging to the family *Passifloraceae* of flowering plants. The industrial production of this monosaccharide is based on the biochemical oxidation (with *A. suboxydans*) of D-glucitol. The water solution of L-sorbose at 31°C contains 93% α-pyranoid, 2% β-pyranoid, 4% α-furanoid, 1% β-furanoid, and a small amount (0.3%) of open-chain ketose forms. Its specific rotation $[\alpha]_D$ in water at 20°C is –43.2. L-Sorbose is an intermediate in the commercial synthesis of L-ascorbic acid (vitamin C — cf., Chapt. "10.3. Vitamins") and is also a microbiological nutrient.

D-Sorbose, D-(–)-sorbose, D-*xylo*-hexulose, D-sorbinose, pseudotagatose. Molar mass 180.16 ($C_6H_{12}O_6$), melting point 165°C, density 1.612 kg/dm³ (17°C).

D,L-Sorbose, D,L-*xylo*-hexulose, D,L-sorbinose. Molar mass 180.16 ($C_6H_{12}O_6$), melting point 162–3°C, density 1.638 kg/dm³ (17°C).

D-Tagatose, D-(–)-Tagatose, D-*lyxo*-hexulose, Tag

CH_2OH
=O
HO–
HO–
–OH
CH_2OH

Molar mass 180.16 ($C_6H_{12}O_6$), melting point 134–5°C.

D-Tagatose occurs as a structural unit in the gum exudate of the tropical tree *Sterculia setigera* and in the lichen *Rocella linearis*. It can be synthesized by alkaline isomerization of D-galactose or from a relatively rare hexitol, D-altritol (also from D-tallitol) by oxidation with *Acetobacter suboxydans*. The water solution of D-tagatose at 27°C contains 75% α-pyranoid, 17% β-pyranoid, 2% α-furanoid, 6% β-furanoid, and a small amount (0.6%) of open-chain ketose forms. Its specific rotation $[\alpha]_D$ in water at 20°C is –2.3.

D-*allo*-Heptulose

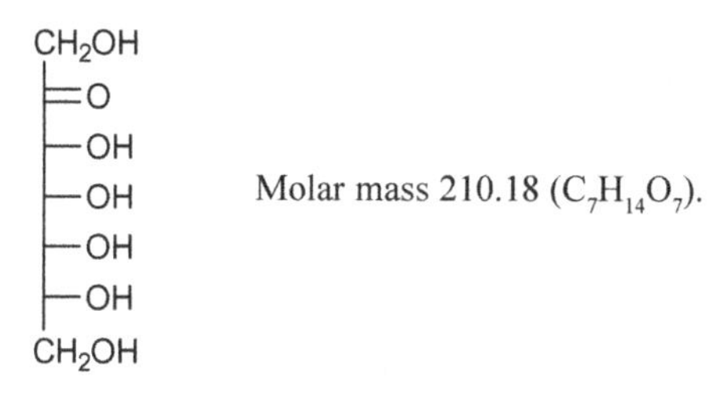

Molar mass 210.18 ($C_7H_{14}O_7$).

D-*altro*-Heptulose, sedoheptulose, sedoheptose

Molar mass 210.18 ($C_7H_{14}O_7$), $[\alpha]_D$ +2.5.

D-*altro*-3-Heptulose

Molar mass 210.18 ($C_7H_{14}O_7$).

D-*manno*-Heptulose, D-mannoketoheptose, D-mannotagatoheptose

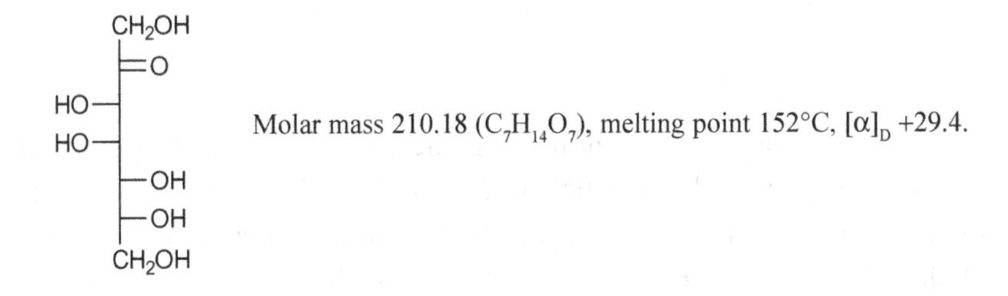

Molar mass 210.18 ($C_7H_{14}O_7$), melting point 152°C, $[\alpha]_D$ +29.4.

D-*talo*-Heptulose

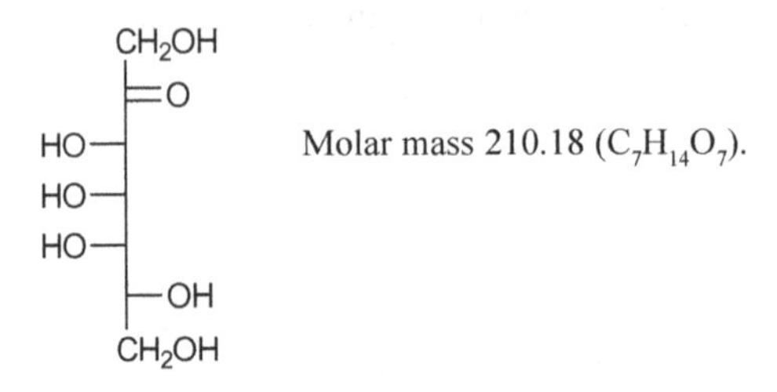

Molar mass 210.18 ($C_7H_{14}O_7$).

L-*gulo*-Heptulose

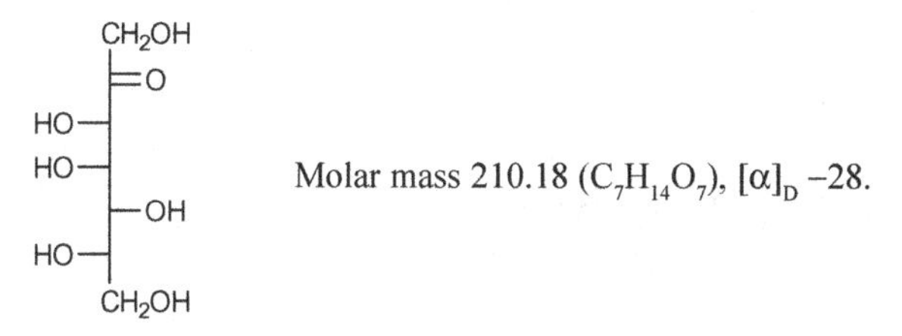

Molar mass 210.18 ($C_7H_{14}O_7$), $[\alpha]_D$ −28.

D-*ido*-Heptulose

Molar mass 210.18 ($C_7H_{14}O_7$), melting point 172°C, $[\alpha]_D$ −34.

L-*galacto*-Heptulose hemihydrate, persculose. Molar mass 219.19 ($C_7H_{17}O_{7.5}$), melting point 110–5°C, $[\alpha]_D$ −80.

D-*glycero*-L-*galacto*-Octulose

Molar mass 240.21 ($C_8H_{16}O_8$), $[\alpha]_D$ −13.4.

D-*glycero*-D-*manno*-Octulose

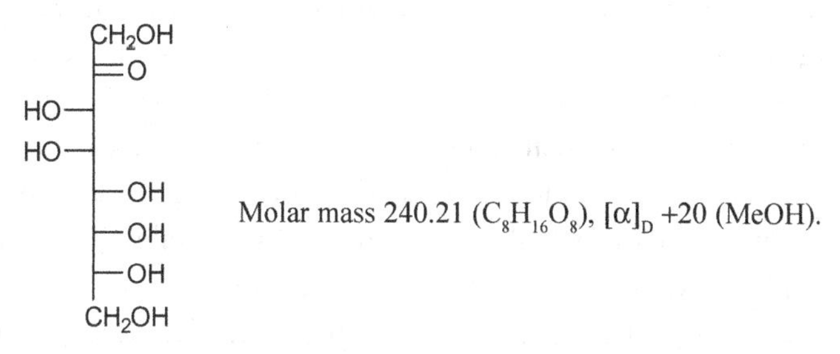

Molar mass 240.21 ($C_8H_{16}O_8$), $[\alpha]_D$ +20 (MeOH).

D-*erythro*-L-*galacto*-Nonulose

CH_2OH
=O
HO—
—OH
—OH
HO—
—OH
—OH
CH_2OH

Molar mass 270.24 ($C_9H_{18}O_9$).

D-*erythro*-L-*gluco*-Nonulose

CH_2OH
=O
HO—
—OH
HO—
HO—
—OH
—OH
CH_2OH

Molar mass 270.24 ($C_9H_{18}O_9$).

9.2. Monosaccharide Derivatives

There are numerous derivatives of sugars whose formation reactions take place both in hydroxyl groups and carbon atoms (cf., Chapt. "11.1. Monosaccharides"). This chapter outlines only briefly the main groups of the most significant derivatives and gives some structure and naming examples of characteristic compounds. The examples comprise, in addition to the important commercial products, also the potential products, the large-scale manufacturing of which might become possible by means of novel biorefinery processes under development (cf., Chapt. "12.1. General Aspects"). However, simple glycosides, such as methyl β-D-glucopyranoside, are not included since typically glycosides are related either to oligo- and polysaccharides or to compounds in which carbohydrates are covalently attached through glycosidic linkages to other organics where the carbohydrate constituent is not an essential feature with respect to chemical and physical properties. Examples of these cases are shown in Chapts. "9.3. Oligosaccharides and Their Derivatives", "9.4. Polysaccharides and Their Derivatives", and "10.1. Glycosides".

9.2.1. *Anhydrosugars*

Anhydrosugars are divided into (i) *intramolecular anhydrides* or *intramolecular ethers* (i.e., non-anomeric anhydrosugars), formally arising by elimination of water from two hydroxy groups (primary or secondary) of a single molecule of a monosaccharide (aldose or ketose) or its derivative (thus resulting in an ether) or into (ii) *monosaccharide anhydrides* or *intramolecular glycosides* (glycosans) (i.e., anomeric anhydrosugars), formally arising by elimination of water from two hydroxyl groups of which one is an anomeric hydroxyl group (thus resulting in a glycoside). The latter compound group is more common; the formation of monosaccharide anhydrides is greatly influenced by the configuration of the monosaccharide. Examples of anhydrosugars are **3,6-anhydro-2,4,5-tri-*O*-methyl-D-glucose** (I), **1,6-anhydro-β-D-glucopyranose** or levoglucosan (II), **1,4-anhydro-β-D-galactopyranose** (III), and **2,7-anhydro-β-D-*altro*-hept-2-ulopyranose** or sedoheptulosan (IV):

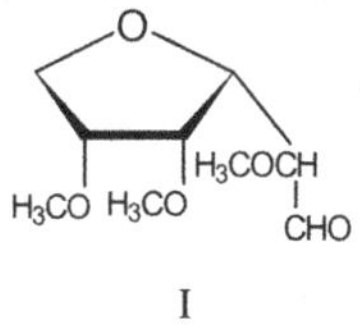

I

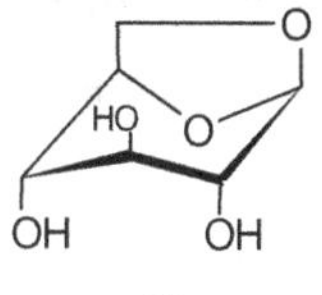

II

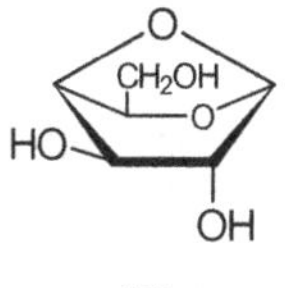

III

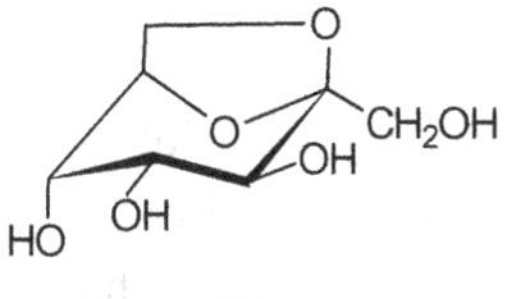

IV

The use of levoglucosan (molar mass 162.14 ($C_6H_{10}O_5$), melting point 180–2°C, $[\alpha]_D$ −65 (20°C)) has been widely studied, since it is formed at a relatively high yield in pyrolysis of cellulose-containing materials or starch (it was first prepared by C. Tanret in 1894). In addition, there exist intermolecular anhydrides formed by the elimination of two water molecules from separate monosaccharides. An example of these derivatives is **α-D-fructopyranose-β-D-fructopyranose 1,2′:1′,2-dianhydride**:

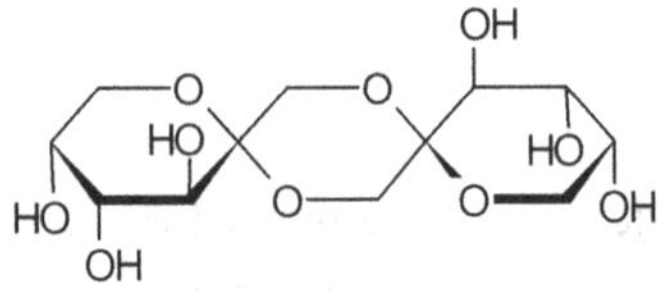

The sugars containing double bonds and which are formed by the elimination of water (i.e., oxygen is removed) can be considered as reduction products. However, these products are rather rare. Examples of these derivatives are **2,3-dideoxy-α-D-*erythro*-hex-2-enopyranose** (I) and **1,5-anhydro-2-deoxy-D-*arabino*-hex-1-enitol** (II):

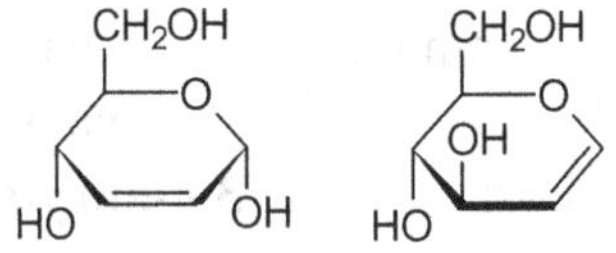

I II

1,6-Anhydro-3,4-dideoxy-β-D-*glycero*-hex-3-enopyranos-2-ulose or levoglucosenone contains both a double bond and an anhydride structure:

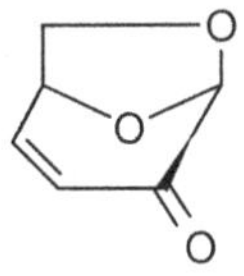

9.2.2. *Deoxy sugars*

Deoxy sugars (previously also known as desoxy sugars) are aldoses or ketoses in which at least one hydroxyl group (i.e., either in a hydroxymethylene or in a hydroxymethyl group) has been replaced by a hydrogen atom, resulting in the formation of a methylene or methyl group, respectively:

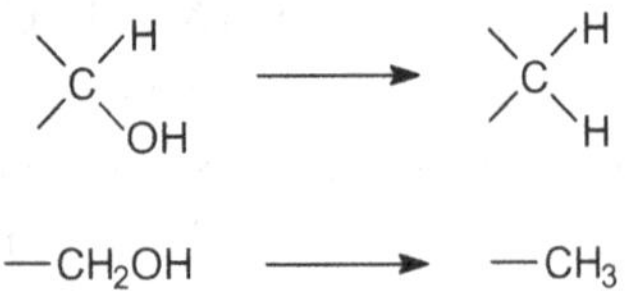

$$-CH_2OH \longrightarrow -CH_3$$

Several deoxy sugars have trivial names established by long usage, especially in cases with a terminal deoxy function at the ω carbon atom (e.g., fucose and rhamnose). However, systematic names are usually preferred for the names of these derivatives, especially where deoxygenation is at a chiral center of the parent sugar. The accepted nomenclature system is simply based on the use of an enumerated prefix deoxy- and the configurations of the remaining asymmetric carbon atoms (need not to be contiguous); for example, **3-deoxy-D-*ribo*-hexose** (I), **5-deoxy-D-*arabino*-3-heptulose** (II), and **6-deoxy-L-mannopyranose** or L-rhamnopyranose (III):

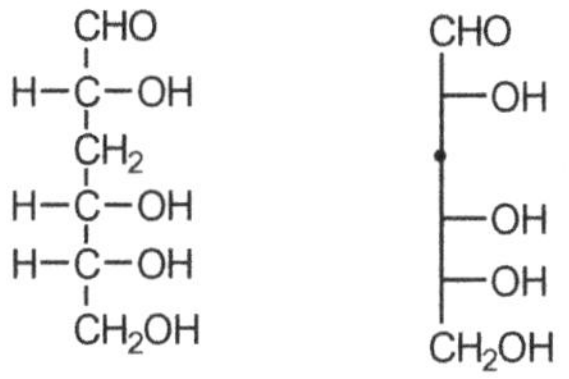

I

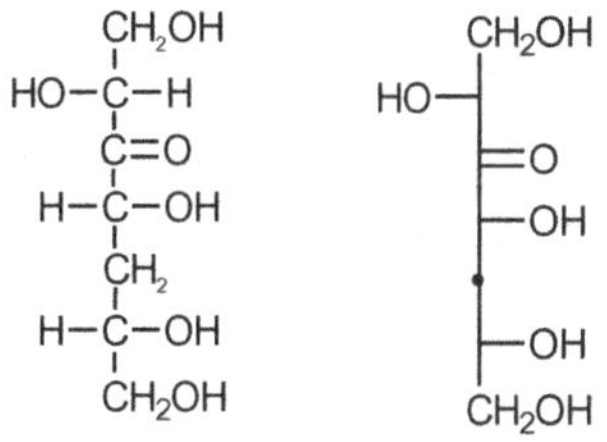

II

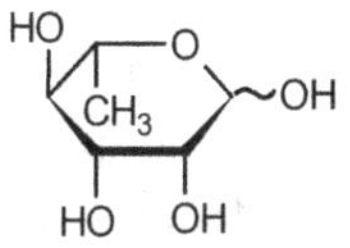

III

In the established cases, a prefix deoxy- can often be used with certain trivial names; for example, **2-deoxy-ribose** (dRib) or 2-deoxy-D-*erythro*-pentose and **2-deoxy-glucose** (2dGlc) or 2-deoxy-D-*arabino*-hexose.

D-Fucose, 6-deoxy-D-galactose, D-galactomethylose, D-rhodeose

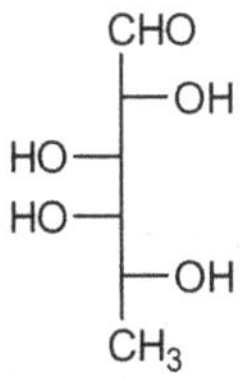

Molar mass 164.16 ($C_6H_{12}O_5$), melting point 140–3°C.

D-Fucose is found in the roots of certain South and Central American plants (*Convolvulaceae*) used as purgatives. These roots give resins of a glycosidic nature, such as convolvulin and jalapin from which it can be separated (together with L-rhamnose and 6-deoxy-D-glucose) by acid hydrolysis. It can be synthesized (6-deoxy-L-altrose is simultaneously formed) from 6-deoxy-6-iodo-1,2:3,4-di-*O*-isopropylidene-D-galactose by dehydrohalogenation. The water solution of D-fucose at 31°C contains 28% α-pyranoid, 67% β-pyranoid, 5% α- and β-furanoid, and a small amount (0.01%) of open-chain aldehyde forms. Its specific rotation $[\alpha]_D$ in water at 20°C is +76.5 (*c* 4, H_2O).

L-Fucose, 6-deoxy-L-galactose, L-galactomethylose, L-rhodeose

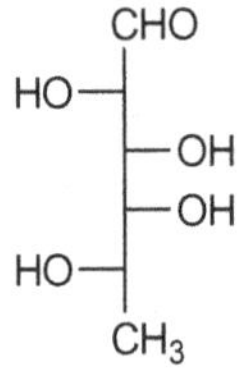

CHO
HO—
—OH
—OH
HO—
CH_3

Molar mass 164.16 ($C_6H_{12}O_5$), melting point 140–1°C.

L-Fucose is found as a constituent of polysaccharides obtained from eggs of the sea urchin (*Echinoidea*), gum tragacanth, frog spawn, and seaweed (*Fucus vesiculosus*). In addition, it is commonly found in glycoproteins, including, for example, mucins and blood-group substances. It can be separated from seaweed by acid hydrolysis and purified by crystallization after deionization and precipitation of impurities with methanol.

D-Rhamnose, 6-deoxy-D-mannose

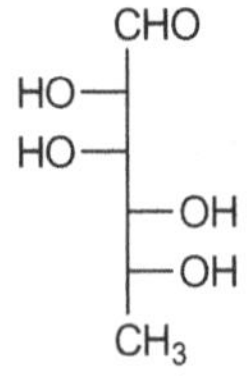

CHO
HO—
HO—
—OH
—OH
CH_3

Molar mass 16416 ($C_6H_{12}O_5$), melting point 90–1°C (monohydrate).

D-Rhamnose is rare in nature. It can be synthesized from 6-*p*-toluenesulfonic acid esters of D-mannose. The water solution of D-rhamnose at 44°C contains 60% α-pyranoid, 40% β-pyranoid, and a small amount (0.01%) of open-chain aldehyde forms. Its specific rotation $[\alpha]_D$ in water at 20°C is –8.2.

L-Rhamnose, 6-deoxy-L-mannose, L-mannomethylose, isodulcit, locaose

```
      CHO
       |
       |-OH
       |-OH
   HO-|
   HO-|
       |
      CH3
```

Molar mass 164.16 ($C_6H_{12}O_5$), melting point 122–6°C.

L-Rhamnose is found as a constituent of many glycosides from which it can be prepared. Especially, it was isolated from extracts of the leaves and blossoms of poison ivy (*Rhus toxicodendron*), poison sumac (*Toxicodendron vernix*), and certain buckthorns (genus *Rhamnus*). In addition, birch xylan, gum arabic, and certain pectins contain this deoxy monosaccharide (cf., Chapt. "9.4.5. Other polysaccharides"). However, the bark of an oak species (*Quercus tinctoria*) provides an excellent source of a glycoside, lemon flavin (quercitrin or quercetin 3-rhamnoside), the acid hydrolysis of which leads to L-rhamnose and the aglycon (quercetin). Its specific rotation $[\alpha]_D$ in water at 20°C is +8.9.

Digitoxose, 2,6-dideoxy-D-*ribo*-hexose

```
      CHO
       |
      CH2
       |-OH
       |-OH
       |-OH
       |
      CH3
```

Molar mass 148.16 ($C_6H_{12}O_4$), melting point 112°C.

Digitoxose is found as a constituent of specific glycosides (digitoxin, gitoxin, and digoxin) from which it can be obtained by mild acid hydrolysis and crystallized from methanol or a mixture of acetone and diethyl ether. Its specific rotation $[\alpha]_D$ in water at 17°C is +46.3.

5-Deoxy-L-arabinose

```
      CHO
       |
       |-OH
   HO-|
   HO-|
       |
      CH3
```

Molar mass 134.13 ($C_5H_{10}O_4$), $[\alpha]_D$ −16.4 (EtOH) (20°C).

6-Deoxy-D-allose

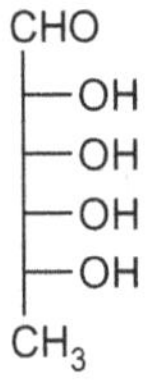

CHO
—OH
—OH
—OH
—OH
CH_3

Molar mass 164.16 ($C_6H_{12}O_5$), melting point 151–2°C, $[\alpha]_D$ +1.2.

6-Deoxy-D-altrose, altromethylose

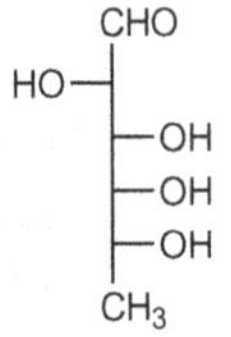

CHO
HO—
—OH
—OH
—OH
CH_3

Molar mass 164.16 ($C_6H_{12}O_5$), $[\alpha]_D$ +16.2.

6-Deoxy-D-glucose, chinovose, isorhodeose, D-glucomethylose, D-isorhamnose, D-epirhamnose

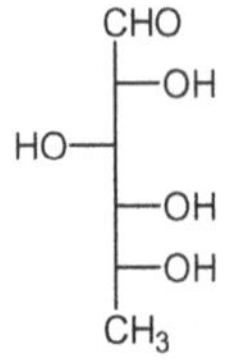

CHO
—OH
HO—
—OH
—OH
CH_3

Molar mass 164.16 ($C_6H_{12}O_5$), melting point 144–6°C, $[\alpha]_D$ +30 (*c* 1.5, H_2O) (28°C).

6-Deoxy-D-gulose, D-antiarose, D-gulomethylose

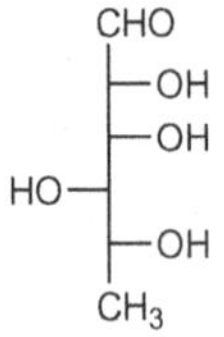

CHO
—OH
—OH
HO—
—OH
CH_3

Molar mass 164.16 ($C_6H_{12}O_5$), melting point 123–6°C, $[\alpha]_D$ –42.4 (*c* 0.5, H_2O) (19°C).

6-Deoxy-D-talose, D-talomethylose

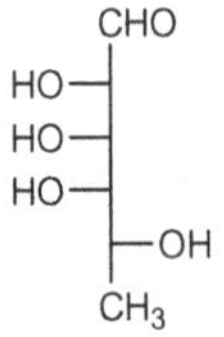

CHO
HO—
HO—
HO—
—OH
CH_3

Molar mass 164.16 ($C_6H_{12}O_5$), melting point 129–31°C, $[\alpha]_D$ +20.6.

7-Deoxy-L-*glycero*-L-*galacto*-heptose

Molar mass 194.18 ($C_7H_{14}O_6$), melting point 186–7°C, $[\alpha]_D$ −62.5 (*c* 2.5, H_2O) (21°C).

7-Deoxy-L-*glycero*-D-*gluco*-heptose

Molar mass 194.18 ($C_7H_{14}O_6$), $[\alpha]_D$ +41.

6-Deoxy-D-tagatose, 6-deoxy-D-*lyxo*-hexulose, D-tagatomethylose

Molar mass 164.16 ($C_6H_{12}O_5$), $[\alpha]_D$ −2 (*c* 2, H_2O) (18°C).

7-Deoxy-D-*altro*-2-heptulose

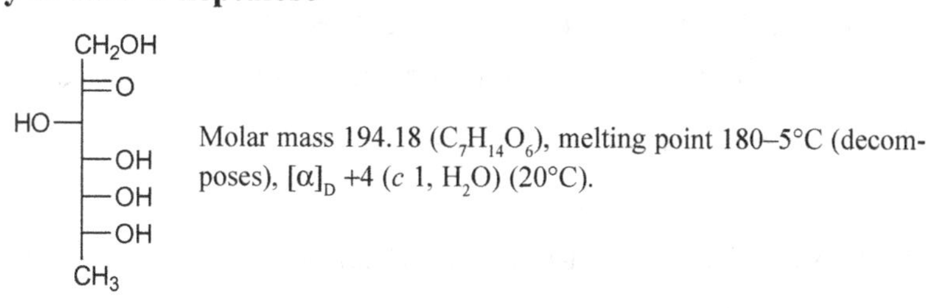

Molar mass 194.18 ($C_7H_{14}O_6$), melting point 180–5°C (decomposes), $[\alpha]_D$ +4 (*c* 1, H_2O) (20°C).

In the following, a few examples are given of rather rare deoxy sugars generally known by their trivial names:

Abequose, 3,6-dideoxy-D-*xylo*-hexose, Abe

Ascarylose, 3,6-dideoxy-L-*arabino*-hexose

Boivinose, 2,6-dideoxy-D-gulose

Colitose, 3,6-dideoxy-L-*xylo*-hexose

Paratose, 3,6-dideoxy-D-*ribo*-hexose

Tyvelose, 3,6-dideoxy-D-*arabino*-hexose, Tyv

Amicetose, 2,3,6-trideoxy-D-*erythro*-hexose

Rhodinose, 2,3,6-trideoxy-L-*threo*-hexose

Deoxy derivatives can also contain one or more organic substituents. In the simplest case, one hydrogen atom, attached to either the asymmetric carbon atom or the oxygen atom bound to this carbon atom, is replaced by a substituent. The location of an organic substituent is shown by adding to the sugar name the number of an asymmetric carbon atom according to the following system: 3-*O*-methyl- (i.e., Me is directly bound to an oxygen atom) and 3-*C*-methyl- (i.e., Me is directly bound to a carbon atom) (cf., Chapt. "7.1. Aldoses"). Examples of such derivatives and their trivial names are as follows:

Chalcose, 4,6-dideoxy-3-*O*-methyl-D-*xylo*-hexose

Cymarose, 2,6-dideoxy-3-*O*-methyl-D-*ribo*-hexose

Diginose, 2,6-dideoxy-3-*O*-methyl-D-*lyxo*-hexose

Mycarose, 2,6-dideoxy-3-*C*-methyl-L-*ribo*-hexose

Oleandrose, 2,6-dideoxy-3-*O*-methyl-D-*arabino*-hexose

Sarmentose, 2,6-dideoxy-3-*O*-methyl-D-*xylo*-hexose

Arcanose, 2,6-dideoxy-3-*C*-methyl-3-*O*-methyl-D-*xylo*-hexose

Cladinose, 2,6-dideoxy-3-*C*-methyl-3-*O*-methyl-L-*ribo*-hexose

Digitalose, 6-deoxy-3-*O*-methyl-D-galactose

Evalose, 6-deoxy-3-*C*-methyl-D-mannose

Mycinose, 6-deoxy-2,3-di-*O*-methyl-D-allose

Noviose, 6-deoxy-5-*C*-methyl-4-*O*-methyl-L-*lyxo*-hexose

9.2.3. *Diols and polyalcohols*

Alditols or "reduced sugars" are polyhydroxy compounds in which an aldehyde group of aldoses or a ketone group of ketoses is, respectively, reduced (i.e., the reduction of a carbonyl functional group) either to a primary or a secondary alcohol. If an alcohol molecule contains two hydroxyl groups, it is called a "*diol*" and the positions of these hydroxyl groups are numbered according to the carbon atoms to which they are attached. Polyhydric alcohols or *polyalcohols* (in the case of sugar alcohols, glycitols or alditols) are called, depending on the number of hydroxyl groups, "triols", "tetritols", "pentitols", "hexitols", "heptitols", etc. Alditols are thus named by changing the suffix *-ose* in the name of the corresponding aldose into the suffix *-itol*. However, many significant polyalcohols have established trivial names, such as glycol, glycerol, xylitol, and sorbitol.

The most common and commercially important polyalcohols in nature are primarily alditols belonging to pentitols or hexitols. Natural alditols, like their aldose precursors, are generally also members of the D series. Since their terminal groups (-CH_2OH) are identical, they can be optically inactive (i.e., they have a plane of symmetry between the "top and bottom halves" of the molecule; e.g., in xylitol), which can be designated by the prefix *meso-* (cf., Chapt. "4.2. The Total Number of Stereoisomers"). This section introduces, besides common alditols, also examples of polyalcohols that can be produced from biomass (cf., Chapt. "12.2. Production of Chemicals"). In addition, some polyalcohols contain double bonds and can exist as *cis*/*trans* isomers (cf., Chapt. "3.7. Geometric Isomerism").

Due to their hydroxyl groups, polyalcohols are polar compounds and readily water soluble. Diols are generally called "glycols" (in Greek, "glykos" or "glykeros" means "sweet"), which describes one of their, and especially polyalcohols', characteristic properties. Distinguished from monoalcohols, polyalcohols are not intoxicating. For this reason, alditols are primarily used as sweetening agents, especially low-molar-mass diols and triols, and also due to their other advantageous properties, mainly as solvents and antifreezing agents.

Ethylene glycol, 1,2-ethanediol, 1,2-dihydroxyethane, glycol, ethylene alcohol

$HOCH_2CH_2OH$

Molar mass 62.07 ($C_2H_6O_2$), melting point −11.5°C, boiling point 198 or 93 (13 mmHg) °C, density 1.109 kg/dm^3 (20°C), refractive index n_D 1.432 (20°C), pK value 14.2 (25°C), flash point 111°C, explosive limits in air 3.2–15.2%, autoignition temperature 413°C.

Ethylene glycol can be produced by many processes, but its principal production is based on ethylene (ethene). It is obtained from ethylene via ethylene chlorohydrin according to the following reaction:

$$H_2C{=}CH_2 + Cl_2 + H_2O \longrightarrow HOCH_2CH_2Cl + HCl$$
$$HOCH_2CH_2Cl + NaHCO_3 \longrightarrow HOCH_2CH_2OH + NaCl + CO_2$$

The present production comprises the synthesis from ethylene via ethylene oxide through the following scheme:

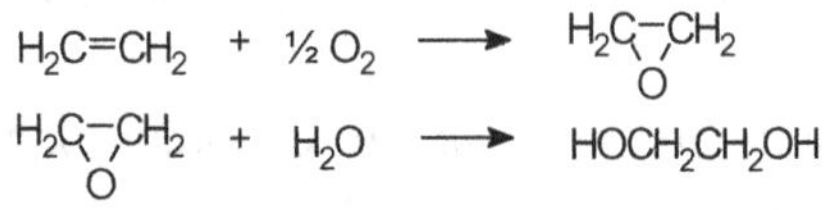

$$H_2C{=}CH_2 + \tfrac{1}{2}O_2 \longrightarrow H_2C\overset{O}{-}CH_2$$
$$H_2C\overset{O}{-}CH_2 + H_2O \longrightarrow HOCH_2CH_2OH$$

The reaction is catalyzed by either acids or bases, or it can also occur at a neutral pH at elevated temperatures. However, under acidic or neutral conditions with a large excess of water, an ethylene glycol yield of about 90% can be achieved, the major by-products being di-, tri-, and tetraethylene glycols, from which the final product is separated by distillation. In addition, ethylene glycol can be directly produced from synthesis gas in the presence of various catalysts. Technical grades normally distill at 190–200°C, but the temperature range may be even wider. The pure ethylene glycol is a colorless, nearly odorless, sweet-tasting, viscous, and hygroscopic liquid that is miscible with water, ethanol, and acetone, and dissolves in lesser amounts in diethyl ether and benzene. It is moderately toxic and can be fatal upon ingestion of sufficient amounts. On heating (it vaporizes slowly at room temperature) the solvent vapors cause headache and nausea. It can react with oxidizing agents. Ethylene glycol is traditionally used as an additive in antifreeze formulations; in this case, some bitter flavoring may be added to it for minimizing the risk of its inadvertent drinking. When mixed with water, the freezing point of the mixture is clearly depressed:

Ethylene glycol/wt.–%	Density/kg/ dm³ (20 °C)	Freezing point/°C
5	1.004	–1.6
10	1.011	–3.4
16	1.019	–5.9
20	1.024	–7.9
24	1.030	–10.3
32	1.041	–16.2
36	1.046	–19.8
40	1.051	–23.8
44	1.057	–28.3
52	1.067	–38.8
60	1.077	–51.2

Ethylene glycol can be also utilized as a solvent, plasticizer, and hydraulic fluid. In addition, it is used as a raw material in the manufacture of polyester fibers (the most significant product, polyethylene terephthalate (PET), is based on ethylene glycol and terephthalic acid) as well as polyurethane fibers and resins. One traditional product has also been ethylene glycol dinitrate (EGDN or nitroglycol, NGc) which was used to lower the freezing point of nitroglycerin to produce dynamite for use in colder weather.

Glycerol, 1,2,3-trihydroxypropane, propane-1,2,3-triol, glycerine, glycerin

CH_2OH
$CHOH$
CH_2OH

Molar mass 92.11 ($C_3H_8O_3$), melting point 20°C, boiling point 182°C (20 mmHg) or 50°C (0.003 mmHg), density 1.261 kg/dm³ (20°C), refractive index n_D 1.475 (20°C), pK value 14.1 (25°C), flash point 160°C, autoignition temperature 393°C.

Glycerol can be produced by several methods, the most significant ones being either hydrolysis (saponification) of natural fats (triglycerides) or synthetic routes from propylene (propene). The latter case primarily involves the chlorination of propylene to allyl chloride, which is converted with hypochlorous acid into dichlorohydrin; this is then hydrolyzed with a strong base via epichlorohydrin to glycerol. In contrast, chlorine-free processes from propylene include the synthesis of glycerol from propylene oxide or acrolein. The commercial products are of variable quality. The anhydrous and pure glycerol is a colorless, odorless, and syrupy liquid with a sweet taste. It is miscible with water and ethanol. Glycerol is hygroscopic and changes rather rapidly from viscous liquid to a fluid one in an open container. It is traditionally used like ethylene glycol as an antifreeze agent for automotive applications and hydraulic fluids. Because of its hygroscopic nature, it is used in medical, pharmaceutical, and personal care preparations, mainly as a non-volatile and non-toxic humectant for improving smoothness and providing lubrication. In food industry, glycerol serves, for example, as a humectant, solvent, sweetener, and thickening agent, and may help preserve foods. In addition, it is used to produce nitroglycerin (i.e., the basic ingredient of many explosives, such as dynamite), alkyd resins, and copying inks.

Erythritol, *erythro*-tetritol, 2*R*,3*S*-1,2,3,4-tetrahydroxybutane

CH_2OH
H–C–OH
H–C–OH
CH_2OH

Molar mass 122.12 ($C_4H_{10}O_4$), melting point 121.5°C, boiling point 329–31 or 294–6 (200 mmHg) °C, density 1.451 kg/dm^3 (20°C).

John Stenhouse (1809–80) discovered erythritol in 1848. It is optically inactive (a *meso* form) and occurs naturally in lichens, algae, yeasts, and in some fruits. This alditol can be produced by fermentation (ribitol and glycerol are formed as by-products) with a yeast *Moniliella pollinis* from D-glucose obtained from the hydrolysis of starch. The synthetic route includes the reduction of D-erythrose (or D-erythronic acid lactone). Erythritol is a crystalline compound that dissolves in water (61.5%), hot ethanol, and pyridine (2.5%, 20–25°C), but not in diethyl ether. It has a sweet taste, does not affect blood sugar or cause tooth decay, and has a low caloric value (<1.6 kJ/g or <0.4 kcal/g; about 90% less than sucrose, 16.2 kJ/g). In addition, in the body, most erythritol is absorbed into the bloodstream in the small intestine, and then excreted for the most part in the urine. Due to these advantageous properties, it is a potential sweetening agent for many commercial products.

Xylitol, *xylo*-pentitol, 2*S*,4*R*-1,2,3,4,5-pentahydroxypentane, xylite, Xylol

```
      CH2OH
      |
   H—C—OH
      |
  HO—C—H
      |
   H—C—OH
      |
      CH2OH
```

Molar mass 152.15 ($C_5H_{12}O_5$), melting point 61.0-1.5 (metastable) and 93.0-4.5 (stable) °C, boiling point 215–7°C (stable).

The most abundant hemicellulose component in hardwoods (or in corncobs), xylan, mainly contains a chemically bound monosaccharide moiety, D-xylose (cf., Chapt. "9.4.5. Other polysaccharides"). Therefore, suitable feedstock materials for producing xylitol are, for example, hydrolysates from the sulfuric acid hydrolysis of birch wood, hydrolysates from the acidic pre-treatment stage of birch kraft pulping, birch spent liquors from the acidic sulfite pulping, or hydrolysates from the steam explosion of birch wood. All these aqueous solutions are fractionated by various chromatographic techniques and the D-xylose-containing fraction obtained is then reduced to xylitol. The large-scale reduction includes catalytic hydrogenation or on a smaller scale, sodium borohydride ($NaBH_4$) can be used for this purpose. In addition, common yeast cells (e.g., *Candida guilliermondii*, *C. boidinii*, *C. tropicalis*, *C. parapsilosis*, *C. mogii*, and *Debaryomyces hansenii*), fungi (e.g., *Petromycens albertensis*), or bacteria (e.g., *Enterobacter liquefaciens*) can produce xylitol from D-xylose. However, fermentation cannot directly be applied to acidic hydrolysates or spent liquors, since the liquor pH and some hydrolysis by-products (e.g., acetic acid and furans) are normally fermentation inhibitors. Consequently, acidic solutions of D-xylose are first treated with alkali (e.g., an aqueous mixture of calcium carbonate and calcium hydroxide) to precipitate impurities and to adjust the pH value generally to a level of 5.0–6.5. The final product, xylitol, can then be purified chromatographically, followed by crystallization.

Xylitol was also one of the key compounds in early carbohydrate chemistry, since in 1890 Emil Fischer together with his research team separated and identified this alditol from the water solution formed in the reduction of D-xylose. However, it was not found in nature until the 1940s. Xylitol is an optically inactive (a *meso* form), white, and crystalline compound that dissolves in water (64.2%, 25°C), ethanol (1.2%, 25°C), and pyridine. It has a sweet taste without any aftertaste and a caloric value of 10.0 kJ/g (2.4 kcal/g) being suitable as a special sweetening agent, especially in chewing gums and pastils. Due to a very low glycemic index (GI) of 7 (the GI of sucrose, glucose, and fructose is 65, 100, and 25, respectively), it absorbs more slowly than typical sugars and thus, has a low effect on a person's blood sugar. In addition, its use appears to have clear dental benefits (found in Finland in the early 1970s) with an inhibitory action on the progress of dental caries (i.e., a cariostatic or an anti-cariogenic effect).

Pentitols similar to xylitol are **ribitol** and **L-arabitol**, both having a sweet taste. Of these stereoisomers, ribitol exists as a *meso* form. It occurs naturally in the plant *Adonis vernalis* and is, for example, a constituent of vitamin B_2 (riboflavin) and flavin mononucleotide (FMN) or riboflavin-5′-phosphate. Ribitol is synthesized by reduction of D- or L-ribose. It is a crystalline compound that dissolves in water and hot ethanol, but not in diethyl ether. L-Arabitol is not found in nature, but it can be synthesized by a reduction of L-arabinose (present in softwood xylan as a monosaccharide moiety) or L-lyxose. It is a crystalline compound that dissolves in water and hot ethanol.

Ribitol, *ribo*-pentitol, adonitol, adonite, adonit, Rib-ol

```
    CH2OH
     |
 H—C—OH
     |
 H—C—OH
     |
 H—C—OH
     |
    CH2OH
```

Molar mass 152.15 ($C_5H_{12}O_5$), melting point 104°C.

L-Arabitol, L-*arabino*-pentitol, L-Ara-ol

```
     CH2OH
      |
  H—C—OH
      |
 HO—C—H
      |
 HO—C—H
      |
     CH2OH
```

Molar mass 152.15 ($C_5H_{12}O_5$), melting point 102–3°C.

D-Glucitol, D-*gluco*-hexitol, D-sorbitol, D-Glc-ol

```
     CH2OH
      |
  H — C — OH
      |
 HO — C — H
      |
  H — C — OH
      |
  H — C — OH
      |
     CH2OH
```

Molar mass 182.18 ($C_6H_{14}O_6$), melting point 110–2 (anhydrous) or 75 (monohydrate) °C, boiling point 295°C (3.5 mmHg), density 1.489 kg/dm³ (20°C), refractive index n_D 1.333 (20°C), $[\alpha]_D$ −2.0 (25°C, H_2O).

D-Glucitol is widespread in nature, being found in many fruits (e.g., apples, pears, peaches, and prunes) and certain berries, but only in insignificant amounts in grapes. It can be synthesized by a reduction (i.e., by electrolytic or high-pressure hydrogenation techniques) of D-glucose and D-fructose but is mostly produced from corn syrup. Commercial products may exist as crystalline product (content 99%), a syrupy solution (content 83–85%) or an aqueous solution (content 70%). The pure, slightly cariogenic D-glucitol is a white, odorless, and crystalline solid with a sweet and cool taste. It is soluble in water (2350 g/L), hot ethanol, and acetone but not in diethyl ether. D-Glucitol can be used as a sugar substitute (about 60% as sweet as sucrose with energy content about 11 kJ/g; for typical sugars about 17 kJ/g), especially in diet foods, a non-stimulant laxative, in modern cosmetics as a humectant and thickener, and as an intermediate in the commercial synthesis of L-sorbose (an intermediate of L-ascorbic acid, vitamin C), or for various esters of fatty acids.

D-Mannitol, D-*manno*-hexitol, Man-ol

```
     CH2OH
      |
 HO — C — H
      |
 HO — C — H
      |
  H — C — OH
      |
  H — C — OH
      |
     CH2OH
```

Molar mass 182.18 ($C_6H_{14}O_6$), melting point 168°C, boiling point 295°C (3.5 mmHg), density 1.489 kg/dm³ (20°C), refractive index n_D 1.333, $[\alpha]_D$ −0.2 (25°C, H_2O).

D-Mannitol is also a widespread sugar alcohol in nature, being found in the exudates of some trees and in various shrubs, grasses, seaweeds, fungi, and lichens. The secretion ("manna") of the flowering manna ash (*Fraxinus ornus*) contains 30–50% of this alditol. It can be synthesized by a catalytic reduction of D-mannose and D-fructose. D-Mannitol crystallizes as white needles or prisms that dissolve in water (16.4%, 15°C or 61.1%, 60°C) and ethanol (0.01%, 15°C or 0.6%, 60°C). It has several industrial uses; mainly the medical use as a filler for producing tablets of medicine. In addition, it is used as a laxative, a sugar substitute, and an intermediate in the explosives industry (i.e., the production of mannitol hexanitrate).

The corresponding stereoisomers (molar mass 182.18; $C_6H_{14}O_6$) with similar properties are, for example, *meso*-galactitol (dulsitol), L-iditol, D-tallitol, and *meso*-allitol:

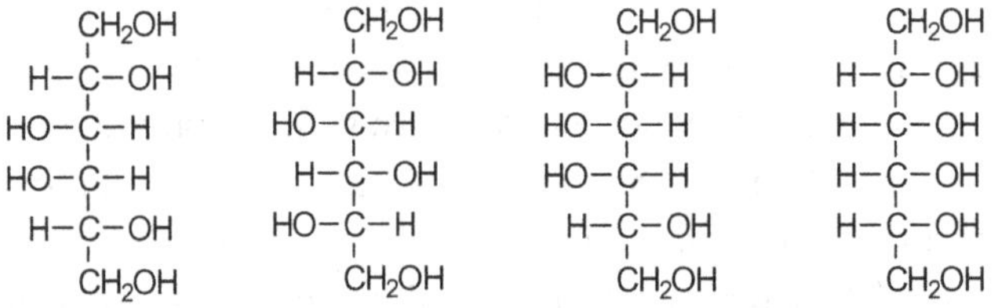

CH_2OH	CH_2OH	CH_2OH	CH_2OH
H–C–OH	H–C–OH	HO–C–H	H–C–OH
HO–C–H	HO–C–H	HO–C–H	H–C–OH
HO–C–H	H–C–OH	HO–C–H	H–C–OH
H–C–OH	HO–C–H	H–C–OH	H–C–OH
CH_2OH	CH_2OH	CH_2OH	CH_2OH
meso-Galactitol	L-Iditol	D-Tallitol	*meso*-Allitol

Heptitols and higher alditols have no significant commercial value. They are found to some extent in nature but are also synthesized.

9.2.4. *Deoxy alditols*

In *deoxy alditols*, a carbonyl group of aldoses or ketoses is replaced by a deoxy structure. In this case, it is possible, as generally in cases where the carbon atoms C_1 and C_ω belong to the same functional group (i.e., the terminal groups are identical) that they are optically inactive. Examples of deoxy alditols are as follows:

L-Rhamnitol, 1-deoxy-L-mannitol, L-Rha-ol

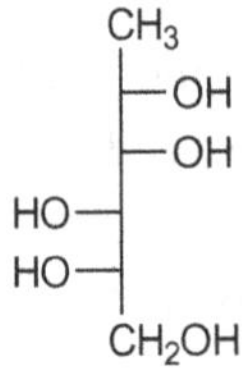

CH_3
–OH
–OH
HO–
HO–
CH_2OH

Molar mass 166.17 ($C_6H_{14}O_5$), melting point 123–4°C, $[\alpha]_D$ −12.0 (*c* 1.0, H_2O) (20°C).

L-Fucitol, 1-deoxy-D-galactitol, L-Fuc-ol

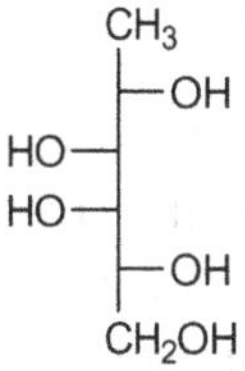

CH_3
–OH
HO–
HO–
–OH
CH_2OH

Molar mass 166.17 ($C_6H_{14}O_5$).

6-Deoxy-D-glucitol, epirhamnitol, isorhamnitol

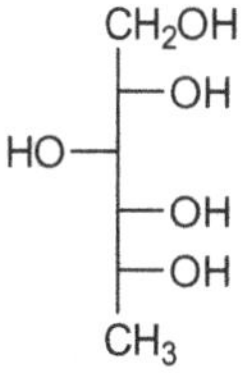

CH_2OH
–OH
HO–
–OH
–OH
CH_3

Molar mass 166.17 ($C_6H_{14}O_5$), $[\alpha]_D$ −10.0 (20°C).

9.2.5. *Other polyalcohols and their derivatives*

In this section, a few polyalcohols and their derivatives are included. They are generally not considered "direct derivatives" of monosaccharides and their present production is mainly based on the petrochemical industry. However, they represent also possibly compounds obtained as biomass conversion products from biorefining (cf., Chapt. "12.2. Production of Chemicals").

Propylene glycol, 1,2-propanediol, 1,2-dihydroxypropane, α-propylene glycol (optically active (+)- and (−)-forms and an optically inactive, racemic (±)-form)

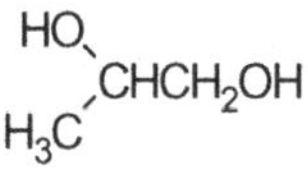

HO
$CHCH_2OH$
H_3C

Molar mass 76.11 ($C_3H_8O_3$), boiling point 187.4 or 96–8 (21 mmHg) °C ((±)), density 1.036 kg/dm³ (20°C (±)), refractive index n_D 1.432 (20°C (±)), flash point 99°C, explosive limits in air 2.6–12.6%, autoignition temperature 421°C.

Propylene glycol is produced analogously with ethylene glycol (see above); in this case, from propylene (propene) originated from various cracking gases of petroleum refining. It is a colorless, nearly odorless, viscous, and hygroscopic liquid with a faintly sweet (slightly bitter) taste and is miscible with water, ethanol, and acetone, and dissolves in lesser amounts in diethyl ether and benzene. Although propylene glycol exists in two enantiomers, the commercial product is often a racemic form. It can react with oxidizing agents. Its solvent properties are similar to those of ethylene glycol and propylene glycol is also able to lower the freezing point of water (cf., its use as an aircraft deicing fluid). In addition, propylene glycol is used, because of its low toxicity, as a humectant, preservative in food products, and chemical feedstock for the production of unsaturated polyester resins, as well as in hydraulic fluids and personal care products.

1,3-Propanediol, 1,3-propylene glycol, trimethylene glycol, β-propylene glycol, 1,3-PD, PDO

$HOCH_2CH_2CH_2OH$

Molar mass 76.11 ($C_3H_8O_2$), melting point −30°C, boiling point 213.5 or 110 (12 mmHg) °C, density 1.060 kg/dm^3 (20°C), refractive index n_D 1.440 (20°C), autoignition temperature 400°C.

1,3-Propanediol has rather similar properties to those of propylene glycol. It is a colorless, odorless, and viscous liquid that is miscible with water and ethanol, and dissolves in lesser amounts in diethyl ether and benzene. 1,3-Propanediol has traditionally been synthesized by hydration of acrolein or by the hydroformylation of ethylene oxide to afford 3-hydroxypropanal (which can be then hydrogenated to 1,3-propanediol), but also, for example, from glycerol (as already described in 1881) by anaerobic fermentation with specific bacteria. It is mainly used as a building block in the production of various polymers. One of the most significant examples is its polyester with terephthalic acid (polytrimethylene terephthalate) which, like many polyesters of 1,3-propanediol, is generally called, in spite of a petrochemical-derived feedstock component, terephthalic acid, also a "green polyester". In addition, it can be formulated into a variety of industrial products or used as a solvent and an antifreeze agent.

1,3-Butanediol, 1,3-butylene glycol, 1,3-dihydroxybutane, β-butylene glycol (optically active (+)- and (–)-forms and an optically inactive, racemic (±)-form)

HO(H$_3$C)CHCH$_2$CH$_2$OH

Molar mass 90.12 ($C_4H_{10}O_2$), melting point <–50°C ((±)), boiling point 207.5 or 103-4 (8 mmHg) °C ((±)), density 1.005 kg/dm^3 (20°C (±) and (+)), refractive index n_D 1.442 ((+)) and 1.441 ((±)) (20°C), flash point 121°C, autoignition temperature 394°C.

1,3-Butanediol is a colorless, odorless, and hygroscopic syrup that dissolves in water, ethanol, ethyl acetate, and acetone, but not in diethyl ether or hydrocarbons. It has rather similar properties to those of ethylene glycol and is commonly used as a solvent for food flavoring agents and a comonomer in certain polyurethane and polyester resins. Its concentration in technical grades is normally about 95% and they distill at 204–208°C. The isomers with similar properties to 1,3-butanediol are **1,4-** and **1,2-butanediol.**

1,4-Butanediol, 1,4-butylene glycol, 1,4-dihydroxybutane, tetramethylene glycol

HOCH$_2$(CH$_2$)$_2$CH$_2$OH

Molar mass 90.12 ($C_4H_{10}O_2$), melting point 20.1°C, boiling point 235 or 120 (10 mmHg) °C, density 1.017 kg/dm^3 (20°C), refractive index n_D 1.446 (20°C), flash point 121°C.

1,2-Butanediol, 1,2-dihydroxybutane, α-butylene glycol (optically active (+)- and (–)-forms and an optically inactive, racemic (±)-form)

HO(CH$_3$CH$_2$)CHCH$_2$OH

Molar mass 90.12 ($C_4H_{10}O_2$), melting point –114°C ((±)), boiling point 190.5 or 96.5 (10 mmHg) °C ((±)), density 1.002 kg/dm^3 (20°C, (±)), refractive index n_D 1.437 ((+)) and 1.438 ((±)) (20°C), flash point 40°C.

2,3-Butanediol, 2,3-butylene glycol, 2,3-dihydroxybutane, dimethylene glycol, *sym*-butylene glycol (optically active (+)- and (–)-forms and an optically inactive, racemic (±)-form)

HO(H$_3$C)CHCH(CH$_3$)OH

Molar mass 90.12 ($C_4H_{10}O_2$), melting point 34.4 (*meso*), 19.7 ((–)) and 7.6 ((±)) °C, boiling point 182.5 or 86 (16 mmHg) °C (±)), density 1.043 (20°C, *meso*), 0.987 (25°C (+)) and 1.003 (20°C (±)) kg/dm^3, refractive index n_D 1.431 (25°C (+) and (±)) and 1.436 (20°C, *meso*), flash point 85°C, autoignition temperature 402°C.

The racemic 2,3-butanediol is found in nature in cocoa butter, sweet corn, rotten mussels, and in the roots of *Ruta graveolens*, and it can react, like other diols, with oxidizing agents. It is a colorless and hygroscopic liquid that is miscible with water and ethanol and dissolves in lesser amounts in diethyl ether and acetone. In addition to the conventional petrochemical-based production, a variety of microorganisms can produce (by "butanediol fermentation") this diol, for example, from glucose. Depending on the fermentation process applied, the simultaneous formation of other products, such as carbon dioxide, hydrogen, formic acid, and glycerol, can be obtained. 2,3-Butane diol is used as a dye solvent and a humectant for wood and textile materials. It is also used together with phthalic anhydride for manufacturing resins.

2-Methyl-1,2-propanediol, isobutylene glycol

$CH_3C(OH)(CH_3)CH_2OH$

Molar mass 90.12 ($C_4H_{10}O_2$), boiling point 178°C, density 1.002 kg/dm^3 (20°C), refractive index n_D 1.435 (20°C).

2-Methyl-1,2-propanediol with a branched carbon chain dissolves in water, ethanol, and diethyl ether.

2,3-Dimethyl-2,3-butanediol, tetramethylethylene glycol, pinacol

$H_3C-C(OH)(CH_3)-C(OH)(CH_3)-CH_3$

Molar mass 118.17 ($C_6H_{14}O_2$), melting point 41.1°C, boiling point 174.4°C, density 0.967 kg/dm^3 (15°C), flash point 77°C.

2,3-Dimethyl-2,3-butanediol can be synthesized from acetone by reduction with magnesium amalgam:

$$2\ (H_3C)_2C{=}O + H_2 \longrightarrow H_3C-C(OH)(CH_3)-C(OH)(CH_3)-CH_3$$

It is a crystalline solid that dissolves in ethanol and diethyl ether and in lesser amounts in water. This diol has been used, besides for research purposes, for military purposes (e.g., the production of a nerve agent soman by reacting pinacolyl alcohol with methylphosphonyl difluoride).

In addition, 2,3-dimethyl-2,3-butanediol forms with six water molecules a hexahydrate, which exists as flaky crystals and has similar properties to those of the anhydrous form.

2,3-Dimethyl-2,3-butanediol hexahydrate. Molar mass 226.27 ($C_6H_{26}O_8$), melting point 45.1°C, density 0.967 kg/dm³ (15°C).

2-Methyl-2,4-pentanediol, hexylene glycol (optically active (+)- and (−)-forms and an optically inactive, racemic (±)-form)

$CH_3C(OH)(CH_3)CH_2CH(OH)CH_3$

Molar mass 188.18 ($C_6H_{14}O_2$), melting point ca. −50°C ((±)), boiling point 198.5°C ((±)), density 0.922 kg/dm³ (20°C (±)), refractive index n_D 1.425 (20°C (±)), flash point about 99°C.

The racemic 2-methyl-2,4-pentanediol is produced by catalytic hydrogenation from diacetone alcohol at low temperatures and high pressures. The boiling range of technical grades is close to that of ethylene glycol and normally about 95% of commercial products distill at 195–199°C. At low temperatures the product is transformed to a semisolid and incrystallizable mass. The pure 2-methyl-2,4-pentanediol is a colorless, slightly sweet-tasting, and hygroscopic liquid that dissolves in water, ethanol, and diethyl ether. It may irritate skin, eyes, and mucous membranes and its ingestion and inhalation at high concentrations may cause narcosis. 2-Methyl-2,4-pentanediol is used as a highboiling solvent in the paint and varnish industries as well as for textile dyeing. It is also suitable, for example, as a humectant and a component in hydraulic brake fluids, printing inks, fuels, and lubricants. In addition, it is used as an intermediate in synthetic chemistry.

2-Ethyl-1,3-hexanediol

$CH_3CH_2CH_2CH(OH)CH(CH_2OH)CH_2CH_3$

Molar mass 146.23 ($C_8H_{18}O_2$), melting point −40°C, boiling point 244°C, density 0.932 kg/dm³ (22°C), refractive index n_D 1.450 (20°C), flash point 127°C, autoignition temperature 360°C.

2-Ethyl-1,3-hexanediol is a colorless, odorless, slightly viscous, and hygroscopic liquid that dissolves in ethanol and diethyl ether and to some extent also in water. It is moderately irritating to eyes and mucous membranes, but not to skin. This diol is used as a vehicle and solvent in printing inks and an insect repellent as well as in cosmetics.

2,2,4-Trimethyl-1,3-pentanediol

$(CH_3)_2CHCH(OH)C(CH_3)_2CH_2OH$

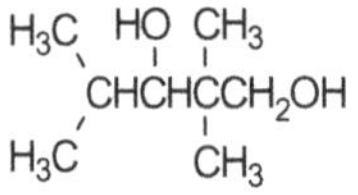

Molar mass 146.23 ($C_8H_{18}O_2$), melting point 51.8–2.2°C, boiling point 234 (737 mmHg) or 81–2 (1 mmHg) °C, density D_{15} 0.937 kg/dm³ (15°C), refractive index n_D 1.451 (15°C), flash point 113°C.

2,2,4-Trimethyl-1,3-pentanediol exists as white and flaky crystals that dissolve in ethanol, diethyl ether, acetone, and hot benzene. It is used like diethylene glycol as a solvent mainly in printing packing inks.

A triol, **2-ethyl-2-hydroxymethyl-1,3-propanediol**, is a colorless solid that is miscible with water and ethanol. It is a widely used block in the polymer industry.

2-Ethyl-2-hydroxymethyl-1,3-propanediol, trimethylol propane, TMP

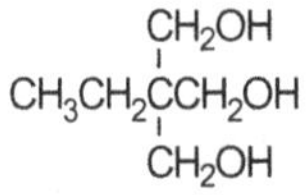

$CH_3CH_2C(CH_2OH)_3$

Molar mass 134.18 ($C_6H_{14}O_3$), melting point 58°C, boiling point 160°C (5 mmHg), density 1.084 kg/dm³, flash point 172°C.

Pentaerythritol, tetramethylol methane, tetrakis(hydroxymethyl) methane, 2,2-bis(hydroxylmethyl)-1,3-propanediol, PETP, THME

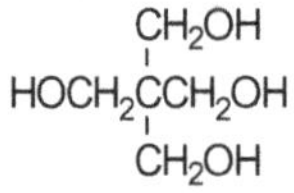

$C(CH_2OH)_4$

Molar mass 136.15 ($C_5H_{12}O_4$), melting point 269°C, boiling point 276°C (30 mmHg), density 1.396 kg/dm³, flash point 200°C.

Pentaerythritol is optically inactive (i.e., it contains no asymmetric carbon atoms) and does not occur in nature. It can be synthesized from acetaldehyde and paraformaldehyde by a two-step mechanism in a calcium hydroxide aqueous suspension, the overall reaction being:

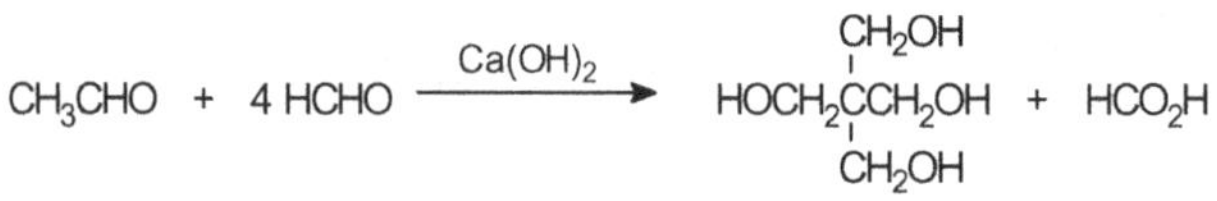

$$CH_3CHO + 4\,HCHO \xrightarrow{Ca(OH)_2} C(CH_2OH)_4 + HCO_2H$$

In this synthesis, dipentaerythritol together with some tri- and even higher polypentaerythritols are formed as by-products. For this reason, the technical product usually contains 85-88% pentaerythritol and 12–15% dipentaerythritol. The pure pentaerythritol is a white and crystalline solid that dissolves in ethanol, ethylene glycol, and only slightly in water (5.6%, 15°C), but not in acetone, diethyl ether or benzene. It is a versatile building block for the preparation of several polyfunctionalized compounds and is mainly used in the manufacture of alkyd-type resins, varnishes, polyvinyl chloride stabilizers, and as an intermediate in the explosives industry (i.e., the production of pentaerythritol tetranitrate).

The following two diols contain four carbon atoms and they are the most significant unsaturated polyalcohols.

2-Butene-1,4-diol

$HOH_2C(H)C{=}C(H)CH_2OH$ — *cis*

$HOH_2C(H)C{=}C(H)CH_2OH$ — *trans*

Molar mass 88.12 ($C_4H_8O_2$), melting point 4 (*cis*) tai 25 (*trans*) °C, boiling point 235 (*cis*) and 131.5 (12 mmHg, *trans*) °C, density 1.070 kg/dm^3 (20°C, *cis* and *trans*), refractive index n_D 1.478 (*cis*) and 1.476 (*trans*) (20°C), flash point 128°C.

2-Butene-1,4-diol can be produced by a reduction of 2-butyne-1,4-diol. It exists as two stereoisomeric forms, but in technical products the most common isomer is the *cis* form (*Z* form). The typical melting and boiling points of these products are 4–7°C and 232–235°C, respectively. The pure isomers are nearly odorless compounds that dissolve readily in water, ethanol, and acetone, but only slightly in benzene. They may irritate skin. In general, 2-butene-1,4-diol is used as an intermediate in synthetic chemistry, for example, in manufacturing alkyd-type resins and stabilizers. It is also used as a fungicide.

2-Butyne-1,4-diol, 1,4-dihydroxy-2-butyne

$HOCH_2C{\equiv}CCH_2OH$ Molar mass 86.09 ($C_4H_6O_2$), melting point 57.5°C, boiling point 238 or 145 (15 mmHg) °C, refractive index n_D 1.480 (20°C).

2-Butyne-1,4-diol can be produced from acetylene and formaldehyde at a high pressure according to the Reppe synthesis. It is a colorless, hygroscopic, and crystalline solid that dissolves in water, ethanol, and acetone, but is insoluble in diethyl ether or benzene. 2-Butyne-1,4-diol is used, for example, as an intermediate in synthetic chemistry, a stabilizer, and a brightener in electroplating.

Diols have only a few commercial halogen derivatives utilized for some specific purposes.

3-Chloro-1,2-propanediol, α-chlorohydrin, glyceryl α-chlorohydrin

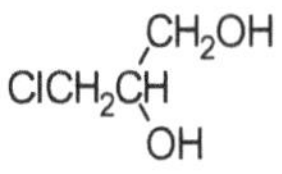

CH_2OH
$ClCH_2CH$
OH

Molar mass 110.54 ($C_3H_7O_2Cl$), boiling point 116°C (11 mmHg), density D_{15} 1.326 kg/dm^3 (18°C), refractive index n_D 1.481 (20°C), flash point 113°C.

3-Chloro-1,2-propanediol is a colorless or pale yellow and hygroscopic liquid that dissolves in water, ethanol, and diethyl ether. It is mainly used as a solvent.

2-Chloro-1,3-propanediol, β-chlorohydrin, glyceryl β-chlorohydrin

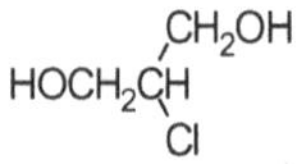

CH_2OH
$HOCH_2CH$
Cl

Molar mass 110.54 ($C_3H_7O_2Cl$), boiling point 146°C (18 mmHg), density 1.322 kg/dm^3 (20°C), refractive index n_D 1.483 (20°C).

2-Chloro-1,3-propanediol (an isomer of α-chlorohydrin) can be synthesized from allyl alcohol and hypochlorous acid (which is gradually formed in aqueous solutions of chlorine):

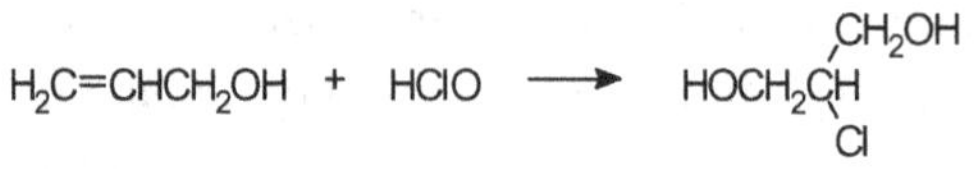

$$H_2C{=}CHCH_2OH + HClO \longrightarrow HOCH_2CH(Cl)CH_2OH$$

It is a liquid that dissolves in water, ethanol, and acetone.

Trichloroacetaldehyde hydrate, chloral hydrate, 2,2,2-trichloro-1, 1-ethanediol, trichloroethylidene glycol

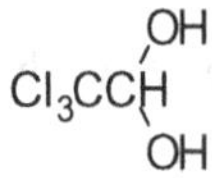

OH
Cl_3CCH
OH

Molar mass 165.40 ($C_2H_3O_2C_3$), melting point 57°C, boiling point 96.3°C (764 mmHg) (decomposes), density 1.908 kg/dm^3 (20°C).

In the exothermic (i.e., with simultaneous evolution of heat) reaction of trichloroacetaldehyde (chloral) with water, a stable diol, trichloroacetaldehyde hydrate or chloral hydrate, is formed:

$$Cl_3CCHO + H_2O \longrightarrow Cl_3CCH(OH)_2$$

The pure product exists as transparent and colorless crystals with an aromatic (slightly bitter) odor and that dissolve readily in water (474 g/100 g H_2O, 17°C) and many common organic solvents. It is a toxic and hypnotic compound which irritates especially eyes, but is also harmful for skin and mucous membranes. Trichloroacetaldehyde hydrate is used as an intermediate in synthetic chemistry (cf., the formerly production of dichlorodiphenyltrichloroethane (DDT)) and a sedative (cf., "knockout drops").

Tribromoacetaldehyde hydrate, bromal hydrate

$Br_3CCH(OH)_2$ Molar mass 298.77 ($C_2H_3O_2Br_3$), melting point 53.5°C, boiling point (decomposes), density 2.566 kg/dm^3 (40°C).

Tribromoacetaldehyde hydrate can be obtained analogously to trichloroacetaldehyde hydrate; however, in this case water is added to tribromoacetaldehyde or bromal. It is a crystalline solid with an odor similar to chloral that dissolves readily in water, ethanol, diethyl ether, and chloroform. The crystals are very hygroscopic and for this reason, tribromoacetaldehyde hydrate should be stored in a tightly closed container.

2,2,3-Trichlorobutanal hydrate, butyl chloral hydrate

$CH_3CHClCCl_2CH(OH)_2$ Molar mass 192.46 ($C_4H_7O_2Cl_3$), melting point 78°C, boiling point (decomposes), density 1.694 kg/dm^3 (20°C).

2,2,3-Trichlorobutanal hydrate is formed in the action of chlorine on paraldehyde. It is a white and crystalline solid that dissolves in ethanol, acetone, and slightly in cold water. 2,2,3-Trichlorobutanal hydrate is used as a medicine (with hypnotic and anticonvulsant effects).

9.2.6. *Cycloalcohols*

Cyclic alcohols are compounds that contain, besides one or more hydroxyl groups, either an aliphatic or aromatic structure of a cyclic hydrocarbon. If a hydroxyl group is attached directly to an aromatic ring, the compound is called a "*phenol*", and these compounds form their own substance group. In contrast, *aliphatic cycloalcohols* are mainly derivatives of cyclohexanol (i.e., a hydroxyl group is attached directly to a cyclohexane ring) or monocyclic and bicyclic terpene alcohols or *terpenoids*. Although these compounds can be named systematically, their nomenclature, especially in the case of terpenoids, is traditionally based on trivial names. It should be noted that the aliphatic cycloalcohols and their simple derivatives shown below are not derivatives of monosaccharides. However, because of the similar properties of aliphatic cycloalcohols and monosaccharides, some examples of cycloalcohols have been included in this chapter.

Inositol, cyclohexane-1,2,3,4,5,6-hexol, cyclohexanehexol, hexahydroxycyclohexane, inosite

Inositol exists, depending on the orientation of hydroxyl groups, in nine possible stereoisomers of which only two (*myo* and *scyllo* forms) are optically inactive. Their nomenclature is simply based on the use of a certain prefix or letter in connection with the parent name "inositol". Four stereoisomers, ***myo*-, D-*chiro*-, L-*chiro*-**, and ***scyllo*-inositols**, are often found in plants and animals representing biologically important compounds.

myo-Inositol is an optically inactive compound which forms white, odorless, and water-soluble (16.3%, 19°C) crystals with a sweet taste. It crystallizes from water (below 50°C) as a dihydrate with a melting point of 224–227°C. The melting point of the corresponding anhydrous form (which loses water of hydration at 100°C) is 215–216°C. This inositol stereoisomer occurs in nature, for example, in citrus fruits and grains as phytic acid (i.e., the hexaphosphate of *myo*-inositol). Its monomethyl ether and two dimethyl ethers are also naturally found. The basic *myo*-inositol is used as a medicine, nutrient, and intermediate in chemical syntheses.

The physical properties of D- and L-*chiro*-inositols are identical; except the former rotates the plane of plane-polarized light to the right and the latter to the left. They are water-soluble (about 40%, 11°C) compounds with a sweet taste and crystallize from water, like the *myo* form, as dihydrates. The monomethyl ether of D-*chiro*-inositol, pinitol or sennitol, is found in small quantities in coniferous trees from which it can be prepared by means of hydrogen iodide. The corresponding monomethyl ether of L-*chiro*-inositol, quebrachitol, occurs in quebracho trees (*Schinopsis* spp) and in the residue obtained after the coagulation of the *Hevea brasiliensis* latex from rubber tapping. *scyllo*-Inositol can be found naturally, for example, in the coconut palm.

myo-Inositol, *cis*-1,2,3,5-*trans*-4,6-cyclohexanehexol, *meso*-inositol, δ-inositol, D,L-inositol

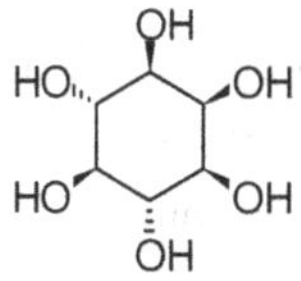

Molar mass 180.16 ($C_6H_{12}O_6$), melting point 224–7°C (dihydrate) or 215–6°C (anhydrous), boiling point 319°C (15 mmHg), density 1.752 kg/dm³ (15°C).

D-*chiro*-Inositol, 1,2,5/3,4,6-inositol, *dextro*-inositol, D-inositol

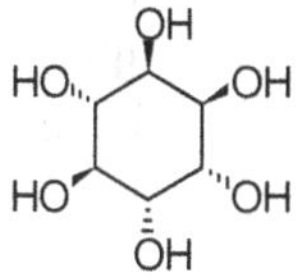

Molar mass 180.16 ($C_6H_{12}O_6$), melting point 247–8°C.

L-*chiro*-Inositol, α-inositol, 1,2,4/3,5,6-inositol, *levo*-inositol, L-inositol

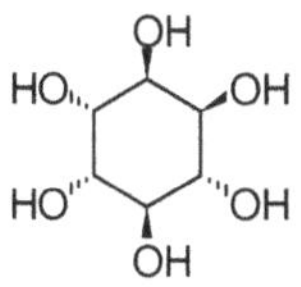

Molar mass 180.16 ($C_6H_{12}O_6$), melting point 247–8°C, density 1.598 kg/dm³ (20°C).

scyllo-Inositol, β-inositol, 1,3,5/2,4,6-hexahydroxycyclohexane, scyllitol, cocositol

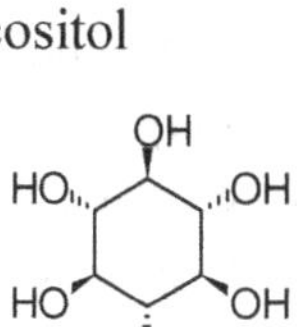

Molar mass 180.16 ($C_6H_{12}O_6$), melting point 349°C.

9.2.7. *Oxidation products — introduction*

The oxidation products of monosaccharides contain, besides an aldehyde group of aldoses and a ketone group of ketoses, other carbonyl group(s). In practice, a primary (terminal) alcohol group $–CH_2OH$ can be oxidized to an aldehyde group –CHO or a carboxylic acid group $–CO_2H$, while a secondary alcohol group >CHOH can be oxidized to a carbonyl group >C=O. The main product groups are *dialdoses*, *diketoses*, *ketoaldoses* (aldoketoses or aldosuloses), *aldonic acids*, *ketoaldonic acids*, *uronic acids*, and *aldaric acids*. In this section, only some examples of commercially important oxidation products are described. However, illustrative structural and naming examples of all these product groups are included.

In *dialdoses*, the carbon atoms C_1 and C_ω of the carbon chain belong to an aldehyde group. Systematic names for individual dialdoses are formed from the systematic stem name of the corresponding aldose, but with the ending *-odialdose* instead of *-ose*, and the appropriate configurational prefix(es). Due to the identical terminal groups –CHO, a choice between the two possible aldose parent names is made as shown in Chapts. "5.1. Fundamental Definitions" and "7.1. Aldoses". Naming examples are L-*threo*-tetrodialdose (I), *galacto*-hexodialdose or *meso-galacto*-hexodialdose (II), and α-D-*gluco*-hexodialdo-1,5-pyranose (III):

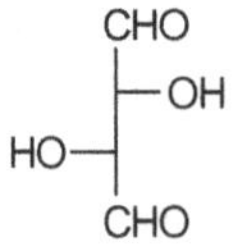

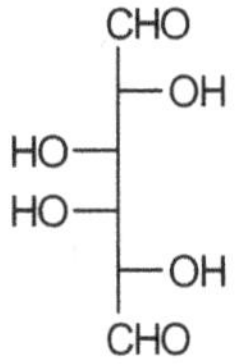

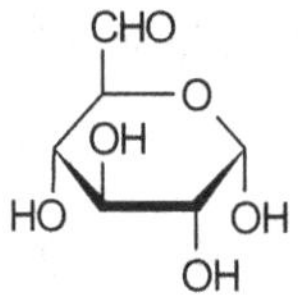

The systematic name of a *diketose* is formed by replacing the terminal *-se* of the stem name by *-diulose*. The locants of the carbonyl groups should be the lowest possible and appear before the characteristic ending. Naming examples are *meso*-pento-2,4-diulose (I), L-*altro*-octo-4,5-diulose (II), and α-D-*threo*-hexo-2,4-diulo-2,5-furanose (III):

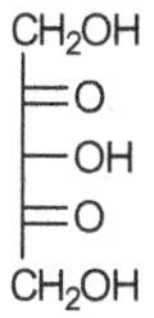

I 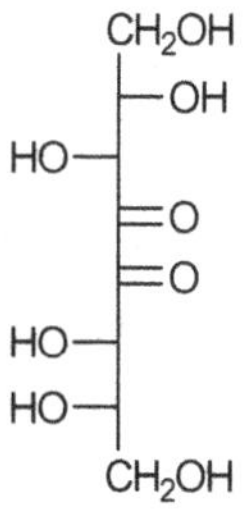 II 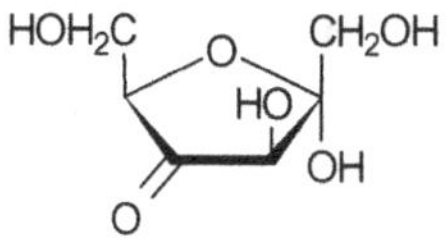 III

Examples of *ketoaldoses* (i.e., the ending *-ulose* instead of *-e*) are D-*arabino*-hexos-3-ulose (I) and methyl α-L-*xylo*-hexos-2-ulo-2,5-furanoside (II):

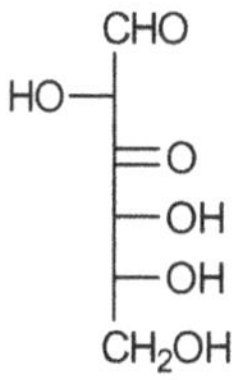

I 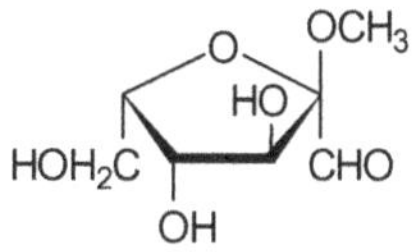 II

9.2.8. *Aldonic acids and lactones*

Aldonic acids are compounds similar, for example, to isosaccharinic acids (cf., "Chapt. 9.2.9. hydroxy monocarboxylic acids and their derivatives") formed from various carbohydrates during kraft pulping (cf., Chapts. "11.2. Polysaccharides" and "12.4.1. Kraft pulping"). They are also aliphatic polyhydroxy monocarboxylic acids. Aldonic acids can be prepared from aldoses by mild oxidants (e.g., aqueous bromine) that convert an aldehyde group of aldoses into a carboxylic acid group. In contrast, stronger oxidants (e.g., nitric acid) convert aldoses into dicarboxylic acids, termed "*aldaric acids*":

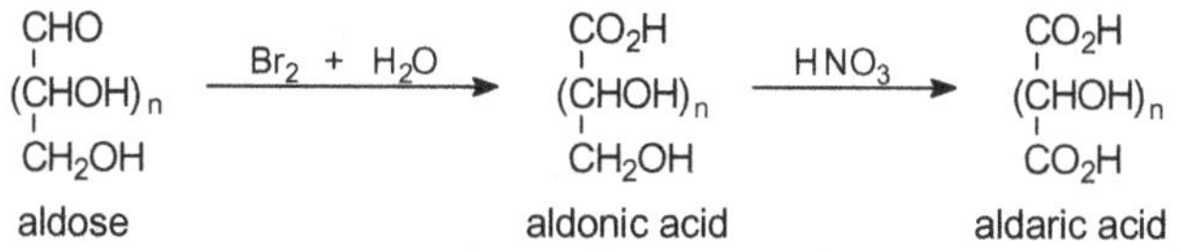

$$\underset{\text{aldose}}{CHO-(CHOH)_n-CH_2OH} \xrightarrow{Br_2\ +\ H_2O} \underset{\text{aldonic acid}}{CO_2H-(CHOH)_n-CH_2OH} \xrightarrow{HNO_3} \underset{\text{aldaric acid}}{CO_2H-(CHOH)_n-CO_2H}$$

where n is generally two to four.

Aldonic acids are usually separated in the form of salts or lactones. Due to several asymmetric carbon atoms, they are also optically active. Aldonic acids are water soluble (pH of their water solutions is about 3) and dissolve only slightly in ethanol. Being carbohydrate-derived compounds, their configurations are shown by the Fischer projection with the prefixes D- and L-. The most significant example of commercial aldonic acids is D-gluconic acid and its 1,4- and 1,5-lactones.

Lactones are chemically cyclic esters formed by intramolecular esterification of the corresponding hydroxy carboxylic acids; i.e., by the elimination of one molecule of water from a hydroxyl and a carboxyl group of an organic acid. They can originate from both aliphatic and aromatic acids and the characteristic structure of the ring is -O-CO-:

$$\overset{4}{HOH_2C}-\overset{3}{CH_2}-\overset{2}{CH_2}-\overset{1}{CO_2H} \xrightarrow{-H_2O}$$

1,4-lactone or γ-lactone

$$\overset{3}{CH}=\overset{2}{CH}-\overset{1}{CO_2H} \xrightarrow{-H_2O}$$

1,5-lactone or δ-lactone

In general, five- or six-membered lactones are thermochemically most stable, but there are also lactones containing smaller and larger rings. In addition, it has been noted that at equilibrium, readily lactonized hydroxy acids exist in aqueous solutions almost entirely as lactones, and their open-chain forms are difficult to produce. Hence, the lactone formation is an essential phenomenon in the chemical behavior and manufacture of polyhydroxy carboxylic acids that cannot be ignored.

Lactones are usually named according to the precursor acid molecule by replacing the ending *-ic acid* by *-olactone* (e.g., butanoic acid or butyric acid → butyrolactone) with the numbers of the carbon atoms that contain the hydroxyl and carboxylic acid groups participating in the ring formation (e.g., glucono-1,4-lactone). The ring size can also be indicated with a Greek letter prefix that specifies the number of carbon atoms in the lactone ring. In this system, the first carbon atom after the one in the carboxylic acid group of the parent compound is labelled α, the second β, etc.; i.e., α-lactone refers to three-membered, β-lactone to four-membered, γ-lactone to five-membered, and δ-lactone to six-membered rings, etc. (e.g., δ-valerolactone and glucono-δ-lactone or glucono-1,5-lactone). Systematic names for the lactones based on aliphatic monohydroxy acids can be formed with the ending *-nolide* instead of *-ic acid* (e.g., 3-propanolide and 5-pentanolide). In addition, general trivial names are often used for certain common heterocyclic compounds, such as coumarin (I), phthalide (II), tetrahydro-2-furanone (III), and tetrahydro-2-pyranone (IV):

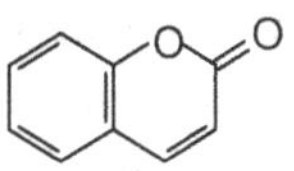

I 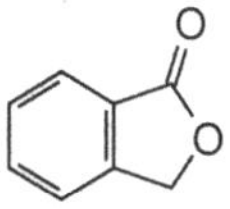 II 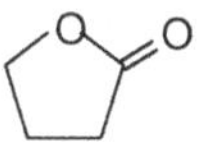 III 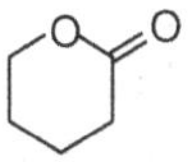 IV

Besides intramolecular esters, α-hydroxy acids undergo during heating bimolecular esterification to "intermolecular substances" with six-membered dilactone rings that are called "*lactides*". Examples are diglycolide or 1, 4-dioxane-2,5-dione (I) and dilactide or 3,6-dimethyl-1,4-dioxane-2,5-dione (II) formed from glycolic acid (hydroxyethanoic acid) and lactic acid (2-hydroxypropanoic acid), respectively:

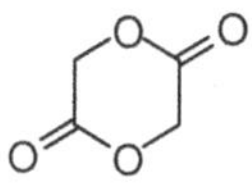

H_3C CH_3

I 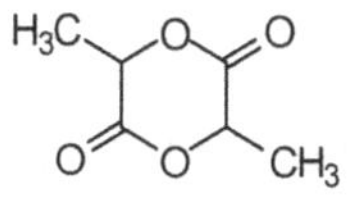 II

In this reaction, linear condensation products of glycolic and lactic acids are also obtained.

D-Gluconic acid, gluconic acid, D-*gluco*-hexonic acid, dextronic acid

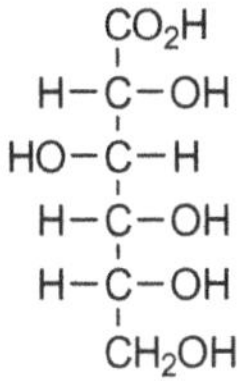

CO_2H
H–C–OH
HO–C–H
H–C–OH
H–C–OH
CH_2OH

Molar mass 196.16 ($C_6H_{12}O_7$), melting point 131°C, pK value 3.9.

Gluconic acid is commonly found in honey, fruit juices, and wine. It can be prepared from glucose by oxidation. Although the oxidation method can be chemical, electrochemical, or biochemical, the industrial production is almost entirely based on an enzymatic process. In this method, the glucose oxidase enzyme produced by the mold *Aspergillus niger* catalyzes the reaction of glucose and oxygen in aqueous nutrient solutions. The formed gluconic acid is neutralized with sodium hydroxide (to almost neutral pH) and the final product, sodium gluconate, is then purified by crystallization (solubility in water 59 g/100 mL). The pure gluconic acid can be produced from its sodium salt by removing the chemically bound sodium with a strong cation exchanger. However, in aqueous solutions gluconic acid exists in equilibrium with the 1,4- and 1,5-lactones, and by evaporation of this aqueous mixture to a concentration of about 90% the latter lactone (i.e., δ-lactone) can be crystallized.

Sodium D-gluconate

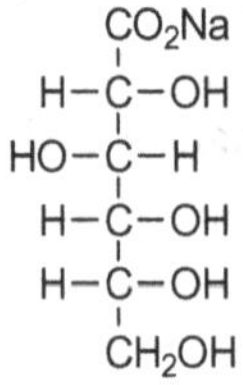

CO_2Na
H–C–OH
HO–C–H
H–C–OH
H–C–OH
CH_2OH

Molar mass 218.14 ($C_6H_{11}O_7Na$).

D-Glucono-1,4-lactone, D-gluconic acid γ-lactone, D-glucono-γ-lactone

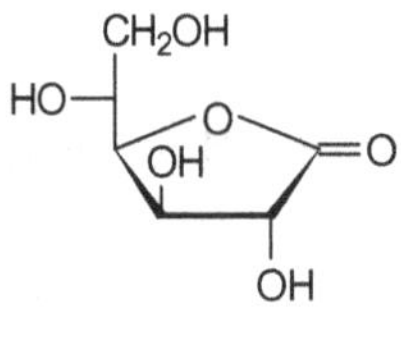

CH_2OH HO O OH =O OH

Molar mass 178.14 ($C_6H_{10}O_6$), melting point 134–6°C.

D-Glucono-1,5-lactone, D-gluconic acid δ-lactone, D-glucono-δ-lactone, GDL

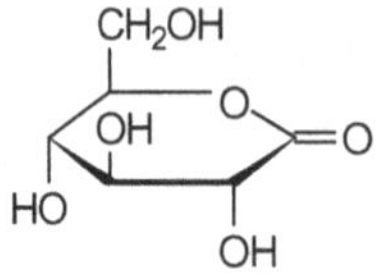

CH_2OH O OH =O HO OH

Molar mass 178.14 ($C_6H_{10}O_6$), melting point 153°C, density 1.610 kg/dm^3 (−5°C).

Sodium gluconate forms odorless white crystals that readily dissolve in water, but only slightly in ethanol. It is widely used, for example, as a chelating agent for cement, a mordant in dyeing, and a cleaning and anticorrosive agent for many purposes. Glucono-1,5-lactone, like the corresponding 1,4-lactone, forms odorless and white needles. It is a food additive mainly used as a sequestrant, an acidifier, or a curing, pickling, or leavening agent. Gluconic acid and its various salts may also be used as chelating agents in pharmaceutical preparations for dosing certain minerals (e.g., containing iron, bismuth, potassium, magnesium, manganese, and calcium), for example, by intramuscular injecttion.

Examples of other salts of D-gluconic acis are the following compounds:

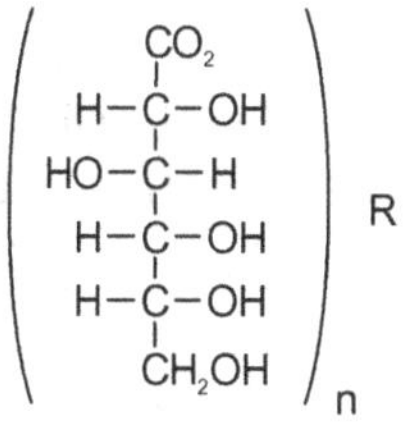

$$\left(\begin{array}{c} CO_2 \\ | \\ H-C-OH \\ | \\ HO-C-H \\ | \\ H-C-OH \\ | \\ H-C-OH \\ | \\ CH_2OH \end{array}\right)_n R$$

Ammonium D-gluconate

$R = NH_4$ $(n = 1)$ Molar mass 213.19 ($C_6H_{15}O_7N$), $[\alpha]_D$ +11.6 (H_2O).

Ammonium D-gluconate is a white powder with a weak odor of ammonia that dissolves in water, but not in ethanol. It is used as an emulsifying agent for cheese and salad dressings.

Calcium D-gluconate monohydrate

$R = Ca \cdot H_2O$ $(n = 2)$ Molar mass 448.40 ($C_{12}H_{24}O_{15}Ca$), melting point 120°C ($-H_2O$).

Calcium D-gluconate is an odorless, practically tasteless, white fluffy power that dissolves slightly in water (3.3 g/100 mL, 15°C), but not in ethanol. It is used as a food additive, a buffer and sequestering agent, and in pharmaceutical tablets.

Barium D-gluconate trihydrate

$R = Ba \cdot 3H_2O$ (n = 2) Molar mass 581.69 ($C_{12}H_{28}O_{17}Ba$), melting point 100°C (–$3H_2O$).

Barium D-gluconate forms flaky prisms that dissolve slightly in water (3.3 g/100 mL, 15.5°C), but not in ethanol.

D-Arabinonic acid, *arabino*-pentonic acid

```
     CO2H
      |
HO – C – H
      |
 H – C – OH
      |
 H – C – OH
      |
     CH2OH
```

Molar mass 166.13 ($C_5H_{10}O_6$), melting point 114–6°C.

The sodium salt of D-arabinonic acid (molar mass 188.11; $C_5H_9O_6Na$) can be prepared from D-glucose by oxidation with oxygen or air in alkaline solutions. In the selective oxidation, an equivalent amount of sodium formate is formed. The free D-arabinonic acid is a crystalline solid that dissolves in water and slightly in ethanol. In aqueous solutions it, like D-gluconic acid, exists in equilibrium with the 1,4- and 1,5-lactones, although in this case, the main component is the 1,4-lactone (i.e., γ-lactone). Sodium D-arabinonate is also able to form complexes with bi- and trivalent metal ions and can be used for the same purposes as sodium D-gluconate, although it is of lesser commercial importance. It is, especially a good stabilizer in iron supplements, since it prevents the reduction of ferric iron (iron(III) or $Fe^{3\oplus}$) to ferrous iron (iron(II) or $Fe^{2\oplus}$) in aqueous solutions. In addition, sodium D-ribonate and further D-ribono-1,4-lactone (a chemical intermediate in the synthesis of riboflavin, vitamin B_2) can be prepared from it by epimerization. The corresponding aldose (D-arabinose) and alditol (D-arabinitol) can also be obtained from this aldonic acid by reduction.

D-Fuconic acid, 6-deoxy-D-galactonic acid

```
     CO2H
      |–OH
HO –|
HO –|
      |–OH
     CH3
```

Molar mass 180.16 ($C_6H_{12}O_6$), melting point 105.5°C, $[\alpha]_D$ –29.1

Examples of other aldonic acids and lactones are:

Compound		Mp./°C
$C_5H_8O_5$		
	L-Arabinono-1,4-lactone	97–9
$C_5H_{10}O_6$		
	L-Arabinonic acid	118–9
	D-Xylonic acid	—
	D-Ribonic acid	112–3
$C_6H_{10}O_6$		
	D-Mannono-1,4-lactone	151
	D-Galactono-1,4-lactone	112
$C_6H_{12}O_7$		
	D-Galactonic acid	122

9.2.9. *Hydroxy monocarboxylic acids and their derivatives*

Hydroxy monocarboxylic acids or *hydroxy acids*, which were earlier called "*oxo-carboxylic acids*" or "*oxo acids*", contain, besides a carboxylic acid group $-CO_2H$, one or more primary $-CH_2OH$, secondary >CHOH or tertiary ≡COH alcohol groups. These carboxylic acid derivatives were among the first organic compounds whose properties were systematically studied and also synthesized. Hydroxy acids exist in nature, for example, as metabolic intermediates in both plants and animals. They are also formed in significant amounts as alkali-catalyzed degradation products of carbohydrates during kraft pulping, where they dissolve in cooking liquor and cause mass loss in pulp (cf., Chapts. "11.2. Polysaccharides" and "12.4.1. Kraft pulping"). Although their chemical structures are similar to those of aldonic acids, hydroxy acids are described here as a separate group, due to their natural appearance and important industrial formation.

The chemical behavior of hydroxy acids is influenced by the location of the hydroxyl group (or groups). This can be often indicated by a Greek letter α, β, γ, and δ, etc., corresponding to the carbon atoms 2, 3, 4, and 5, etc. by counting the carboxyl carbon as C_1. In addition to the systematic and aliphatic carboxylic acid-derived names, hydroxy acids generally have trivial names.

The acidity of α-hydroxy acids (AHAs) and also β-hydroxy acids is enhanced relative to the corresponding unsubstituted acids because of the proximity of the polar hydroxyl group (cf., the inductive effect of -OH). Due to the possibility of forming intermolecular hydrogen bonds, hydroxy acids have exceptionally high melting and boiling points and dissolve more readily in water (but less in diethyl ether) than the corresponding unsubstituted acids with the same chain length. In addition, hydroxy acids do not generally exhibit toxic properties and some of them are widely used in the food industry.

Like aldonic acids containing four and five carbon atoms, γ- and δ-hydroxy acids can form stable lactones by intramolecular esterification with the simultaneous elimination of water, leading to five- and six-membered cyclic structures. However, the formation of four- and six-membered lactones is thermochemically not very favorable and they can be obtained only in special cases. In aqueous solutions, the hydroxy acids that can form stable lactones exist almost entirely as lactones even at room temperature; thus, the open-chained forms are not present. In contrast, in alkaline solutions hydroxy acids exist as their neutral salts, and the formation of lactones is not possible until the acidity of this solution is adjusted to the level where these acids are liberated (cf., pK values).

As mentioned above, a substantial amount of hydroxy monocarboxylic acids (together with formic and acetic acids as well as some hydroxy dicarboxylic acids) is formed during alkaline pulping when cellulosic fibers are chemically liberated from the wood matrix. For example, in the kraft pulping of birch and pine chips about 30% of the feedstock carbohydrates (mostly from hemicelluloses, roughly half of the total wood hemicelluloses degrade) is converted by various degradation reactions into aliphatic carboxylic acids, which dissolve in cooking liquor as their sodium salts (cf., Chapt. "12.4.1. Kraft pulping"). In spite of this huge amount, even a partial recovery of these acids presents a complicated separation problem and has so far not been accomplished on an industrial scale. After cooking, the spent cooking liquor (black liquor) is separated by washing from the pulp and is concentrated to 65–80% solids content in multiple-effect evaporators and then combusted in the recovery furnace for the recovery of cooking chemicals and the generation of energy. The main acid components in black liquor are, besides formic (methanoic acid) and acetic (ethanoic acid) acids, the "low-molar-mass" hydroxy acids (glycolic, lactic, and 2-hydroxybutanoic acids) and the following "high-molar-mass" hydroxy acids, which exist (except 3,4-dideoxy-pentonic acid) after liberation from their sodium salts as 1,4-lactones ((γ-lactones):

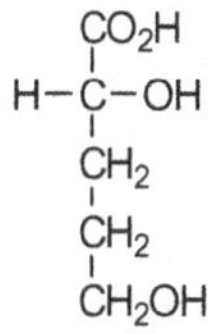

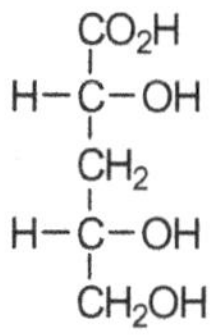

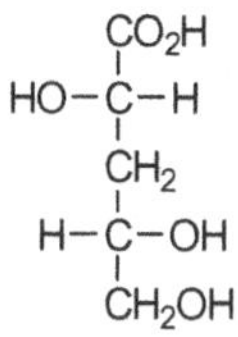

CO_2H–CH(OH)–CH_2–CH_2–CH_2OH

3,4-dideoxy-pentonic acid

CO_2H–C(H)(OH)–CH_2–C(H)(OH)–CH_2OH and CO_2H–C(HO)(H)–CH_2–C(H)(OH)–CH_2OH

3-deoxy-*erythro*- and *threo*-pentonic acid

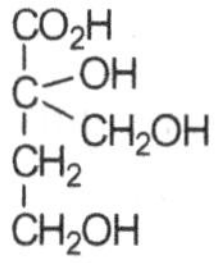

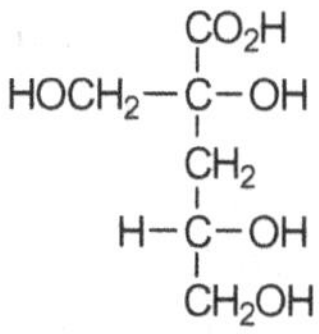

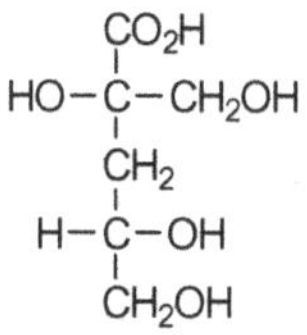

CO_2H–C(OH)(CH_2OH)–CH_2–CH_2OH

xyloisosaccharinic acid

CO_2H–C($HOCH_2$)(OH)–CH_2–C(H)(OH)–CH_2OH and CO_2H–C(HO)(CH_2OH)–CH_2–C(H)(OH)–CH_2OH

α- and β-glucoisosaccharinic acid

These compounds are chiral and hence exhibit optical activity. Due to the structures of feedstock carbohydrates (glucomannan, xylan, and cellulose), they are also "ordinarily" members of the D series.

Glycolic acid, hydroxyethanoic acid, hydroxyacetic acid

$HOCH_2CO_2H$ Molar mass 76.05 ($C_2H_4O_3$), melting point 79–80°C, boiling point 100°C (decomposes), density 1.49 kg/dm^3 (25°C), pK value 3.83 (25°C), flashing point >300°C.

Glycolic acid or hydroxyethanoic acid is the smallest hydroxy acid. It occurs naturally, for example, in sugar beets and unripe grapes. Glycolic acid was first synthesized in 1848 by treating glycine (aminoacetic acid) with nitric acid. Before the 1940s this compound was produced by hydrolysis of monochloroacetic acid or by reduction of oxalic acid (ethanedioic acid). Today, its commercial production is mainly based on an acid catalyzed reaction of formaldehyde (methanal) with carbon monoxide (i.e., carbonylation of formaldehyde) under high pressures at elevated temperatures. It can also be obtained by acid hydrolysis of glycolonitrile, oxidation of ethylene glycol, or by an enzymatic biochemical process (i.e., from fermentable carbon sources). The pure glycolic acid is a colorless, odorless, hygroscopic, and crystalline solid that dissolves in water and ethanol, but only slightly in ethyl acetate. It is commercially available as about 70% water solutions. This α-hydroxy acid is rather unstable during heating and, for example, various condensation products, such as diglycolic acid, diglycolic anhydride, glycyl glycolic acid, diglycolide, and higher polymerization products are formed via the elimination of water during distillation or gradually in water solutions. Recently, especially glycolic acid-derived and biodegradable polymers (poly(glycolic acid) PGA) and co-polymers (poly(lactic acid-glycolic acid) PLGA) are of increasing interest. The monomeric acid is slightly poisonous if ingested and is a strong irritant for skin abrasions. It is used for many purposes including leather dyeing and tanning, textile dyeing, cleaning and polishing of metals, soldering, copper pickling, and electroplating, and is also used in various skin care products (i.e., as an "anti-aging product" to smooth wrinkles, fade hyperpigmentation, and sun damage). In addition, glycolic acid and its derivatives have many special uses.

A glycolate or a hydroxyethanoate is a salt or ester of glycolic acid. Significant examples of such derivatives are as follows:

Sodium glycolate, sodium hydroxyethanoate

$HOCH_2CO_2Na$ Molar mass 98.03 ($C_2H_3O_3Na$).

Zinc glycolate, zinc hydroxyethanoate

$(HOCH_2CO_2)_2Zn$ Molar mass 215.47 ($C_4H_6O_6Zn$).

Methyl glycolate, methyl hydroxyethanoate

$HOCH_2CO_2CH_3$ Molar mass 90.08 ($C_3H_6O_3$), boiling point 151.1°C, density 1.168 kg/dm^3 (18°C).

Ethyl glycolate, ethyl hydroxyethanoate

$HOCH_2CO_2CH_2CH_3$ Molar mass 104.11 ($C_4H_8O_3$), boiling point 160 or 69 (25 mmHg) °C, density 1.083 kg/dm³ (23°C), refractive index n_D 1.418 (20°C).

Lactic acid, 2-hydroxypropanoic acid, 2-hydroxypropionic acid, α-hydroxypropanoic acid, α-lactic acid, milk acid (optically active (+)- and (–)-forms and an optically inactive, racemic (±)-form)

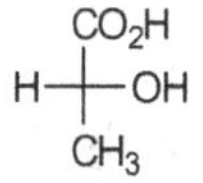

CO_2H / H–C–OH / CH_3 D-(–)- CO_2H / H–C–OH / CH_3 L-(+)-

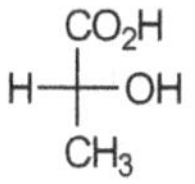

Molar mass 90.08 ($C_3H_6O_3$), melting point 52.8 ((–)) and 18 ((±)) °C, boiling point 103 (2 mmHg, (–)) and 122 (15 mmHg (±)) °C, density 1.206 kg/dm³ (25°C (±)), refractive index n_D 1.439 (20°C (±)), pK value 3.08 (100°C) or 3.86 (25°C).

Lactic acid or 2-hydroxypropanoic acid is the smallest optically active hydroxy acid, and this chiral α-hydroxy acid exists in three different forms: (i) optically active L-(+)-lactic acid (*S*-lactic acid) formed during metabolism and exercise; i.e., glucose is degraded and oxidized to pyruvate or 2-oxopropanoic carboxylate, and L-lactate is then produced from this intermediate via the enzyme lactate dehydrogenase (LDH) faster than the body can process it, thus causing lactate concentrations to rise, (ii) D-(–)-lactic acid (*R*-lactic acid) formed in the lactic acid fermentation, and (iii) optically inactive D,L-lactic acid. Lactic acid occurs naturally and Carl Wilhelm Scheele (1742–86) first refined it in 1780 from sour milk. Jöns Jacob Berzelius (1779–1848) discovered in 1808 that lactic acid is produced in muscles during exertion and in 1873 Johannes Wislicenus (1835–1902) established its structure. Lactic acid is present in many foods both naturally and as a product of in situ microbial fermentation (as in sauerkraut, yogurt, kefir, buttermilk, and sourdough bread). In industry, lactic acid is produced by fermenting feedstocks that contain carbohydrates (e.g., starch, cane and beet sugar, molasses, potatoes, and milk whey) by homolactic organisms, such as *Lactobacillus delbrückii*, *L. amylophilus*, *L. bulcarcius*, *L. leichmanii*, and *L. helveticus*. It has also been conventionally prepared by hydrolysis of lactonitrile, a by-product of the acrylonitrile technology. The commercial production of lactic acid started in 1895. The crude acid product (excess calcium hydroxide/carbonate is added to the fermenters to maintain the pH at five to six and the calcium lactate-containing broth should first be acidified with sulfuric acid to liberate the acid) can be purified via esterification (i.e., it is converted into ethyl lactate) and then by hydrolysis of this ester after its separation by distillation. Another possibility is to use electrodialysis for the separation and liberation of lactic acid from the broth.

The purity of technical grades of lactic acid varies in the range of 20–80%, and they are yellowish or brownish in color and contain mainly a racemic form. The pure lactic acid is a colorless or yellowish, odorless, hygroscopic, and crystalline solid or syrupy liquid that dissolves in water, ethanol, and slightly in diethyl ether. The pure isomers can be prepared from the racemic lactic acid, but they can also be obtained directly from fermentation. Like glycolic acid, this α-hydroxy acid is also rather unstable during heating and, for example, various condensation products, such as dilactic acid, dilactic anhydride, lactyl lactic acid, dilactide, and higher polymerization products are formed via the elimination of water during distillation or gradually in water solutions. The monomeric acid is used, besides of many special purposes, as a food additive (mainly as an acidulant), an active ingredient in pharmaceutical and cosmetic applications, and a mordant in dyeing wool. Technical-grade lactic acid has traditionally been used in the leather tanning industry as an acidulant for deliming hides and in vegetable tannin. It is also a suitable raw material for producing many significant chemicals (e.g., propylene glycol, propylene oxide, and acrylic acid) and its derivatives (e.g., its esters can be used as lacquer solvents and plastic additives). In addition, various biodegradable polymers (e.g., poly(lactic acid) PLA) can be used in packaging and bone surgery.

Salts and esters of lactic acid are called "lactates" or "2-hydroxypropanoates" and are often based on a racemic acid. Significant examples of such derivatives are as follows:

$$(CH_3CH(OH)CO_2)_nR$$

Ammonium lactate, ammonium 2-hydroxypropanoate

$R = NH_4$ $(n = 1)$ Molar mass 107.11 ($C_3H_9O_3N$), density 1.19–1.21 kg/dm^3 (15°C).

Ammonium lactate is a colorless or yellowish, hygroscopic, and syrupy liquid that is miscible with water and ethanol. It is unstable during heating and for this reason, it should be kept cool. Ammonium lactate is used in leather tanning and electroplating.

Sodium lactate, sodium 2-hydroxypropanoate

R = Na (n = 1) Molar mass 112.06 ($C_3H_5O_3Na$), melting point 17°C, boiling point 140°C (decomposes), density 1.313 kg/dm^3 (25°C, 60% water solution), refractive index n_D 1.421 (25°C, 60% water solution).

It is difficult to produce pure sodium lactate, since it is a very hygroscopic solid that readily dissolves in water and ethanol. The purity of commercial technical grades generally varies in the range of 50–60% and they are yellowish or brownish, syrupy liquids. Sodium lactate is used as an effective humectant for many purposes, as a corrosion inhibitor in alcoholic antifreeze formulations, and a food additive (acting as a preservative, acidity regulator, and bulking agent). In addition, it can be given intravenously for preventing or controlling mild to moderate metabolic acidosis in patients with restricted oral intake.

Calcium lactate pentahydrate

R = Ca · 5 H_2O (n = 2) Molar mass 308.30 ($C_6H_{20}O_{11}Ca$), melting point 100–20°C (−3H_2O).

Calcium lactate pentahydrate crystallizes as needles or is a white and nearly odorless powder that dissolves in hot water and in lesser amounts in tepid water (3.1 g/100 mL, 0°C, 5.4 g/100 mL, 15°C, or 7.9 g/100 mL, 30°C), but is insoluble in ethanol and diethyl ether. In medicine it is commonly used as an antacid and a blood clotting agent as well as a treatment in calcium deficiencies. It also has a strong cariostatic effect (i.e., it is added to sugar-free foods as an effective anti-caries agent to prevent tooth decay, see xylitol above). Calcium lactate is also added to fresh-cut fruits to keep them firm and extend their shelf life. In addition, it is used in foods as an ingredient in baking powder and powdered (dried) milk, as well as a gelatinizing agent in jellies.

Aluminum lactate, aluminum 2-hydroxypropanoate

R = Al (n = 3) Molar mass 294.20 ($C_9H_{15}O_9Al$).

Aluminum lactate is a colorless or yellowish powder that dissolves in water. It is used in medicine.

Antimony lactate, antimony 2-hydroxypropanoate

R = Sb (n = 3) Molar mass 388.96 ($C_9H_{15}O_9Sb$).

Antimony lactate is an odorless and tan-colored solid that dissolves in water. It is a highly toxic compound if inhaled, swallowed or absorbed through the skin. This lactate is used as a mordant in fabric dyeing.

Copper(II) lactate dihydrate

R = Cu· 2 H_2O (n = 2) Molar mass 277.71 ($C_6H_{14}O_8Cu$).

Copper lactate dihydrate or cupric lactate forms greenish blue crystals or is a granular powder that dissolves in water (17 g/100 mL, 20°C or 45 g/100 mL, 100°C) and slightly in ethanol. It is used as a source of copper in electroplating.

Iron(II) lactate trihydrate

R = Fe · 3 H_2O (n = 2) Molar mass 287.96 ($C_6H_{16}O_9Fe$), melting point (decomposes).

Iron(II) lactate trihydrate or ferrous lactate is a crystalline, deliquescent solid with a slight peculiar odor. It dissolves in water (2.1 g/100 mL, 10°C or 8.5 g/100 mL, 100°C) and in lesser amounts in ethanol, but not in diethyl ether. Iron(II) lactate may cause irritation and some health symptoms including both acute and delayed nausea if ingested. This lactate is used as a catalyst and in foods as an acidity regulator and a color retention agent, and is also used to fortify foods with iron.

Strontium lactate trihydrate

R = Sr · 3 H_2O (n = 2) Molar mass 319.81 ($C_6H_{16}O_9Sr$), melting point 120°C (–3H_2O).

Strontium lactate trihydrate forms white crystals or is an odorless, granular powder that readily dissolves in water (25 g/100 mL, 20°C or 200 g/100 mL, 100°C) and in lesser amounts in ethanol.

Methyl lactate, methyl 2-hydroxypropanoate, lactic acid methyl ester

R = CH_3 (n = 1) Molar mass 104.12 ($C_4H_8O_3$), melting point −66°C ((±)), boiling point 40 (13 mmHg, (+)) and 144.8 ((±)) °C, density 1.086 (26°C, (+)) and 1.093 (20°C, (±)) kg/dm^3, refractive index n_D 1.414 (20°C (±) and (−)), flashing point 49°C, explosive limits in air 2.2–% (100°C), autoignition temperature 385°C.

Methyl lactate is the most important ester of lactic acid. It can be produced by a direct, acid-catalyzed esterification of lactic acid with methanol (see lactic acid above), although its condensation polymer is a better raw material for this purpose. It can also be synthetized from methyl acrylate by acetylation followed by pyrolysis (550°C). The pure methyl lactate is a clear and colorless liquid with a mild odor. It is miscible with water and dissolves in ethanol and diethyl ether. This ester forms an azeotrope with water containing 25% ester and having a boiling point of 99°C. Methyl lactate is used, for example, as a solvent for cellulose acetate, nitrocellulose, and stains.

Ethyl lactate, ethyl 2-hydroxypropanoate, lactic acid ethyl ester

R = CH_2CH_3 (n = 1) Molar mass 118.13 ($C_5H_{10}O_3$), melting point −26°C ((±)), boiling point 154,5 ((±)) and 58 (20 mmHg (+)) °C, density 1.030 (20°C (±)) and 1.031 (20°C, (−)) kg/dm^3, refractive index n_D 1.412 (20°C (±) and (+)), flashing point 46°C, explosive limits in air 1.5–30% (100°C), autoignition temperature 399°C.

Ethyl lactate can be produced by a direct, acid-catalyzed esterification of lactic acid with ethanol (see lactic acid, above). It is also obtained from petrochemical stocks by combining acetaldehyde (ethanal) with hydrogen cyanide to form acetaldehyde cyanohydrin that is then converted into ethyl lactate by treatment with ethanol and an inorganic acid. About 95% of the commercial product distills within the temperature range of 150–160°C. The pure ethyl lactate is a colorless liquid with a mild odor. It is miscible with water and also readily dissolves in ethanol and diethyl ether. Ethyl lactate is used as a solvent for cellulose acetate, nitrocellulose, and cellulose ethers. Since it easily biodegradable, it is usually considered a "green solvent". Due to its relatively low toxicity, ethyl lactate is also commonly utilized in pharmaceutical preparations, cosmetic formulations, and food additives. In addition, it can be used as a chemical intermediate in the preparation of higher esters of lactic acid by transesterification.

Butyl lactate, butyl 2-hydroxypropanoate, lactic acid butyl ester

$R = CH_2CH_2CH_2CH_3$ (n = 1) Molar mass 146.19 ($C_7H_{14}O_3$), melting point −43°C ((±)), boiling point 187 or 83 (13 mmHg) °C ((±)), density 0.980 kg/dm^3 (32°C (±)), refractive index n_D 1.422 (20°C (±)), flashing point 76°C, autoignition temperature 382°C.

Butyl lactate can be produced from methyl lactate with butanol by transesterification. About 95% of the commercial product distills within the temperature range of 185–200°C. The pure butyl lactate is a colorless liquid with a mild odor. It is miscible with ethanol and diethyl ether and dissolves in lesser amounts in water (4.4 g/100 mL, 25°C). It is used as a solvent, for example, in lacquers, varnishes, inks, and perfumes.

Pentyl lactate, pentyl 2-hydroxypropanoate, amyl lactate, lactic acid amyl ester

$R = CH_2(CH_2)_2CH_2CH_3$ (n = 1) Molar mass 160.22 ($C_8H_{16}O_3$), boiling point 202°C ((±)), density 0.965 kg/dm^3 (20°C (±)), refractive index n_D 1.424 (25°C (±)), flashing point 79°C.

Pentyl lactate can be produced from fusel oil (crude amyl alcohols) and lactic acid. The commercial products are usually yellowish liquids, about 95% of which distill within the temperature range of 190–220°C and mainly contain isomeric pentyl esters of lactic acid. The pure pentyl lactate is a colorless liquid with a brandy-like odor. It dissolves in ethanol and diethyl ether, but is insoluble in water. It is used as an industrial solvent especially for cellulose esters and as a plasticizer.

In addition to all these esters, many higher esters of lactic acid with higher boiling points are known. They are principally used as plasticizers for thermoplastic vinyl resins and cellulose-based plastics.

3-Hydroxypropanoic acid, 3-hydroxypropionic acid, β-hydroxypropanoic acid, β-lactic acid

R = $CH_2CH_2CH_2CH_3$ (n = 1) Molar mass 90.08 ($C_3H_6O_3$), melting point 143°C, boiling point (decomposes), refractive index n_D 1.449 (20°C), pK value 4.5.

3-Hydroxypropanoic is an optically inactive β-hydroxy acid, since it does not contain any asymmetric carbon atoms. It can be prepared catalytically from formaldehyde and lead acetate, but it is also formed from ethylene chlorohydrin or β-bromopropanoic acid by hydrolysis. In addition, 3-hydroxypropanoic acid is obtained from acrylic acid during an alkaline treatment. The product is a hygroscopic syrup that is miscible with diethyl ether and dissolves in water and ethanol. It is not of significant commercial importance and cannot be used directly, for example, as a raw material for condensation polymers, since upon distillation acrylic acid is readily formed from it via the elimination of water, which is typical for β-hydroxy acids. Therefore, it is used for producing various acrylates. In spite of this, a recent method to produce a biodegradable polyester, poly(3-hydroxypropanoic acid) (3-HP) has been developed.

3-Propanolide, β-propanolactone, β-propiolactone, propiolactone, 3-hydroxypropanoic acid lactone

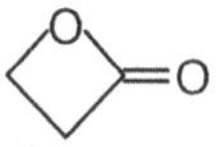

Molar mass 72.06 ($C_3H_4O_2$), melting point −33.4°C, boiling point 155 (decomposes) or 51 (10 mmHg) °C, density 1.146 kg/dm^3 (20°C), refractive index n_D 1.410 (20°C), flashing point 75°C, explosive limits in air 2.9–%.

3-Propanolide can be prepared from ketene and formadehyde:

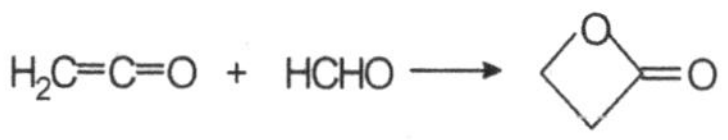

$$H_2C{=}C{=}O + HCHO \longrightarrow$$

It is a colorless liquid with a pungent odor and is chemically stable during storage at 5–10°C. 3-Propanolide is miscible with diethyl ether and dissolves in chloroform, but it decomposes in water and ethanol. It is a strong irritant to skin and is expected to be carcinogenic. This lactone is used as a chemical intermediate and disinfectant.

Glyceric acid, 2,3-dihydroxypropanoic acid (optically active (+)- and (−)-forms and an optically inactive, racemic (±)-form)

$HOCH_2CH(OH)CO_2H$ Molar mass 106.08 ($C_3H_6O_4$), boiling point (decomposes).

Glyceric acid or 2,3-dihydroxypropanoic acid occurs particularly in the form of phosphorylated derivatives in the body as an intermediate in glycolysis. It can be syntheticcally prepared by oxidation of glycerol with nitric acid. It is a syrupy liquid that is miscible with water, ethanol, and acetone, but does not dissolve in diethyl ether. Its salts and esters are known as "glycerates" or "2,3-dihydroxypropanoates". The most significant example is the methyl derivative, **methyl glycerate**.

Methyl glycerate, methyl 2,3-dihydroxypropanoate, glyceric acid methyl ester (optically active (+)- and (−)-forms and an optically inactive, racemic (±)-form)

$HOCH_2CH(OH)CO_2CH_3$ Molar mass 120.12 ($C_4H_8O_4$), boiling point 239–44 or 119–20 (14 mmHg) °C (±)), density (D_{15}) 1.281 kg/dm^3 (15°C (±)), refractive index n_D 1.450 (20°C (±)).

Almost all the normal aliphatic hydroxy monocarboxylic acids occur in nature. Especially, the occurrence and biochemical behavior of hydroxybutanoic acids have been studied. In contrast, the significance of other open-chain monohydroxy monoacids is negligible and they are prepared only for research purposes.

2-Hydroxybutanoic acid, α-hydroxybutanoic acid, 2-hydroxybutyric acid (optically active (+)- and (−)-forms and an optically inactive, racemic (±)-form)

$CH_3CH_2CH(OH)CO_2H$ Molar mass 104.12 ($C_4H_8O_3$), melting point 44.0–4.5°C ((±)), boiling point 260 (decomposes) or 140 (14 mmHg) °C ((±)), density 1.125 kg/dm^3 (20°C (±)).

2-Hydroxybutanoic acid can be prepared by heating α-bromobutanoic acid in an aqueous potassium carbonate solution. In addition, it is formed from trichloropropene oxide via a more complicated reaction route. 2-Hydroxybutanoate, the conjucate base of 2-hydroxybutanoic acid, is produced in mammalian tissues (principally hepatic) that catabolize L-threonine or synthesize glutathione.

3-Hydroxybutanoic acid, β-hydroxybutanoic acid, 3-hydroxybutyric acid (optically active (+)- and (−)-forms and an optically inactive, racemic (±)-form)

$CH_3CH(OH)CH_2CO_2H$ Molar mass 104.12 ($C_4H_8O_3$), melting point 48–50°C ((±)), boiling point 130 (12–4 mmHg) or 94–6 (0.1 mmHg) °C ((±)), refractive index n_D 1.442 (20°C, (±)).

The oxidized and polymeric derivatives of 3-hydroxybutanoic acid occur widely in nature. It can be synthetically prepared by oxidation of β-hydroxybutanal. In addition, it is formed during the reduction of ethyl acetoacetate. In humans, 3-hydroxybutanoic acid is formed in the liver from acetyl-CoA and the catalyst of this biosynthesis is the enzyme β-hydroxybutyrate dehydrogenase.

4-Hydroxybutanoic acid, γ-hydroxybutanoic acid, 4-hydroxybutyric acid, GHB

$HOCH_2CH_2CH_2CO_2H$ Molar mass 104.12 ($C_4H_8O_3$), melting point −17°C, boiling point 178–80°C (decomposes).

Alexander Mikhaylovich Zaytsev (1841–1910) first synthesized 4-hydroxybutanoic acid in 1874. It occurs naturally and is found in the human central nervous system (in small amounts in almost all animals) and can accumulate in certain diseases in the blood. In addition, it is produced as a result of fermentation and for this reason, it exists in small quantities in some beers and wines. 4-Hydroxybutanoic acid is categorized as an illegal drug (it has been labeled as a date rape drug or a predator drug) in many countries and has many "street names" (e.g., liquid ecstasy, liquid E or liquid X). It is a central nervous system depressant used as an intoxicant although it produces a stimulant effect at lower doses. However, when death is associated with this compound, it is usually in conjunction with other drugs, such as alcohol or depressants. 4-Hydroxybutanoic acid has traditionally been used as a sleeping agent and an anesthetic in childbirth, but newer drugs have recently led to a decrease in its legitimate medical use. Nowadays, the common medical applications for it are in the treatment of narcolepsy and more rarely alcoholism. In this case, it is generally used as a sodium salt, **sodium 4-hydroxybutanoate** or sodium oxybate (the trade name "Xyrem") with a molar mass of 126.09 ($C_4H_7O_3Na$), a white, crystalline powder. Other salts include **potassium 4-hydroxybutanoate** with a molar mass of 142.19 ($C_4H_7O_3K$).

4-Butanolide, γ-butanolactone, γ-butyrolactone, butyrolactone, 4-hydroxybutanoic acid lactone, GBL

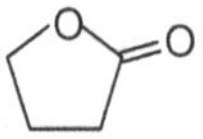

Molar mass 86.09 ($C_4H_6O_2$), melting point −42°C, boiling point 206 or 89 (12 mmHg) °C, density D_0 1.129 kg/dm³ (16°C), refractive index n_D 1.434 (20°C), flashing point 98°C.

4-Butanolide is a naturally occurring component in some wines and similar foodstuffs. It can be prepared from butane-1,4-diol by a catalytic oxidation:

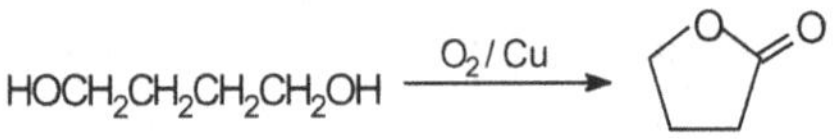

$$HOCH_2CH_2CH_2CH_2OH \xrightarrow{O_2/Cu}$$

or by a high-pressure reaction between acetylene and methanol. 4-Butanolide is a colorless and hygroscopic liquid with a weak characteristic odor. It is miscible with water and dissolves readily in ethanol, diethyl ether, acetone, and benzene. In humans, it is rapidly converted into 4-hydroxybutanoic acid by paraoxonase enzymes, thus acting as a prodrug for this acid. For this reason, it is used as a recreational intoxicant with effects similar to alcohol. In addition, it is used as a common solvent, a cleaning agent, a superglue remover, and a flavoring agent. Many butanoic acid derivatives, such as methionine or 2-amino-4-(methylthio)butanoic acid, and polyvinylpyrrolidone can be produced from it.

4-Pentanolide, γ-valerolactone, 5-methyldihydrofuran-2-one, 4-hydroxypentanoic acid lactone, GVL (optically active (+)- and (−)-forms and an optically inactive, racemic (±)-form)

Molar mass 100.12 ($C_5H_8O_2$), melting point −31°C ((±)), boiling point 206 ((±)), 83–4 (13 mmHg (±)) or 86–90 (14 mmHg (+)) °C, density 1.046 kg/dm³ (25°C (±)), refractive index n_D 1.433 (20°C (±)), flashing point 96°C.

4-Pentanolide can be prepared by isomerization of allyl acetic acid with mild sulfuric acid. It is also readily obtained by hydrogenation of α- and β-angelica lactones that can be formed from levulinic acid. In general, hexosans and hexoses are converted under acidic conditions into 5-(hydroxymethyl)furfural, which via the elimination of formic acid forms levulinic acid. 4-Pentanolide is a colorless liquid that is miscible with water and dissolves in ethanol and acetone. It is a potential liquid fuel and a "green solvent" for many purposes. In addition, because of its herbal odor, it can be used in perfume and flavor industries.

5-Pentanolide, δ-pentanolactone, δ-valerolactone, tetrahydro-2-pyrone, 5-hydroxy-pentanoic acid lactone

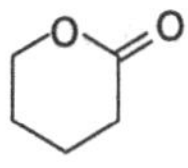

Molar mass 100.12 ($C_5H_8O_2$), melting point −12.5°C, boiling point 218–20 or 113–4 (14 mmHg) °C, density 1.079 kg/dm^3 (20°C), refractive index n_D 1.450 (20°C).

5-Pentanolide is a colorless liquid that is miscible with ethanol and diethyl ether and dissolves in lesser amounts in water. It is used as a chemical intermediate, for example, in the production of polyesters.

D,L-Xyloisosaccharino-1,4-lactone, D,L-3-deoxy-2-*C*-hydroxymethyl-tetrono-1,4-lactone, XISAL

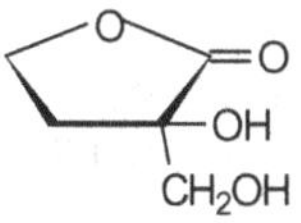

OH
CH_2OH

Molar mass 132.12 ($C_5H_8O_4$), melting point 95.5–6.5°C, boiling point 125°C (0.2 mmHg), pK value 3.0.

The sodium salt of xyloisosaccharinic acid (two isomers) is one of the major alkali-catalyzed degradation products of the hemicellulose component xylan and is formed in significant amounts especially during the kraft pulping of hardwoods (cf., Chapt. "12.4.1. Kraft pulping"). In acidic solutions, this hydroxy acid undergoes lactonization to generate xyloisosaccharino-1,4-lactone.

α-D-Glucoisosaccharino-1,4-lactone, 3-deoxy-2-*C*-hydroxymethyl-D-*erythro*-pentono-1,4-lactone, α-GISAL

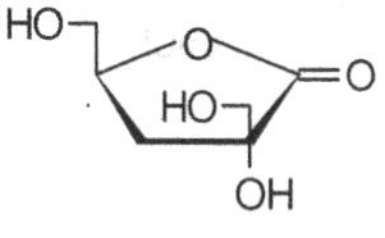

HO
HO
OH

Molar mass 162.14 ($C_6H_{10}O_5$), melting point 94–5°C, boiling point 200°C (0.5 mmHg), pK value 3.6.

The calcium salt of α-D-glucoisosaccharinic acid can be prepared together with its β-isomer by the action of calcium hydroxide on lactose and other similar carbohydrates. However, although the β-form is a more prominent product, the calcium salt of α-form is very crystalline and can be readily crystallized from ethyl acetate. The sodium salts of both acids are also among the major alkali-catalyzed degradation products of cellulose and the hemicellulose component glucomannan. They are formed in significant amounts, especially during the kraft pulping of softwoods (cf., Chapt. "12.4.1. Kraft pulping"). In acidic solutions these hydroxy acids undergo lactonization to generate the corresponding glucoisosaccharino-1,4-lactones.

β-D-Glucoisosaccharino-1,4-lactone, 3-deoxy-2-*C*-hydroxymethyl-D-*threo*-pentono-1,4-lactone, β-GISAL

Molar mass 162.14 ($C_6H_{10}O_5$).

2-Hydroxy-2-phenylacetic acid, phenylhydroxyacetic acid, phenylglycolic acid, mandelic acid, amygdalic acid (optically active (+)- and (–)-forms and an optically inactive, racemic (±)-form)

Molar mass 152.16 ($C_8H_8O_3$), melting point 133–5 ((–)) and 121–3 ((±)) °C, boiling point (decomposes), density 1.300 kg/dm³ (20°C, (±)).

The (–)-form of phenylhydroxyacetic acid or mandelic acid (in German, "Mandel" means "almond") was discovered while heating amygdalin (cf., Chapt. "10.1. Glycosides"), which is an extract of bitter almond, with diluted hydrochloric acid. The racemic mandelic acid can be prepared by first reacting benzaldehyde with sodium bisulfite, next making mandelonitrile with sodium cyanide (i.e., the cyanohydrin of benzaldehyde) from the adduct obtained, and finally converting this nitrile derivative into the end product by acid-catalyzed hydrolysis:

$$C_6H_5CHO \xrightarrow{NaHSO_3} C_6H_5CH(OH)SO_3Na \xrightarrow{NaCN} C_6H_5CH(OH)CN \xrightarrow{H_2O/H^{\oplus}} C_6H_5CH(OH)CO_2H$$

Mandelic acid is an aromatic α-hydroxy acid that forms big, transparent crystals, or a white crystalline solid with a faint odor. It darkens when affected by light. It is soluble in water and polar organic solvents and is toxic if ingested. This compound is used as a useful precursor to various drugs and has been traditionally used in medical applications as an anti-bacterial, particularly in the treatment of urinary tract infections. It has been also used, along with other α-hydroxy acids, as a component of "chemical face peels".

The salts and esters of mandelic acid are called "phenylhydroxy-acetates" or "mandelates".

Calcium phenylhydroxyacetate, calcium mandelate, calcium dimandelate

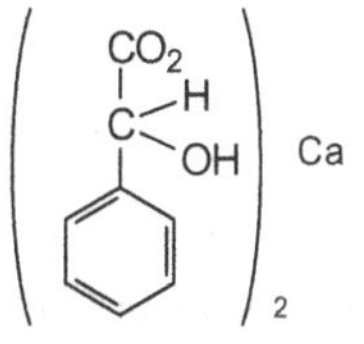

Molar mass 342.41 ($C_{16}H_{14}O_6Ca$).

Calcium mandelate is used as a urinary antiseptic, for example, for the treatment of pyelitis, a fairly common disease that usually can be diag-nosed and cured without great difficulty.

9.2.10. *Hydroxy polycarboxylic acids and their derivatives*

Naturally occurring organic acids include, besides hydroxy mono-carboxylic acids, also *hydroxy dicarboxylic acids*. These dicarboxylic acids are still normally prepared syntheticcally for commercial purposes. Like hydroxy monocarboxylic acids, the diacids may contain one or more alcohol groups and can be chiral with different stereoisomers.

In *aldaric acids*, both terminal carbon atoms of an aldose (carbon atoms C_1 and C_ω) are oxidized to carboxyl groups. Names of individual aldaric acids are formed by replacing the ending *-ose* of the systematic or trivial name of the parent aldose by *-aric acid*. One example of aldaric acids related to carbohydrates is the earlier briefly described tartaric acid (cf., Chapts. "3.5. Racemic Modifications" and "4.2. The Total Number of Stereoisomers", see also below). As examples, the chemical structures of *meso*-xylaric acid (I), 1-methyl D-galactaric acid (1-methyl hydrogen D-galactarate) (II), and D-mannaro-1,4:6,3-dilactone (III) are given:

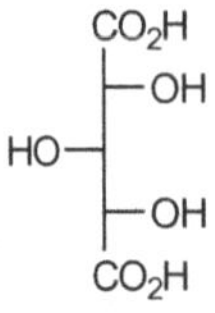

I 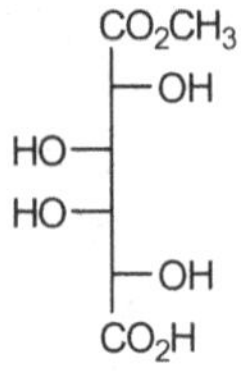 II 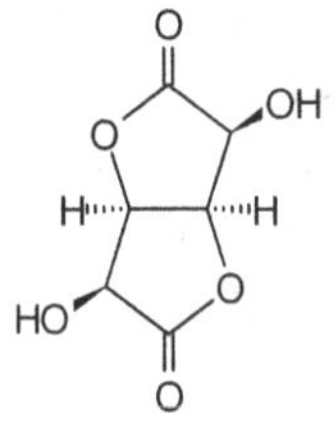 III

Tartronic acid, hydroxypropanedioic acid, 2-hydroxypropanedioic acid, 2-tartronic acid, hydroxymalonic acid, 2-hydroxymalonic acid

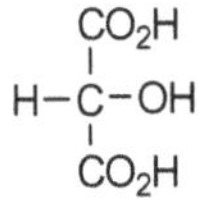

Molar mass 120.06 ($C_3H_4O_5$), melting point 156–8°C (decomposes).

Hydroxypropanedioic acid can be obtained by heating bromopropanedioic acid (bromomalonic acid) with a gold oxide suspension in water or by reduction of mesoxalic acid (ketomalonic acid or oxomalonic acid — molar mass 118.05; $C_3H_2O_5$). It is also formed from malonic acid (propanedioic acid) by ozonation. This dicarboxylic acid is synthesized only for laboratory purposes. The pure hydroxypropanedioic acid is a crystalline solid that dissolves in water and ethanol and slightly in diethyl ether. It decomposes during heating resulting in the formation of CO_2 and polyglycol $(C_2H_2O_2)_n$. Hydroxypropanedioic acid is known as a reactant in the catalytic oxidation with air to form mesoxalic acid. It also forms chelates as well as salts and esters, of which the latter ones are called "tartronates" or "hydroxypropanedioates".

Malic acid, hydroxybutanedioic acid, 1-hydroxy-1,2-ethanedicarboxylic acid, apple acid, hydroxysuccinic acid, deoxy-tetraric acid (optically active (+)- and (−)-forms and an optically inactive, racemic (±)-form)

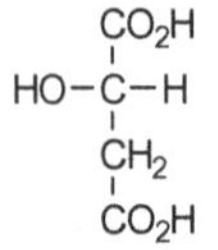

S-(−)- (*l*) *R*-(+)- (*d*)

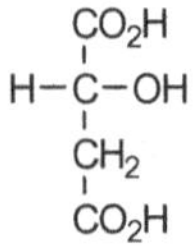

Molar mass 134.09 ($C_4H_6O_5$), melting point 100 ((+)) and 128 ((±)) °C, boiling point 140 ((−), decomposes) and 150 ((±), decomposes) °C, density 1.595 ((−)) and 1.601 ((±)) kg/dm^3 (20°C), pK_1 value 3.40 and pK_2 value 5.11 (25°C).

The *R,S* system (cf., Chapt. "5.2. Relative and Absolute Configuration") is often used to describe the configuration of asymmetric carbon atom C_2 of malic acid. The levorotatory isomer, *S*-(−)-malic acid (L-malic acid), is a natural constituent and common metabolite of animals and plants; the malate anion is an intermediate in the citric acid cycle. It occurs in low concentrations in many fruits including apples, cherries, plums, and watermelons (97, 94, 98, and 100% of total acids, respectively), thus making its isolation from natural sources impractical and expensive. It confers a tart taste to wine and contributes to the sourness of green apples, although its amount decreases with increasing fruit ripeness. Its taste is very clear and pure in rhubarb. Malic acid was first isolated from apple juice by Carl Wilhelm Scheele in 1785. For this reason, in 1787 Antoine Laurent Lavoisier proposed for it the name "apple acid" (in French "acide malique" and in Latin, "malum" means "apple"). Its optical activity changes with dilution; a 34% water solution at 20°C is optically inactive, while dilution results in increasing levo rotation and more concentrated solutions show dextro rotation. In contrast, *R*-(+)-malic acid (D-malic acid) is only a laboratory chemical. Malic acid was first produced synteticcally in 1923, and its commercial production was started in the 1960s. In its synthesis, benzene is first catalytically oxidized to maleic anhydride which is then converted to malic acid by heating (180°C) with steam under pressure (about 1 MPa). The synthetic product, the racemic *R,S*-malic acid (D,L-malic acid or *d,l*-malic acid), is optically inactive. It is a colorless, odorless, hygroscopic, and crystalline solid with a sour taste. It readily dissolves in water (58 g/100 g, 25°C and 80 g/100 g, 75°C) and ethanol (39 g/100 g, 25°C) and in lesser amounts in diethyl ether (1.4 g/100 g, 25°C). Malic acid undergoes many of the characteristic reactions of dicarboxylic acids and monoalcohols. On heating at about 180°C, it decomposes to fumaric acid and maleic anhydride, which sublimes on further heating. When heated, it does not form an anhydride, but two molecules of malic acid form both a linear malomalic acid and a cyclic dilactone (malide). It closely resembles both citric and tartaric acids in its physical and chemical properties and does not exhibit toxic properties. Malic acid is used as a chemical intermediate in the synthesis of its esters, amide derivatives, and salts. It is used primarily as an ingredient (an acidulant) in hard candies and other sweets, jams, jellies, and various canned fruits and vegetables. In addition, malic acid and its derivatives have many other uses as chelating or flavoring agents, as well as for purposes, such as cleaning and polishing metals.

Salts and esters of malic acid are called "malates" or "2-hydroxybutanedioates". Significant examples of such derivatives are:

$$\begin{array}{l} CO_2R \\ | \\ C\!\!<^{H}_{OH} \\ | \\ CH_2 \\ | \\ CO_2R' \end{array}$$

Ammonium hydrogen malate, ammonium hydrogen 2-hydroxybutanedioate

$R = NH_4$, $R' = H$ Molar mass 151.08 ($C_4H_9O_5N$), melting point 161°C, boiling point (decomposes), density 1.15 kg/dm^3.

Ammonium hydrogen malate is a crystalline solid that dissolves in water (32 g/100 mL, 16°C) and in lesser amounts in ethanol.

Dipotassium malate, dipotassium 2-hydroxybutanedioate

$R = R' = K$ Molar mass 210.28 ($C_4H_4O_5K_2$).

Dipotassium malate is a crystalline solid that dissolves in water. It is used as a food additive, mainly as an acidity regulator (acidifier), for example, in canned soups and soft drinks. In addition, it acts as an antioxidant and a food flavor.

Calcium malate, calcium 2-hydroxybutanedioate

$R = R' = ½ Ca$ Molar mass 208.18 ($C_4H_8O_7Ca$, dihydrate) and 226.20 ($C_4H_{10}O_8Ca$, trihydrate).

The hydrates of calcium malate are colorless and crystalline solids or powders that dissolve slightly in water (dihydrate: 0.8 g/100 mL, 0°C or 1.2 g/100 mL, 37.5°C and trihydrate: 0.3 g/100 mL, 0°C or 0.4 g/100 mL, 37.5°C). They are used as food additives and in pharmaceutical products.

Dimethyl malate, dimethyl 2-hydroxybutanedioate, malic acid dimethyl ester

R = R' = CH_3

Molar mass 162.14 ($C_6H_{10}O_5$), boiling point 104–8 (1 mmHg) and 242°C, density 1.223 kg/dm³ (20°C), refractive index n_D 1.442 (20°C).

Diethyl malate, diethyl 2-hydroxybutanedioate, malic acid diethyl ester

R = R' = CH_2CH_3

Molar mass 190.20 ($C_8H_{14}O_5$), boiling point 253°C, density 1.128 kg/dm³ (20°C), refractive index n_D 1.436 (20°C).

Dipropyl malate, dipropyl 2-hydroxybutanedioate, malic acid dipropyl ester

R = R' = $CH_2CH_2CH_3$

Molar mass 218.25 ($C_{10}H_{18}O_5$), melting point 10.5°C, boiling point 151°C (10 mmHg), density 1.075 kg/dm³ (20°C), refractive index n_D 1.438 (20°C).

Tartaric acid, 2,3-dihydroxybutanedioic acid, 2,3-dihydroxysuccinic acid (optically active (+)- and (−)-forms and an optically inactive, racemic (±)-form)

CO_2H HO−C−H H−C−OH CO_2H	CO_2H H−C−OH HO−C−H CO_2H	CO_2H H−C−OH H−C−OH CO_2H
D-(−)- (*l*) *SS*- *S*-*R**, *R**-	L-(+)- (*d*) *RR*- *R*-*R**, *R**-	*meso*- *RS*-

Molar mass 150.09 ($C_4H_6O_6$), melting point 169–70 ((+)) or 205–6 ((±)) °C, density 1.760 ((+)), 1.788 ((±)) or 1.666 (*meso*) kg/dm³ (20°C), refractive index n_D 1.495 ((+)), pK_1 value 3.0 ((+)) or 3.2 (*meso*) or pK_2 value 4.3 ((+)) or 4.8 (*meso*) (25°C), flashing point 210°C, autoignition temperature 428°C.

Tartaric acid or 2,3-dihydroxybutanedioic acid is a dihydroxy dicarboxylic acid with two chiral centers. Consequently, it exists in three stereoisomeric forms; D-(−)-tartaric acid (dextrotartaric acid), L-(+)-tartaric acid (levotartaric acid), and the achiral form (*meso*-tartaric acid or mesotartaric acid). In addition, it has a racemic form (traditionally known as "paratartaric acid" or "racemic acid"), which is typically indicated by the alternative prefixes D,L-, (±)-, *dl*-, *R*,*S*- or 2*R*,3*S*-. The naturally occurring

L-tartaric acid occurs particularly in grapes and bananas mostly as its acid potassium salt (cream of tartar or potassium hydrogen tartrate). In the fermentation of wine (tartaric acid is one of the main acids found in wine), this salt forms deposits (due to racemization, they also contain small amounts of D and D,L forms) in the vats, a precipitation which Carl Wilhelm Scheele first processed in 1769. However, evidence shows that Islamic alchemist Jabir ibn Hayyan (known as Geber, ca. 721–815) was aware of its separation from potassium tartrate in ca. 800. Tartaric acid played also an important role in the discovery of chemical chirality (cf., Chapts. "2.2. The Time After the 1700s" and "5.2. Relative and Absolute Configuration"). Due to its typical formation, L-tartaric acid is obtained from lees, a solid by-product of fermentations, while the commercial D,L-tartaric acid can be prepared catalytically in a multistep oxidative (with H_2O_2) reaction from maleic acid anhydride. The purity of commercial products varies from technical grades to pharmacopoeial ones. Pure tartaric acid forms colorless crystals, or it is a white, crystalline powder with a sour taste that is stable in air. Like its other stereoisomers, L-tartaric acid readily dissolves in water and ethanol (20 g/100 g, 18°C), but only slightly in diethyl ether (0.3 g/100 g, 18°C). Their solubilities (g/100 g) in water are the following:

Temperature/°C	L and D forms	D,L form	*meso* form
20	139	20	125
40	176		
60	218		
80	273		
100	343		

The lower water solubility and higher melting point and density of D,L-tartaric acid than those of L-tartaric acid (or D-tartaric acid) are due to its more compact crystal structure. When heated, both pure enantiomers racemize with great ease to the D,L form that crystallizes from water as a dihydrate. The corresponding anhydrous form can be obtained by heating this dihydrate at 100°C. *Meso*-tartaric acid does not exist naturally, but it is formed from the pure enantiomer via thermal isomerization under alkaline conditions. It crystallizes from water as a monohydrate and loses water of crystallization at 110°C. Tartaric acid is recognized as nontoxic and is poorly absorbed by the intestines. Its metabolism is different from that of citric acid in that tartaric acid is only slightly oxidized. Tartaric acid is used in baking powder (i.e., it reacts with baking soda or sodium bicarbonate to release carbon dioxide), as an antioxidant, or is generally added to foods to give a sour taste. Due to its ability to form complexes, this acid and its salts are utilized for cleaning and polishing of metals. In addition, these compounds are suitable as a mordant and as chemicals in leather tanning, silver mirrors, photography, textile dyeing, and printing inks.

Salts and esters of tartaric acid are known as "tartrates" or "dihydroxybutanedioates". Depending on their structure some of them can be optically active. Significant examples of tartrates are:

$$RO_2CCH(OH)CH(OH)CO_2R'$$

Ammonium hydrogen tartrate, ammonium bitartrate, monoammonium tartrate, ammonium hydrogen dihydroxybutanedioate

$R = NH_4$, $R' = H$ Molar mass 167.12 ($C_4H_9O_6N$), melting point (decomposes), density 1.636 kg/dm^3 ((±)).

Ammonium hydrogen tartrate or acid ammonium tartrate forms white crystals that dissolve in water ((±)-form: 2.3 g/100 mL, 15°C or 3.2 g/100 mL, 25°C) but are insoluble in ethanol. It is used to make baking powder and to detect calcium.

Ammonium tartrate, diammonium tartrate, diammonium dihydroxybutanedioate

$R = R' = NH_4$ Molar mass 184.15 ($C_4H_{12}O_6N_2$), melting point (decomposes), density 1.601 kg/dm^3 ((±)).

Ammonium tartrate forms white crystals that readily dissolve in water ((±)-form: 58 g/100 mL, 15°C or 81 g/100 mL, 60°C) and in lesser amounts in ethanol. It is used for chemical analyses.

Sodium tartrate dihydrate, disodium tartrate, disodium dihydroxybutanedioate, sal tartar

R = R' = Na · 2 H_2O — Molar mass 230.08 ($C_4H_8O_8Na_2$), melting point 150°C (–2H_2O), density 1.818 kg/dm^3 ((+)).

Sodium tartrate dihydrate forms white crystals or granules that readily dissolve in water ((+)-form: 29 g/100 mL, 6°C or 66 g/100 mL, 66°C) but are insoluble in ethanol. It is used as food additive, a stabilizer, and a chemical reagent.

Sodium hydrogen tartrate monohydrate, sodium bitartrate, sodium hydrogen dihydroxybutanedioate

R = H , R' = Na · H_2O — Molar mass 190.09 ($C_4H_7O_7Na$), melting point 100°C (–H_2O), boiling point 234 ((+)) and 219 ((±)) °C (decomposes), $[\alpha]_D$ +21 (1% in H_2O, 20°C).

Sodium hydrogen tartrate monohydrate or acid sodium tartrate forms white crystals or it is a white powder that dissolves in water ((+)-form: 6.7 g/100 mL, 18°C or 9.2 g/100 mL, 30°C and (±)-form: 8.9 g/100 mL, 19°C). It is used as a chemical reagent and in effervescing mixtures.

Potassium tartrate dihydrate, dipotassium tartrate, potassium dihydroxybutanedioate, argol

R = R' = K · 2 H_2O — Molar mass 262.31 ($C_4H_8O_8K_2$), melting point 100°C (–2H_2O), density 1.984 kg/dm^3 ((±)).

Potassium tartrate dihydrate is a crystalline solid that readily dissolves in water ((±)-form: 100 g/100 mL, 25°C). This compound can also crystallize as a hemihydrate (1/2H_2O, molar mass 235.28; $C_4H_5O_{6.5}K_2$). In this case, it loses water of crystallization at 155°C and decomposes at 200–220°C. Its density is 1.98 kg/dm^3 (20°C, (+)) and it readily dissolves in water ((+)-form: 150 g/100 mL, 14°C or 278 g/100 mL, 100°C). Potassium tartrate is used as a food additive, a laboratory reagent, and a medicine (cathartic).

Potassium hydrogen tartrate, potassium bitartrate, potassium hydrogen dihydroxybutanedioate, cream of tartar

R = K , R' = H — Molar mass 188.18 ($C_4H_5O_6K$), density 1.984 (18°C (+)) and 1.954 ((±)) kg/dm^3.

Potassium hydrogen tartrate is formed during the fermentation of wine (see tartaric acid, above). It can be produced from the impure tartar (in German "Weinstein") either by making first potassium sodium tartrate with sodium carbonate or by saturating the aqueous solution of tartar directly with sulfur dioxide. When purified it is an odorless, white, and crystalline powder with a pleasant acidulous taste. It dissolves in water ((+)-form: 0.6 g/100 mL, 20°C or 6.1 g/100 mL, 100°C, (±)-form: 0.4 g/100 mL, 15°C or 7.0 g/100 mL, 100°C, and *meso* form: 16.7 g/100 mL, 25°C) and slightly in ethanol. Potassium hydrogen tartrate is used for many purposes, for example, as a leavening agent in baking powders and an additive in medicines, or for cleaning and polishing metals.

Calcium tartrate tetrahydrate, calcium tartrate, calcium dihydroxybutanedioate

R = R' = ½ Ca · 4 H_2O Molar mass 260.21 ($C_4H_{12}O_{10}Ca$), melting point 200°C (−4H_2O, (±)).

Calcium tartrate was generally prepared by acidification of the impure tartar with hydrochloric acid and by neutralization of the liberated free acid with calcium hydroxide. Nowadays, it is produced from the impure potassium hydrogen tartrate according to the following method:

$$2\,KHC_4H_4O_6 + Ca(OH)_2 + CaSO_4 \longrightarrow 2\,CaC_4H_4O_6 + K_2SO_4 + 2\,H_2O$$

The impure product is a grey or reddish powder. When purified, it forms white crystals that dissolve in dilute acids and slightly in water ((+)-form: 0.03 g/100 mL, 0°C or 0.07 g/100 mL, 37.5°C, (±)-form: 0.003 g/100 mL, 0°C or 0.008 g/100 mL, 37.5°C, and *meso* form: 0.03 g/100 mL, 20°C) and ethanol. Calcium tartrate is used as a food preservative, an antacid, and a chemical intermediate for producing tartaric acid and its salts.

Potassium sodium tartrate tetrahydrate, potassium sodium tartrate, potassium sodium dihydroxybutanedioate, Rochelle salt, Seignette's salt

R = K , R' = Na · 4 H_2O Molar mass 282.23 ($C_4H_{12}O_{10}KNa$), melting point 70–80°C ((+)), boiling point 215°C (−4H_2O, (+)), density 1.790 kg/dm^3 ((+)).

Pierre Seignette (1660–1719) prepared first potassium sodium tartrate in ca. 1675. It can be produced from the impure tartar with sodium carbonate:

$$2\,KHC_4H_4O_6 + Na_2CO_3 \longrightarrow 2\,KNaC_4H_4O_6 + CO_2 + H_2O$$

The purity of commercial products varies from technical grades to laboratory reagents. The pure potassium sodium tartrate forms colorless, transparent, and efflorescent crystals or it is a white powder with a cool, saline taste. It readily dissolves in water ((+)-form: 26 g/100 mL, 0°C or 66 g/100 mL, 100°C) and is insoluble in ethanol. Potassium sodium tartrate is used in baking powders, medicinally as a laxative, and in the process of silvering mirrors. In addition, it is used as an ingredient of Fehling's solution (see below) that is conventionally used for the determination of reducing sugars in solutions.

Antimony potassium tartrate hemihydrate, antimony potassium tartrate, antimony potassium dihydroxybutanedioate, antimonyl potassium tartrate, tartrated antimony, emetic tartar

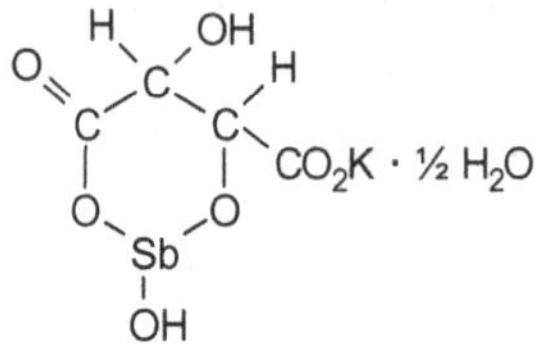

Molar mass 333.93 ($C_4H_5O_{7.5}KSb$), melting point 100°C (–1/2H_2O), density 2.607 kg/dm^3.

Antimony potassium tartrate can be prepared by refluxing an aqueous solution of potassium hydrogen tartrate and antimony trioxide followed by crystallization. It forms transparent, odorless, and efflorescent crystals or it is a white powder with a sweetish, metallic taste. It dissolves in water (8.7 g/100 mL, 25°C or 35.7 g/100 mL, 100°C) and glycerol (6.7 g/100 mL), but is insoluble in ethanol. Antimony potassium tartrate has been known since the Middle Ages as a powerful emetic and was also used later in the treatment of schistosomiasis and leishmaniasis. It is chemically used as a textile and leather mordant and an insecticide.

Fehling's solution was developed in 1849 by Hermann von Fehling (1812–85) for the detection of reducing sugars in solutions. It is prepared by combining immediately prior to the determination (i.e., the compound to be tested is added to Fehling's solution, and the mixture is heated) equal volumes of two separate solutions (I and II). The colorless solution I contains 173 g potassium sodium tartrate tetrahydrate (Rochelle salt, see above) and 5 g sodium hydroxide in 500 mL water. Solution II is an aqueous solution (500 mL) of copper(II) sulfate (34.7 g), and its color is deep blue. The blue active ingredient in Fehling's solution is the bis(tartrate) complex of $Cu^{2\oplus}$. In the reaction, aldehydes (hemiacetals) reduce the cuprate(II) salt complex to red copper(I) oxide (cuprous oxide, Cu_2O) and are oxidized to carboxylic acids (aldonic acids). In general, only aldehydes (aldoses) are oxidized giving a positive result, but ketones (ketoses) do not react, unless they are α-hydroxy ketones. A similar phenomenon can also be observed with the traditional Tollen's reagent consisting of a solution of silver nitrate and ammonia. In this case, a positive result is indicated by the precipitation of elemental silver, often producing a characteristic "silver mirror" on the reaction vessel surface. In addition, there are many other methods available for this purpose including, for example, the use of dinitrosalicylic acid (DNS) as a reagent.

The neutral esters of tartaric acid with low-molar-mass alcohols are liquids or easily melting solids. These esters are primarily used as solvents and plasticizers. Like salt derivatives, some of them can be optically active. The most significant examples of ester derivatives are as follows:

$$RO_2CCH(OH)CH(OH)CO_2R'$$

Diethyl tartrate, diethyl dihydroxybutanedioate, diethyl ester of tartaric acid

$R = R' = CH_2CH_3$ Molar mass 206.20 ($C_8H_{14}O_6$), melting point 18.7°C ((+) and (±)), boiling point 280 ((+) and (±)), 142 (8 mmHg, (+)), and 158 (18 mmHg (±)) °C, density 1.204 ((+)) and 1.205 ((±)) kg/dm³ (20°C), refractive index n_D 1.447 ((+)) and 1.444 ((±)) (20°C), flashing point 93°C.

Diethyl tartrate is a colorless, thick, and oily liquid that dissolves in water, alcohol, diethyl ether, and acetone. It is used as a plasticizer for automobile lacquers and a solvent for nitrocellulose, gums, and resins.

Dimethyl tartrate, dimethyl dihydroxybutanedioate, dimethyl ester of tartaric acid

R = R' = CH_3 — Molar mass 178.4 ($C_6H_{10}O_6$), melting point 48 ((+)), 90 ((±)), and 114 (*meso*) °C, boiling point 280 or 166 (12 mmHg (+)), 282 or 169 (20 mmHg), and 98 (0.01 mmHg, *meso*) °C, density 1.306 (45°C, (+)) and 1.260 (90°C, (±)) kg/dm^3.

Dimethyl tartrate is a solid compound that readily dissolves in ethanol and chloroform, but in lesser amounts in water, diethyl ether, and acetone.

Dibutyl tartrate, dibutyl dihydroxybutanedioate, dibutyl ester of tartaric acid

R = R' = $CH_2CH_2CH_2CH_3$ — Molar mass 262.31 ($C_{12}H_{22}O_6$), melting point 22.0–2.5 ((+)) and 21 ((±)) °C, boiling point 320 or 178 (12 mmHg, (+)) and 312 or 185 (12 mmHg (±)) °C, density 1.091 (20°C (+)) and 1.086 (25°C (±)) kg/dm^3, refractive index n_D 1.445 ((+)) and 1.446 ((±)) (20°C), flashing point 90°C, autoignition temperature 284°C.

Dibutyl tartrate is a light tan and almost odorless liquid that dissolves in water, ethanol, and acetone. It is used as a solvent and plasticizer, for example, for cellulose esters and ethers, lacquers, and transfer inks.

Dipentyl tartrate, dipentyl dihydroxybutanedioate, diamyl tartrate, dipentyl ester of tartaric acid

R = R' = $CH_2(CH_2)_2CH_2CH_3$ — Molar mass 290.36 ($C_{14}H_{26}O_6$), boiling point ca. 400°C, density 1.040–1.048 kg/dm^3 (15°C), refractive index n_D 1.45 (20°C).

D-Glucaric acid, 2*R*,3*S*,4*S*,5*S*-2,3,4,5-tetrahydroxyhexanedioic acid, 2,3,4,5-tetrahydroxyadipic acid, saccharic acid

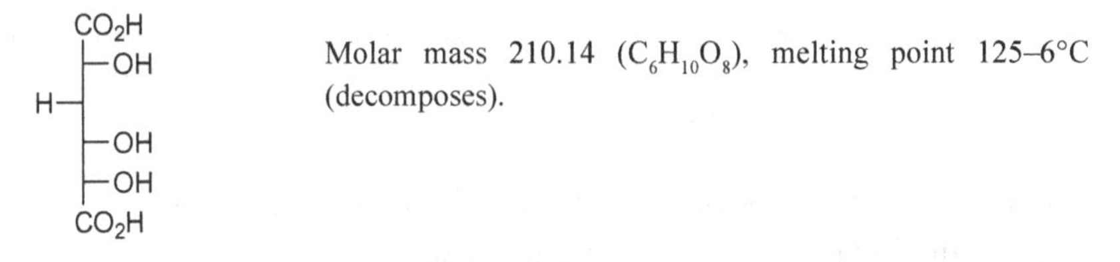

Molar mass 210.14 ($C_6H_{10}O_8$), melting point 125–6°C (decomposes).

D-Glucaric acid can be prepared from D-glucose, sucrose, or starch by oxidation with nitric acid. It forms white, hygroscopic needles or is a syrupy liquid that readily dissolves in water and ethanol, but only slightly in diethyl ether. Its specific rotation $[\alpha]_D$ in water at 19°C is +20.6.

Galactaric acid, *meso*-galactaric acid, 2*R*,3*S*,4*R*,5*S*-2,3,4,5-tetrahydroxy-hexanedioic acid, mucic acid

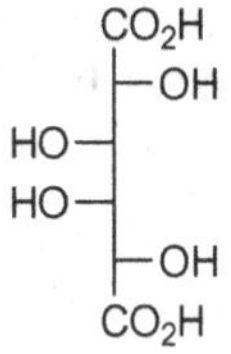

CO_2H / –OH / HO– / HO– / –OH / CO_2H

Molar mass 210.14 ($C_6H_{10}O_8$), melting point about 210°C (decomposes).

Galactaric acid can be prepared from D-galactose or lactose by oxidation with nitric acid. It is a white, crystalline powder that sparsely dissolves in cold water (1 g/300 mL) and hot water (1 g/60 mL), but is insoluble in ethanol and diethyl ether. Galactaric acid is used to replace tartaric acid in some applications, as a sequesterant for metal ions (calcium and iron), an intermediate chemical for the synthesis of heterocyclic compounds (pyrroles), and for retarding hardening of concrete.

D-Mannaric acid, 2*S*,3*S*,4*S*,5*S*-2,3,4,5-tetrahydroxyhexanedioic acid

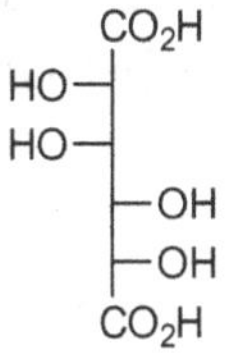

CO_2H / HO– / HO– / –OH / –OH / CO_2H

Molar mass 210.14 ($C_6H_{10}O_8$), melting point 128.5°C.

Citric acid, 2-hydroxypropane-1,2,3-tricarboxylic acid, 3-carboxy-3-hydroxypentane-dioic acid

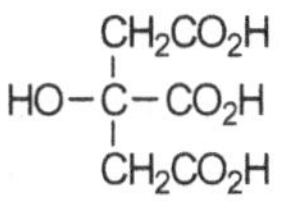

CH_2CO_2H / HO–C–CO_2H / CH_2CO_2H

Molar mass 192.14 ($C_6H_8O_7$), melting point 153°C, boiling point (decomposes) °C, density 1.665 kg/dm³ (18°C), pK_1 value 3.1, pK_2 value 4.8, and pK_3 value 6.4 (25°C).

Citric acid or 2-hydroxypropane-1,2,3-tricarboxylic acid is widely found in plant and animal cells. It exists in a variety of fruits and vegetables, most notable in citrus fruits; 4.0–8.0% of the dry solids (citron), 1.2–2.1% (grapefruit), and 0.6–1.0% (orange). Several berries (red currant 0.7–1.3%, black currant 1.5–3.0%, raspberry 1.0–1.3%, gooseberry 1.0%, strawberry 0.6–0.8%, and apple 0.008%), vegetables (potato 0.3–0.5%, asparagus 0.08–0.2%, cabbage 0.016%, and eggplant 0.01%), as well as many flowers and plants contain this compound. It is also an essential compound in animal cells and body fluids; for example, 15 ppm (blood), 25 ppm (blood plasma), 10 ppm (red blood cells), 500–1250 ppm (milk), 100–750 ppm (urine), 20 ppm (kidneys), and 7500 ppm (bones). In addition, citric acid is one of a series of compounds involved in the physiological oxidation of carbohydrates, fats, and proteins to carbon dioxide and water with the simultaneous liberation of energy for vital activities (cf., the Krebs cycle, Chapt. "8.1. Photosynthesis"). Citric acid was first crystallized from lemon juice in 1784 by Carl Wilhelm Scheele although the knowledge of the acidic nature of lemon and lime juices was already recorded in ca. 1200 in one famous encyclopedia, the Great Mirror (in Latin "Speculum Maius") written by the alchemist Vincent of Beauvais (Vincentius Bellovacensis or Vincentius Burgundus ca. 1190–1264). However, the Islamic alchemist Jabir Ibn Hayyan discovered the presence of citric acid in citrus fruits as early as in the 8th century. Citric acid has been traditionally isolated from various fruit juices. In 1893, Carl Wehmer (1858–1935) reported that *Penicillum glaucum* could produce it from sugars as a by-product of the metabolism of this mold. In 1917, James N. Currie discovered that certain strains of the mold *Aspergillus niger* could be effective citric acid producers. In the present day, most citric acid is still produced by fungal (*A. niger*) fermentation of different carbohydrates-containing juices. Moreover, many other microorganisms, such as fungi and bacteria, can produce this compound.

In the present day, the product is precipitated from the broth solution as a sparingly soluble calcium salt. Citric acid can be also synthetically prepared from rather common chemical intermediates, including the derivatives of maleic and fumaric acids, acetone, and ketene, although this production is not of importance. The purity of commercial products varies from technical grades to pharmacopoeial ones. The pure anhydrous citric acid forms colorless crystals that dissolve in water (69 g/100 mL, 40°C, 73 g/100 mL, 60°C or 84 g/100 mL, 100°C) and ethanol (38 g/100 g, 25°C), and in lesser amounts in diethyl ether (1.0 g/100 g, 25°C) and pentyl acetate (4.2 g/100 g, 25°C). It is insoluble in chloroform, tetrachloromethane, benzene, toluene, and carbon disulfide. Citric acid is nontoxic and easily oxidized in the human body. Because of its palatability, low toxicity, high water solubility, as well as its buffering and chelating abilities, it is used in food, biotechnical, pharmaceutical, and chemical industries. It is used, for example, as an acidifier, a dispersing agent, an antioxidant, a sequestering agent, a water-conditioning agent, and a detergent builder. It is also used as a mordant, in removal of sulfur dioxide from smelter waste gases, and cleaning and polishing stainless steel and other metals. In addition, many commercial salts as well as esters suitable for plasticizers are produced from this tricarboxylic acid.

2-Hydroxypropane-1,2,3-tricarboxylic acid monohydrate. Molar mass 210.14 ($C_6H_{10}O_8$), density 1.542 kg/dm^3 (18°C), refractive index n_D 1.498.

The anhydrous citric acid crystallizes from hot water, while the monohydrate forms when citric acid is crystallized from colder (<37°C) water. The monohydrate forms crystals that are stable in air. It can be converted into the anhydrous form by slow heating above 70–75°C and it melts in the temperature range of 135–152°C. When heated rapidly, the monohydrate first melts at 100°C and then solidifies to the anhydrous form with a melting point of 153°C. Like the pure anhydrous citric acid, the monohydrate dissolves in water (54 g/100 mL, 10°C or 64 g/100 mL, 30°C) and ethanol (50 g/100 g, 25°C), and in lesser amounts in diethyl ether (2.2 g/100 g, 25°C), ethyl acetate (5.3 g/100 g, 25°C), pentyl acetate (6.0 g/100 g, 25°C), and chloroform (0.007 g/100 g, 25°C).

Salts and esters of citric acid are known as "citrates" or "2-hydroxypropane-1,2,3-tricarboxylates". Significant examples of citrates are as follows:

$$\begin{array}{c} CH_2CO_2R \\ | \\ HO-C-CO_2R' \\ | \\ CH_2CO_2R'' \end{array}$$

Triammonium citrate, ammonium citrate, *tert*-ammonium citrate

R = R' = R'' = NH_4 Molar mass 243.22 ($C_6H_{17}O_7N_3$), melting point (decomposes).

Triammonium citrate is a white and deliquescent powder or a granular solid that readily dissolves in water and ethanol. It is used as a buffering agent, a mordant in cotton fabric dyeing, an anticorrosive agent, a chelating agent, and an analytical reagent.

Diammonium citrate, *sec*-ammonium citrate

R = R'' = NH_4 , R' = H   Molar mass 226.19 ($C_6H_{14}O_7N_2$), melting point (decomposes), density 1.48 kg/dm^3.

Diammonium citrate is a white powder or a granular solid that readily dissolves in water and ethanol. It is used for similar purposes as triammonium citrate. Especially, it is used for determining phosphorus content in fertilizers and for cleaning and polishing metals (particularly the removal of aluminum oxide). In addition, like triammonium citrate, it is an effective chelate for iron.

Trilithium citrate tetrahydrate, lithium citrate

R = R' = R'' = Li · 4 H_2O Molar mass 281.98 ($C_6H_{13}O_{11}Li_3$), melting point 105°C (−4H_2O).

Trilithium citrate forms colorless needles or it is a white powder that dissolves readily in water (74.5 g/100 mL, 25°C or 66.7 g/100 mL, 100°C) and in lesser amounts in ethanol and diethyl ether. It is used in carbonated beverages and pharmaceutical products and as a mood stabilizer.

Trisodium citrate, sodium citrate

R = R' = R'' = Na   Molar mass 258.07 ($C_6H_5O_7Na_3$).

The anhydrous trisodium citrate is a white and granular powder that dissolves readily in water (57 g/100 mL, 25°C).

Trisodium citrate dihydrate. Molar mass 294.10 ($C_6H_9O_9Na_3$), melting point 150°C (−2H_2O), boiling point (decomposes).

The dihydrate of trisodium citrate forms white and odorless crystals or it is a powder that dissolves readily in water (72 g/100 mL, 25°C or 167 g/100 mL, 100°C) and in lesser amounts in ethanol.

Trisodium citrate pentahydrate. Molar mass 348.15 ($C_6H_{15}O_{12}Na_3$), melting point 150°C (−5H_2O), boiling point (decomposes), density 1.857 kg/dm^3 (23.5°C).

The pentahydrate of trisodium citrate is a crystalline solid that readily dissolves in water (93 g/100 mL, 25°C or 250 g/100 mL, 100°C) and in lesser amounts in ethanol.

Sodium citrate possesses a saline, mildly tart taste and is chemically stable in air. It is the most significant salt of citric acid that is widely used as a food additive, usually for flavor or as a preservative or an acidity regulator (acidifier). It can also perform as a buffering agent in many applications and in medical uses as an anticoagulant in blood transfusions or an antacid, especially prior to anaesthesia for caesarian section procedures. In addition, it is an effective agent for removal of carbonate scale from boilers and for cleaning automobile radiators.

Tripotassium citrate monohydrate, potassium citrate

R = R' = R" = K · H_2O Molar mass 324.41 ($C_6H_7O_8K_3$), melting point 230°C (decomposes).

The monohydrate of tripotassium citrate is a white, hygroscopic, and crystalline solid that readily dissolves in water (167 g/100 mL, 15°C or 199 g/100 mL, 31°C) and in lesser amounts in ethanol. It is used as a food additive and in medicines and pharmaceutical products as a stabilizer and a buffering agent. Medicinally, it is also widely used to treat urinary calculi (kidney stones) and as an alkalizing agent in the treatment of mild urinary tract infections such as cystitis. The molar mass of the anhydrous form is 306.40 ($C_6H_5O_7K_3$).

Triethyl citrate, ethyl citrate, triethyl ester of citric acid

R = R' = R" = CH_2CH_3 Molar mass 276.29 ($C_{12}H_{20}O_7$), boiling point 294 or 185 (17 mmHg) °C, density 1.137 kg/dm^3 (20°C), refractive index n_D 1.445 (20°C), flashing point 150°C.

Triethyl citrate is a colorless, odorless, and oily liquid with a bitter taste. It is insoluble in water, but dissolves in ethanol and diethyl ether. It is used as a food additive and a solvent and a plasticizer for nitrocellulose, natural resins, polyvinyl chloride, and similar plastics. In addition, it is used in pharmaceutical coatings and plastics as well as in paint removers.

Tributyl citrate, butyl citrate, tributyl ester of citric acid

R = R' = R" = $CH_2CH_2CH_2CH_3$

Molar mass 360.44 ($C_{18}H_{32}O_7$), melting point −20°C, boiling point 232.5 (22.5 mmHg) or 169–70 (1 mmHg) °C, density D_{25} 1.042 kg/dm^3 (25°C), refractive index n_D 1.445 (20°C), flashing point 157°C, autoignition temperature 368°C.

Tributyl citrate is a colorless, odorless, pale yellow, and non-volatile liquid that is insoluble in water. It is used as a plasticizer, an antifoam agent, and a solvent for nitrocellulose.

Trimethyl citrate, methyl citrate, trimethyl ester of citric acid

R = R' = R" = CH_3

Molar mass 234.21 ($C_9H_{14}O_7$), melting point 78.5–9.0°C, boiling point 287 (decomposes) or 176 (16 mmHg) °C.

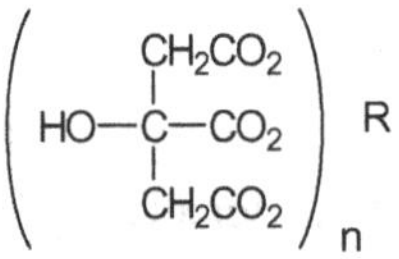

$$\left(HO-\underset{CH_2CO_2}{\overset{CH_2CO_2}{\overset{|}{\underset{|}{C}}}}-CO_2 \right)_n R$$

Calcium citrate tetrahydrate, calcium citrate, tricalcium citrate, lime citrate

R = 3 Ca · 4 H_2O (n = 2)

Molar mass 570.51 ($C_{12}H_{18}O_{18}Ca_3$), melting point 100–20°C (−4H_2O).

Calcium citrate is a chemical intermediate in the separation of citric acid from the industrial fermentation process where the citric acid in the broth solution is neutralized by calcium hydroxide. The pure product is a white, odorless powder or a crystalline solid that only slightly dissolves in water (0.085 g/100 mL, 18°C) and ethanol (0.0065 g/100 mL, 18°C). It loses most water of crystallization at 100°C and at 120°C the anhydrous form can be obtained. Calcium citrate is used as a food additive, a calcium supplement, and a buffering agent.

Aluminum citrate

R = Al (n = 1)

Molar mass 216.08 ($C_6H_5O_7Al$).

Aluminum citrate is a white powder that usually contains about 10% water and readily dissolves in hot water. This salt of citric acid is used in medicines and as a mordant in textile dyeing.

Iron(III) citrate pentahydrate, iron(III) citrate, ferric citrate

R = Fe · 5 H_2O (n = 1) Molar mass 335.03 ($C_6H_{15}O_{12}Fe$).

Iron(III) citrate forms reddish-brown scales that dissolve in water but are insoluble in ethanol. It is a light sensitive compound and is used, like a double salt ammonium ferric citrate, in blueprint papers. It has also some medical uses as an iron supplement.

Lead(II) citrate trihydrate, lead(II) citrate

R = Pb · 3 H_2O (n = 2) Molar mass 1053.82 ($C_{12}H_{16}O_{17}Pb_3$).

Lead(II) citrate forms white and water-soluble crystals. It is toxic by ingestion and inhalation. Its high concentrations may cause acute lead poisoning.

9.2.11. *Ketoaldonic acids*

The structures of *ketoaldonic acids* contain, besides a carboxylic acid group, a ketone group. Names of individual compounds are formed by replacing the ending *-ulose* of the corresponding ketose by *-ulosonic acid*, preceded by the locant of the ketone group. The anion takes the ending *-ulosonate*. The carbon atom C_1 belongs to a carboxylic acid group. Examples are D-*erythro*-pent-2-ulosonic acid (I), α-D-*arabino*-hex-2-ulopyranosonic acid (II), and ethyl (α-D-*arabino*-hex-2-ulopyranosid)onate (III):

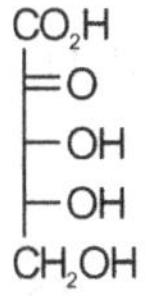

I 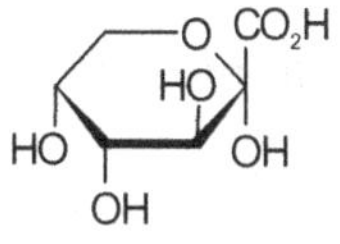 II 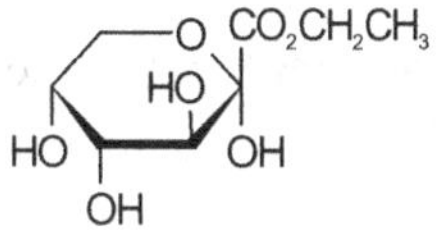 III

9.2.12. *Uronic acids*

Uronic acids are monocarboxylic acids derived by oxidation of the terminal -CH_2OH group (carbon atom C_ω) of an aldose to a carboxylic acid group. Since an aldehyde group of aldoses can be readily oxidized to a carboxylic acid group even by mild oxidants, exclusive oxidation of the primary alcohol group can be accomplished only with the use of suitable protecting groups. The names of the individual uronic acid are formed in different ways by replacing (i) the *-ose* of thc systematic or trivial name of an aldose by *-uronic acid* or (ii) the *-oside* of the name of the glycoside by *-osiduronic acid*, or (iii) the *-osyl* of the name of glycosyl group by *-osyluronic*. The carbon atom of the aldehyde group (and not that of the carboxylic acid group as in normal systematic nomenclature) is numbered C_1. Uronic acids can also form intermolecular esters. However, when the primary alcohol group of a ketose is oxidized to a carboxylic acid group, the product is considered an oxidized aldonic acid (ketoaldonic acid, see above) and is named according to this group. In nature, uronic acids occur especially as structural units of hemicelluloses and pectins (cf., Chapt. "9.4.5. Other polysaccharides"). In the following, only some examples of the natural uronic acids and their naming are given. The chemical structures of sodium (methyl α-L-glucofuranosid)uronate (I), methyl α-D-glucofuranosidurono-6,3-lactone (II), and 4-deoxy-L-*threo*-hex-4-enopyranuronic acid (III) are as follows:

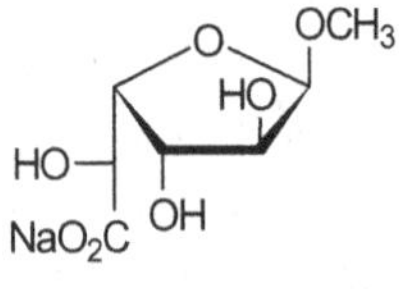

I 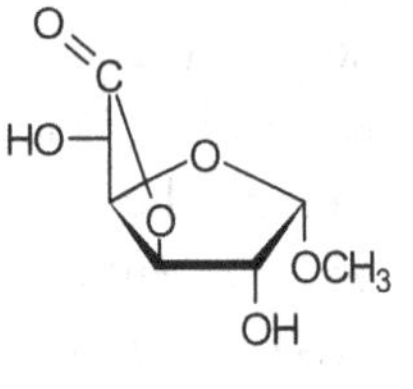 II 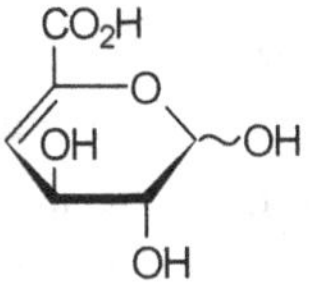 III

D-Glucuronic acid, β-D-glucopyranuronic acid, GlcA, GlcU

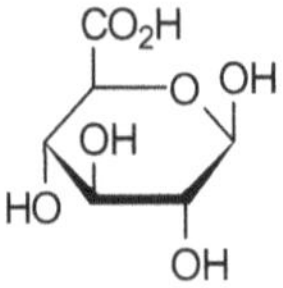

Molar mass 194.14 ($C_6H_{10}O_7$), melting point 165°C.

D-Glucuronic acid or β-D-glucopyranuronic acid occurs widely in nature either as a common building block of certain polymers (including polysaccharides, proteoglycans, and glycoglycerolipids) or as glycosides (glucuronides) and as its salts and esters (glucuronates). The formation of D-glucuronic acid from D-glucose takes place in the liver of all animals, including humans and other primates. The formation of various glycolsides via glucuronidation in the human body detoxicates or inactivates certain substances (e.g., alcohols, phenols, carboxylic acids, mercaptans, primary and secondary aliphatic amines, and many specific poisons) by making them more water-soluble and, in this way, allowing for their subsequent excretion from the body in the urine. D-Glucuronic acid was also first isolated from urine. It crystallizes as needles from ethanol or ethyl acetate. Its specific rotation $[\alpha]_D$ at 24°C is +36.3 (*c* 6, H_2O). This compound is used for determining urinary steroids and steroid conjugates in blood and is also a precursor of ascorbic acid (vitamin C).

4-*O*-Methyl-α-D-glucopyranuronic acid (molar mass 208.17; $C_7H_{12}O_7$) units exist in hemicelluloses (e.g., xylan):

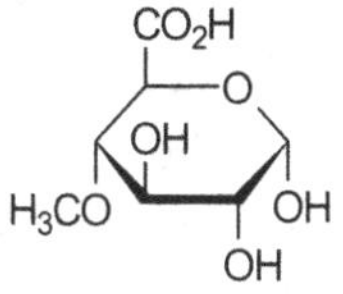

These side-groups of xylan are partly converted via the elimination of methanol to 4-deoxy-4-hexenuronic acid (hexenuronic acid; HexA or HexU) groups during kraft pulping (cf., Chapt. "12.4.1. Kraft pulping"). The hexenuronic acid groups are unreactive in the subsequent alkaline oxygen delignification and peroxide bleaching stages, but react, due to their "ene" functionality, with several other bleaching chemicals, such as chlorine dioxide, ozone, and peracids, thus consuming these expensive chemicals. However, the hexenuronic acid groups can be liberated and converted into furan derivatives (i.e., 2-furoic acid and 5-carboxy-2-furaldehyde), for example, prior to bleaching by mild acid hydrolysis, thus reducing the consumption of bleaching chemicals.

D-Galacturonic acid, α-D-galactopyranuronic acid, GalA, GalU

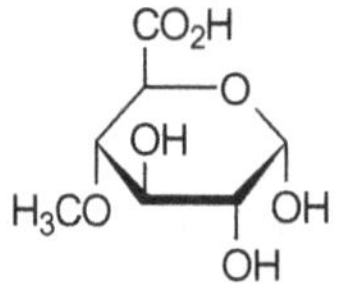

Molar mass 194.14 ($C_6H_{10}O_7$), melting point 159°C.

D-Galacturonic acid or α-D-galactopyranuronic acid is an oxidized form of D-galactose. It exists as a structural component of polymer pectin (polyuronide), from which it can be liberated by acid hydrolysis. The monohydrate of this uronic acid crystallizes from ethanol as needles having a specific rotation $[\alpha]_D$ of +50.9 in water at 20°C. It is used for biochemical research.

Naturally occurring other uronic acids are **L-iduronic acid** and **D-mannuronic acid**, both having a molar mass of 194.14 ($C_6H_{10}O_7$).

9.2.13. *Nitrogen compounds*

The substance group of different nitrogen-containing monosaccharides forms a significant class of compounds. When the carbonyl oxygen or it and the substituents attached to its adjacent carbon atom are replaced by nitrogen-containing substituent(s), the following types of compounds can be obtained (Ph is a phenyl group) (cf., Chapt. "11.1. Monosaccharides"):

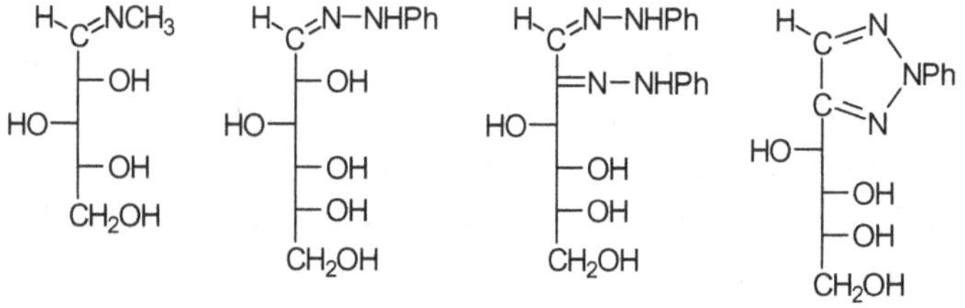

I II III IV

The names of these compounds are: 1-deoxy-1-(methylimino)-D-xylitol (I), D-glucose phenylhydrazone (II), D-*arabino*-hexos-2-ulose bis(phenylhydrazone) or D-glucose phenylosazone or D-*arabino*-hex-2-ulose phenylosazone (III), and D-*arabino*-hexos-2-ulose phenylosotriazole or 1*R*-1-(2-phenyl-2*H*-1,2,3-triazol-4-yl)-D-erythritol (IV). If an excess of phenylhydrazine (e.g., 1-methyl-1-phenylhydrazine) is used in the reaction, each carbon atom of the formed monosaccharide derivative (alkazone) binds chemically the group =N-NPhCH_3.

The replacement of an alcohol group of a monosaccharide by an amino group -NH_2 is envisaged as a substitution of the appropriate hydrogen atom of the corresponding deoxy monosaccharide by the amino group, leading to *amino sugars*. The stereochemistry at the carbon atom carrying the amino group is designated by regarding the amino group as equivalent to -OH. In addition, the characteristic ending *-osamine* is used for 2-amino derivatives. In the systematic names, the compounds are specified by using a combination of prefixes deoxy- and amino- with a stem name.

D-Glucosamine, 2-amino-2-deoxy-D-glucopyranose, chitosamine

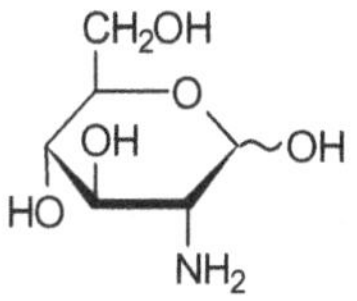

Molar mass 179.17 ($C_6H_{13}O_5N$), melting point 110 (β-form, decomposes) and 86 (α-form) °C, $[\alpha]_D$ +47.5 (20°C).

D-Glucosamine or chitosamine is a prominent biochemical precursor (mainly in the form of D-glucosamine-6-phosphate) in the biochemical synthesis of nitrogen-containing sugars, glycosylated proteins, and lipids. It is a structural unit in mucoproteins (consisting primarily of mucopolysaccharides) as well as in chitosan and chitin, which both compose the exoskeletons of crustaceans and other arthropods. Georg Ledderhose (1855–1925) first prepared it in 1876 by acid hydrolysis of chitin and Walter Haworth finally determined the stereochemistry of this compound in 1939. The compound crystallizes generally as needles that readily dissolve in water, slightly in methanol and ethanol, but are insoluble in diethyl ether and chloroform. D-Glucosamine is used as a dietary supplement for adults and in biochemical research. It is a strong base that forms with acids various salts; for example, an easily crystallizable hydrochloride (molar mass 215.63; $C_6H_{14}O_5NCl$).

D-Galactosamine, 2-amino-2-deoxy-D-galactopyranose

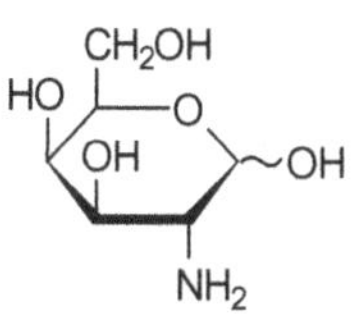

Molar mass 179.17 ($C_6H_{13}O_5N$), $[\alpha]_D$ +93.

D-Galactosamine is, together with galactose and glucose, a constituent of some glycolprotein hormones, such as follicle-stimulating hormone (FSH) and lutropin or luteinizeing hormone (LH). It is a hepatotoxic (liver-damaging) agent; it is used in animal models of liver failure (hepatitis) as well as in biochemical research. The hydrochloride of this compound is a crystalline solid with molar mass 215.63 ($C_6H_{14}O_5NCl$), melting point 180°C (α form, decomposes), and $[\alpha]_D$ +91–6 (20°C).

The corresponding other monosaccharide derivatives are as follows:

D-Mannosamine, 2-amino-2-deoxy-D-mannopyranose

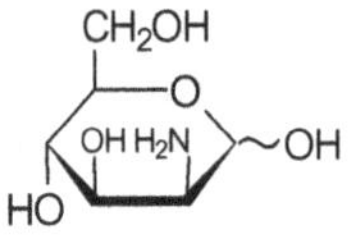

CH2OH O OH H2N OH HO

Molar mass 179.17 ($C_6H_{13}O_5N$).

D-Fucosamine, 2-amino-2,6-dideoxy-D-galactopyranose

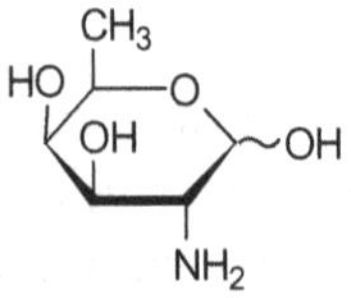

CH3 HO O OH OH NH2

Molar mass 163.17 ($C_6H_{13}O_4N$).

Neuraminic acid, 5-amino-3,5-dideoxy-D-*glycero*-D-*galacto*-non-2-ulosonic acid, Neu

Molar mass 267.24 ($C_9H_{17}O_8N$).

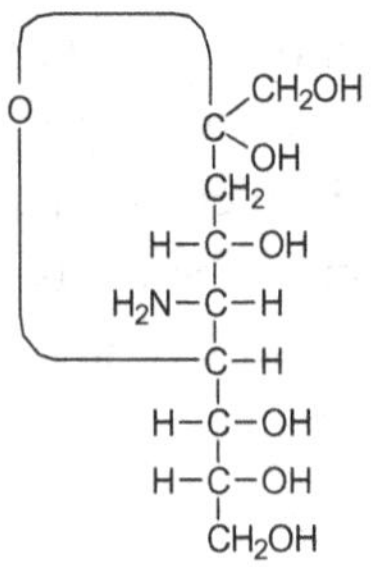

O CH2OH C OH CH2 H–C–OH H2N–C–H C–H H–C–OH H–C–OH CH2OH

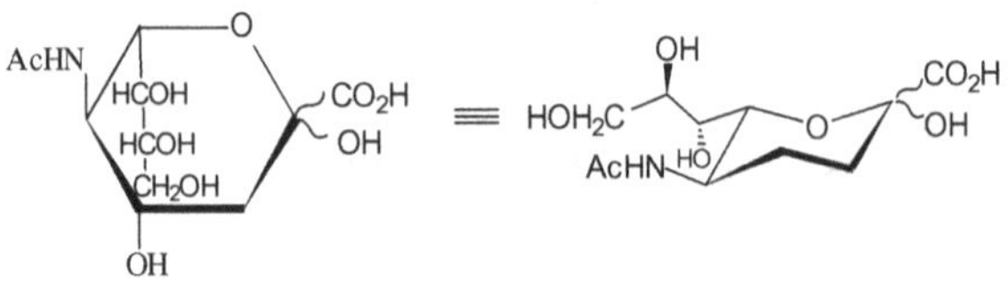

AcHN O HCOH HCOH CH2OH OH CO2H OH ≡ HOH2C OH AcHN HO O CO2H OH

Muramic acid, 2-amino-3-*O*-[*R*-1-carboxyethyl]-2-deoxy-D-glucose, Mur

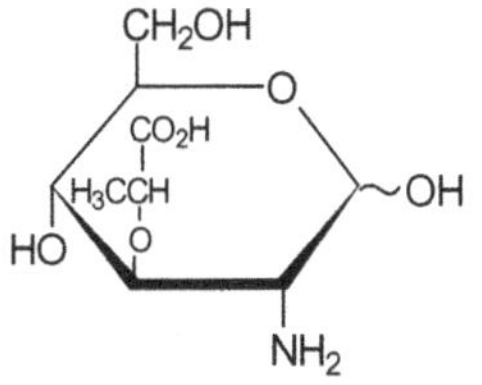

Molar mass 251.24 ($C_9H_{17}O_7N$).

Neosamine B, 2,6-diamino-2,6-dideoxy-β-L-*ido*-pyranose

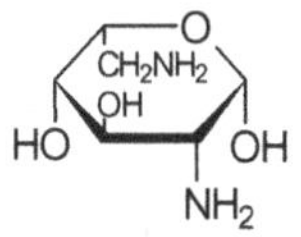

Molar mass 178.19 ($C_6H_{14}O_4N_2$).

An amino group can also chemically bind one further substituent. For example, in the naming of "acyl derivatives" the prefix acetamido- is used (e.g., 2-acetamido-2-deoxy-D-glucose) or for "alkyl derivatives" the name is, for example, 2-butylamino-2-deoxy-D-glucose. Other examples are 2-acetamido-2-deoxy-D-galactopyranose or *N*-acetyl-D-galactosamine (I), 2-deoxy-2-sulfoamino-D-glucopyranose or *N*-sulfo-D-glucosamine (II), *N*-glycolyl-α-neuraminic acid (α-Neu5Gc) (III), and 5-*N*-acetyl-4,8,9-tri-*O*-acetyl-α-neuraminic acid (α-Neu4,5,8,9Ac$_4$) (IV):

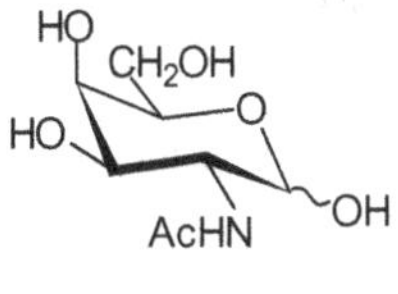

I

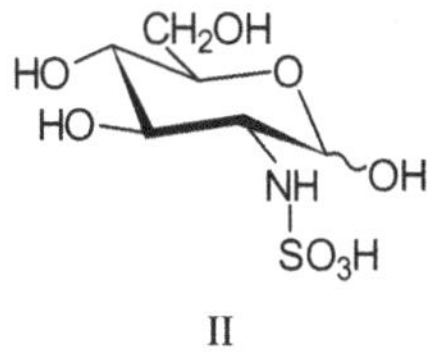

II

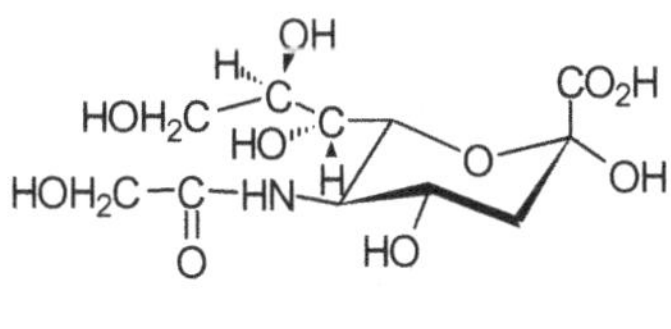

III

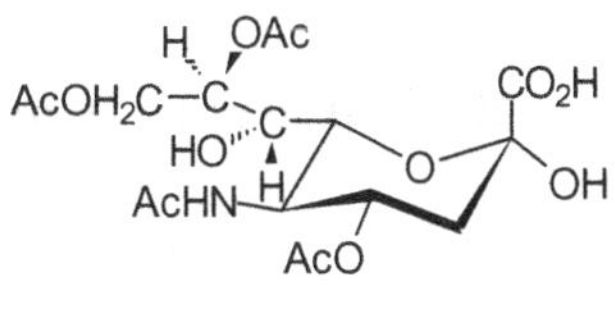

IV

N-Methyl-L-glucosamine

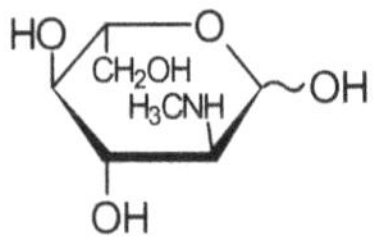

Molar mass 177.20 ($C_7H_{15}O_4N$).

Desosamine, 3,4,6-trideoxy-3-dimethyl-β-D-glucose

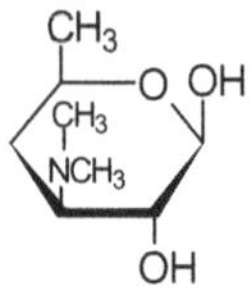

Molar mass 175.23 ($C_8H_{17}O_3N$).

1-Amino-1-deoxyalditols belong to this class of compounds and are used as intermediate chemicals for producing pharmaceuticals, cosmetics, and detergents.

1-Amino-1-deoxy-D-arabinitol

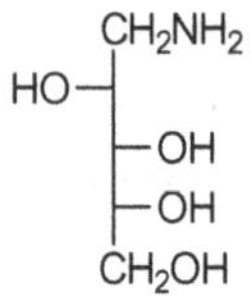

Molar mass 151.16 ($C_5H_{13}O_4N$), melting point 95–7°C, $[\alpha]_D$ +3.8° (20°C).

1-Amino-1-deoxy-L-arabinitol hydrochloride

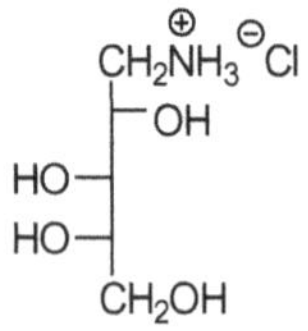

Molar mass 187.62 ($C_5H_{14}O_4NCl$), melting point 134–6°C, $[\alpha]_D$ −13.2 (20°C).

1-Amino-1-deoxy-D-xylitol hydrochloride

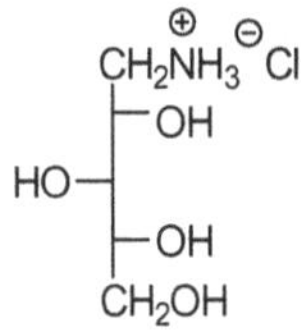

Molar mass 187.62 ($C_5H_{14}O_4NCl$), melting point 139–40°C, $[\alpha]_D$ −12.3 (20°C).

1-Amino-1-deoxy-D-lyxitol hydrobromide

Molar mass 232.07 ($C_5H_{14}O_4NBr$), melting point 103–4°C, $[\alpha]_D$ +3.5 (20°C).

1-Amino-1-deoxy-D-galactitol

Molar mass 181.20 ($C_6H_{15}O_5N$), melting point 148–50°C, $[\alpha]_D$ −1.5 (20°C).

1-Amino-1-deoxy-D-glucitol

Molar mass 181.20 ($C_6H_{15}O_5N$), melting point 131–3°C, $[\alpha]_D$ −7.2 (20°C).

1-Amino-1-deoxy-D-mannitol

Molar mass 181.20 ($C_6H_{15}O_5N$), melting point 136–7°C, $[\alpha]_D$ −2.5 (20°C).

The corresponding *1-deoxy-1-nitroalditols* are as follows:

1-Deoxy-1-nitro-D-mannitol

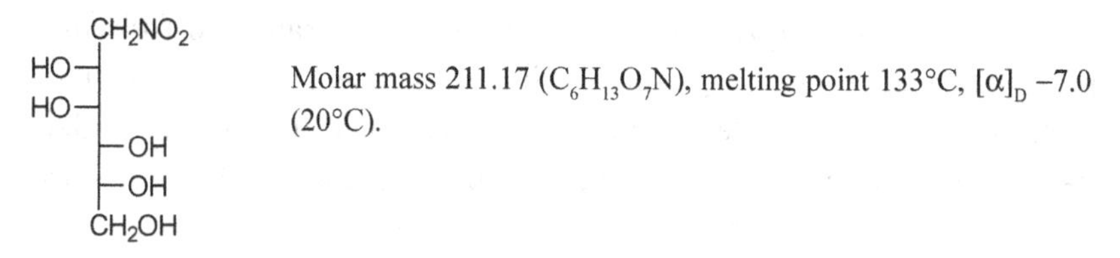

Molar mass 211.17 ($C_6H_{13}O_7N$), melting point 133°C, $[\alpha]_D$ −7.0 (20°C).

1-Deoxy-1-nitro-L-mannitol

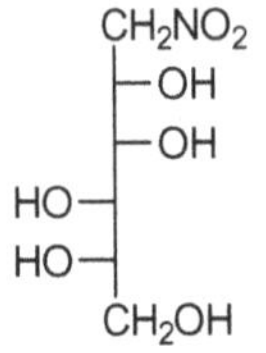

CH_2NO_2
—OH
—OH
HO—
HO—
CH_2OH

Molar mass 211.17 ($C_6H_{13}O_7N$), melting point 133°C, $[\alpha]_D$ −7.0 (20°C).

1-Deoxy-1-nitro-L-galactitol

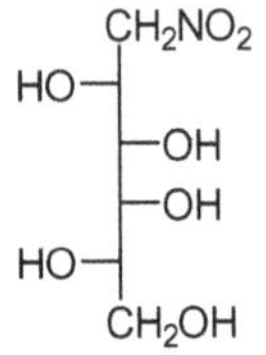

CH_2NO_2
HO—
—OH
—OH
HO—
CH_2OH

Molar mass 211.17 ($C_6H_{13}O_7N$), melting point 142°C, $[\alpha]_D$ −1.7 (20°C).

1-Deoxy-1-nitro-D-iditol hemihydrate

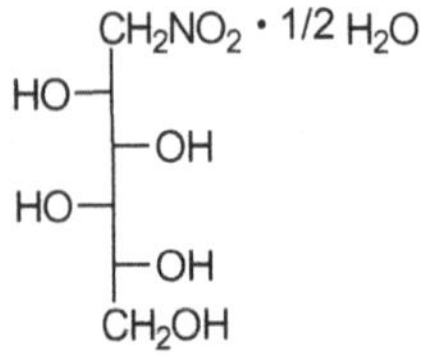

$CH_2NO_2 \cdot 1/2\ H_2O$
HO—
—OH
HO—
—OH
CH_2OH

Molar mass 220.18 ($C_6H_{14}O_{7.5}N$), melting point 89°C.

1-Deoxy-1-nitro-L-iditol hemihydrate

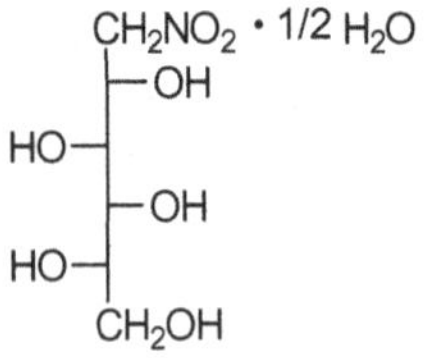

$CH_2NO_2 \cdot 1/2\ H_2O$
—OH
HO—
—OH
HO—
CH_2OH

Molar mass 220.18 ($C_6H_{14}O_{7.5}N$), melting point 89°C.

9.2.14. *Other monosaccharide derivatives*

A typical example of *free phosphates* is D-glucopyranose 6-(dihydrogen phosphate) or 6-*O*-phosphone-D-glucopyranose (I), of *phosphonates* methyl-β-ribofuranoside-5-(hydrogen phosphonate) or methyl 5-deoxy-β-D-ribofuranosid-5-yl hydrogen phosphonate (II), of *sulfates* α-D-galactopyranose 2-sulfate or 2-*O*-sulfonato-α-D-galactopyranose (III), of *thiohemiacetals* 1*S*-D-glucose ethyl dithiohemiacetal (IV), and of *dithioacetals* D-glucose propane-1,3-diyl dithioacetal (V):

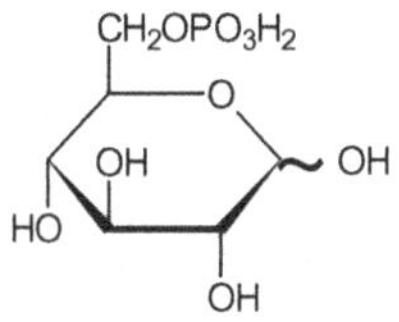

I 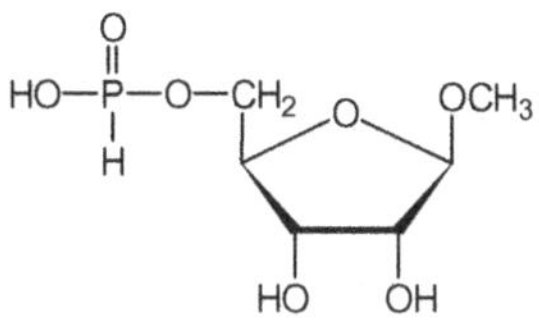 II

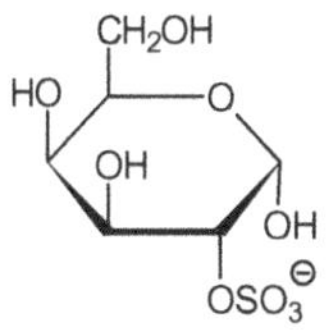

III 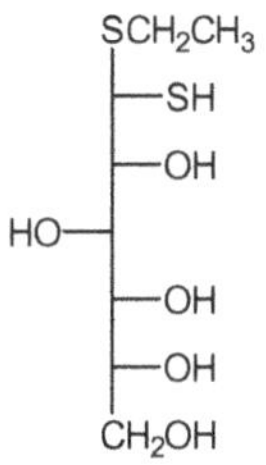 IV 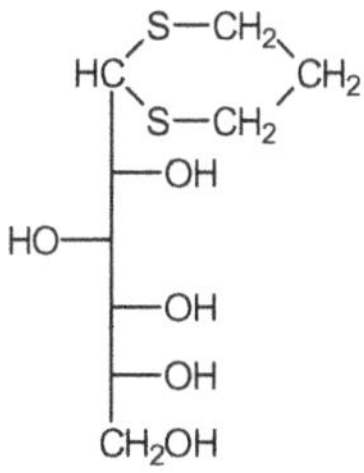 V

Finally, examples of other types of sugar derivatives are given; for example, D-mannononitrile (I), *E*-1,2,3,4,5-penta-*O*-acetyl-D-*erythro*-pent-1-enol (II), α-D-manno-pyranosyl bromide (III), 3-deoxy-3-methyl-D-glucose (IV), 3-deoxy-3,3-dimethyl-D-*ribo*-hexose (V), 4-*C*-(hydroxymethyl)-D-*erythro*-pentose (VI), uridine 5′-(α-D-glucopyranosyl diphosphate) (UDP-Glc) (VII) (cf., Chapt. "8.2. Polysaccharides"), 5-thio-β-D-glucopyranose (VIII), and methyl 4-seleno-α-D-xylofuranoside (IX):

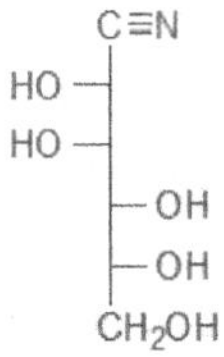

I 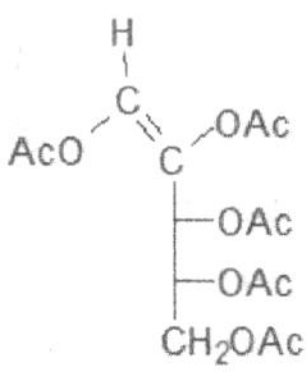 II 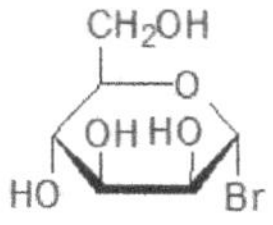 III

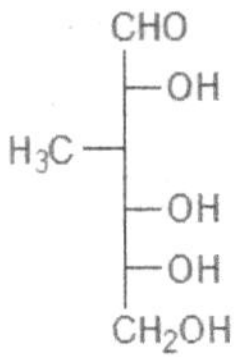

IV

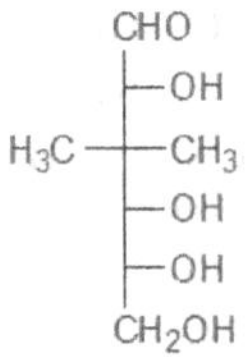

V

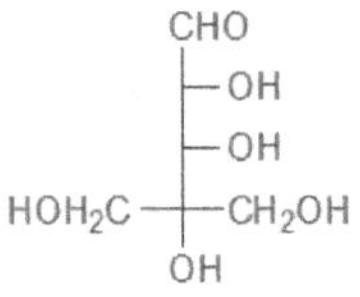

VI

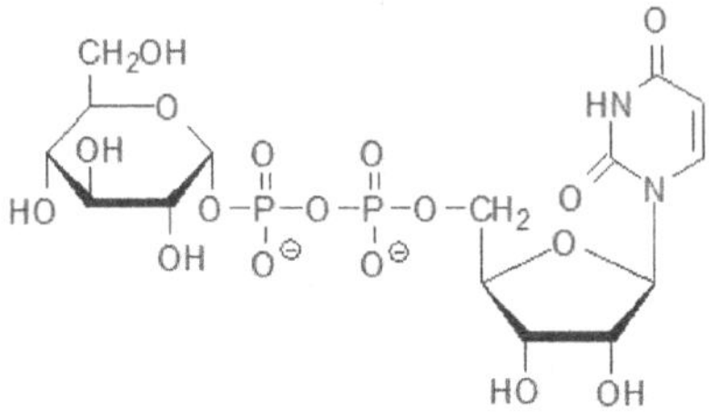

VII

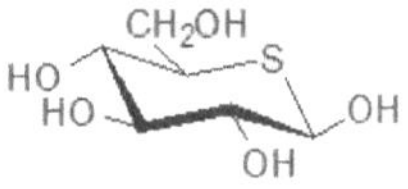

VIII

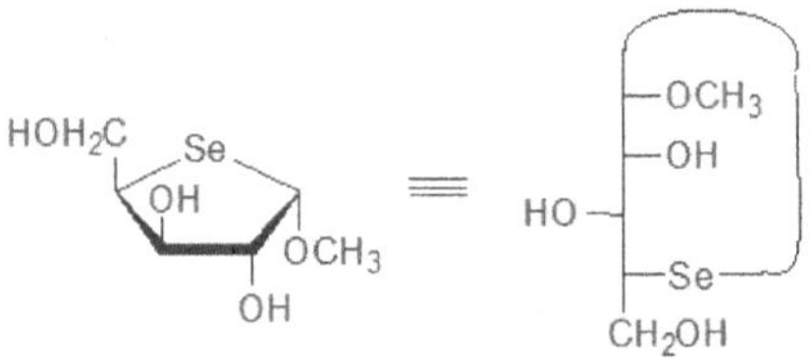

IX

9.2.15. *Furans and their derivatives*

Heterocyclic compounds are cyclic compounds that have at least one ring structure containing, besides carbon atom(s), one or more atoms of different elements or heteroatoms (in Greek, “heteros” means “different”) as members of their rings. The most common heteroatoms are oxygen, nitrogen, sulfur, and phosphorus. The naming of heterocyclic compounds is normally based on the trivial names of simple ring structures (e.g., furan, pyridine, pyrrole, thiophene, and xantene). These names can be used together when naming more complicated derivatives according to a certain priority order. Several compounds of this substance group are significant industrial chemicals, but they can also occur at low concentrations as structural units of different natural substances. Typical examples of such compounds are the pyrimidine and purine base components (e.g., cytosine and adenine) of nucleosides in nucleic acids (cf., Chapt. "10.2. Nucleic Acids"). Since, besides mono- and disaccharides, some furans are formed as by-products of acid hydrolysis of oligo- and polysaccharides, these compounds and their simple upgrading products are also described in this connection. On the other hand, the furan compounds can also be the desired main products of various carbohydrate resources.

Furan, 1,4-epoxy-1,3-butadiene, oxole

Molar mass 68.08 (C_4H_4O), melting point −85.6°C, boiling point 31.4°C, density 0.951 kg/dm^3 (20°C), refractive index n_D 1.421 (20°C), flashing point −36°C, explosive limits in air 2.3–14.3%.

Heinrich Limprich (1827–1909) first prepared furan (in Latin, "furfur" means "bran") in 1870. It is produced commercially by catalytic decarbonylation of furfural, but is also formed in decarboxylation of 2-furoic acid, in oxidation of butane dial (succinaldehyde), or as a by-product in manufacturing of 1,4-dicyanobutane (adiponitrile). It is a colorless and highly flammable liquid with an ethereal odor, miscible with ethanol and diethyl ether, but only slightly in water (1%, 25°C). Furan turns brown on standing in air or light and this color change is retarded if a small amount of water is added. It is absorbed by the skin and its vapors are poisonous. Unstabilized furan products slowly form unstable peroxides on exposure to air. Furan is used as a chemical intermediate in producing tetrahydrofuran, thiophene, and pyrrole.

Tetrahydrofuran, 1,4-epoxybutane, oxolane, tetramethylene oxide, diethylene oxide, THF

Molar mass 72.12 (C_4H_8O), melting point −65°C, boiling point 65.4°C, density 0.889 kg/dm^3 (20°C), refractive index n_D 1.405 (20°C), flashing point −15°C, explosive limits in air 1.5–11.8%, autoignition temperature 321°C.

Tetrahydrofuran is obtained by acid-catalyzed dehydration of 1,4-butanediol or by catalytic hydrogenation of furan. It is a colorless and highly flammable liquid with an ethereal odor and it readily dissolves in ethanol, diethyl ether, acetone, and benzene and in lesser amounts in water. Tetrahydrofuran penetrates the skin and has a tendency to form highly explosive peroxides on storage in air. When vaporized, it has a physiological effect similar to diethyl ether and is also toxic by inhalation and ingestion. Tetrahydrofuran is widely used as a solvent for many purposes. It is a popular solvent for hydroboration reactions and generally for organometallic compounds, such as organolithium and Grignard reagents (cf., Chapt. "11.1. Monosaccharides"). In addition, this furan is used as a chemical intermediate for producing many renewable platform chemicals.

Furfural, 2-furaldehyde, furaldehyde, furfuraldehyde, furan-2-carboxaldehyde, furan-2-carbaldehyde, fural

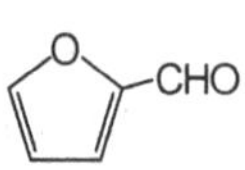

CHO

Molar mass 96.09 ($C_5H_4O_2$), melting point −38.7°C, boiling point 161.7 or 90 (65 mmHg)°C, density 1.159 kg/dm^3 (20°C), refractive index n_D 1.526 (20°C), flashing point 62°C, explosive limits in air 2.1–19.3%, autoignition temperature 315°C.

Johan Wolfgang Döbereiner (1780–1824) first isolated furfural or 2-furaldehyde in 1821 by as "yellow oil" from the reaction mixture formed as a by-product in the preparation of formic acid from carbohydrates with sulfuric acid and manganese dioxide. After this discovery, during the years 1835–40, furfural was prepared (primarily by John Stenhouse, 1809–80) from a wide range of vegetable materials including, for example, corn, oats, sawdust, and bran, by boiling finely divided raw materials with aqueous sulfuric acid or other acids. The same product that resulted in all cases had a characteristic aroma and was named in 1845 by George Fownes (1815–49) as "bran oil" according to its principal origin (see furan, above). In 1901, the chemical structure of furfural was deduced by Carl Dietrich Harries (1866–1923) when this compound gradually became comercially important. Nowadays, furfural is still produced from annually renewable sources, such as non-food residues of food crops (e.g., corncobs, various hulls, and cereal grasses), bagasse (a by-product of sugarcane harvesting), and pentosan-rich forest and wood industrial waste materials (e.g., hardwood sawdust and harvesting residues). It is produced in batch or continuous digesters, where the pentosans (primarily xylan) are first hydrolyzed typically under elevated temperature (>160°C) by mineral acid hydrolysis to pentoses (primarily xylose), which are then subsequently cyclodehydrated to furfural (see the production of xylose and xylitol). The product can be isolated by steam distillation. The pure furfural is a colorless liquid with a strong odor of almonds (i.e., similar to benzaldehyde) that quickly darkens (first to yellow and finally to reddish-brown) upon exposure to air and light. However, this phenomenon can be prevented by adding a small amount (0.25%) hydroquinone or paraldehyde. Furfural is miscible with ethanol and diethyl ether and dissolves readily in acetone and in lesser amounts in benzene, chloroform, and water (8.3%, 20°C). When vaporized, it might be slightly toxic at higher concentrations, irritates eyes, skin, and mucous membranes, and may cause allergic reactions. Furfural forms the following binary azeotropes:

Other component	Furfural/wt.-%	Bp./°C
o-Xylene	13	141
Pinene	38	143
Camphene	40	147
o-Chlorotoluene	35	155
Cyclohexanol	6	156

Furfural is used as a specialty and selective solvent for many purposes. It is also a reactive solvent and contributes low viscosity to resin formulations. In addition, it is an important renewable platform chemical for a number of monomeric compounds and resins. For example, its oxidation results in 2-furoic acid and its hydrogenation provides furfuryl alcohol (FA), which may be further hydrogenated to tetrahydrofurfuryl alcohol (THFA). Examples of other utilization include its use as a weed killer, a fungicide, an analytical reagent, and a wetting agent in the manufacture of abrasive wheels and brake linings.

5-Hydroxymethyl-2-furaldehyde, 5-(hydroxymethyl)furfural, hydroxymethylfurfural, 5-hydroxymethyl-2-formylfurane, HMF

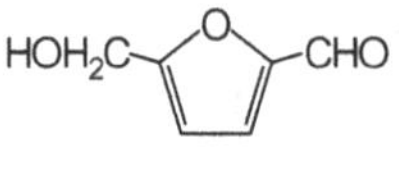

HOH_2C — CHO

Molar mass 126.11 ($C_6H_6O_3$), melting point 35.0–5.5°C, boiling point 114–6°C (0.5 mmHg), density 1.206 kg/dm^3 (25°C), refractive index n_D 1.563 (18°C).

Louis Camille Maillard (1878–1936) first prepared 5-hydroxymethyl-2-furaldehyde or hydroxymethylfurfural in 1912 from inulin using oxalic acid. It can be found in low amounts, for example, in honey (in fresh honey less than 15 mg/kg), fruit juices, coffee, and ultra-heat-treated (UHT) milk. In practice, it is absent in fresh food, but is naturally generated in sugar-containing food during heat treatments, such as drying or cooking, thus being used as an indicator for excess heat treatment. Hydroxymethylfurfural is produced in the same way as furfural, although in this case, the raw materials include hexose-based polysaccharides or six-carbon atoms-containing monosaccharides (especially fructose). Compared to furfural, this compound cannot be isolated by steam distillation and its separation and purification are carried out by solvent extraction and distillation. In addition, it readily degrades on heating via the elimination of formic acid into levulinic acid which further cyclicizes to α- and β-angelica lactones:

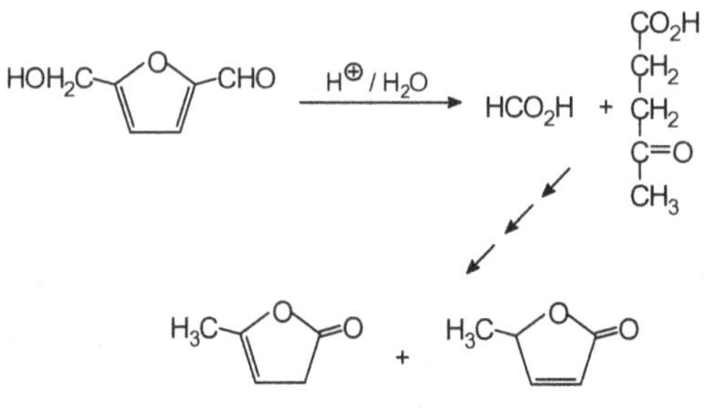

HOH_2C — CHO $\xrightarrow{H^{\oplus} / H_2O}$ HCO_2H + CO_2H–CH_2–CH_2–C=O–CH_3

H_3C — =O + H_3C — =O

Hydroxymethylfurfural forms needles which dissolve in water, ethanol, and benzene, and slightly in diethyl ether. It is used as a chemical intermediate for producing many significant chemicals.

Furfuryl alcohol, 2-furylmethanol, 2-(hydroxymethyl)furan, 2-furanmethanol, 2-furancarbinol, furyl carbinol, FA

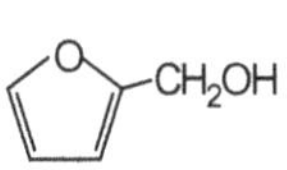

CH_2OH

Molar mass 98.10 ($C_5H_6O_2$), melting point −14.6°C, boiling point 170 or 68–9 (20 mmHg) °C, density 1.130 kg/dm^3 (20°C), refractive index n_D 1.487 (20°C), flashing point 65°C, explosive limits in air 1.8–16.3%, autoignition temperature 391°C.

Furfuryl alcohol or 2-furylmethanol or is manufactured industrially by employing both liquid-phase and vapor-phase hydrogenation of 2-furaldehyde according to the Cannizzaro reaction:

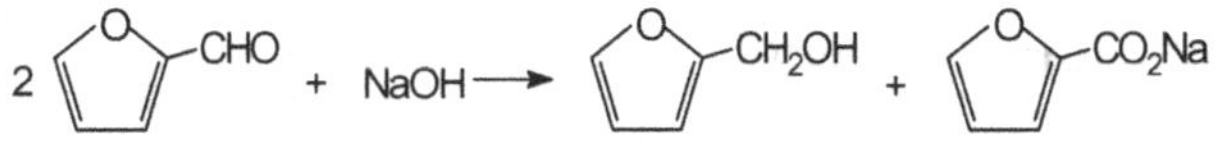

The pure product is a colorless and mobile liquid with a faint burning odor and a bitter taste. On exposure to air and light, it becomes amber colored (finally brown to dark-red) upon prolonged standing. It also autopolymerizes with acid catalysts, often with explosive violence, to a black and chemically highly stable polymer, poly(furfuryl alcohol), which eventually becomes crosslinked and insoluble in the reaction medium. Furfuryl alcohol is miscible with water, ethanol, diethyl ether, acetone, and ethyl acetate, although it is unstable in water. When vaporized, it is toxic by inhalation and is absorbed by the skin. Furfuryl alcohol is widely used as a monomer in manufacturing resins and as an excellent solvent for a variety of synthetic resins. Its industrial value is a consequence of its low viscosity and high reactivity. In addition, due to the low molar mass of this compound, it can impregnate the cells of wood, where it can be polymerized and bonded with the wood by heat, radiation or catalysts. This kind of treated wood has improved moisture-dimensional stability and hardness as well as decay and insect resistance.

Tetrahydrofurfuryl alcohol, 2-tetrahydrofuranmethanol, tetrahydrofuryl carbinol, THFA (optically active (+)- and (−)-forms and an optically inactive, racemic (±)-form)

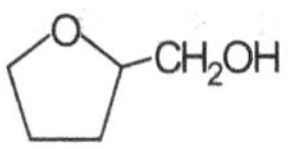

Molar mass 102.13 ($C_5H_{10}O_2$), melting point <−80°C, boiling point 177–8 or 80–2 (20 mmHg) °C, density 1.054 kg/dm^3 (20°C), refractive index n_D 1.452 (20°C), flashing point 75°C, explosive limits in air 1.5–9.7%, autoignition temperature 282°C.

Tetrahydrofurfuryl alcohol or 2-tetrahydrofuranmethanol is commercially produced by vapor-phase catalytic hydrogenation of furfuryl alcohol. It is a colorless liquid with a mild, pleasant odor and is miscible with water and dissolves in lesser amounts in diethyl ether and acetone. Tetrahydrofurfuryl alcohol is used as an ingredient in proprietary stripping and lacquer formulations. In addition, it is used as a solvent for many purposes and a chemical intermediate for the pharmaceutical and chemical specialities sector.

Tetrahydrofurfuryl benzoate

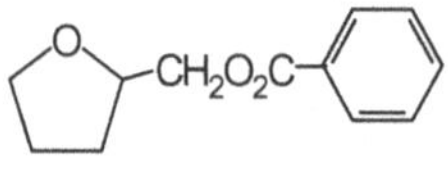

Molar mass 206.24 ($C_{12}H_{14}O_3$), boiling point 300–2 or 138–40 (2 mmHg) °C, density (D_0) 1.137 kg/dm^3 (20°C).

Tetrahydrofurfuryl laurate

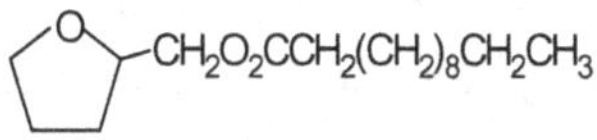

Molar mass 332.47 ($C_{17}H_{32}O_3$), density 0.930 kg/dm^3 (25°C).

Tetrahydrofurfuryl oleate

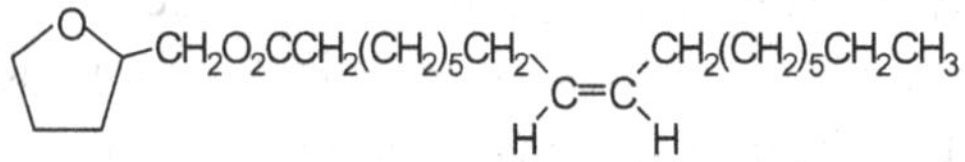

Molar mass 366.58 ($C_{23}H_{42}O_3$), melting point −30°C, boiling point 240°C (5 mmHg), density 0.923 kg/dm^3 (25°C), flashing point 165°C.

Tetrahydrofurfuryl phthalate

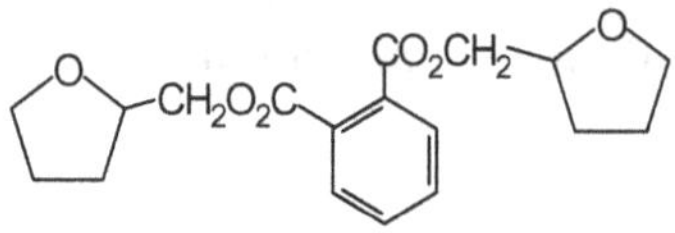

Molar mass 334.37 ($C_{18}H_{22}O_6$), melting point <15°C, density 1.194 kg/dm^3 (25°C).

2,5-Tetrahydrofurfuryl dimethanol

Molar mass 132.16 ($C_6H_{12}O_3$), melting point <–50°C, boiling point 265°C, density (D_0) 1.172 kg/dm^3 (4°C).

2-Furoic acid, 2-furancarboxylic acid, furan-2-carboxylic acid

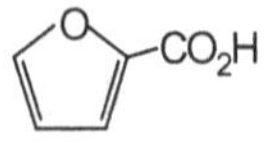

Molar mass 112.09 ($C_5H_4O_3$), melting point 133–4°C, boiling point 230–2 or 141–4 (20 mmHg) °C (sublimes).

2-Furoic acid or 2-furancarboxylic acid was the first furan derivative described in 1780 by Carl Wilhelm Scheele. It is produced by the chemical or biochemical oxidation of furfuryl alcohol or furfural and is purified by sublimation or crystallization from hot water. It forms colorless needles or flaky crystals that dissolve in hot water, ethanol, and diethyl ether and slightly in cold water. 2-Furoic acid is used as a preservative and a flavoring agent in food products, in textile processing, and in the field of optic technology. The salts and esters of this compound are known as furoates.

Ethyl 2-furoate, ethyl furoate, 2-furoic acid ethyl ester

$-CO_2CH_2CH_3$ Molar mass 140.15 ($C_7H_8O_3$), melting point 34–5°C, boiling point 196.8 or 128 (95 mmHg) °C, density 1.117 kg/dm^3 (21°C).

2-Furoyl chloride, 2-furancarboxylic acid chloride, 2-furancarbonyl chloride

$-COCl$ Molar mass 130.53 ($C_5H_3O_2Cl$), melting point −2°C, boiling point 173 or 66 (10 mmHg) °C.

2-Furoyl chloride or 2-furancarboxylic acid chloride is produced by the treatment of 2-furoic acid with phosphorus pentachloride. It is a colorless and corrosive liquid that dissolves in diethyl ether and chloroform, but decomposes in water. 2-Furoyl chloride is a powerful lachrymator and a strong irritant to eyes and skin. This compound is used as a pharmaceutical and chemical intermediate and a substituent for chloropicrin in disinfecting grain elevators.

3-Furoic acid, furan-3-carboxylic acid, 3-furancarboxylic acid

CO_2H Molar mass 112.09 ($C_5H_4O_3$), melting point 122–3°C, boiling point 105–10°C (12 mmHg, sublimes).

Other simple furan derivatives are as follows:

2-Chlorofuran, 2-furyl chloride

$-Cl$ Molar mass 102.52 (C_4H_3OCl), boiling point 77.5°C (744 mmHg), density 1.192 kg/dm^3 (20°C), refractive index n_D 1.457 (20°C).

2-Bromofuran, 2-furyl bromide

Molar mass 146.98 (C_4H_3OBr), boiling point 102°C (744 mmHg), density 1.650 kg/dm^3 (20°C), refractive index n_D 1.498 (20°C).

2-Furonitrile, 2-cyanofuran, 2-furyl cyanide, furan-2-carbonitrile, 2-furancarbonitrile

Molar mass 93.08 (C_5H_3ON), boiling point 146°C, density 1.082 kg/dm^3 (20°C).

2-Acetylfuran, acetylfuran, 2-furyl methyl ketone, 1-(2-furanyl)ethanone

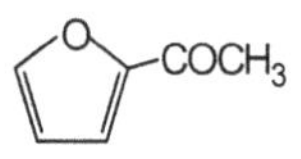

Molar mass 110.11 ($C_6H_6O_2$), melting point 33°C, boiling point 173 or 67 (10 mmHg) °C, density 1.098 kg/dm^3 (20°C), refractive index n_D 1.502 (20°C).

2-Ethoxyfuran, ethyl-2-furyl ether

Molar mass 112.14 ($C_6H_8O_2$), boiling point 125–6°C, density 0.985 kg/dm^3 (23°C), refractive index n_D 1.450 (23°C).

2-Furanacetic acid, 2-furanethanoic acid, 2-furylacetic acid, 2-furylethanoic acid, 2-(furan-2-yl)acetic acid

Molar mass 126.11 ($C_6H_6O_3$), melting point 68–9°C, boiling point 102–4°C (0.4 mmHg).

2-(Aminomethyl)furan, 2-furylmethylamine, 2-furfurylamine, α-furfurylamine, furfurylamine, 2-furanmethylamine

Molar mass 97.12 (C_5H_7ON), melting point –70°C, boiling point 145–6 or 80 (84 mmHg) °C, density 1.050 kg/dm^3 (25°C), refractive index n_D 1.491 (20°C), flashing point 37°C.

2-(Aminomethyl)furan can be produced from furfural and ammonia. This liquid is miscible with water and dissolves in lesser amounts in ethanol and diethyl ether. 2-(Aminomethyl)furan is used as a an anti-corrosive agent, a chemical intermediate, and an additive in soldering fluids.

2-(Aminomethyl)tetrahydrofuran, 2-tetrahydrofurfurylamine

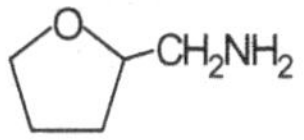

Molar mass 101.15 ($C_5H_{11}ON$), boiling point 150–6°C, density (D_{20}) 0.977 kg/dm^3 (20°C), refractive index n_D 1.455 (20°C).

2-(Aminomethyl)tetrahydrofuran is a colorless or slightly yellowish liquid. It is used as a chemical intermediate and an accelerating agent in vulcanization.

9.3. Oligosaccharides and Their Derivatives

9.3.1. *Disaccharides*

Disaccharides are formed when two monosaccharides are joined together via a glycosidic bond after the elimination of a molecule of water (i.e., by a dehydration reaction). If two anomeric hydroxyl groups of monosaccharides are reacted with one another, the product is called a "non-reducing disaccharide" (cf., Chapt. "1. Introduction"). However, most common disaccharides belong to "reducing disaccharides". In such a disaccharide one glycosyl unit has replaced the hydrogen atom of an alcoholic hydroxyl group of the other, thus having a free anomeric hydroxyl group either in a hemiacetal or hemiketal structure (i.e., permiting two possible orientations of the uninvolved hydroxyl group). The non-reducing disaccharides are systematically named by using the endings *-osyl* and *-oside*; for example, sucrose or β-D-fructofuranosyl α-D-glucopyranoside. Using the recommended abbreviations for monosaccharides (e.g., fructose Fru and glucose Glc — cf., Chapts. "7.1. Aldoses" and "7.2. Ketoses"), the abbreviated name for sucrose is (β-D-Fru*f*-(2↔1)-α-D-Glc*p*). In this formula, the letters *f* and *p* refer to a furanoid and pyranoid structure, respectively (cf., Chapt. "7.3. Cyclic forms"). In addition, the numbering of carbon atoms participating in the formation of a glycosidic bond is shown and the arrowhead is pointed to an anomeric carbon atom. In contrast, in the names of reducing disaccharides, the endings *-osyl* and *-ose* are used; for example, α-lactose or 4-*O*-β-D-galactopyranosyl-α-D-glucopyranose or β-D-galactopyranosyl-(1→4)-α-D-glucose (β-D-Gal*p*-(1→4)-α-D-Glc*p*). However, many of the naturally occurring disaccharides have well established trivial names. For clarity, in this chapter, the chemical structures of sugars are mainly depicted in the Haworth projections without taking into consideration the "actual stereochemical structures" (e.g., an actual ring conformation). The structures of sucrose and lactose shown by these two systems are given as follows:

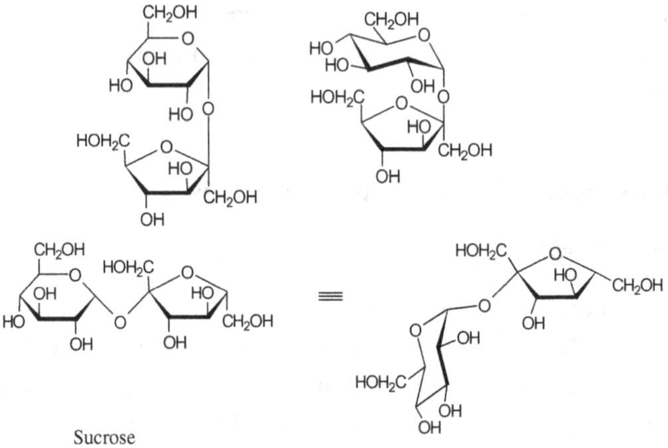

Sucrose

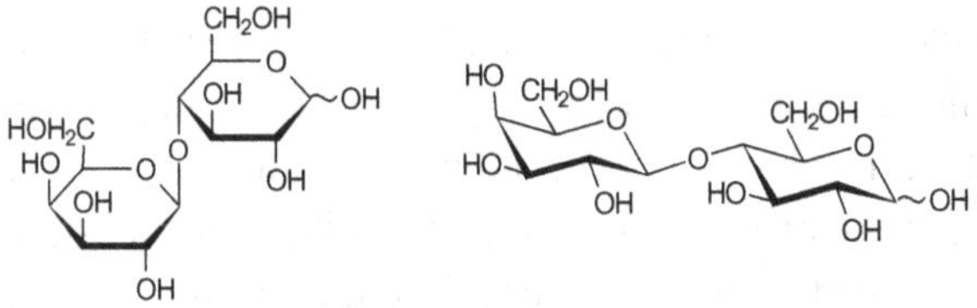

Lactose

In nature, only a few free — but industrially significant — disaccharides occur.

Sucrose, β-D-fructofuranosyl α-D-glucopyranoside, β-D-Fru*f*-(2↔1)-α-D-Glc*p*, saccharose, cane sugar, beet sugar, table sugar, sugar, Suc

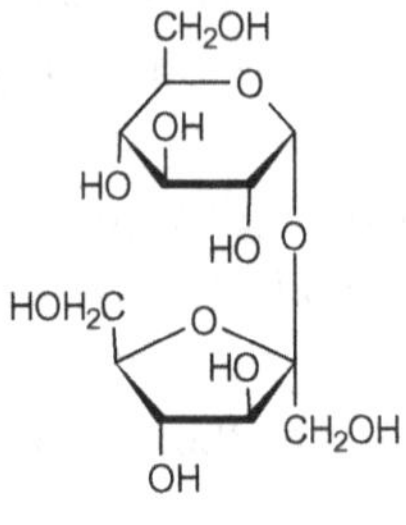

Molar mass 342.30 ($C_{12}H_{22}O_{11}$), melting point 160–86°C (decomposes), density 1.588 kg/dm^3.

Sucrose is found naturally in seeds, leaves, fruits, flowers, and roots of many plants. Honey, which is a sweet food made by bees using nectar from flowers, also contains high percentages of this disaccharide and its hydrolysis product, invert sugar syrup (i.e., a mixture of glucose and fructose) similar to high-fructose corn syrup. The biosynthesis of sucrose proceeds via the precursors UDP-glucose and fructose-6-phosphate and is catalyzed by the enzyme sucrose-6-phosphate synthase (cf., Chapt. "8.2. Polysaccharides"). It is an energy source in the metabolism and a carbon source in the biosynthesis. The industrial production of table sugar (annually about 185 million tons) consisting of a variety of commercial products is based on its separation from two sugar crops: sugar cane (*Saccharum officinarum*) and sugar beet (*Beta vulgaris*) which are also indicated by their common trivial names, cane sugar and beet sugar, respectively. Sucrose forms odorless, white, and very hard crystals or is a powder that dissolves at room temperature readily in water (200 g/100 mL) and slightly in ethanol (0.6 g/100 mL). It becomes hygroscopic when the relative humidity (RH) of air rises to 82%, and at an RH of 95%, the absorption of water is so powerful that the final product is sugar liquor. Sucrose combusts to carbon dioxide and water and at elevated temperatures it gradually changes into a brown solid, caramel (caramelization). It is fermented by common yeasts and its specific rotation $[\alpha]_D$ at 20°C is +66.5 (*c* 26, H_2O). Sucrose has a natural sweet taste and is the most significant sweetener in foods and soft drinks. However, studies have indicated obvious links between the consumption of free sugars including sucrose and health hazards, such as obesity, tooth decay, and the development of a metabolic syndrome (i.e., the development of type 2 diabetes mellitus). Hence, it has been recommended that especially the consumption of sugar-containing soft drinks should be limited. Sucrose is also used to some extent for the manufacture of its salts ("saccharinates") and as a hydrophilic component, for example, in surface active compounds, such as the sucrose polyesters (SPEs) containing between five to eight fatty acids esterified to a sucrose molecule. In addition, other carbohydrates (e.g., isomaltulose) can be biotechnically prepared from it by the bacterium *Protaminobacter rubrum*).

The concept of "sweetness" is a complex with respect to the sense of taste perceived in the mouth on different parts of the tongue and what molecular characteristics are actually essential for sweetness. This taste is universally regarded as pleasurable and simple carbohydrates, such as sucrose, are those most commonly associated with it. Sweetness has also very ancient evolutionary beginnings for human beings who learned to regularly consume the milk of other mammals. It is plausible to think that, for example, the liking of sweetness in newborn human infants is connected with the sweetness of the initial lactose-containing food, breast milk. The early milk from mammals is called "colostrum"; it contains antibodies that provide protection to the newborn baby as well as nutrients and growth factors. On the other hand, the growing consumption of sugars can be simply considered as the productive and technological development of the food industry.

In 1967, a general theory of sweetness (the AH-B theory) was proposed by Robert Sands Shallenberger and Terry E. Acree for the relationships between the typical structural features of a compound and the taste of sweetness. According to this original theory (which has been extended later), for example, the property of sugar sweetness requires the presence of two adjacent hydroxyl groups (i.e., an α-glycolic group) together with a suitable spatial orientation. Of four possible three-dimensional possibilities of α-glycolic groups, only two can cause the sensation of sweetness. In addition, one prerequisite for a compound to be sweet is that a hydrogen bond is formed between the "key group" (AH-B, i.e., a covalently bound H-bonding proton (AH) and an electronegative group (B)) of a sweetener and the receptor site on the tongue. In this case, the "strength" of the sweetness can be said to be partly in proportion to the strength of the interaction caused by the formed hydrogen bond although the sweetness cannot be determined directly as such by physical methods. Consequently, one potential way is to use standard taste panel methods to evaluate the taste characteristics of intense sweeteners and sugars. These tests are conducted using individual panelists trained to recognize various characteristics and able to describe and distinguish between different products presented to them. In general, the sweetness intensity of various compounds is compared to that of the standard sweetener, sucrose, which enables the ranking of the *relative sweetness values* for the samples studied on a certain scale. Normally, varying water solution concentrations of a sweetener are compared to those of sucrose with known concentrations. For example, if a 10% sucrose solution and a 15% solution of a sample possess an equal taste at room temperature (a majority of the panel members should have the same opinion), the calculated relative sweetness value on a dry mass basis is (10/15) × 100 or 67 (i.e., the value for sucrose is 100). The results can be also indicated by using a value of 1.0 for sucrose. However, it is evident that factors, such as pH and the use of other ingredients, also influence the sensation of sweetness. Examples of the guiding relative sweetness values for common sugars, sugar alcohols, sugar mixtures, and some artificial sweeteners are the following:

Compound	Relative sweetness value
Fructose	1.2–1.8
Xylitol	1.0
Sucrose	1.0
Glucose	0.7
Glucitol	0.6
Mannitol	0.5
Maltose	0.3
Lactose	0.2
Invert sugar	1.0–1.3
Starch syrups	0.3–0.8
Glycerol	0.5
Saccharin	300
Cyclamate	30
Aspartame	180

In practice, based on these values, the same sensation of sweetness as that with sucrose can be obtained by using about double the amount of glucitol or mannitol. *Lactose intolerance*, which is the inability of adults and children to digest lactose, causes side effects, such as abdominal pains. In modern lactose-free products (e.g., "HYLA" products, acronym for "hydrolyzed lactose"), this disaccharide has been cleaved into its monosaccharide moieties, galactose and glucose. The degradation also results in the formation of sweeter products, which is understandable because of the relative sweetness values for these carbohydrates. The heating value of common carbohydrate-based sweeteners is about 16 kJ/g (about 4 kcal/g), thus indicating that the most effective ratio of sweetness-to-heat content can be obtained with fructose.

Sugar cane (*Saccharum officinarum*) is one of the several species of tall perennial true grasses (height 6–8 m), native in tropical and subtropical regions, but effectively cultivated for sugar production. Its mature stalk (about 75% of the entire plan) is typically composed of 12–16% sucrose, 63–73% water, and 13–19% other materials (mainly organic fibers). The cultivation of sugarcane was known in China and India long before Christ, from where it gradually spread to the Mediterranean countries and the West Indies (cf., Chapt. "2.1. The Era Before the 1800s"). Before the recovery of sucrose, the leaves and the root, which altogether contain significant amounts of impurities, should be removed from the stalk. Sugar cane is harvested by hand or mechanically. In tradetional hand harvesting, the field is first set on fire to burn dry leaves and chase away or kill venomous snakes without harming the stalks. Harvesters then cut the cane just above ground-level using cane knives or machetes. Once cut, harvesters should commence the recovery process of sucrose within one day since sugar cane begins to lose its sugar content rapidly, especially if slower hand cutting is applied. In mechanical harvesting, the remains (i.e., the top of the sugar cane and the dead leaves) left in the field by the machine serve as mulch for the next round of planting.

The other source of sucrose is the cultivated sugar beet (*Beta vulgaris*), which was bred from the "White Silesian" fodder beet by Andreas Sigismund Marggraf (1709–82) and Franz Karl Achard (1753–1821) in the late 18th century. However, sugar beet-based production did not significantly increase until during the wars of Napoleon Bonaparte (cf., the continental system) in the early 19th century. In contrast to sugar cane, which grows in tropical and subtropical zones, sugar beet grows exclusively in the temperate zone. The root of the beet with an average mass of 800 g contains, due to purposeful development, 12–20% sucrose together with about 75% water. In harvesting, beet roots are mechanically separated from the crown and leaves (suitable for cattle fodder) and transported to a factory. If the beets are to be left for later delivery, they are formed into clamps, which are shielded from the weather. In this case, provided the clamp is well-built with the right amount of ventilation, no significant loss in the sucrose content takes place. Since the beet load coming to a factory may contain even 50% of impurities, the first important process stage is washing the beets with water. After this stage, the beet roots are mechanically sliced into thin strips (called "cossettes" with a thickness of about 3.5 mm), and passed to a diffuser to extract the sucrose into a water solution. Diffusers are tall vessels (about 20 m, the residence time of cossettes about 1.5 h) in which the beet slices move in one direction, while hot water (70°C) flows in the opposite direction. The color of the liquid ("raw juice") exiting the diffuser varies from black to dark red; it contains 14–15% sucrose and is introduced to further refining. The molasses from the refining process (see below) is added to the extracted cossettes (pulp) with a low sucrose content, and after drying this product is sold as animal feed.

The major part (80%) of the worldwide production of sucrose is from sugar cane and is used as raw material in human food industries or is fermented to produce bioethanol, especially in Brazil. The sucrose-containing liquid ("cane juice") is separated from sugarcane by pressing, and the residual dry material ("bagasse") is generally burned for energy. This residue is also used as an additive in concrete or a feedstock material for producing pulp and chemicals. The recovery of sucrose can also be done by a similar extraction as applied in beet root processing. However, regardless of the original raw material, the pressed or extracted dark-colored raw juice still contains a lot of impurities and is normally clarified by the addition of the hot milk of lime (a suspension of calcium hydroxide in water) and separately carbon dioxide. This defecation treatment neutralizes (pH from 5.5 to almost neutral, thus also preventing the hydrolysis of sucrose) and precipitates a number of impurities, including low-molar-mass organic acids, large organic molecules, and colored substances. After filtration, the thin juice (dry solids content about 15%) is concentrated via multiple-effect evaporation to produce a thick juice (dry solids content about 70%) which is fed to the crystallizers. In this stage, the liquor is concentrated further by boiling in vacuum pans, and sucrose gradually crystallizes from the supersaturated solution, "mother liquor". Fine sucrose crystals are added as seed crystals to accelerate the process, and the resulting sucrose crystals and syrup mix (in French, "massecuite" means "cooked mass") are finally passed through a centrifuge. Several crystallization stages are usually carried out, and the final mother liquor is called "molasses". During this process, it is also possible to produce so-called "brown sugar", which contains certain impurities and is a commercial product.

Granulated white sugar (table sugar) is mainly produced directly in beet sugar-based factories. In contrast, in sugar cane-based factories, only raw sugar with a yellowish-brown or sometimes even reddish color is usually produced. This brown sugar with a significant molasses content is then further refined separately in a process called "affination" in which the syrup and crystals are first separated in a spinning centrifugal basket. The refining after this stage is similar to that used in sugar beet-based factories, resulting, for example, in granulated and lump sugar as well as coarse-grain sugar. All these products are pure sucrose. In contrast, about 1.5% tricalcium phosphate is added (to prevent clumping) to powdered sugar (confectioner's sugar or icing sugar) produced by grinding granulated sugar.

Lactose, 4-*O*-β-D-galactopyranosyl-D-glucopyranose, β-D-galactopyranosyl-(1→4)-D-glucopyranose, β-D-Gal*p*-(1→4)-D-Glc*p*, milk sugar, lactobiose

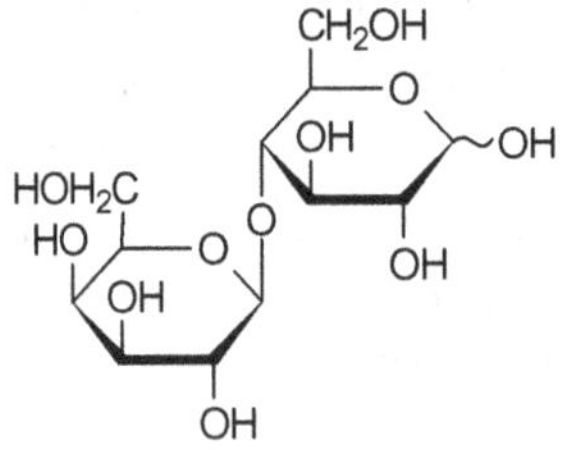

Molar mass 342.30 ($C_{12}H_{22}O_{11}$), melting point 202 (α form, monohydrate) or 252 (β form, anhydrous) °C, density 1.525 kg/dm^3 (α form, monohydrate).

Fabrizio Bartoletti (1576–1630) first reported the isolation of lactose in 1633 and in 1780, Carl Wilhelm Scheele identified it as a sugar. Lactose occurs free or combined with other carbohydrates and their derivatives (e.g., 2-acetamido-2-deoxy-D-glucose, L-fucose, D-galactose, and 5-acetamido-3,5-dideoxy-D-*glycero*-D-*galacto*-nonulopyranosonic acid or lactamic acid) in milk (human, 6–7% and cow, 4–5%), from which its trivial name, milk sugar (in Latin, "lac" or "lactis" means "milk"), is derived. It is also found as a constituent of many oligosaccharides in milk.

In addition, it exists at low concentrations, for example, in the pollen extract of forsythia (e.g., *Forsythia suspense* in the family *Oleaceae*) and the fruits of sapodilla (*Achras sapota*). The monohydrate of its α form can be crystallized by evaporating the whey obtained as a by-product of manufacture of cheese or casein. In general, whey is the liquid remaining after milk has been curdled and strained. The corresponding anhydrous β form can be obtained by carrying out the crystallization at higher temperatures (95°C). Lactose can also be synthesized as a chemically pure product. α-Lactose loses water of crystallization at 120°C and is not fermented by common yeasts. It is a white, crystalline mass or powder that slightly dissolves at room temperature in water (17%) and its specific rotation $[\alpha]_D$ is +52.6 (*c* 8, H_2O) (α form) or +55.4 (*c* 4, H_2O) (β form). Lactose has major applications in food industry and pharmaceutical industry.

Maltose, 4-*O*-α-D-glucopyranosyl-D-glucopyranose, α-D-glucopyranosyl-(1→4)-D-glucopyranose, α-D-Glc*p*-(1→4)-D-Glc*p*, maltobiose, malt sugar

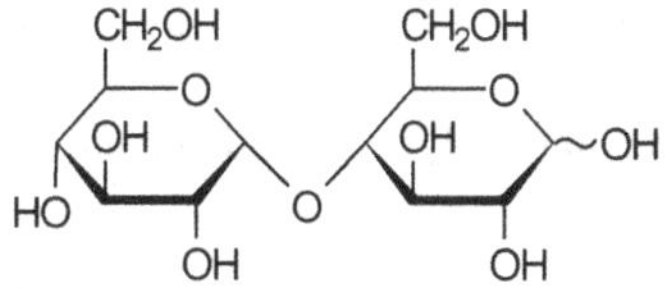

Molar mass 342.30 ($C_{12}H_{22}O_{11}$), melting point 103°C (β form, monohydrate).

Cornelius O'Sullivan (1841–1907) discovered maltose in 1872; it occurs in low concentrations in some plants. Its name comes from "malt" representing germinated cereal grains (mostly barley) that have been dried in a process known as "malting". In this process (cf., brewing), the enzymes (mainly α-amylase, a diastase) are developed that modify the grain's starches and break them into carbohydrates, such as maltose. This disaccharide can be produced at high yields (80%) enzymatically from starch. Maltose is fermented by common yeasts in the presence of D-glucose. It forms colorless and combustible crystals that dissolve in water and slightly in alcohol, but are insoluble in diethyl ether. Its (β form) specific rotation $[\alpha]_D$ is +130 (*c* 4, H_2O). Maltose is used as a nutrient, a sweetener, and a stabilizer for polysulfides.

Isomaltose, 6-*O*-α-D-glucopyranosyl-D-glucopyranose, α-D-glucopyranosyl-(1→6)-D-glucose, α-D-Glc*p*-(1→6)-D-Glc*p*

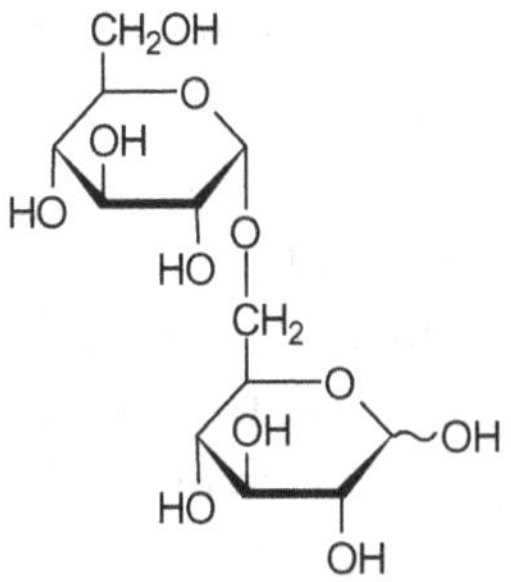

Molar mass 342.30 ($C_{12}H_{22}O_{11}$), melting point 120°C (β form).

Isomaltose is found as a constituent of starch and glycogen and as a regularly repeating unit in many dextran type bacterial polysaccharides. Isomaltose is produced by treating high maltose syrup with the enzyme transglucosidase (TG). Its (β form) specific rotation $[\alpha]_D$ is +122 (*c* 2, H_2O).

Cellobiose, 4-*O*-β-D-glucopyranosyl-D-glucopyranose, β-D-glucopyranosyl-(1→4)-D-glucopyranose, β-D-Glc*p*-(1→4)-D-Glc*p*

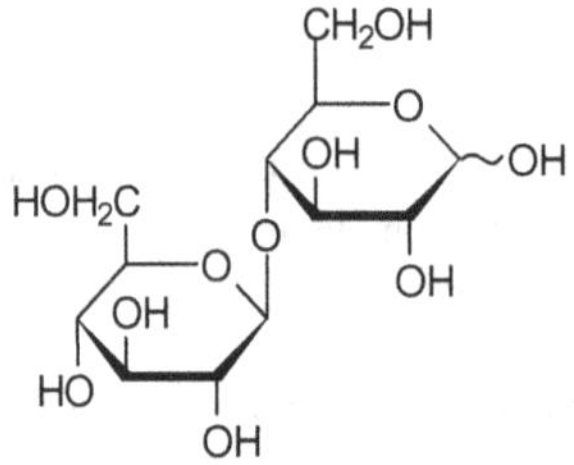

Molar mass 342.30 ($C_{12}H_{22}O_{11}$), melting point 225°C (decomposes).

Cellobiose is not found as such in nature, but it can be considered to be a repeating constituent of cellulose. In addition, it occurs similarly in the structures of many polysaccharides and certain plant glycosides. Cellobiose can be obtained by enzymatic or acid hydrolysis (under mild conditions) of cellulose or cellulose-rich materials, such as cotton, jute, and paper. It is not fermented by brewer's yeast. Cellobiose forms colorless crystals that dissolve in water and slightly in ethanol, but not in acetone. Its specific rotation $[\alpha]_D$ is +34.9 (*c* 1, H_2O). Cellobiose is used in bacteriology.

α,α-Trehalose, α-D-glucopyranosyl-α-D-glucopyranoside, α-D-Glc*p*-(1↔1)-α-D-Glc*p*, mycose, tremalose

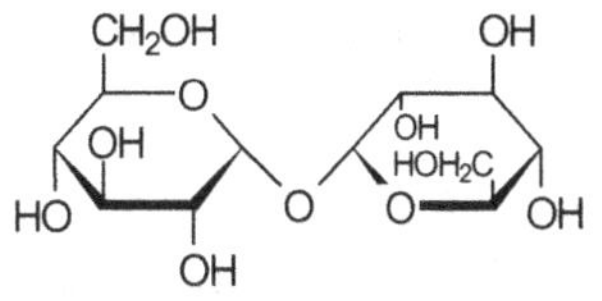

Molar mass 342.30 ($C_{12}H_{22}O_{11}$), melting point 96.5–7.5 (dihydrate) or 203 (anhydrous) °C.

α,α-Trehalose occurs widely in nature, especially in fungal colonies, whose certain spores may contain 5% of this compound. H.A.L. Wiggers discovered it in 1832 in an ergot of rye. This compound can also be found in low concentrations in some yeasts, lichens, and algae. However, α,α-trehalose is found in high concentrations (20–30%) in a cocoon substance (trehala manna) secreted by true weevils (genus *Larinus*) living on thorn bushes (*Echinops pesicus*) in the Middle East. It can be extracted from trehala manna or starch with hot 75% ethanol, but can also be prepared synthetically. α,α-Trehalose forms white crystals as the dihydrate and at about 130°C an anhydrous form (regains readily moisture) is obtained. It is fermented by common yeasts. Its specific rotation $[\alpha]_D$ is +178 (*c* 2, H_2O). α,α-Trehalose is an antioxidant, has a high water retention capacity, and is used as a preservative in food and medicines.

α,β-Trehalose, β-D-glucopyranosyl-α-D-glucopyranoside, α-D-Glc*p*-(1↔1)-β-D-Glc*p*, neotrehalose

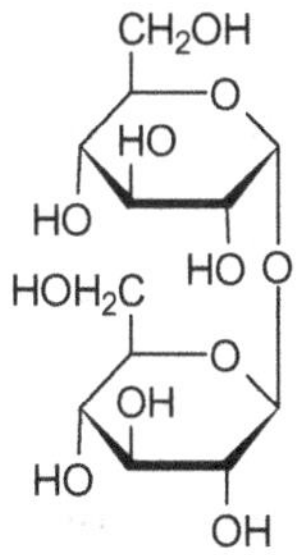

Molar mass 342.30 ($C_{12}H_{22}O_{11}$), melting point 210–20°C, $[\alpha]_D$ +95.

β,β-Trehalose, β-D-glucopyranosyl-β-D-glucopyranoside, β-D-Glc*p*-(1↔1)-β-D-Glc*p*, isotrehalose

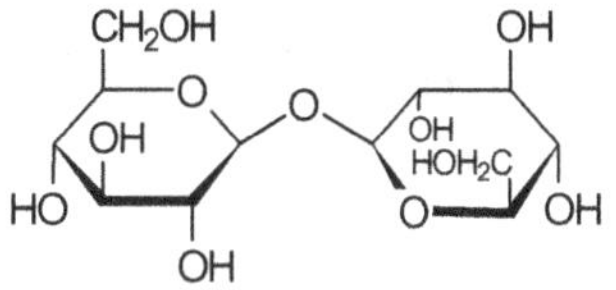

Molar mass 342.30 ($C_{12}H_{22}O_{11}$), melting point 135–40°C, $[\alpha]_D$ –40.

Gentiobiose, 6-*O*-β-D-glucopyranosyl-D-glucopyranose, β-D-glucopyranosyl-(1→6)-D-glucopyranose, β-D-Glc*p*-(1→6)-D-Glc*p*

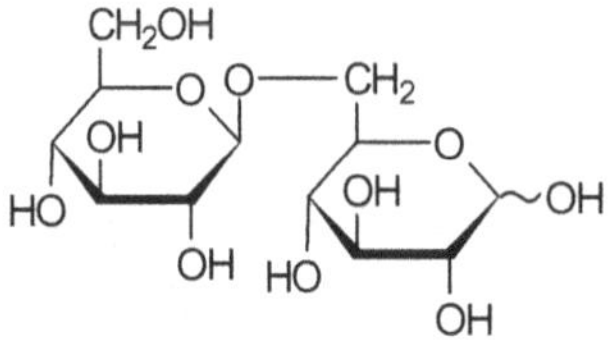

Molar mass 342.30 ($C_{12}H_{22}O_{11}$), melting point 86 (α form, contains two methanol molecules) or 190 (β form, solvent free) °C.

Gentiobiose is rare in nature. It occurs as a constituent of many glycosides, of which the most significant ones are amygdalin (cf., Chapt. "10.1.2. Cyanogenic glycosides") and α-crocin. In addition, it is a structural unit of several polysaccharides. Gentiobiose is produced by partial acid hydrolysis or enzymatically from **gentianose** (a trisaccharide, β-D-fructofuranosyl-β-D-glucopyranosyl-(1→6)-α-D-glucopyranoside, β-D-Glc*p*-(1→6)-α-D-Glc*p*-(1↔2)-β-D-Fru*f* — molar mass 504.44 ($C_{18}H_{32}O_{16}$), melting point 211°C, and $[\alpha]_D$ +33 (*c* 2, H_2O)). This disaccharide can also be prepared synthetically. Its specific rotation $[\alpha]_D$ is +87 (*c* 5, H_2O) (α form) or +10 (*c* 5, H_2O) (β form).

Xylobiose, 4-*O*-β-D-xylopyranosyl-D-xylopyranose, β-D-xylopyranosyl-(1→4)-D-xylopyranose, β-D-Xyl*p*-(1→4)-D-Xyl*p*

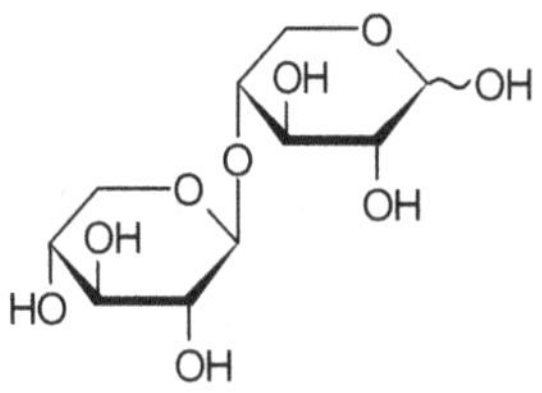

Molar mass 282.25 ($C_{10}H_{18}O_9$), melting point 185–6°C, $[\alpha]_D$ −25 (*c* 2, H_2O).

Laminarabiose, 3-*O*-β-D-glucopyranosyl-D-glucopyranose, β-D-glucopyranosyl-(1→3)-D-glucopyranose, β-D-Glc*p*-(1→3)-D-Glc*p*

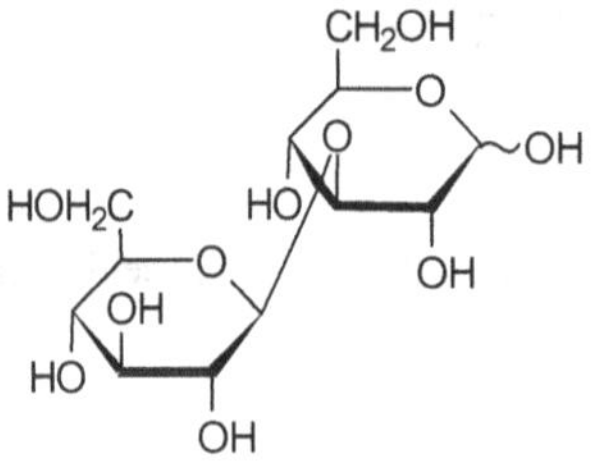

Molar mass 342.30 ($C_{12}H_{22}O_{11}$), melting point 205°C, $[\alpha]_D$ +18 (*c* 2, H_2O).

Cellobiulose, 4-*O*-β-D-glucopyranosyl-D-fructofuranose, β-D-glucopyranosyl-(1→4)-D-fructofuranose, β-D-Glc*p*-(1→4)-D-Fru*f*

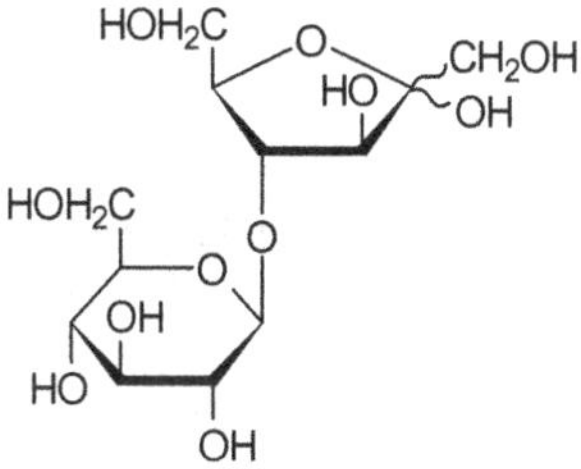

Molar mass 342.30 ($C_{12}H_{22}O_{11}$), $[\alpha]_D$ −60.1 (*c* 2.4, H_2O).

Inulobiose, 1-*O*-β-D-fructofuranosyl-D-fructofuranose, β-D-fructofuranosyl-(1→2)-D-fructofuranose, β-D-Fru*f*-(1→2)-D-Fru*f*

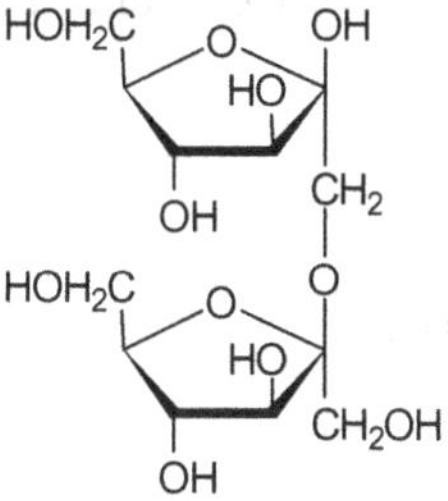

Molar mass 342.30 ($C_{12}H_{22}O_{11}$), $[\alpha]_D$ −72 (*c* 3, H_2O).

Mannobiose, 4-*O*-β-D-mannopyranosyl-D-mannopyranose, β-D-mannopyranosyl-(1→4)-D-mannopyranose, β-D-Man*p*-(1→4)-D-Man*p*

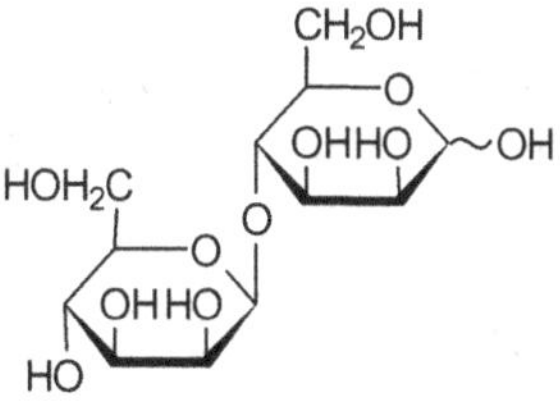

Molar mass 342.30 ($C_{12}H_{22}O_{11}$), melting point 194°C, $[\alpha]_D$ −2 (*c* 1, H_2O).

Maltulose, 5-*O*-α-D-glucopyranosyl-D-fructopyranose, α-D-glucopyranosyl-(1→5)-D-fructopyranose, α-D-Glc*p*-(1→5)-D-Fru*p*

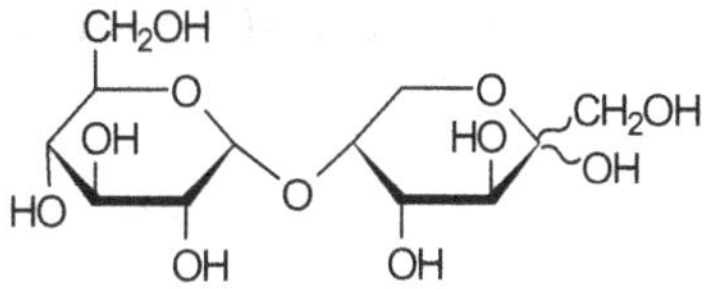

Molar mass 342.30 ($C_{12}H_{22}O_{11}$), melting point 113–5°C, $[\alpha]_D$ +64.

Isomaltulose, 6-*O*-α-D-glucopyranosyl-D-fructofuranose, α-D-glucopyranosyl-(1→6)-D-fructofuranose, α-D-Glc*p*-(1→6)-D-Fru*f*

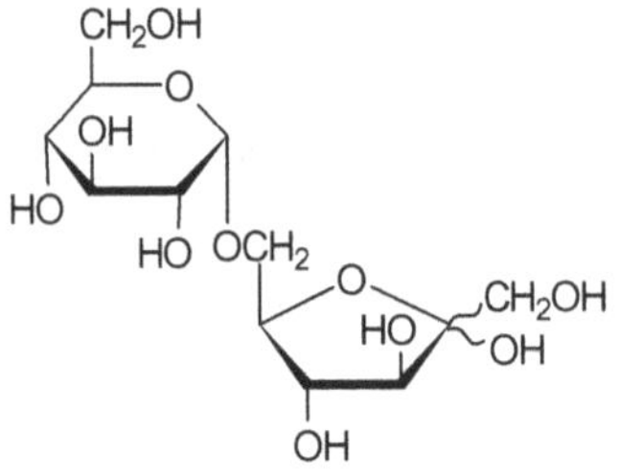

Molar mass 342.30 ($C_{12}H_{22}O_{11}$), $[\alpha]_D$ +97.2.

Planteobiose, 6-*O*-α-D-galactopyranosyl-D-fructofuranose, α-D-galactopyranosyl-(1→6)-D-fructofuranose, α-D-Gal*p*-(1→6)-D-Fru*f*, melibiulose

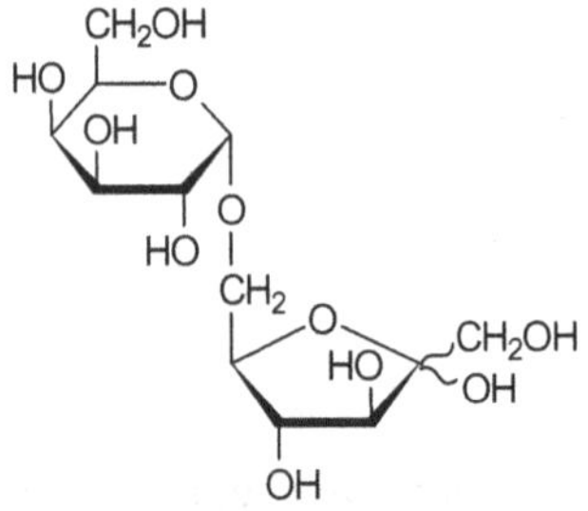

Molar mass 342.30 ($C_{12}H_{22}O_{11}$).

Examples of other disaccharides are as follows:

Melibiose, α-D-galactopyranosyl-(1→6)-D-glucose, 6-*O*-α-D-galactopyranosyl-D-glucose, α-D-Gal*p*-(1→6)-D-Glc. Molar mass 342.30 ($C_{12}H_{22}O_{11}$).

Primeverose, β-D-xylopyranosyl-(1→6)-D-glucose, 6-*O*-β-D-xylopyranosyl-D-glucose, β-D-Xyl*p*-(1→6)-D-Glc. Molar mass 312.27 ($C_{11}H_{20}O_{10}$).

Turanose, α-D-glucopyranosyl-(1→3)-D-fructose, 3-*O*-α-D-glucopyranosyl-D-fructose, α-D-Glc*p*-(1→3)-D-Fru. Molar mass 342.30 ($C_{12}H_{22}O_{11}$).

Nigerose, α-D-glucopyranosyl-(1→3)-D-glucose, α-D-Glc*p*-(1→3)-D-Glc. Molar mass 342.30 ($C_{12}H_{22}O_{11}$).

Sophorose, β-D-glucopyranosyl-(1→2)-D-glucose, β-D-Glc*p*-(1→2)-D-Glc. Molar mass 342.30 ($C_{12}H_{22}O_{11}$).

Arabinopyranobiose, β-L-Ara*p*-(1→3)-L-Ara. Molar mass 282.25 ($C_{10}H_{18}O_9$).

Arabinofuranobiose, β-L-Ara*f*-(1→3)-L-Ara. Molar mass 282.25 ($C_{10}H_{18}O_9$).

Galactobiose, β-D-Gal*p*-(1→3)-D-Gal. Molar mass 342.30 ($C_{12}H_{22}O_{11}$).

Glucosylgalactose, β-D-Glc*p*-(1→6)-D-Gal. Molar mass 342.30 ($C_{12}H_{22}O_{11}$).

Kojibiose, α-D-Glc*p*-(1→2)-D-Glc. Molar mass 342.30 ($C_{12}H_{22}O_{11}$).

Vikianose, β-L-Ara*p*-(1→6)-D-Glc. Molar mass 312.27 ($C_{11}H_{20}O_{10}$).

9.3.2. *Other oligosaccharides*

Higher oligosaccharides than disaccharides (trisaccharides, tetrasaccharides, etc.) are found in nature only in low concentrations. The most significant examples belong to trisaccharides. The compounds are not of high commercial importance.

Raffinose, β-D-fructofuranosyl-α-D-galactopyranosyl-(1→6)-α-D-glucopyranoside, α-D-Gal*p*-(1→6)-α-D-Gal*p*-(1↔2)-β-D-Fru*f*, melitose, melitriose, gossypose, α-D-galactosylsucrose

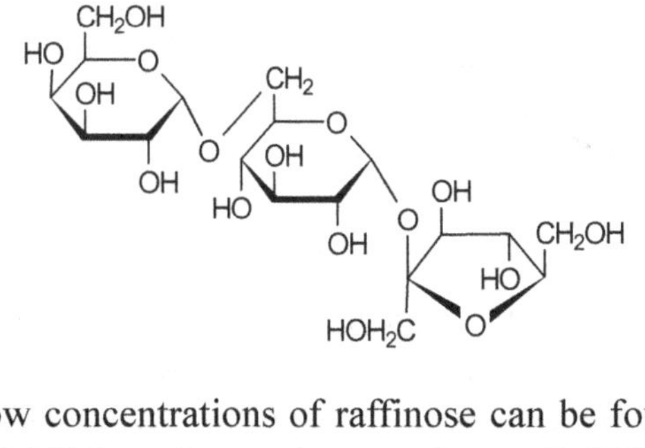

Molar mass 504.44 ($C_{18}H_{32}O_{16}$), melting point 78 (pentahydrate) or 118 (anhydrous) °C.

Low concentrations of raffinose can be found in vegetables, sugar beet (<0.05%), cottonseeds, soya beans (1.9%), and manna (see mannose), which is a sweet-tasting sugary extract (hardens readily in air to granules) from the sap of manna ash or South European flowering ash (*Fraxinus ornus*). Raffinose is a non-reducing trisaccharide corresponding to the structure of *O*-galactosylsucrose. It can be crystallized from the by-product molasses (see sucrose) or extracted from cottonseeds with water followed by precipitation with calcium or barium hydroxide. Its treatment with enzymes (invertase or raffinase) leads to melibiose and sucrose. In addition, it is hydrolyzed to galactose and sucrose by the enzyme α-galactosidase (α-GAL). Since humans and other monogastric animals (e.g., poultry and pigs) do not possess this enzyme in the main intestines, raffinose passes undigested through the stomach and upper intestine. However, it is fermented in the lower intestine by gas-producing bacteria that possess the α-GAL enzyme resulting in the flatulence (i.e., the formation of carbon dioxide and methane) commonly associated with eating beans and other vegetables. Raffinose is fermented by low fermentation yeast and partly by top yeast (baker's yeast). It is a white, combustible, and crystalline powder with a sweet taste and dissolves in water (1 g/7 mL) and slightly in ethanol. Its specific rotation $[\alpha]_D$ is +123 (*c* 2, H_2O). Raffinose is used in bacteriology and in the preparation of other carbohydrates.

Planteose, α-D-galactopyranosyl-(1→6)-β-D-fructofuranosyl-α-D-glucopyranoside, α-D-Gal*p*-(1→6)-β-D-Fru*f*-(2↔1)-α-D-Glc*p*

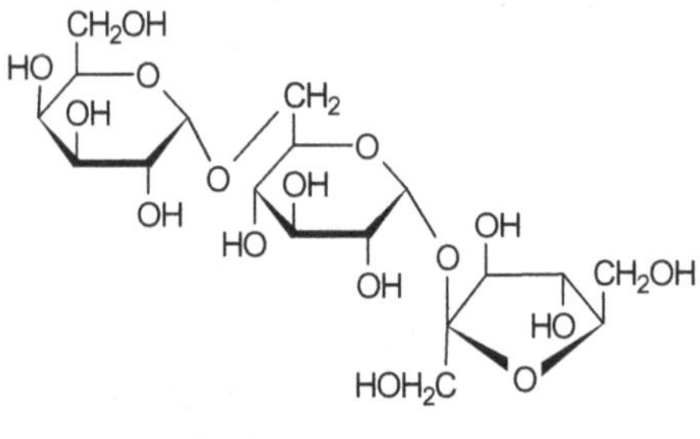

Molar mass 504.44 ($C_{18}H_{32}O_{16}$), melting point 124°C.

Planteose is found, for example, in the seeds of broadleaf plantain or greater plantain (*Plantago major*) and blond plantain (*P. ovata*). It can be extracted from these seeds with methanol, after which sucrose is removed from the extracted sugars by fermentation, and the product is finally purified by chromatography. Planteose is a non-reducing trisaccharide and its specific rotation $[\alpha]_D$ is +130 (*c* 5, H_2O).

Meletsitose, α-D-glucopyranosyl-(1→3)-β-D-fructofuranosyl-α-D-glucopyranoside, α-D-Glc*p*-(1→3)-β-D-Fru*f*-(2↔1)-α-D-Glc*p*

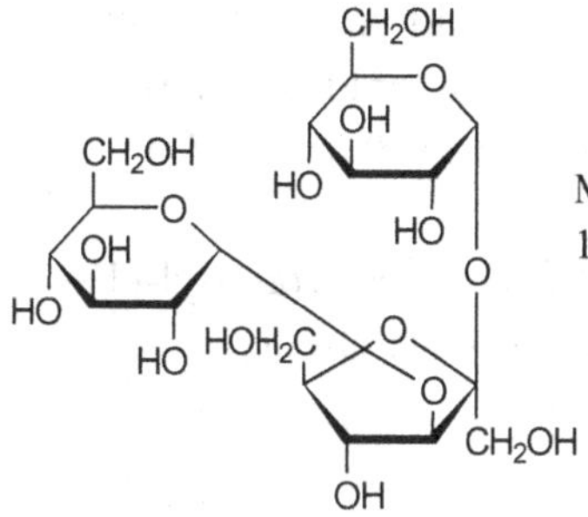

Molar mass 504.44 ($C_{18}H_{32}O_{16}$), melting point 148°C (dihydrate).

Meletsitose is found in in the sap of many trees (e.g., limes and poplars) and manna, which is secreted to injured spots of certain trees caused by insects. It is primarily obtained from honey; in this procedure, honey is first diluted with ethanol and then meletsitose is separated by centrifugation. However, honey with a high concentration of this trisaccharide cannot be used as food of bees, because meletsitose is not hydrolyzed by the invertase enzymes. It is a non-reducing trisaccharide and its specific rotation $[\alpha]_D$ is +88.2 (*c* 4, H_2O).

Maltotriose, α-D-glucopyranosyl-(1→4)-α-D-glucopyranosyl-(1→4)-D-glucopyranose, α-D-Glc*p*-(1→4)-α-D-Glc*p*-(1→4)-D-Glc*p*, amylotriose

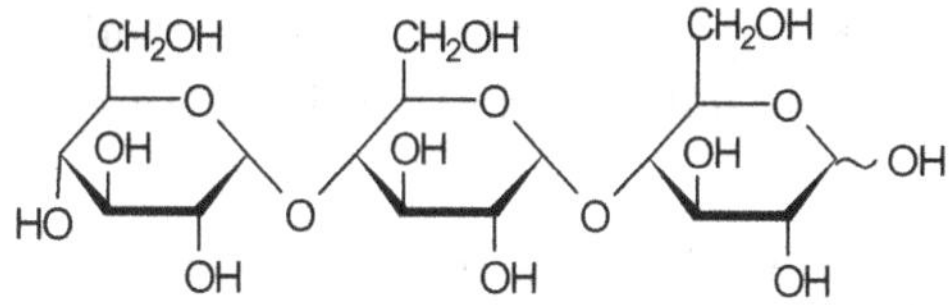

Molar mass 504.44 ($C_{18}H_{32}O_{16}$), melting point 148°C (dihydrate).

Maltotriose cannot be found in nature, although it can be accumulated to some extent in certain animal tissues as a result of the influence of α-amylase (a common enzyme in human saliva) on glycogen. Since it can be considered to be a structural unit of starch (cf., amylose and amylopectin parts), the acid hydrolysis of starch leads to a product from which this trisaccharide can be separated from other glycosyl oligosaccharides by chromatography. The treatment of amylose with α-amylase results in maltose, maltotriose, and maltotetraose from which the latter one is gradually converted into maltose. On the other hand, by stopping the reaction at an appropriate stage, it is possible to convert amylose mainly to maltotriose and maltose, which can be removed by fermentation. The final product is then purified by chromatography. Maltotriose is gradually fermented by common yeasts and its specific rotation $[\alpha]_D$ is +160 (*c* 1, H_2O).

Cellotriose, β-D-glucopyranosyl-(1→4)-β-D-glucopyranosyl-(1→4)-D-glucopyranose, β-D-Glc*p*-(1→4)-β-D-Glc*p*-(1→4)-D-Glc*p*

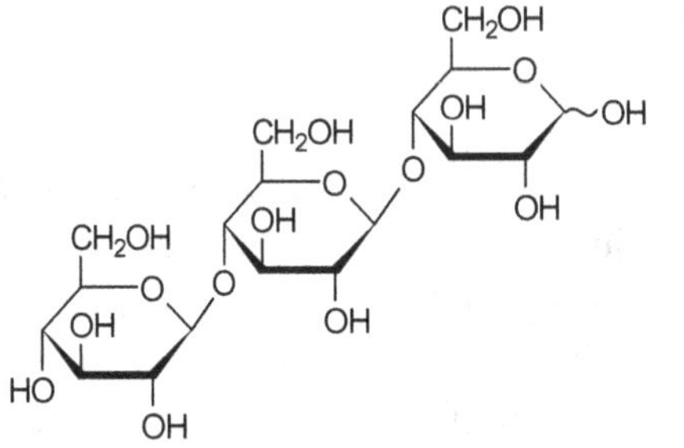

Molar mass 504.44 ($C_{18}H_{32}O_{16}$), melting point 209°C.

The hendecaacetate of cellotriose can be obtained from cellulose by acetolysis and the product can be purified by chromatography. The deacetylation of this derivative leads to cellotriose. It is also formed by a partial acid hydrolysis of cellulose with fuming hydrochloric acid, or enzymatically. The synthetic preparation of this trisaccharide is based on the Koenigs-Knorr reaction (Wilhelm Koenigs 1851–1906 and Eduar Knorr 1867–1926) starting from 2,2′,3,3′,4,6,6′-hepta-*O*-acetyl-α-cellobiosylbromide and 1,2,3,4-tetra-*O*-acetyl-D-glucose. Its specific rotation $[\alpha]_D$ is +22 (*c* 4, H_2O).

Panose, α-D-glucopyranosyl-(1→6)-α-D-glucopyranosyl-(1→4)-D-glucopyranose, α-D-Glc*p*-(1→6)-α-D-Glc*p*-(1→4)-D-Glc*p*

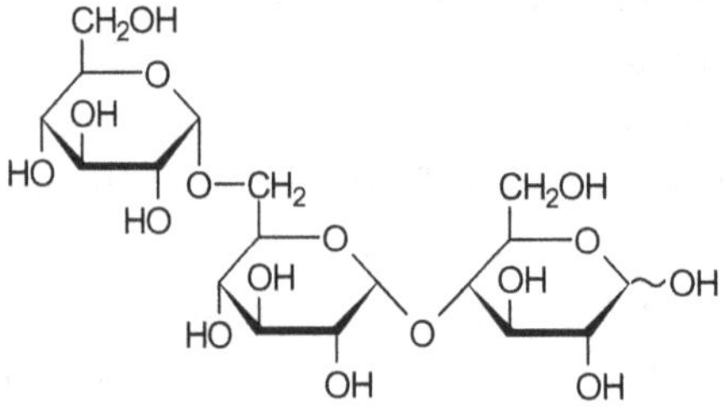

Molar mass 504.44 ($C_{18}H_{32}O_{16}$), melting point 213°C (decomposes).

Panose is a structural unit of starch and glycogen and has been separated, for example, from wort and beer. It is mainly prepared synthetically from maltose by enzymatic (D-glucosyltransferase) treatment and the reaction product is purified by various fermentation, precipitation, and chromatographic methods. Its specific rotation $[\alpha]_D$ is +154 (*c* 2,H_2O).

Examples of other oligosaccharides are as follows:

Cellotetraose, β-D-glucopyranosyl-[(1→4)-β-D-glucopyranosyl]$_2$-(1→4)-D-glucopyranose, β-D-Glc*p*-[(1→4)-β-D-Glc*p*]$_2$-(1→4)-D-Glc*p*

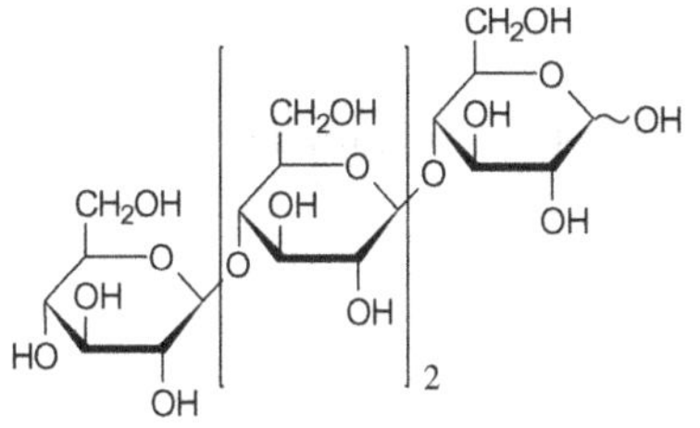

Molar mass 666.58 ($C_{24}H_{42}O_{21}$), melting point 252°C (decomposes), $[\alpha]_D$ +16.5 (*c* 3.4, H_2O).

Maltotetraose, α-D-glucopyranosyl-[(1→4)-α-D-glucopyranosyl]$_2$-(1→4)-D-glucopyranose, α-D-Glc*p*-[(1→4)-α-D-Glc*p*]$_2$-(1→4)-D-Glc*p*, amylotetraose

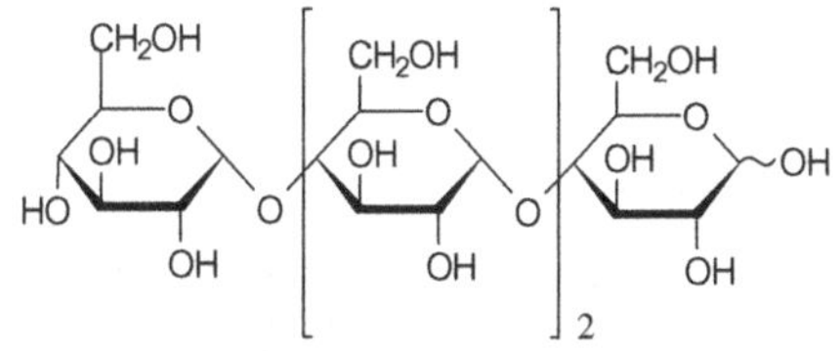

Molar mass 666.58 ($C_{24}H_{42}O_{21}$), $[\alpha]_D$ +165.5 (*c* 5, H_2O).

Stachyose, β-D-fructofuranosyl-α-D-galactopyranosyl-(1→6)-α-D-galactopyranosyl-(1→6)-α-D-glucopyranoside, α-D-Gal*p*-(1→6)-α-D-Gal*p*-(1→6)-α-D-Glc*p*-(1↔2)-β-D-Fru*f*

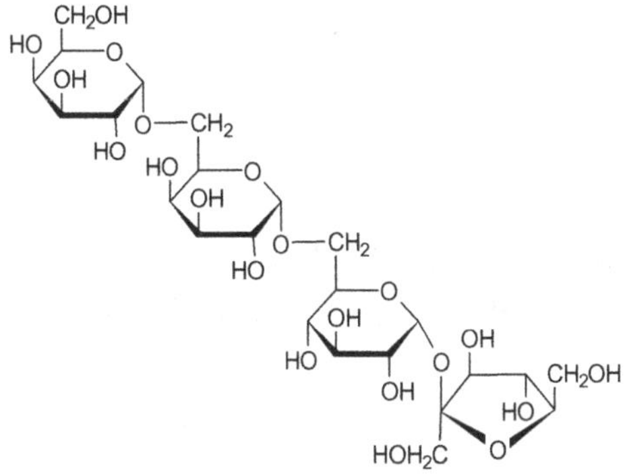

Molar mass 666.58 ($C_{24}H_{42}O_{21}$), melting point 101°C (hydrate).

Stachyose occurs in several plants often together with sucrose and raffinose. It has been separated, for example, from the rootstocks of the plants (e.g., crosne or Chinese artichoke, *Stachys affinis*) belonging to the genus *Stachys*, from which its trivial name is derived. It can also be isolated from soya beans (*Glycine max* or *Soja hispida*), sap of manna ash or South European flowering ash (*Fraxinus ornus*), seeds (lupin beans) of annual yellow lupin or European yellow lupin (*Lupinus luteus*), and sprigs of winter jasmine or Indian jasmine (*Jasminum multiflorum*). Its specific rotation $[\alpha]_D$ is +133 (*c* 4.5, H_2O).

Cellopentaose, β-D-glucopyranosyl-[(1→4)-β-D-glucopyranosyl]$_3$-(1→4)-D-glucopyranose, β-D-Glc*p*-[(1→4)-β-D-Glc*p*]$_3$-(1→4)-D-Glc*p*

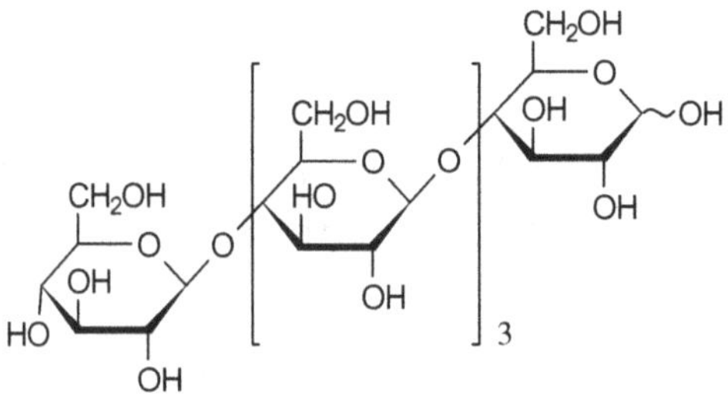

Molar mass 828.73 ($C_{30}H_{52}O_{26}$).

Cellohexaose, β-D-glucopyranosyl-[(1→4)-β-D-glucopyranosyl]$_4$-(1→4)-D-glucopyranose, β-D-Glc*p*-[(1→4)-β-D-Glc*p*]$_4$-(1→4)-D-Glc*p*

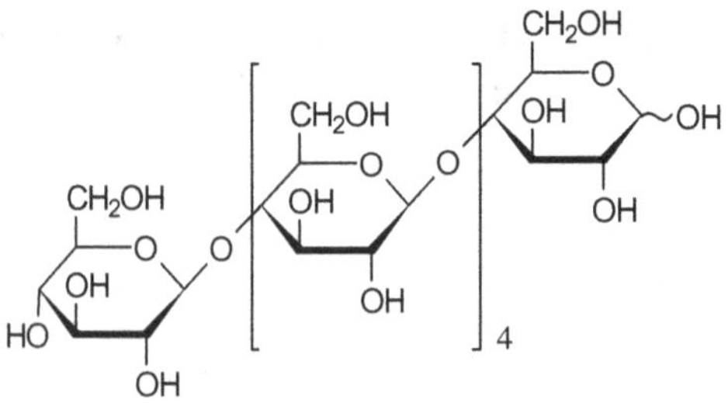

Molar mass 990.87 ($C_{36}H_{62}O_{31}$), melting point 275–8°C (decomposes), $[\alpha]_D$ +10 (*c* 1.2, H_2O) (30°C).

Celloheptaose, β-D-glucopyranosyl-[(1→4)-β-D-glucopyranosyl]$_5$-(1→4)-D-glucopyranose, β-D-Glc*p*-[(1→4)-β-D-Glc*p*]$_5$-(1→4)-D-Glc*p*

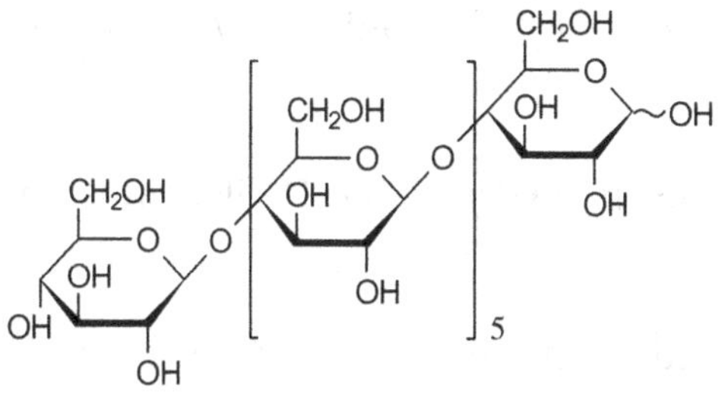

Molar mass 1153.01 $C_{42}H_{72}O_{36}$), melting point 283–6°C (decomposes), $[\alpha]_D$ +7 (*c* 0.1, H_2O) (30°C).

Isomaltotriose, α-D-glucopyranosyl-(1→6)-α-D-glucopyranosyl-(1→6)-D-glucopyranose, α-D-Glc*p*-(1→6)-α-D-Glc*p*-(1→6)-D-Glc*p*. Molar mass 504.44 ($C_{18}H_{32}O_{16}$).

Isopanose, α-D-glucopyranosyl-(1→4)-α-D-glucopyranosyl-(1→6)-D-glucopyranose, α-D-Glc*p*-(1→4)-α-D-Glc*p*-(1→6)-D-Glc*p*. Molar mass 504.44 ($C_{18}H_{32}O_{16}$).

Manninotriose, α-D-galactopyranosyl-(1→6)-α-D-galactopyranosyl-(1→6)-D-glucopyranose, α-D-Gal*p*-(1→6)-α-D-Gal*p*-(1→6)-D-Glc*p*. Molar mass 504.44 ($C_{18}H_{32}O_{16}$).

Umbelliferose, β-D-fructofuranosyl-α-D-galactopyranosyl-(1→2)-α-D-galactopyranoside, α-D-Gal*p*-(1→2)-α-D-Gal*p*-(1↔2)-β-D-Fru*f*. Molar mass 504.44 ($C_{18}H_{32}O_{16}$).

Verbascose, β-D-fructofuranosyl-α-D-galactopyranosyl-[(1→6)-α-D-galactopyranosyl]$_2$-(1→6)-α-D-glucopyranoside, α-D-Gal*p*-[(1→6)-α-D-Gal*p*]$_2$-(1→6)-α-D-Glc*p*-(1↔2)-β-D-Fru*f*. Molar mass 828.73 ($C_{30}H_{52}O_{26}$), melting point 220°C, $[\alpha]_D$ +169 (*c* 5, H_2O).

Hystose, β-D-fructofuranosyl-(2→1)-β-D-fructofuranosyl-(2→1)-β-D-fructofuranosyl-α-D-glucopyranoside, β-Fru*f*-(2→1)-β-D-Fru*f*-(2→1)-β-D-Fru*f*-(2↔1)-α-D-Glc*p*. Molar mass 666.58 ($C_{24}H4_{42}O_{21}$).

Isokestose, β-D-Fru*f*-(2→1)-β-D-Fru*f*-(2↔1)-α-D-Glc*p*. Molar mass 504.44 ($C_{18}H_{32}O_{16}$).

Kestose, β-D-Fru*f*-(2→6)-β-D-Fru*f*-(2↔1)-α-D-Glc*p*. Molar mass 504.44 ($C_{18}H_{32}O_{16}$).

Neokestose, β-D-Fru*f*-(2→6)-α-D-Glc*p*-(1↔2)-β-D-Fru*f*. Molar mass 504.44 ($C_{18}H_{32}O_{16}$).

Laminaratiose, β-D-Glc*p*-(1→3)-β-D-Glc*p*-(1→3)-D-Glc*p*. Molar mass 504.44 ($C_{18}H_{32}O_{16}$).

Cyclodextrins, sometimes called "cycloamyloses" or "cyclomaltooligosaccharides", are a family of compounds made up of α-D-glucopyranose units bound chemically together in a ring. They are enzymatically prepared from starch by cyclodextrin glycosyltransferase (CGTase) employed along with α-amylase. Cyclodextrins are composed of five or more sugar units linked together by (1→4)-glucosidic bonds and the toroid structures have interiors that are considerably less hydrophilic than the aqueous environment and are able to host hydrophobic molecules. In contrast, the exterior surfaces are sufficiently hydrophilic to ensure the water solubility of the toroids. Cyclodextrins are easily crystallized and are typically constituted by six, seven, or eight glucopyranose units. The trivial names of these well-established structures are α-, β-, and γ-cyclodextrin, respectively, and they are also traditionally known by the symbols α-CD, β-CD, and γ-CD:

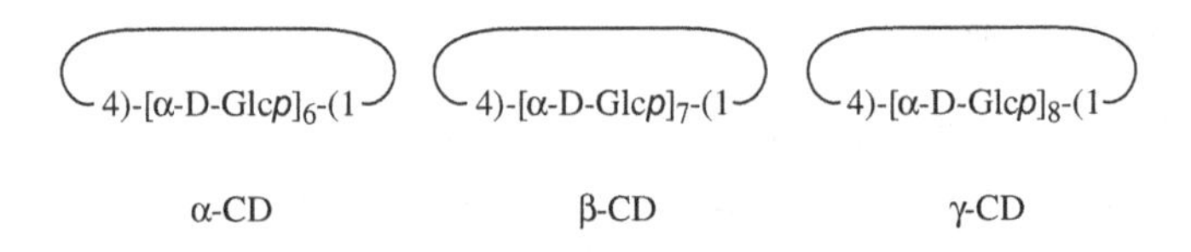

Since cyclodextrins can form host-guest complexes with hydrophobic molecules (i.e., certain organic molecules or metals), they have found a number of interesting applications in a wide range of fields including foodstuff, fertilizer, and pharmaceutical applications, as well as environmental protection and synthetic chemistry. They are generally less volatile, less reactive, and more soluble than their uncomplexed components alone. In addition, they are crystalline, even if the complexed compound is liquid. The chemical structure of the most common cyclodextrin, β-cyclodextrin, is as follows:

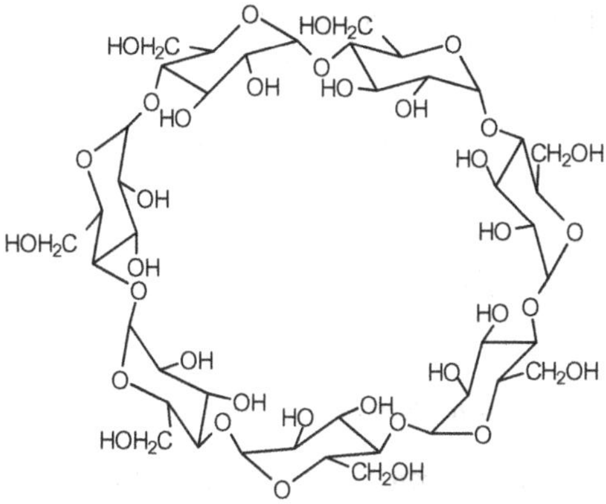

9.3.3. *Deoxy disaccharides*

Examples of rare *deoxy disaccharides* generally known as their trivial names are as follows:

Rutinose, α-L-rhamnopyranosyl-(1→6)-D-glucose, 6-*O*-α-L-rhamnopyranosyl-D-glucose

Cacotriose, α-L-rhamnopyranosyl-(1→2)-α-L-rhamnopyranosyl-(1→4)-D-glucose

Solatriose, α-L-rhamnopyranosyl-(1→2)-β-D-glucopyranosyl-(1→3)-D-galactose

9.3.4. *Disaccharide alcohols*

Examples of reduced disaccharides (*disaccharide alcohols*) suitable for sweetener agents are as follows:

Maltitol, 4-*O*-(α-D-glucopyranosyl)-D-glucitol

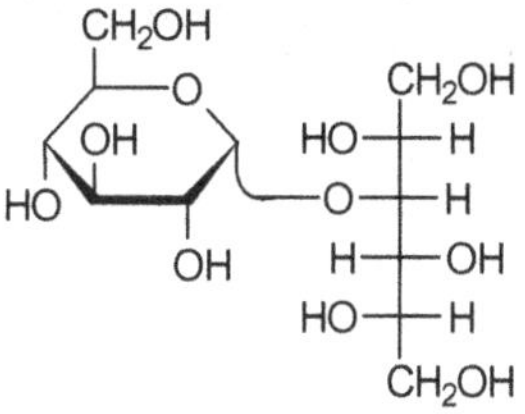

Molar mass 344.32 ($C_{12}H_{24}O_{11}$).

Lactitol, 4-*O*-(β-D-galactopyranosyl)-D-glucitol

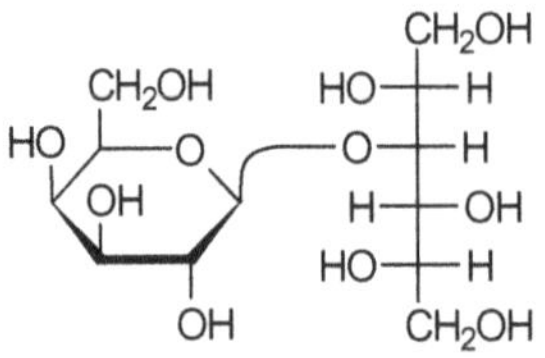

Molar mass 344.32 ($C_{12}H_{24}O_{11}$).

9.3.5. *Disaccharide uronic acids*

Examples of *disaccharide uronic acids* are as follows:

Cellobiouronic acid, 4-*O*-β-D-glucopyranuronosyl-D-glucose

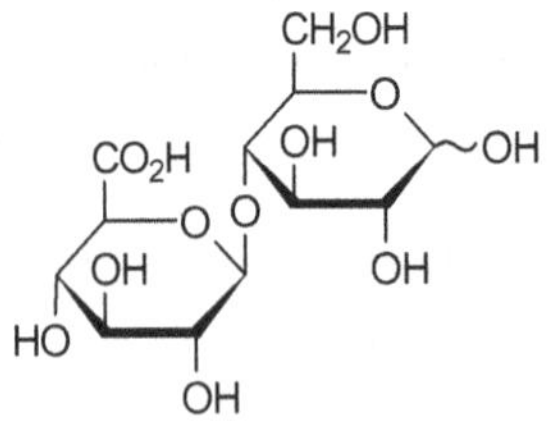

Molar mass 356.28 ($C_{12}H_{20}O_{12}$), melting point 189°C, $[\alpha]_D$ +7 (*c* 2, H_2O).

2-*O*-(4-*O*-Methyl-α-D-glucopyranuronosyl)-D-xylopyranose

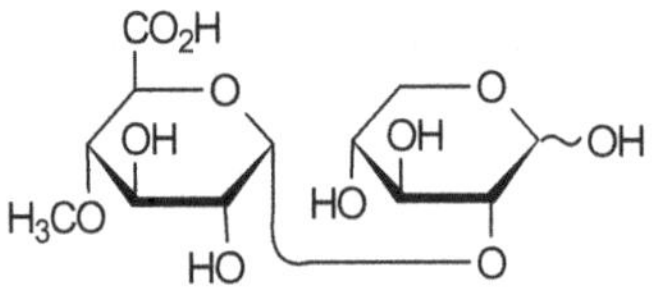

Molar mass 340.28 ($C_{12}H_{20}O_{11}$), $[\alpha]_D$ +110 (*c* 1, H_2O).

6-*O*-β-D-Glucopyranuronosyl-D-galactopyranose

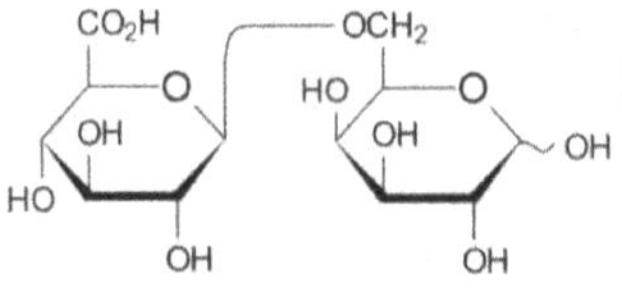

Molar mass 356.28 ($C_{12}H_{20}O_{12}$), melting point 116°C, $[\alpha]_D$ −8.3 (*c* 5, H_2O).

4-*O*-α-D-Galactopyranuronosyl-D-galactopyranuronic acid

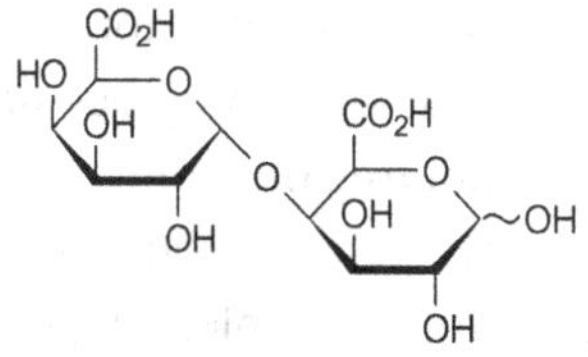

Molar mass 370.27 ($C_{12}H_{18}O_{13}$).

9.3.6. *Disaccharide nitrogen derivatives*

Examples of the nitrogen-containing derivatives of disaccharides are as follows:

2-Amino-2-deoxy-4-*O*-(α-D-glucopyranosyl)-D-glucopyranose

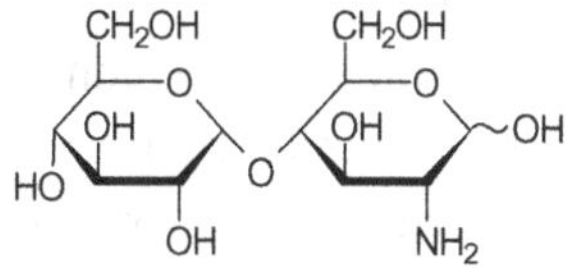

Molar mass 341.31 ($C_{12}H_{23}O_{10}N$).

Hyalobiouronic acid, 2-amino-2-deoxy-3-*O*-(β-D-glucopyranuronosyl)-D-glucopyranose

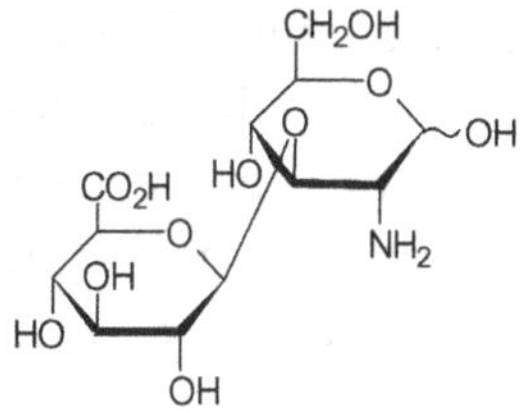

Molar mass 355.30 ($C_{12}H_{21}O_{11}N$), melting point 190°C (decomposes), $[\alpha]_D$ +30 (*c* 1, 0.1 N HCl).

Condrosine, 2-amino-2-deoxy-3-*O*-(β-D-glucopyranuronosyl)-D-galacto-pyranose

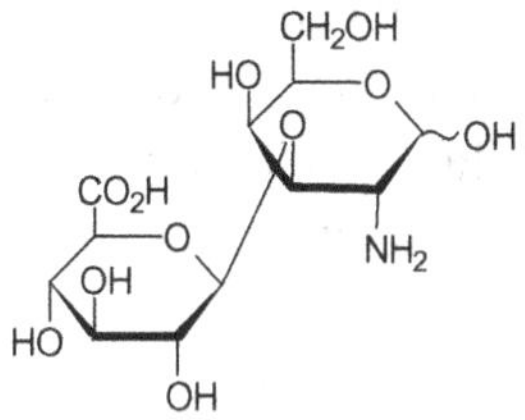

Molar mass 355.30 ($C_{12}H_{21}O_{11}N$), $[\alpha]_D$ +42 (*c* 2, H_2O).

9.4. *Polysaccharides* and Their Derivatives

Polysaccharides are complex carbohydrates that consist of more than nine monosaccharide units covalently bound together by glycosidic linkages (cf., Chapt. "1. Introduction"). Their structures may range from linear to highly branched polymers. These polymeric carbohydrates occur naturally as such or as structural parts, for example, of peptidoglycans and lipopolysaccharides (cf., Chapt. "10.5.1. Hormones"). They can be roughly divided into plant (e.g., cellulose, hemicelluloses, starch, pectins, and gums) and animal (e.g., glycogen and chitin) polysaccharides. In addition, polysaccharides produced by algae, fungi, and microbes are often separately classified. The trivial names of polysaccharides typically indicate their origin.

The general term "glycan" (or "homoglycan" or "homopolysaccharide") refers to polysaccharides that contain only one type of monosaccharide unit (i.e., glycose). For example, a glucan means a glycan with a repeating unit of glucose. Similarly, established names are normally used for arabinan, xylan, galactan, mannan, and fructan. It is also possible that the accurate configuration of the monosaccharide unit, as well as the carbon atoms participating in the formation of a glycosidic bond, are shown; for example, cellulose $[\rightarrow 4)$-β-D-Glc*p*-$(1\rightarrow]_n$, is alternatively called "(1→4)-β-D-glucan" or "(1→4)-β-D-glucopyranan". There are also many similar specific names, such as that of pectin or (1→4)-α-D-galacturonan with repeating α-D-galacturonic acid units. Polysaccharides that contain more than one type of monosaccharide unit are generally called "heteroglycans" or "heteropolysaccharides". Examples of these heterogeneous polysaccharides are natural hemicelluloses.

It has been estimated that over 90% of the carbohydrate mass in nature consists of polysaccharides; they also form a significant part of the industrial utilization of carbohydrates (cf., Chapt. "12.1. General Aspects"). Commercially, the most important polysaccharides are cellulose (a water-insoluble material), starches (only swell in water), and water-soluble gums. In addition, a wide range of industrial cellulose derivatives exist. The most abundant polysaccharides and their derivatives are briefly described in this section.

9.4.1. *Cellulose and its derivatives*

Chemical structure of **cellulose** can be presented by stereochemical (1), abbreviated (2), Haworth perspective (3), and Mills (4) formulas:

1)

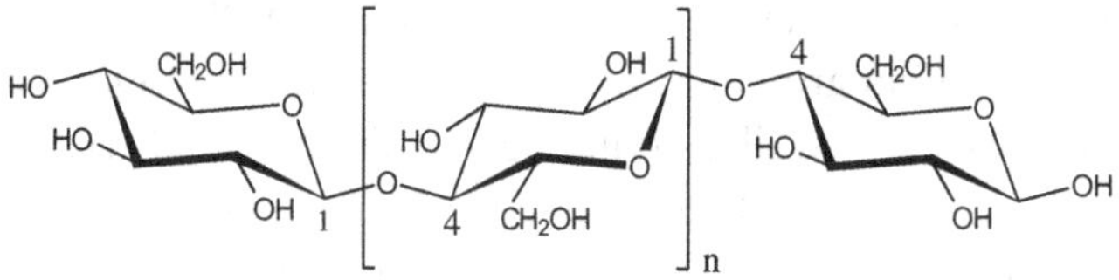

2) β-D-Glc*p*-(1 → [4)-β-D-Glc*p*-(1 →]$_n$ 4)-β-D-Glc*p*

3)

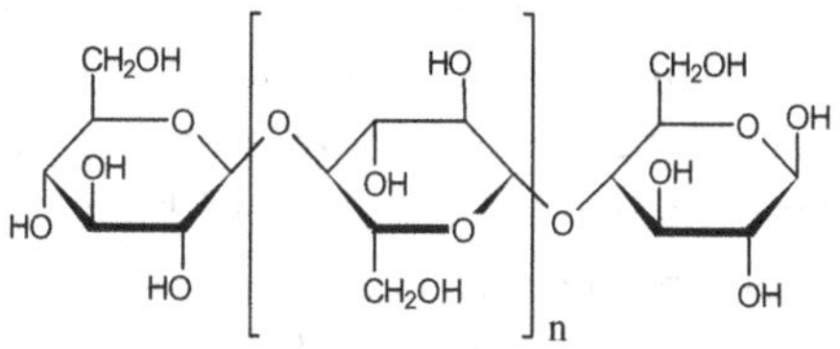

4)

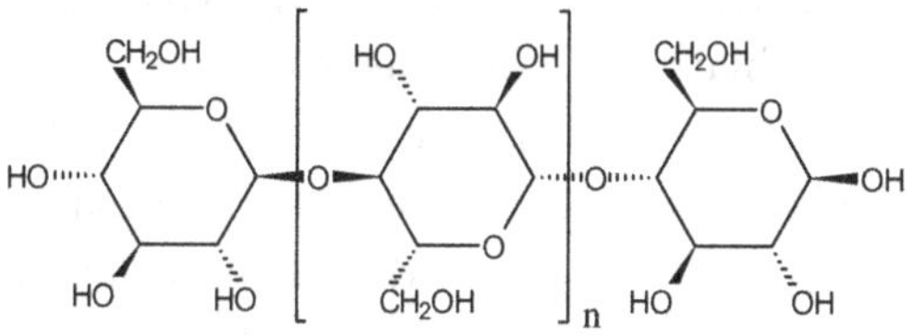

Cellulose is the world's most abundant natural biopolymer and thus, the major organic component in most biomass materials. It has been estimated that almost half of the biomass formed in the photosynthesis consists of this polysaccharide; globally about 10^{12} tons of carbon dioxide are chemically bound by photosynthesis each year (cf., Chapt. "8.1. Photosynthesis"). Almost all cellulose is formed by photosynthesis (i.e., "plant cellulose"), but there are also microbial extracellular carbohydrates, such as "biocellulose" or "bacterial cellulose" (BC) that are synthesized by various bacteria. The most common bacteria responsible for the biosynthesis of BC belong to the genera *Acetobacter*, *Achromobacter*, *Aerobacter*, *Agrobacterium*, *Alcaligenes*, *Pseudomonas*, *Rhizobium*, *Sarcina*, and *Zoogloea*. However, its most effective producer is a

Gram-negative bacterium *Gluconacetobacter xylinus* (formerly known as *Acetobacter xylinum*), which alone produces enough cellulose to justify commercial interest. Plant cellulose and BC have the same chemical structure, but somewhat different physical properties.

Cellulose is distributed in its native form throughout the plant kingdom. The radial growth of wood begins in the cambium, composed of a single layer of thin-walled living cells (initials) filled with protoplasm. On division, the initial cell produces a new initial and a xylem mother cell, which in turn gives rise to daughter cells; each of the latter is capable of further division, and so on. More cells are produced towards the xylem on the inside than towards the phloem on the outside. For this reason, trees always contain much more wood than bark. During the following phase of cell development (i.e., cell wall thickening), the formation of different cell wall layers is initiated. At the same time, the formation of lignin that contains aromatic units begins mainly from pectic substances (i.e., lignification). Thus, lignin supports the tissues of higher plants (as trees) and some algae. The cell architecture is basically composed of two separate layers: the relatively thin primary wall (P, the outer layer, thickness 0.05–0.1 μm) and the thick secondary wall (S), which can be further divided into three sublayers termed "the outer layer of the secondary wall" (S_1 or S1, thickness 0.1–0.3 μm), "the middle layer of the secondary wall" (S_2 or S2, the thickest one, 1–8 μm), and "the inner layer of the secondary wall" (S_3 or S3, thickness <0.1 μm). All these layers differ from one another in the different orientation of their cellulose-containing microfibrils and, to some extent, in their chemical composition. A wood cell consists mainly of celluloses, hemicelluloses, and lignin. A simplified picture is that cellulose forms a skeleton (i.e., microfibril-containing lamellas) that is surrounded by other substances functioning as matrix (hemicelluloses) and encrusting materials (see lignin, below).

The major part (about 90 and 65 vol.-% in softwood and hardwood xylem, respectively) of the wood cells (called "tracheids" or "fibers") are responsible for support (in softwoods and hardwoods) or water conduction (in softwoods). All cell development phases take place during a few weeks, resulting in the death of the cell after lignification with the simultaneous formation of a wood fiber or a pulp fiber. The central cavity of this hollow fiber is termed "lumen" (L). The middle lamella

(ML, contains 60–70% lignin, thickness 0.2–1.0 μm) is located between the P walls of adjacent cells and serves the function of binding the cells together. Since it is difficult to distinguish ML from the two P walls on either side, the term "compound middle lamella" (CML) is generally used to designate the combination of ML with the two adjacent P walls. In general, the cell wall thickening depends on the time of year (earlywood or latewood, S_2 thicknesses 1–4 μm or 3–8 μm, respectively) and/or the function of the cell. Based on detailed microscopy observations, the cell was discovered in 1663 by Robert Hook (1635–1703), who named this biological unit for its resemblance to cells (in Latin, "cella" means "small room") inhabited by Christian monks in a monastery.

The content of cellulose in plant materials and technical products varies depending on origin. Examples of the constitution of different materials (as a percentage of the dry solids) are as follows:

Material	Cellulose	Hemicelluloses	Lignin
Softwoods	40–45	25–30	25–30
Hardwoods	40–45	30–35	20–25
Fiber plants (e.g., flax, cotton, hemp, and sisal)	70–95	5–25	<5
Natural-growing reeds and agricultural residues	25–45	25–50	10–30
Chemical pulps	65–80	20–30	<5

These materials also contain extractives (3–10%) and, especially fiber plants and agricultural residues (e.g., maize husks, and stalks) 5–20% proteins and inorganic compounds (primarily SiO_2).

Elucidation of the chemical origin of cellulose dates back to 1837–42, when the French botanist Anselme Payen noted that most plant materials contain a relatively resistant fraction with essentially the same elemental composition (i.e., $C_6H_{10}O_5$). This fibrous material isolated as a residue in the nitric acid treatment of plant materials was named according to his suggestion "cellulose" (i.e., the basic constituent of the cell wall of plants, see above) and this name was "confirmed" in 1839 by the French Academy. The incrusting material (in French "les matières encrustantes" — $C_{35}H_{24}O_{10}$) of cellulose was dissolved and isolated with solutions of nitric acid and sodium hydroxide and was named "lignin" (in Latin, "lignum" means "wood") in 1865 by Franz Eilhard Schulze (1840–1921). This term was probably used in a similar context already in 1819 by Augustin Pyramus de Candolle (1778–1841).

Cellulose is a polydispersed, completely linear homopolysaccharide that consists of repeating β-D-glucopyranose (β-D-Glc*p*) moieties (in a $^4C_1$ conformation) linked together by (1→4)-glycosidic (or glucosidic) bonds. In the $^4C_1$ conformation, all substituents (C_1-OR, C_2-OH, C_3-OH, C_4-OR, and C_5-CH_2OH) of the β-D-glucopyranose chain units are equatorially oriented, making the chain very stable due to the minimized interaction between the pyranose ring substituents (cf., Chapt. "6.3. Naming of Conformations"). It should be noted that the cellulose chain has both reducing (C_1-OH) and non-reducing (C_4-OH) units in its molecular structure.

The degree of polymerization (DP) of native wood cellulose is of the order of about 10,000 and is lower than that of cotton cellulose (about 15,000). These DP values correspond to molar masses of 1.6 million Da and 2.4 million Da, and to molecular lengths of 5.15 μm and 7.73 μm, respectively. In technical processes, such as chemical pulping (cf., "12.4. Chemical Delignification"), the DP of cellulose can decrease to 500–2000, corresponding to molecular lengths of 0.26 μm and 1.03 μm, respectively (in regenerated cellulose DP is 200–300). The polydispersity (M_w/M_n) of cellulose is rather low (<2), indicating that the weight average molar mass (M_w) and the number average molar mass (M_n) do not deviate much from each other. The biosynthesis of cellulose is briefly discussed in Chapt. "8.2. Polysaccharides".

Because of the strong tendency for intra- and intermolecular hydrogen bonding, bundles of cellulose molecules aggregate to microfibrils and further to fibrils (and finally lamellas in the cell wall) that contain both highly ordered crystalline (60–75% of the total cellulose and 50–150 nm in length) and less ordered (disordered) amorphous regions (25–50 nm in length). Several different crystalline structures of cellulose are also known: celluloses I, II, III, and IV. The detailed crystalline structures of these polymorphous lattices of cellulose have been determined from X-ray diffraction data. The unit cell of native cellulose (i.e., cellulose I) generally consists of four β-D-glucopyranose residues and, for example, in the chain direction the repeating unit is a cellobiose residue (1.03 nm in length). Furthermore, cellulose I contains two coexisting phases: triclinic cellulose I_α and monoclinic cellulose I_β. In practice, cellulose produced by bacteria and algae is enriched in the I_α form, while I_β is the major form in higher plants. The regenerated cellulose (i.e., monoclinic cellulose II) has a slightly different and more stable crystalline structure than cellulose I. This form is obtained whenever the lattice of cellulose I is destroyed, for example, on swelling with strong alkali (see mercerization, below) or on dissolution of cellulose. The conversion of cellulose I to cellulose II is irreversible. It is possible to produce celluloses III and IV from celluloses I or II with specific chemical treatments.

Microfibrils are relatively stable against mechanical and chemical stress. In nature, lignocellulosic raw materials are rather resistant against, for example, enzymatic digestion; the rate and extent of biochemical degradation of these materials is influenced not only by the effectiveness of the biomass degrading enzymes (secreted by many microorganism, mostly bacteria and fungi) but also by the chemical, physical, and morphological characteristics of the heterogeneous substrate. Efficient degradation of cellulose requires a mixture of different enzymes (cellulases) acting sequentially or in concert; they primarily turn cellulose into hydrolysis products, such as cellobiose and glucose. Endoglucanases (EGs) cleave the cellulose chains internally, exoglucanases (cellobiohydrolases, CBHs) degrade cellulose starting from free chain ends (producing cellobiose), and β-glucosidase, which is needed for total cellulose hydrolysis, hydrolyzes short cellooligosaccharides to glucose. Cellulolytic enzymes produced by various microbes are termed "glycoside hydrolases" (GHs), often including some lignin-modifying catalysts. On the other hand, specific enzymes catalyzing the hydrolysis of hemicelluloses are, for example, endoxylanases ((1→4)-β-D-xylan xylanohydrolases), β-xylosidase ((1→4)-β-D-xyloside xylohydrolase), endomannanases ((1→4)-β-D-mannan mannanohydrolases), and β-mannosidase ((1→4)-β-D-mannoside mannohydrolase), plus many accessory enzymes, such as α-glucuronidase, α-arabinosidase, α-galactosidase, and esterases.

Naturally occurring microbial respiration and decomposition of lignocellulosic materials are significant reactions that release the carbon bound in various plants to the carbon cycle of the Earth. In addition, herbivores (e.g., ruminants and termites) can digest cellulose due to the microorganisms in their gastrointestinal tract. In contrast, humans are not able to degrade cellulose into sugars to utilize it as a source of energy. This is due to the fact that the glucosidic bonds (β-(1$\rightarrow$4)-bonds) in cellulose differ from those in starch (α-(1$\rightarrow$4)-bonds) and the starch-hydrolyzing enzymes (see amylases, below) present in the saliva of humans and some other mammals cannot affect cellulose.

In general, cellulose is relatively inert against chemical treatments; it is also soluble only in a few solvents, although some of them can be used consequently to improve its reactivity and accessibility. Different solvents may cause interfibrillar or intrafibrillar swelling or dissolution of cellulose. The extent of swelling depends on the solvent, as well as on the nature of the cellulose used (e.g., the degree of chemical and mechanical treat-ments performed and the content of hemicelluloses and lignin). The prerequisite for swelling is breaking the internal hydrogen bonds between cellulose chains mostly in the amorphous regions (i.e., intercrystalline swelling) and, more effectively, both in the amorphous and crystalline regions (i.e., intracrystalline swelling). When the penetrating agent causes intracrystalline swelling, a cellulose-swelling agent complex is formed. This can be accomplished by concentrated solutions of strong bases or acids in addition to some salts. In the case of limited swelling, swelling agents only combine with the ordered cellulose but do not completely destroy the interfibrillar hydrogen bonding. In contrast, in unlimited swelling the swelling agents are bulky and form complexes with cellulose, resulting in the breakage of the adjacent hydrogen bonds and separation of the cellulose chains with gradual dissolution.

Cellulose swells in electrolyte solutions due to the penetration of hydrated ions that require more space than the water molecules. The water retention of cellulose fibers at a given relative humidity (i.e., sorption of water from the vapor state) varies depending on whether the equilibrium has occurred by sorption or desorption (i.e., hysteresis). In addition, the water uptake continuously decreases after repeating moistening and drying of the fibers. When dry cellulose fibers are exposed to humidity, they adsorb water and the cross section of fibers increases to some extent; at 20, 40, 60, 80, and 100% relative humidity the corresponding water contents (sorption) are roughly 3, 4, 6, 10, and 23% of cellulose, respectively. In desorption the corresponding values are 4, 5, 8, 12, and 24% of cellulose, respectively. Under varying humidity, in contrast to the change in the fiber diameter, the dimensional change in the longitudinal direction is very small.

The most important swelling complex of cellulose is that formed with sodium hydroxide (NaOH as well as other hydroxides; e.g., KOH, LiOH, CsOH, and RbOH), although corresponding "addition compounds" (ROH · MOH, however, "chemically" more likely as hydrates of alkoxides $ROM \cdot H_2O$) are also formed with other inorganic (e.g., LiCl, $ZnCl_2$, LiCSN, and $Ca(CSN)_2$) compounds and organic bases (e.g., tri- and tetraalkyl ammonium hydroxides). These "alkali celluloses" have certain pseudo-stoichiometric relations between alkali and cellulose. The alkaline pre-treatment is extremely important in making intermediates (see the production of cellulose xanthate, below) which are, due to an increased penetration of the reagents into the swollen cellulose structure, more reactive than the original cellulose. For this reason, the activation of cellulose by alkali (NaOH or NH_3) increases its accessibility, particularly when the reaction is accompanied by strong swelling of the cellulose structure (i.e., the subsequent reactions proceed in an alkaline medium). As a strong swelling agent, NaOH also causes changes in the cellulose crystalline structure (i.e., changes in the polymorphous lattices).

The traditional process for making alkali celluloses (i.e., the treatment with a 16–18% NaOH solution at room temperature for varying times) was developed already in 1844 and named "mercerization" after its inventor John Mercer (1791–1866). It should be noted that since the solutions are alkaline, cellulose can be to some extent depolymerized in the presence of oxygen. It is also known that, in addition to making different "cellulosates", extensive swelling can be achieved in solutions of various acids and salts, although the formation of definite complexes is not straightforward. However, when using strong acids (H_2SO_4 or HCl), some hydrolysis of cellulose may occur depending on the pretreatment used. In contrast, H_3PO_4 and HNO_3 do not hydrolyze cellulose, but HNO_3 reacts to some extent with it (see the formation of nitrocellulose below). There are also other activation pre-treatments, including solvent exchange, use of structure-loosening additives or even mechanical action that can be utilized to increase the reactivity of cellulose.

A new area of cellulose chemistry started almost 50 years ago when non-aqueous solvent systems for cellulose, including aprotic ones, were discovered. Solvents for cellulose are central in the preparation of cellulose derivatives, but they are also needed in laboratory work. Those conventional solvents cannot be recovered and reused. For this reason, novel solvents for cellulose have been systematically sought for industrial purposes since the 1970s, especially for rayon and cellophane industries (i.e., for preparing regenerated cellulose fibers). The dissolution of cellulose with proper solvents allows a complete modification of cellulose in a homogeneous system. Some advanced derivatives of cellulose synthesized with new methods cannot even be prepared under heterogeneous conditions.

Solvents for cellulose generally fall into two main categories: (i) non-derivatizing solvents and (ii) derivatizing solvents. Both groups of solvent systems can be established in aqueous or non-aqueous media. The term "non-derivatizing solvent" denotes systems dissolving cellulose only by intermolecular interactions, whereas derivatizing solvents comprise all the systems where the dissolution of cellulose occurs in combination with the covalent derivatization into an unstable ester, ether, or acetal. The cellulose derivative formed in the latter case is readily decomposed to regenerated cellulose by changing the medium or the pH of the system.

The most significant aqueous complex-forming solvents (i.e., solubility via complex formation) for cellulose are ("en" refers to ethylenediamine — $HNCH_2CH_2NH$ — and "T" to tartrate $C_4H_3O_6^{2\ominus}$) as follows:

Abbreviation	Structure	Color
Cuoxam Schweizer's solution	$[Cu(NH_3)_4](OH)_2$	Violet
Nioxam	$[(Ni(NH_3)_6](OH)_2$	Dark blue
Cuen or CED	$[Cu(en)_2](OH)_2$	Violet
Cadoxen	$[Cd(en)_3](OH)_2$	Transparent
Cooxen	$[Co(en)_3](OH)_2$	Dark red
Nioxen	$[Ni(en)_3](OH)_2$	Violet
Zincoxen	$[Zn(en)_3](OH)_2$	Transparent
EWNN	$[FeT_3]Na_6$	Greenish

Thus, this group of non-derivatizing solvents contains mostly inorganic complex-forming compounds: cuprammonium hydroxide (Cuoxam or Schweizer's solution), nickel-ammonium hydroxide (Nioxam), cupriethylenediamine hydroxide (Cuen or CED), cadmium-ethylenediamine hydroxide (Cadoxen), cobalt-ethylenediamine hydroxide (Cooxen), nickel-ethylenediamine hydroxide (Nioxen), zinc-ethylenediamine hydroxide (Zincoxen), and iron sodium tartrate (the sodium salt of ferric tartaric acid or EWNN, in German "Eisen-Weinsäure-Natrium Komplex").

Other similar compounds are ("en" is ethyleneamine, "pp" 1,3-propylenediamine, "tren" tris(2-aminoethyl)amine, and "dien" diethylenetriamine) $Pd(en)(OH)_2$, $Cu(pp)_2(OH)_2$, $Cd(tren)(OH)_2$, $Ni(tren)(OH)_2$, and $Zn(dien)(OH)_2$. Molten salt hydrates such as lithium perchlorate ($LiClO_4{\cdot}3H_2O$) and lithium thiocyanate ($LiSCN{\cdot}2H_2O$), as well as aqueous solutions, such as saturated ammonium thiocyanate (NH_3/NH_4SCN) or calcium thiocyanate ($Ca(SCN)_2$) can dissolve limited amounts of cellulose and were used in the past. In addition, the following organic liquid/salt-solvent systems belong to the first group of the solvents: dimethylamine/lithium chloride (DMA/LiCl), *N,N*-dimethylacetamide/lithium chloride (DMAc/LiCl), dimethyl sulfoxide/tetrabutylammonium fluoride ($\cdot 3H_2O$) (DMSO/TBAF), ethylenediamine/potassium thiocyanate (EDA/KSCN), 1,3-dimethyl-2-imidazolidone/lithium chloride (DMI/LiCl), *N*-methyl-2-pyrrolidone/lithium chloride (NMP/LiCl), and aqueous poly(ethylene glycol)/sodium hydroxide (PEG/NaOH) (Fig. 9.1). An important example of non-derivatizing and non-aqueous solvents for cellulose is *N*-methylmorpholine *N*-oxide (NMMO or NMNO — cf., the commercial Lyocell process).

Examples of aqueous and non-aqueous derivatizing and solubilizing solvents systems for cellulose are as follows. Phosphoric acid/water (H_3PO_4 (>85%)/H_2O, cellulose derivative Cell-O-PO_3H_2), formic acid/zinc chloride ($HCO_2H/ZnCl_2$, Cell-O-C(O)H), trifluoroacetic acid/trifluoroacetic acid anhydride ($CF_3CO_2H/CF_3(CO)_2O$, Cell-O-C(O)CF_3), nitrogen tetroxide/*N,N*-dimethylformamide (N_2O_4/DMF, Cell-O-N=O), trimethylchlorosilane/pyridine ($(CH_3)_3SiCl$/pyridine, Cell-O-$Si(CH_3)_3$), dimethyl sulfoxide/paraformaldehyde (DMSO/PF, Cell-O-CH_2OH), trichloroacetaldehyde/dimethyl sulfoxide/tetraethylammonium chloride (CCl_3CHO/DMSO/TEA, Cell-O-CH(OH)CCl_3, and carbon disulfide/sodium hydroxide/water ($CS_2/NaOH/H_2O$, Cell-O-C(S)SNa).

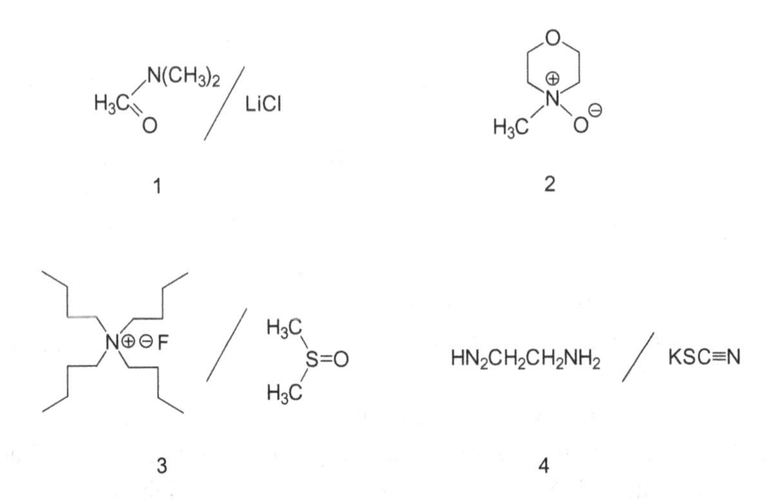

Fig. 9.1. Examples of non-derivatizing solvent systems for cellulose. *N,N*-dimethylacetamide/lithium chloride (DMAC/LiCl) (1), *N*-methylmorpholine *N*-oxide (NMMO) (2), dimethyl sulfoxide/tetrabutylammonium fluoride (DMSO/TBAF) (3), and ethylenediamine/potassium thiocyanate (EDA/KSCN) (4).

Ionic liquids (ILs, Fig. 9.2) are non-aqueous solvents that are organic salts existing as liquids at relatively low temperatures of <100°C and being mostly involatile with high thermal stability. They represent new solvents for cellulose in the preparation of regenerated cellulose and the subsequent synthesis of cellulose derivatives, as well as in the fractionation of lignocellulosic materials. In general, they are non-derivatizing solvents ("green solvents") with strong intra- and intermolecular interactions. Due to the diversity and exceptional properties of ILs, they are currently in the focus of versatile scientific interest for enhancing the efficiency of many electrochemical, synthetic, and analytical processes, and they may still extend their range of applications. It should be noted that as early as 1934 Charles Graenacher suggested that molten *N*-ethylpyridinium chloride in the presence of nitrogen-containing bases could be used to dissolve cellulose. However, this molten salt system was considered at the time somewhat esoteric and, due to its relatively high melting point (118°C), also a novelty of small practical value.

The most important ILs are 1-ethyl-3-methylimidazolium $[C_2mim]^{\oplus}$ chloride (EMIMCl), 1-butyl-3-methylimidazolium $[C_4mim]^{\oplus}$ chloride

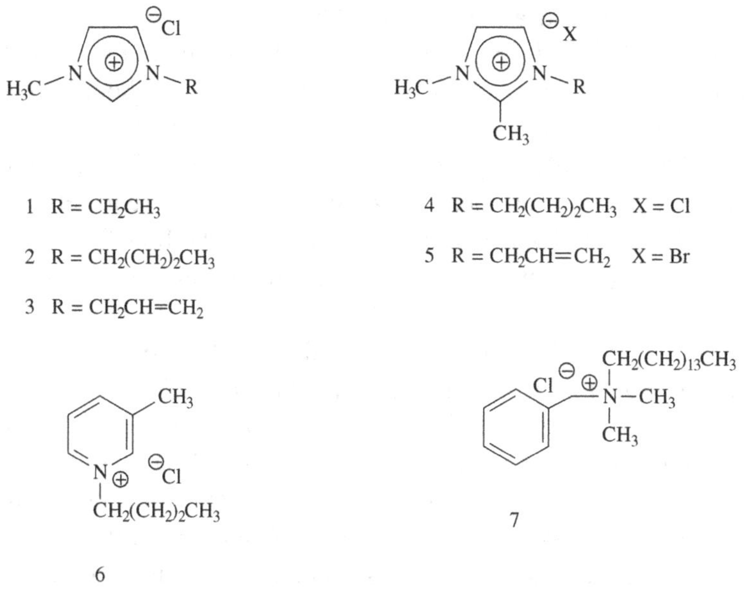

1 $R = CH_2CH_3$

2 $R = CH_2(CH_2)_2CH_3$

3 $R = CH_2CH{=}CH_2$

4 $R = CH_2(CH_2)_2CH_3$ $X = Cl$

5 $R = CH_2CH{=}CH_2$ $X = Br$

6

7

Fig. 9.2. Examples of typical ionic liquids. 1-Ethyl-3-methylimidazolium chloride (1), 1-butyl-3-methylimidazolium chloride (2), 1-allyl-3-methylimidazolium chloride (3), 1-butyl-2,3-dimethylimidazolium chloride (4), 1-allyl-2,3-dimethylimidazolium bromide (5), 1-*N*-butyl-3-methylpyridinium chloride (6), and benzyldimethyl(tetradecyl)ammonium chloride (7).

(BMIMCl) (or tetrafluoroborate BMIM-BF_4), 1-butyl-2,3-dimethylimidazolium $[C_4dmim]^{\oplus}$ chloride (BDMIMCl), 1-allyl-3-methylimidazolium chloride (AMIMCl), 1-allyl-2,3-dimethylimidazolium $[Admim]^{\oplus}$ bromide, 1,3-dimethylimidazolium dimethyl phosphate (DMIMDMP), 1-ethyl-3-methylimidazolium dicyanamide, 1-*N*-butyl-3-methylpyridinium chloride, 1-*N*-ethyl-3-methylpyridinium ethyl sulfate, benzyldimethyl(tetradecyl) ammonium chloride, and 1-*N*-ethyl-3-hydroxymethylpyridinium ethyl sulfate.

The primary driving forces behind the investigation of novel solvents (especially ILs) for cellulose include environmental concerns (i.e., a "green chemistry" approach) and, on the other hand, the ability to create liquid crystals in the new solvent. Room-temperature ILs (or those utilizing slight heating in a microwave oven) can replace conventional organic solvents. The most effective ILs are those containing anions that are strong hydrogen bond acceptors. In addition, it has been claimed that cellulose can be easily regenerated from its IL solutions by adding, for example, water, ethanol or acetone, although other methods have been applied as well. Further research dealing with the economical production of ILs and the physical behavior (viscous or elastic) of different cellulose IL solutions, as well as their utilization possibilities (e.g., effects on cellulose), recyclability, and toxicity (cf., the toxic DMSO/TBAF) is still needed. All these factors are important when promoting industrial application of ILs.

Almost pure cellulose can be prepared by first removing extractives from cotton wool followed by the removal of hemicelluloses with alkaline extraction. Wood-derived cellulose preparations can be obtained by extracting hemicellulose materials from the total polysaccharide fraction of wood meal (termed "holocellulose") with various alkaline solutions. Holocellulose is prepared from the extractive-free wood meal by removing lignin from it; the possible methods include treatments with chlorine-ethanolamine or acidic sodium chlorite (at pH 4). There is also a wide range of other methods for making celluloserich preparations. The products can be characterized with a parameter called "alpha-cellulose". It is the pulp fraction resistant to consecutive extractions with 17.5% and 9.45% sodium hydroxide solutions at 25°C. In this determination, "beta-cellulose" is the soluble fraction reprecipitated on acidification of the solution (mainly degraded cellulose) and "gamma-cellulose" is the fraction remaining in the solution (mainly hemicelluloses).

Nanotechnology involves the manipulation of materials to create products (e.g., nanocomposites) where at least one of the constituent phases has one dimension <100 nm (i.e., between 1 nm and 100 nm, 1 μm = 1000 nm); at 1 nm or below, quantum physics rules and at 100 nm or above, classical physics and chemistry dictate properties of matter. For this reason, "hybrid matter" within a nanoscale dimension area of 1–100 nm can behave differently and provide materials with improved physical and chemical properties such as higher strength.

Cellulose itself has a nanofibrillar structure; it self-assembles into well defined architectures at multiple scales ranging from the nanoscale to the macroscale. Cellulose, as an abundant and renewable feedstock, has thus, great potential to be utilized in nanotechnology. However, the opportunities offered by cellulose and cellulose-containing materials are only recently being realized. The research is still in its early phase and significant research and development investments in the science and engineering are needed to fully exploit this opportunity. The main challenges include, in addition to the development and characterization of new products and their applications, the development of production processes transferable to an industrial scale.

The degree of crystallinity of cellulose can be manipulated during its regeneration and a product with "microcrystallinity" is obtained under different regeneration conditions. The terminology used for varying types of cellulose microfibrils is slightly confusing. **Microfibrillated cellulose** (MFC) was discovered in the 1960s; it can be produced by liberating and degrading substructural fibrils and microfibrils from wood pulp fibers under mechanical action and heat (e.g., the subsequent pre-treatment; disintegration and homogenization). The microfibrils, visible only in an electron microscope, are 10–50 nm wide and even several μm long; they can thus be regarded as "**nanofibrils**" or "**nanocellulose**".

If the product originates from highly quality wood pulp and the crystalline regions are partially isolated from the amorphous regions with mild acid hydrolysis, for example, with hydrochloric acid (i.e., the structure is only moderately degraded), the term "**microcrystalline cellulose**" (MCC) is regularly used. This is also an established term for a commercially available material in the pharmaceutical industry and for materials generally produced by chemical hydrolysis, often in a combination with mechanical milling of some kind. It still seems appropriate to use the descriptive term "MFC", for example, in the context of composite materials.

The properties of MCC (or MFC) depend on the raw material (wood pulp, algae, or bacterial cellulose) used, as well as on the manufacturing conditions. In general, it is a white, odorless, tasteless, and free-flowing powder with advantageous physical properties. For example, MCC produced as dispersion in water or other polar solvents has the appearance of an extremely stable, viscous, and thixotropic gel at a concentration of a few percent. At this concentration it is an excellent suspending medium for other solids and an emulsifying base for organic liquids; this opens commercial uses in foods, paints, and cosmetics. MCC has also been widely used (first known under the trade name "Avicel MCC") as an additive in the direct compression of tablets and capsules of pharmaceutical products (i.e., as a drug carrier or a processing aid) because of its good flowability, compactibility, and compressibility. Data from humans and animals provided no evidence that the ingestion of MCC can cause toxic effects in humans (i.e., it is metabolically inert) when used in foods according to good manufacturing practices. However, toxic effects of nanoparticles (especially in inhalation) are still an open question.

Mineral acid hydrolysis (with HCl or H_2SO_4) of MCC or cellulose fibers (followed by vigorous agitation of the slurry and spray drying) results in highly crystalline, rod-shaped small particles through selective and intensive degradation of the readily accessible material. This product, developed in the early 1990s, is called "**nanocrystalline cellulose**" (NCC) or "**crystalline nanocellulose**" (CNC); its further processing, for example, by ultrasonic treatment and differential centrifugation to needle-like "**NCC whiskers**" or "**cellulose nanowhiskers**" (CNW) has attracted significant attention during the last decade. CNW, representing the smallest subunits of cellulose, are not yet commercial products and are today only made on a pilot scale. Their size depends on the source; the wood-derived products are typically 5–10 nm wide and about 200 nm long. The abbreviation "CNXLs" refers to "**cellulose nanocrystals**", which in practice are identical with CNW.

Due to the rod-like shape of the CNXL particles, the suspensions of it display "liquid crystalline behavior" above a critical concentration. At low CNXL concentrations the suspensions are isotropic with a random arrangement of rod-shaped particles, while at high concentrations the suspensions are anisotropic with the particles packed in a chiral nematic (cholesteric) arrangement. The same phenomenon can be observed when high concentrations of cellulose (>10%) are dissolved in BMIMCl or in NMMO (see above). In these cases, liquid crystalline solutions of cellulose are formed that are optically anisotropic between crossed polarizing filters and display birefringence. In fact, oriented suspensions of MCC were observed already in 1959, although spontaneously anisotropic molecular solutions of cellulose derivatives were not discovered until the mid seventies. At that time, it was noted that a concentrated aqueous solutions of HPC (see below) displays iridescent colors that change with concentration and viewing angle. Later, a wide range of cellulose derivatives (especially cellulose esters) has been found to form both lyotropic and thermotropic cholesteric liquid crystals in different solvents. The discovery of these liquid crystalline derivatives has provided considerable stimulus to the cellulose industry. Possible applications range from their high ability to form fibers of high modulus and tensile strength to their versatile use in modern liquid-crystal display devices and screens.

Composite materials are today used in an enormous amount of applications due to their versatility and wide applicability. NCC and CNW have potential for many such applications as reinforcements in nanocomposites or, after chemical modification of their surfaces, for various other purposes. The properties of these polymers can be improved by changes in the composition of the reinforcing and matrix phases. The very large specific area of CNW results in increased interaction with the matrix polymer on molecular level, which leads to materials with new properties. The average microfibril aspect ratio (fiber length divided by diameter) and fibrillar structure of NCC (and MCC) control the reinforcement effects. Since NCC has, compared to chemical pulps, a very high tensile strength and Young's modulus, this material is a good reinforcing filler for various composite materials. The properties of NCC are different from those of MCC and offer further opportunities.

As already discussed in Chapt. "2.2. The Time After the 1700s", cellulose is utilized after the chemical separation of lignin, extractives, and most hemicelluloses from wood by kraft (sulfate) or various sulfite processes, predominantly as **chemical pulp** (annually about 130 million tons) in the form of paper and board. The annual amounts of high-yield pulps (i.e., **semichemical pulp**), **mechanical pulps**, and non-wood-derived pulps are about 45 million tons and about 20 million tons, respectively. At present, about 90% of chemical pulps are produced by the dominant kraft process (cf., Chapt. "12.4.1. Kraft pulping"). The main reasons for this dominance are excellent pulp strength properties and low demands on wood species and wood quality, as well as a well-established recovery of cooking chemicals, energy, and by-products. In this conventional process, delignification takes place at elevated temperature and pressure (160–170°C and 7 bar) under strong alkaline conditions with an aqueous solution of NaOH and Na_2S. Due to the lack of selectivity in kraft pulping, roughly one half of the wood substance degrades and dissolves into the cooking liquor (black liquor) that thus contains, besides degraded lignin fragments, also a huge amount of carbohydrate-derived materials (mainly aliphatic carboxylic acids).

Unlike the kraft process, the sulfite process covers the whole range of pH reaching from acidic conditions (pH 1–2) to alkaline conditions (pH 9–13) (cf., Chapt. "12.4.2. Sulfite pulping"). For this reason a high flexibility in pulp yields and properties is possible. The active sulfur-containing species in the sulfite process are, in acid sulfite and bisulfite cooking, sulfur dioxide (SO_2) and hydrogen sulfite ions ($HSO_3^{\ominus}$), and in neutral and alkaline sulfite cooking, sulfite ($SO_3^{2\ominus}$). The active base ($Ca^{2\oplus}$, $Mg^{2\oplus}$, $H_4N^{\oplus}$, and $Na^{\oplus}$) used depends on the pH range; acid sulfite at 1–2 ($Ca^{2\oplus}$, $Mg^{2\oplus}$, $H_4N^{\oplus}$, and $Na^{\oplus}$), bisulfite at 3–5 ($Mg^{2\oplus}$, $H_4N^{\oplus}$, and $Na^{\oplus}$), neutral sulfite (NSSC) at 6–9 ($H_4N^{\oplus}$ and $Na^{\oplus}$), and anthraquinone (AQ) alkaline sulfite at 9–13 ($Na^{\oplus}$). Also in the acid sulfite and bisulfite cooking, roughly one half of the wood substance degrades and dissolves into the cooking liquor (spent liquor) that contains degraded lignin fragments (lignosulfonates) and a considerable amount of carbohydrate-derived materials, such as mono-, oligo-, and polysaccharides and aliphatic carboxylic acids (mainly acetic acid and aldonic acids).

Chemical pulps are normally further purified in the bleaching stage where the residual lignin in pulp can be removed rather selectively and, for example, also the residual extractives are destroyed. The pulp products are used as raw materials in the manufacture of paper and board. In general, they still contain (20–30%) hemicelluloses necessary for suitable strength properties. A relatively small amount of chemical pulps (about 3%, corresponding to less than about 4 million tons annually) with flexible fibers from dissolving pulping (i.e., based mainly on acidic sulfite, multistage sulfite, or pre-hydrolysis kraft methods) is used as a source for purified cellulose (**dissolving pulp**, alpha-cellulose content typically 80–90%) for producing mainly **cellulose derivatives** (e.g., cellulose esters and ethers) and **regenerated celluloses** (e.g., rayon fibers, annual production about 3 million tons worldwide) used mainly for the production of threads (textile fibers) and foils. Regenerated cellulose has almost the same DP and polydispersity as the initial cellulose in dissolving pulp, but its morphology is changed and its microfibrils are fused into a relatively homogeneous macrostructure. Regenerated cellulose fibers have long been produced by different processes, resulting in fibers with a wide range of mechanical properties. These products, as well as cellulose derivatives, are known under many trade names.

The properties of cellulose derivatives are influenced by the amount and distribution of the substituents in the molecule. As indicated above, all β-D-glucopyranose units within the cellulose chain have three different reactive hydroxyl groups: two secondary groups attached to carbon atoms C_2 and C_3 and one primary group in the hydroxymethyl group at C_5. The rate and degree of conversion ("reactivity") strongly depends on the availability of these hydroxyl groups and this parameter is called "accessibility". In other words, accessibility means the relative ease of the reactants reaching the hydroxyl groups. There are various methods to determine the accessibility of cellulose to different reagents under varying conditions. Because of the insolubility of cellulose, the chemical reactions of hydroxyl groups proceed, at least in their initial phase, in a heterogeneous medium. In these reactions with cellulose, chemical conversion preferentially proceeds at the surface and in the low-ordered parts of the cellulose structure. It is possible that during several common reactions the partially and/or totally substituted products are dissolved and further reaction steps may then occur homogeneously.

Several factors affect the accessibility and reactivity of cellulose. One prerequisite for etherification (see below) is the ionization of hydroxyl groups and the formation of reactive alkoxides ($C\text{-}O^{\ominus}$). Owing to the inductive effects of neighboring substituents, the tendency towards this kind of dissociation (which depends on acidity) decreases along the series $C_2\text{-}OH > C_3\text{-}OH > C_6\text{-}OH$. For this reason, C_2-OH is more readily etherified than the other hydroxyl groups. However, after the substitution of C_2-OH, the acidity of C_3-OH usually increases resulting in its higher reactivity. On the other hand, in esterification (see below) the primary hydroxyl group C_6-OH possesses the highest reactivity. Being sterically least hindered, C_6-OH groups show higher reactivity toward bulky substituents than the other hydroxyl groups.

The topochemistry and morphology, the ratio of crystalline regions to amorphous ones, are important factors in the chemical reactivity and accessibility of cellulose. The hydroxyl groups located in the amorphous regions are accessible and readily react, while those in crystalline regions with close packing and strong internal hydrogen bonding can be completely inaccessible. For this reason, pre-treatment (pre-swelling) of cellulose is needed in etherification (by alkalis) and esterification (by acids). It may also be possible to increase accessibility by biochemical pre-treatments, but they all result in the creation of new accessible hydroxyl groups. The degree of crystallinity largely depends on the origin of the cellulose preparation, and substantial variations are seen by alternative determination techniques even for the same sample.

Differences in the degree and rate of conversion between regions of high and low degree of order in cellulose are most distinct in reactions accompanied by only a small or medium amount of swelling. In the strongly swollen or soluble state of cellulose, all hydroxyl groups are accessible to the reactants. The fraction of hydroxyl groups available for reactions can be only about 10% in highly crystalline cellulose, but as much as about 90% in decrystallized one. However, due to the random nature of the reaction, a homogeneous product can only be obtained by a complete substitution of hydroxyl groups. In this case, the maximum DS of 3 is reached. At any DS lower than this maximum value, the reaction leads to random sequences of units consisting of unreacted, monosubstituted, disubstituted, and fully substituted moieties. Experiments indicate that moderate differences in reactivity do not significantly influence the overall distribution of these moieties at varying DS values. For example, when DS is 2, the distribution of unreacted, monosubstituted, disubstituted, and fully substituted moieties is, respectively, 5, 20, 45, and 30%.

Cellulose is a polyalcoholic compound, and its functionalization is principally based on the typical reactions of hydroxyl groups. Cellulose derivatives can be divided into *cellulose esters* and *cellulose ethers* (cf., Chapts. "9.4.2. Cellulose esters" and "9.4.3. Cellulose ethers"). However, cellulose was used as a starting material for technically important derivatives even before its polymeric nature was fully understood. For example, Henri Braconnot (1780–1855) first prepared the oldest cellulose derivative of commercial importance, cellulose nitrate, as early as in 1832. The early history of this inorganic ester is related to the militaries of most European nations during the second half of the 19th century. In 1846, Christian Friedrich Schönbein (1799–1868) developed the preferred method of highly nitrated cellulose ("cotton powder" or "gun cotton" with a nitrogen content of 12.0–13.6%) using HNO_3-H_2SO_4 mixtures, and in 1863, Frederick Abel (1827–1902) developed a method of safely handling cellulose nitrate, making possible its use as an explosive. Later in 1868, John Wesley Hyatt (1837–1920) discovered semisynthetic thermoplastic ("plastic"), "celluloid", by combining cellulose nitrate having a nitrogen content of 9–11% ("collodion") with camphor (about 20% of the product) and a minor amount of other plasticizers (e.g., dibutyl sulfate). This discovery was made in a public competition with the objective of finding substitute materials for ivory in producing billiard balls. Thus, cellulose nitrate gradually became the progenitor of industries of explosives, plastics, lacquers, protective coatings, photographic films, and cements. The synthesis of cellulose nitrate formed the basis for the first industrial process for "artificial silk", for example, by spinning cellulose nitrate in acetic acid into a precipitation bath of cold ethanol.

The largest part of the cellulose-based artificial fibers, "viscose rayon" (in French, "rayon" means "honeycomb" or "department"), the first semisynthetic fiber product, is manufactured by the so-called "viscose process" originally invented in 1892 by Charles Frederick Cross (1855–1935), Edward John Bevan (1856–1921), and Clayton Beadle (1899–1976). This product was commercialized in 1894 for textile purposes. Today, rayon is used in many clothing articles; it may be the most useful synthetic fiber or filament to human beings. The process ("viscose route to artificial silk") is based on the formation of cellulose xanthate (cellulose xanthogenate), which is then decomposed by spinning in an acid bath. Cellulose xanthate is a water-soluble unstable anionic ester obtained by reacting cellulose with carbon disulfide (CS_2, "sulfocarbide") in an aqueous solution of sodium hydroxide. The formation of cellulose from cellulose nitrate through denitration, dissolution, spinning, and regeneration was already realized by Hilaire de Chardonnet (1839–1924) in 1885 (full-scale production in 1899). This invention can even be traced back to 1850, when Joseph Wilson Swan (1828–1914) made electric lamp filaments by extrusion, and to 1855, when Georges Audemars prepared nitrocellulose-based artificial silk by a method impractical for commercial use. Due to the high flammability of this nitrogen-containing artificial silk, it was quickly taken off the market. Chardonnet's artificial silk was exhibited at the Paris Exhibition of 1899, and it gained the Grand Prix. At that time in the 1890s, there were also other methods for producing similar materials; Matthias Eduard Schweizer (1818–60) invented the so-called "cuprammonium method" (the cupron process). In this method, regenerated cellulose filaments are obtained by spinning an alkaline cellulose solution containing tetraamine-copper dihydroxide into an aqueous hardening bath that removes the copper and ammonia and neutralizes the caustic soda (NaOH).

During the 1890s, a development was stimulated, not only by scientific curiosity, but also by a more practical intention to chemically transform water-insoluble cellulose into a dissolved state. The ultimate aim was artificial silk prepared from an endless cellulose thread. By using a slit instead of a normal spinning jet (spinneret), and by adding glycerol (to prevent excess crystallinity) to the solution, "**cellophane**" plastic (film) can be produced. Of the other cellulose derivatives of great importance, cellulose acetate (see below), formed by reacting cellulose, acetic acid, and acetic anhydride with an acid catalyst, was first available in the 1930s. This important cellulose ester was the first ester derivative of cellulose; it was produced already in 1865 by heating cotton and acetic anhydride at 180°C by Paul Schützenberger (1829–97). Cellulose chemistry as an individual branch of polymer chemistry can be traced back to fundamental experiments in the 1920s and 1930s on acetylation and deacetylation of cellulose ("acetyl cellulose" or "cellulose acetate" and "acetate fiber" or "acetate rayon"), leading to the concept of polymer-analogous reactions.

It is claimed that Charles Goodyear (1800–60) was the first American to modify on purpose a semi-synthetic polymer when he vulcanized natural rubber (i.e., latex containing the *cis*-polymer of isoprene or 2-methyl-1,3-butadiene with a molar mass of 100 kDa to 1000 kDa) in 1839, and opened this natural polymer to industrial utilization. However, ancient Mesoamericans had already used a similar procedure in about 1600 B.C. to cure natural rubber when processing it to balls and other objects. Gutta-percha (in Malay, "getah perca" means "percha sap") is latex that contains the *trans*-polymer of isoprene; it is more rigid than the common latex elastomer, natural rubber. Gutta-percha can be obtained from trees of the genus *Palaquium* (e.g., *Palaquium gutta*) found in Sumatra, Peninsular Malaysia, Singapore, and Borneo. In 1842, London was introduced to this material by a British surgeon, William Montgomerie, who noted that gum from these trees was used by native woodsmen to make handles for their machetes. He considered that there was also potential for its use, both as knife handles and for various medical devices.

9.4.2. *Cellulose esters*

Cellulose acetate Cell-O-(CO)CH_3 is the most significant organic ester of cellulose. Its main production is still based on a solution process ("solution acetylation") that includes a reaction between cellulose and acetic anhydride dissolved in acetic acid with sulfuric acid as a catalyst:

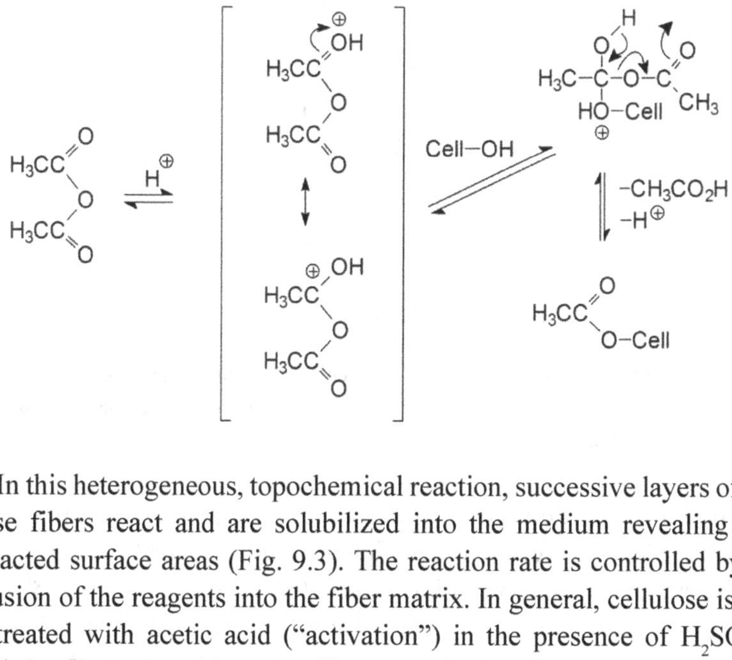

In this heterogeneous, topochemical reaction, successive layers of cellulose fibers react and are solubilized into the medium revealing new unreacted surface areas (Fig. 9.3). The reaction rate is controlled by the diffusion of the reagents into the fiber matrix. In general, cellulose is first pre-treated with acetic acid ("activation") in the presence of H_2SO_4 to swell the fibers, to achieve a uniform reaction, and to adjust the DP to a suitable lower level (typically 350–500). Acetylation is then performed (for several hours at 50°C) with a mixture of acetic anhydride/acetic acid. A fully acetylated product with a DS of 3 is necessary to secure its complete solubility. The use of H_2SO_4 as a catalyst enables the production at lower temperature (up to 50°C). Typical specifications for acetylation-grade dissolving pulps include a low ash (<0.08%), metal (Fe <10 ppm),

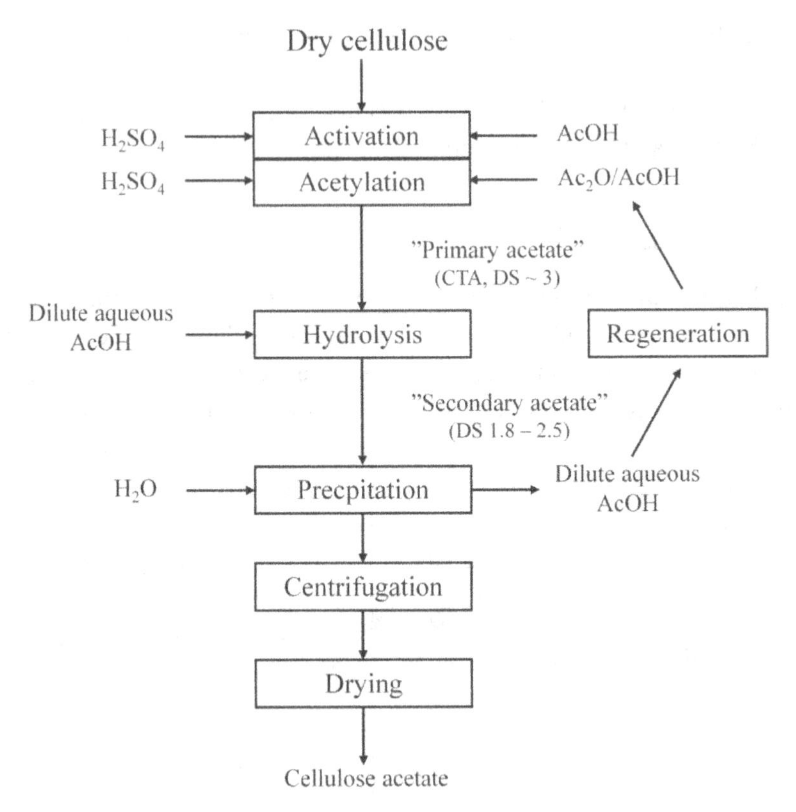

Fig. 9.3. Simplified flow sheet of the conventional cellulose acetate process (Alén 2011c). AcOH is acetic acid, Ac_2O acetic anhydride, CTA cellulose triacetate, and DP degree of substitution.

extractives (<0.15%, diethyl ether-extractable compounds), and pentosans (<2.1%) content, as well as a high alpha-cellulose content (>95.6%).

The end product CTA ("primary acetate", acetate content 44.8% and melting point about 300°C) is usually partially deacetylated without isolation in an aqueous acetic acid solution (for some hours at 40–80°C) to an acetone-soluble product ("secondary acetate") with a DS of about 2.5 ("cellulose 2.5 acetate", acetate content about 40% and melting point about 230°C), or most commonly about 2 ("diacetate" or simply "acetate", acetate content about 35%). In this controlled hydrolysis, sulfate ions (i.e., degradation products of sulfate half ester groups) and a sufficient number of acetyl groups are removed and a homogeneous distribution of the residual acetyl groups is obtained. The product is then precipitated and washed with water to remove acid impurities, and the water is removed by centrifugation or pressing and finally by drying. The product is readily soluble in acetone (i.e., it is suitable for extrusion) and it can be converted into filaments or films by the so-called "dry spinning process".

Both CTA and cellulose acetate are white, odorless, and nontoxic substances that tolerate a wide range of solvents. Cellulose acetate is typically used in textiles (see "acetate fiber" or "acetate rayon", above), composite fabrics, plastics, films, cigarette filters, lacquers, insulating foils, and reverse osmosis membranes. As an acetate fiber it has many beneficial properties; for example, special dyes have been developed for it since it does not accept dyes used for cotton and rayon. "Fortisan" is a trade name for hightenacity cellulose fiber manufactured by partial saponification of stretched (i.e., handled in steam under pressure to improve orientation) cellulose acetate.

Besides cellulose acetate, **cellulose propionate** and **cellulose butyrate** (**cellulose formate** is unstable) are produced to some extent. They have some good properties compared to those of cellulose acetate. Some mixed thermoplastic ester derivatives, such as **cellulose acetate propionate** or cellulose acetopropionate (CAP), and **cellulose acetate butyrate** or cellulose acetobutyrate (CAB), are prepared in one step from cellulose with acetic anhydride and the corresponding acid (propanoic or butanoic acid) in the presence of H_2SO_4. These mixed esters have desirable properties not exhibited by cellulose acetate or CTA. They find commercial applications, for example, in lacquers, sheetings, molding plastics, film products, and hot melt coatings.

Due to cellulose acetate's sensitivity to moisture, its limited compatibility with other synthetic resins, and its relatively high processing temperature, a great number of cellulose esters of higher molar mass have been prepared (Table 9.1), but only a few have attained commercial use. They can be easily prepared with procedures similar to those used for cellulose acetate. In a series with increasing acyl chain length, for example, from C_2 to C_6, the melting point, density, and tensile strength generally decrease, while resistance to moisture and solubility in nonpolar solvents

Table 9.1. Some properties of cellulose triesters (Malm *et al.* 1951)

Cellulose ester		Melting point/°C	Density/g/cm^3	Tensile strength/MPa	Moisture regain/% (95% rh)
Acetate	C_2	306	1.28	71.6	7.8
Propionate	C_3	234	1.23	48.0	2.4
Butyrate	C_4	183	1.17	30.4	1.0
Valerate	C_5	122	1.13	18.6	0.6
Caproate	C_6	94	1.10	13.7	0.4
Heptylate	C_7	88	1.07	10.8	0.4
Laurate	C_{12}	91	1.00	5.9	0.3

increases. In addition, some mixed esters (e.g., **cellulose propionate isobutyrate** and **cellulose propionate valerate**) have found use in plastic composites when good grease- and water-repelling properties are required. Due to their ordered arrangement in solution, some cellulose esters dissolved in appropriate solvents show liquid crystalline characteristics similar to those of other rigid chain polymers.

In contrast, the preparation of cellulose esters of aromatic acids has so far acquired little commercial interest; it has mainly been limited to special cases (e.g., **cellulose cinnamate**, **cellulose salicylate**, **cellulose phthalate**, and **cellulose terephthalate**). However, some mixed cellulose esters containing a dicarboxylic moiety (e.g., **cellulose acetate phthalate**) have technically useful properties, such as solubility in alkalis and excellent film forming characteristics. In addition, various cellulose esters with organic acids carrying sulfonic or phosphonic acid groups have been prepared. In all these cases, a homogeneous esterification in different solutions (e.g., in DMA/LiCl solution) is possible (Table 9.2).

Cellulose can be dissolved by several methods and solidified back as regenerated cellulose with properties (mainly its DP and crystalline structure) different from those of the initial cellulose. High quality regenerated cellulose, **viscose fiber**, can be obtained by the viscose process (Fig. 9.4),

Table 9.2. Typical examples of aromatic cellulose derivatives from homogeneous esterification

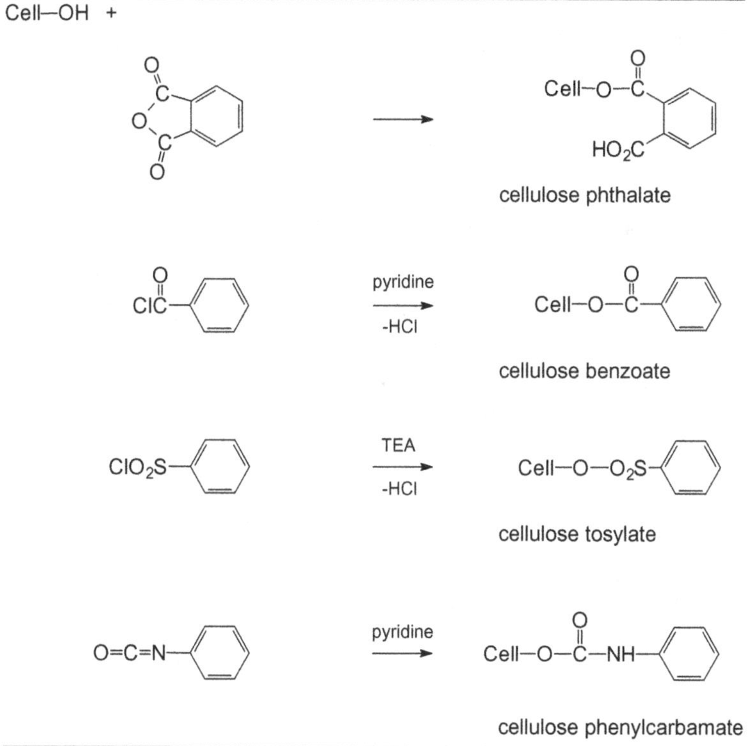

Cell—OH +		
phthalic anhydride	→	Cell—O—C(=O)— (HO_2C) cellulose phthalate
ClC(=O)—	pyridine → -HCl	Cell—O—C(=O)— cellulose benzoate
ClO_2S—	TEA → -HCl	Cell—O—O_2S— cellulose tosylate
O=C=N—	pyridine →	Cell—O—C(=O)—NH— cellulose phenylcarbamate

where the cellulose feedstock material (cellulose content between 91% and 96% and ash content <0.1%) is first mercerized (see above) to convert it to alkali cellulose. After this the cellulose mass is pressed between rollers to remove excess liquid (to a cellulose and NaOH content of 32–36% and 15–16%, respectively) followed by shredding or crumbling (the product is called "white crumb") and oxidative depolymerization ("aging" or "preripening") to obtain an appropriate DP level of 200–400 for the production of **cellulose xanthate** Cell-O-CS_2 with suitable properties. In the alkaline oxidation stage, the cellulose is exposured to ambient air or pressurized oxygen at 20–50°C. Thus, the treatment of alkali cellulose with carbon disulfide (CS_2) results in the formation of cellulose xanthate:

Formation of cellulose xanthate

$$\text{Cell–O}^{\ominus} + S{=}C{=}S \rightleftharpoons \text{Cell–O–C}(=S)S^{\ominus}$$

Regeneration of cellulose from viscose

$$\text{Cell–O–C}(=S)\text{SNa} + H^{\oplus} \rightleftharpoons \text{Cell–O–C}(=S)\text{SH} + Na^{\oplus}$$

$$\text{Cell–O–C}(=S)\text{SH} \rightarrow \text{Cell–OH} + CS_2$$

$$2\,\text{Cell–O–C}(=S)\text{SNa} + Zn^{2\oplus} \rightleftharpoons [\text{Cell–O–C}(=S)\text{–S}]_2\text{Zn} + 2Na^{\oplus}$$

$$[\text{Cell–O–C}(=S)\text{–S}]_2\text{Zn} \xrightarrow{H^{\oplus}} 2\,\text{Cell–O–C}(=S)\text{SH} + Zn^{2\oplus}$$

$$2\,\text{Cell–O–C}(=S)\text{SH} \rightarrow 2\,\text{Cell–OH} + 2CS_2$$

The xanthation (at 25–30°C for about three hours) yields an orange-yellow product ("yellow crumb", DS about 0.5) that is not completely soluble, but is dissolved in dilute aqueous NaOH, usually under high-intensity mechanical action, resulting in the formation of a viscous yellow solution (Fig. 9.4). This viscose solution contains, besides cellulose xanthate (about 8%) and free NaOH (6–7%), trithiocarbonate ($CS_3^{2\ominus}$) and carbonate ($CO_3^{2\ominus}$) at the 1% level, sulfide ($S^{2\ominus}$) and perthiocarbonate ($CS_4^{2\ominus}$) at the 0.1% level, and small amounts of thiosulfate ($S_2O_3^{2\ominus}$), dithiocarbonate ($COS_2^{2\ominus}$), and monothiolcarbonate ($CO_2S^{2\ominus}$). The purpose of

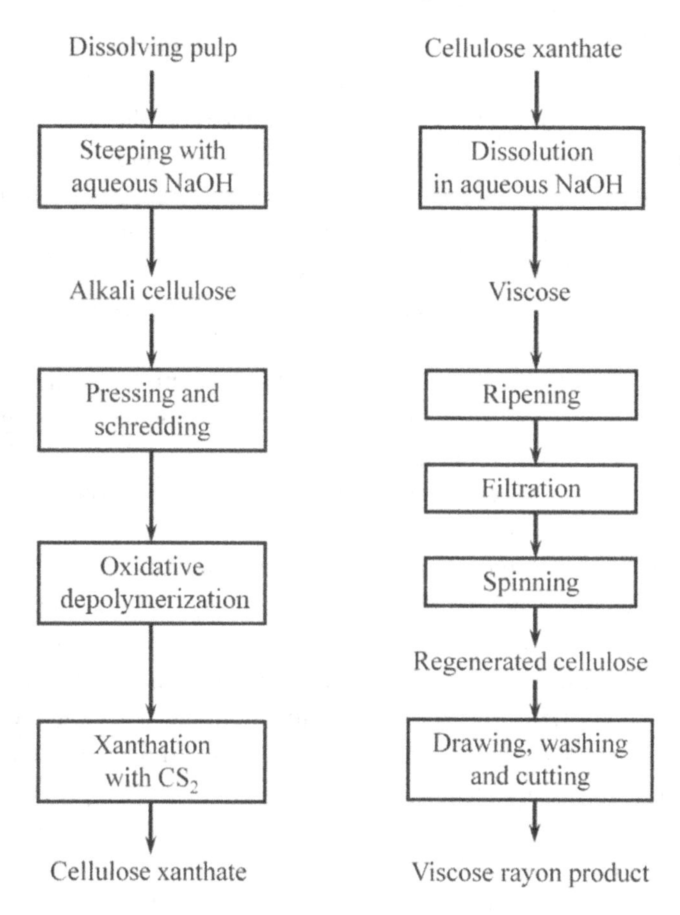

Fig. 9.4. Simplified scheme of the viscose process (Alén 2011c). For explanations, see the text.

the viscose aging ("ripening", approximately at room temperature for 1–3 days) is to adjust the viscosity of the solution to the level desired for the spinning process. Cellulose chains have to be short enough to give manageable viscosities in the spinning solution, but still long enough to impart good physical properties to the fiber product. The DS value is also reduced to some extent during this ripening, but the xanthate groups are redistributed more uniformly along the cellulose chains. In the conventional spinning process, the viscose solution is pressed through corrosion-resistant spinnerets having, depending on the product, 100–1000 holes (hole diameter of 50–100 μm) into a bath of aqueous H_2SO_4 and Na_2SO_4 at about 40°C. Finally, the fine rayon filaments are mechanically treated to modify their properties suitable for textile fibers and washed to remove salts and other water-soluble impurities.

The chemistry of the viscous process is rather complicated; only about 70% of the CS_2 input is consumed in the reaction of cellulose, while the rest is consumed in various side reactions under alkaline conditions. Furthermore, the conversion of the alkaline cellulose solution to filaments of cellulose by spinning in the presence of transition metal cations (e.g., $Zn^{2\oplus}$, $Hg^{2\oplus}$, or $Ag^{\oplus}$) in a combination with special additives (e.g., amines and/or polyethylene oxides) is a complex chemical process that is strongly affected by the previous steps of "xanthation", as well as the dissolution of cellulose xanthate. The regeneration reaction (the viscose is predominantly coagulated by metal ions and decomposed by the acid) takes place in two basic steps with different velocity constants (see above): the formation of cellulose xanthogenic acid (a fast reaction) and then the decomposition of this acid (a slow reaction) with the simultaneous regeneration of cellulose.

The mechanism behind this process is simply the introduction of a sufficient number of anionic dithiocarbonate groups to the cellulose chain to convert it to a water- or alkali-soluble polymer. This means that the cellulose is chemically converted into a homogeneous soluble derivative. This aqueous solution (viscose) is then passed through a spinneret (the production of artificial fibers, rayon) or a slit (the production of cellophane) directly into an acid bath for the regeneration ("solidifying") of almost pure cellulose. Nowadays, several comercial rayon fibers are named according to the process used to convert the cellulose into a soluble polymer and then regenerated. The dominating step is still the viscose process. In spite of certain shortcomings (e.g., the "excess consumption" of CS_2, as well as the need to handle sulfur-containing discharges together with some slow unit operation stages), the main reasons for the clear dominance of this process are its versatility and adaptability to end-use requirements, as well as a long engineering experience and development that has led to many recent improvements. In summary, this process converts the short wood-derived fibers of cellulose into endless filaments or stable fibers, and also into films utilized primarily in food packaging.

Rayon fibers of varying grades are used in a great variety of applications for textiles in apparel and home furnishing, but they also have significant uses, for example, as non-woven materials and many other industrial products. They are generally easy to dye, soft, resistant to wrinkles (they drape well), and highly absorbent. Depending on the process, it is also possible to adjust the product properties within a wide range; for example, a drawing stage, when applied, results in products that have about twice the strength and two-thirds of the stretch of regular rayon.

Cellulose fibers produced by a direct solvent process have the generic name of "lyocell". As shown above, many attempts have been made to develop new solvents to directly dissolve cellulose. One of the most potential solvents for this purpose is NMMO monohydrate (Fig. 9.1); a commercial process called "solvent spinning" has been introduced to utilize an aqueous NMMO solution in the production of wood-based regenerated cellulose fibers ("lyocell fibers"). In spite of good product properties, there are still problems before a "real breakthrough" of this process will happen. In recent years, **cellulose carbamates** have received increasing attention as new alkali-soluble intermediates with a DS of 0.2–0.3 for the production of regenerated cellulose fibers. These cellulose derivatives are formed at elevated temperatures from cellulose and urea via isocyanic acid as an active intermediate.

Cellulose nitrate Cell-O-NO_2 was the first cellulose derivative produced on an industrial scale when manufacturing cotton powder and an artificial silk (see above). The current industrial production is still based on the typical heterogeneous equilibrium reaction between cellulose and nitric acid (HNO_3) in the presence of H_2O and H_2SO_4 ("nitrating acid"). The reaction is retarded by water; the removal of water with sulfuric acid forces the reaction to completion. The first reaction step involves the generation of the nitronium ion ($NO_2^{\oplus}$):

$$HNO_3 + 2H_2SO_4 \rightleftharpoons NO_2^{\oplus} + 2HSO_4^{\ominus} + H_3O^{\oplus}$$

Variations of the nitrating acid, as well as the reaction time and temperature, determine product qualities of the resulting cellulose nitrate. Industrial-scale nitration of high-quality dissolving pulp or cotton linters can be performed either in a batch process or in a continuous reactor. In the classical former case, a rather small quantity of cellulose compared to the liquid phase of HNO_3-H_2O-H_2SO_4 (1:20–1:50), is vigorously agitated for 20–30 minutes. A proper ratio of HNO_3 to H_2SO_4 is chosen to yield the desired degree of nitration and to help maintain a fibrous product structure. At a target DS of 1.9, 2.4, and 2.7 the relative percentage composition of the system HNO_3:H_2O:H_2SO_4 is, respectively, 25.0:19.3:55.7, 25.0:15.2: 59.8, and 25.0:8.5:66.5.

After nitration, most of the excess acid is separated by centrifuging and recycled. "Pre-stabilization" of the cellulose nitrate consists of a series of washing and cooking with water to remove last traces of adhering acid. With low- and medium-nitrated products, subsequent digestion under pressure at 130–150°C reduces the chain length of the molecules (i.e., adjusts its viscosity) and equalizes the distribution of NO_2-subtituents ("stabilization" or "post-stabilization"). Highly nitrated products for explosives require careful stabilization to avoid uncontrolled decomposition. Finally, for further processing, water is displaced by alcohol in the case of lacquers or celluloid nitrates; the latter products are available in the form of fibers or flakes. Products can also be gelatinized by using softeners, such as phthalic acid esters; it is also possible to make aqueous dispersions of the softened cellulose nitrates for their further use in coatings. In general, at a DS of 1.8–2.7 the product mass is 125–150% of the initial mass of cellulose. The continuous process, originally from the 1960s, has clear advantages, such as higher product uniformity, shorter processing time, and higher safety compared to the conventional batch process. The DS range of commercial cellulose nitrates is between 1.8 and 2.8 (i.e., a content of nitrogen between 10.5% and 13.7%, respectively) — typical solvents suitable for each derivative are indicated in parentheses:

DS	Nitrogen content/%	Use examples
1.8–2.0	10.5–11.0	Plastics, lacquers (ethanol)
1.9–2.1	10.9–11.3	Lacquers (ethanol/diethyl ether)
2.2–2.3	11.8–12.2	Lacquers, films, and coatings (esters)
2.2–2.8	12.0–13.6	Explosives (acetone)

The products are white, odorless, transparent, tasteless, and rather hydrophobic substances whose physical and chemical properties depend on DS; for example, their density varies from about 1.5 g/cm^3 to above 1.7 g/cm^3. Commercial cellulose nitrates can be plasticized with a variety of conventional softeners; in this they are compatible with a large number of synthetic polymers.

Some other inorganic esters of cellulose, such as **cellulose sulfates** Cell-O-SO_3H and **cellulose phosphates** Cell-O-PO_3H_2, have been systematically investigated, but so far they have not achieved commercial significance (Table 9.3). Cellulose sulfates can be prepared by a variety of reagent combinations; the straightforward treatment of cellulose with aqueous sulfuric acid results only in a very low conversion (maximum DS 1.5) together with the formation of a large bulk of different degradation products. The stable sodium salt of cellulose sulfate is a white, odorless, and tasteless powder that is completely water-soluble at a DS of above 0.2–0.3 and exhibits good thermal stability up to 100°C for a long time. Cellulose sulfates have found limited specialty use, for example, as thickeners for lacquers, cosmetics, pharmaceuticals, and food and as printing inks, coatings of photographic films, detergents, and oil drilling fluids. They react with hydrogen halides to yield deoxyhalogenated products with excellent flame-retardant properties. Furthermore, various phosphorus-containing cellulose esters (i.e., phosphates, phosphites, and phosphinates) have been of considerable interest because of their flame

Table 9.3. Examples of the inorganic esters of cellulose and their main routes of production

Cellulose nitrates			
Cell–OH	+ HNO_3/H_2SO_4	→→	Cell–O–NO_2
Cellulose sulfates			
Cell–OH	+ SO_3 + H_2SO_4/SO_3 + $ClSO_3H/SO_3$	→→	Cell–O–SO_3H
Cellulose phosphates			
Cell–OH	+ H_3PO_4	→→	Cell–O–PO_3H_2
Cellulose borates			
Cell–$(OH)_3$	+ H_3BO_3 + $B(OR)_3$	→→	Cell–O_3B
Cellulose nitrites			
Cell–OH	+ N_2O_4	→→	Cell–O–NO

retarding properties and potential use in textiles, as well as their ion-exchange capability (as weak cation exchangers). In addition, there are many organic flame retardants added to materials, such as cotton and viscous fibers to inhibit, suppress, or delay the production of flames. However, the needed amounts are often high, generally 10–30% of fire-safe consumer products.

9.4.3. *Cellulose ethers*

Cellulose ethers comprise a class of cellulose derivatives, which contains many commercially important members. The most prominent ethers are water-soluble; they are widely used in many water-based formulations as thickeners for adjusting the rheology of solutions (i.e., the control of thickening, viscosity, and flow behavior) mainly in food applications, cosmetics, pharmaceuticals, drilling muds, building materials, and latex paints. The most important properties of cellulose ethers are their solubility together with their water-binding (absorbency and retention), nontoxicity, and chemical stability. The introduction of ether groups into a cellulose molecule at very low DS levels results in the swelling ability or solubility of the product, even in cold water. The type of constituents, including the combination of ether groups as well as DS and the uniformity of substitution, control the water solubility and/or organic solvent solubility of cellulose ethers. For example, for methylcellulose and ethylcellulose (see below), the DS range typically needed for solubility in water is 1.5–2.0 and 0.7–1.7, and in organic solvents >2.5 and >2.2, respectively.

Cellulose ethers can be prepared by treating alkali cellulose with a number of various reagents. Among the synthesis routes under heterogeneous conditions only three are of commercial importance (Table 9.4): (i) the reaction of hydroxyl groups with an alkyl or aryl chloride in the presence of sodium hydroxide ("the alkali-consuming process", 1 mol NaOH/1 mol reagent) and often also an inert diluent according to the Williamson reaction, (ii) the reaction of hydroxyl groups with an alkylene oxide ("the ring-opening reaction" without alkali consumption), and (iii) the reaction of hydroxyl groups with α,β-unsaturated compounds activated by electron-attracting groups (the Michael addition reaction).

Table 9.4. Examples of typical cellulose ethers obtained by different methods

A. Reaction with alkyl or aryl chlorides		
Carboxymethylcellulose	Cell–O–CH_2CO_2H	CMC
Methylcellulose	Cell–O–CH_3	MC
Ethylcellulose	Cell–O–CH_2CH_3	EC
Propylcellulose	Cell–O–$CH_2CH_2CH_3$	PC
Benzylcellulose	Cell–O–$CH_2C_6H_5$	BC
B. Reaction with an alkylene oxide		
Hydroxyethylcellulose	Cell–O–CH_2CH_2OH	HEC
Hydroxypropylcellulose	Cell–O–$CH_2CH(OH)CH_3$	HPC
Hydroxybutylcellulose	Cell–O–$CH_2CH(OH)CH_2CH_3$	HBC
C. Reaction with an α,β-unsaturated compounds		
Cyanoethylcellulose	Cell–O–CH_2CH_2CN	
Carbamoylethylcellulose	Cell–O–$CH_2CH_2CONH_2$	
Carboxyethylcellulose	Cell–O–$CH_2CH_2CO_2H$	CEC

In the first case, the purposes of the inert diluent are to suspend/disperse the cellulose raw material, to provide heat transfer, to moderate reaction kinetics and to facilitate recovery of the products. Reactions are typically conducted at elevated temperatures (50–140°C) and, if needed, under nitrogen atmosphere to avoid oxidative degradation reactions of cellulose. However, in all cases a great variety of side reactions take place. After an etherification reaction crude grades are simply dried, ground, and packed out, but purified grades require the removal of various by-products in a separate operation prior to drying. There are also some commercially important cellulose ethers, which contain several types of functional groups; these are called "mixed cellulose ethers".

Carboxymethylcellulose Cell-O-CH_2CO_2H or **sodium carboxymethylcellulose** Cell-O-CH_2CO_2Na (CMC or NaCMC, respectively) is commercially the most important cellulose ether with an annual production of over 300,000 tons worldwide. The basic chemistry of carboxymethylation has been well known for a long time (invented in 1918); recent efforts have mainly been directed towards optimization and rationalization of the process. In general, it is prepared from alkali cellulose with sodium monochloroacetate as reagent (in the "semi-dry process" the solvent is either ethanol or methanol and in the "slurry process" isopropanol or 2-propanol):

$$\text{Cell-O}^{\ominus} + \text{ClCH}_2\text{CO}_2^{\ominus} \longrightarrow \text{Cell-O-CH}_2\text{CO}_2^{\ominus} + \text{Cl}^{\ominus}$$

The side reaction is the formation of sodium glycolate ($HOCH_2CO_2Na$) and NaCl. The product is the sodium salt (NaCMC), but usually designated simply as CMC. Another commercial form of CMC is the product with calcium as the counter ion (CaCMC). This product is not water soluble, but it swells substantially in aqueous media and hence, is useful as a tablet disintegrant.

In the process, monochloroacetic acid (MCA) is added to the reaction slurry containing sufficient excess NaOH to neutralize the MCA and to promote its reaction. The reaction requires at least 0.8 mol of NaOH per mol of an anhydroglucose unit (AGU) in the cellulose molecule to remain at the proper alkalinity throughout etherification (e.g., at 60–80°C for 90 min). The product consists of purified CMC powder (>95% CMC, about one third of the total production) and various unpurified, crude technical grade powders (55–75% CMC). Commercial CMCs are normally water-soluble (DP 200–1000) with DS values ranging from 0.4 to about 1.4, typically 0.5–0.8. At a lower DS (0.05–0.25), the product is soluble only in aqueous NaOH (4–10%), whereas at a DS of 0.4–1.4, a good water solubility is attained, and at a DS of >2.2 the products are soluble in polar organic solvents.

Since its commercial introduction, CMC has found use in an ever-increasing number of applications. It can be used in a variety of products, such as detergents (e.g., as a soil antiredeposition aid), food products (mainly as a thickener, but also as a stabilizer and water-binder, as well as a whipping agent in ice-creams and jellies), oil drilling muds (as a dispersion-stabilizing aid and viscosity-adjusting aid), textiles (e.g., as a warp size), paper products (as a papermaking chemical), paper coatings, pharmaceuticals (as a thickener, water-binder, stabilizer, and granulation aid), latex paints, ceramics, and cosmetics (e.g., as a thickener and suspension aid in toothpastes and shampoos). CMC absorbs many times its own mass of water to form a stable colloidal mass. Because of the carboxylic acid groups, CMC is also a polyelectrolyte (pK_a 4–5 depending on the DS).

The other commercial *hydroxyalkylcelluloses* are **hydroxyethylcellulose** Cell-$(OCH_2CH_2)_nOH$ (HEC) and **hydroxypropylcellulose** Cell-$(OCH_2CH(CH_3))_nOH$ (HPC). They are obtained in the reaction of alkali cellulose with gaseous alkene oxides or their corresponding liquid chlorohydrins. Thus, HEC (an annual production about 60,000 tons) is typically produced from alkali cellulose and ethylene oxide:

$$\text{Cell}-O^{\ominus\oplus}Na + n\,H_2C\underset{O}{-}CH_2 \longrightarrow \text{Cell}-(OCH_2CH_2)_nOH + NaOH$$

Ethylene glycol and the oligomeric glycol side product are formed in the competing reaction that consumes ethylene oxide:

$$NaOH + H_2C\underset{O}{-}CH_2 \longrightarrow HOCH_2CH_2O^{\ominus\oplus}Na \dashrightarrow HO(CH_2CH_2O)_nH + NaOH$$

Hydroxyalkylation with epoxides is a base-catalyzed substitution reaction that does not require a stoichiometric, but in principle only a catalytic amount of $HO^{\ominus}$ ions for the cleavage of the epoxy ring. However, hydroxyalkylation is not limited only to the hydroxyl groups originally present in alkali cellulose, but can proceed further at the newly formed primary hydroxyl groups, resulting in hydroxyalkyl chains of varying length. Thus, in the case of hydroxylethylation, pendant oxyethylene chains are formed. Because the ethylene oxide used can react with both the original hydroxyl groups of cellulose and/or the hydroxyethyl substituents of the already substituted cellulose (i.e., forming an oxyethylene chain), the term "molar substitution" (MS) is used to describe the degree of reaction and the product itself; i.e., it denotes the average number of alkylene oxide molecules added per AGU. DS is lower than MS in the products and the ratio MS/DS can be used as a measure of the relative length of the side chains. In addition, the alkali-catalyzed hydroxyethyl ether formation is accompanied by a reaction of $HO^{\ominus}$ ions with ethylene oxide to glycol and further to polyglycols (see above). Usually, only about one half of the ethylene oxide reacts with alkali cellulose while the other half is consumed in side reactions.

HEC and HPC are white, odorless, and nontoxic powders, whose solubility depends highly on the kind of substitution; products with a MS of 0.05–0.5 are only soluble in alkalis, while those with a MS of 1.5 or more are water soluble. HPC is more hydrophobic than HEC and it can be extruded without a softener at 160°C, while HEC is not thermoplastic and decomposes in aqueous solutions already above 100°C. The commercial HEC (total annual production about 60,000 tons, MS 1.8–3.5) is used as a thickener, protective colloid, binder, stabilizer, and suspending agents in a variety of industrial applications including latex paints, construction substances, ceramics, paper chemicals, cosmetics, pharmaceuticals, textiles, and aids in polymerization. It can be also used in coatings on food-contact surfaces of metal, paper, or board. HPC is used for similar purposes (especially in the production of PVC) as HEC, although its use is more limited. In addition, HEC and particularly HPC have also a tendency to form liquid-crystalline aqueous systems that have been comprehensively studied in recent years. Besides, **hydroxybutylcellulose** (HBC) Cell-$(OCH_2CH(CH_2CH_3))_nOH$ is produced to some extent.

Further modification of cellulose ether with another functional group can enhance or produce novel properties. **Carboxymethylhydroxy-ethylcellulose** (CMHEC) is one commercial example of "mixed cellulose ethers". This product is an anionic modification of HEC; it is manufactured by a reaction of alkali cellulose, either simultaneously or sequentially, with ethylene oxide and sodium monochloroacetate. Various grades are available and primarily used in oil recovery applications. Examples of other commercial mixed cellulose ethers include **methylhydroxyethyl-cellulose** (MHEC) or hydroxyethylmethylcellulose (HEMC), **ethylhy-droxyethylcellulose** (EHEC) or hydroxyethylethylcellulose (HEEC), **hydrophobically modified hydroxyethylcellulose** (HMHEC), **cationic hydroxyethylcellulose** (cationic HEC), **methylhydroxypropylcellulose** (MHPC) or hydroxypropylmethylcellulose (HPMC) and **hydroxybutyl-methylcellulose** (HBMC). In addition, simple cellulose ethers with both methyl and ethyl groups have been manufactured.

The most common and simplest representatives of *alkylcelluloses* or cellulose alkyl ethers are **methylcellulose** Cell-OCH_3 (MC) and

ethylcellulose Cell-O-CH_2CH_3 (EC). They are mainly produced (total annual production about 100,000 tons) by a reaction ("alkylation") of alkali cellulose with alkyl chlorides:

$$\text{Cell-O}^{\ominus} + \text{R-Cl} \rightleftharpoons \text{Cell-O-R} + \text{Cl}^{\ominus}$$

$$R = CH_3 \text{ or } CH_2CH_3$$

The Williamson reaction proceeds according to the S_N2 mechanism (bimolecular nucleophilic substitution). Under these alkaline conditions, methanol or ethanol (R-OH) is also formed as by-products that can react further with alkyl chloride (R-Cl) to form dimethyl ether or diethyl ether (R-O-R). This by-product formation accounts for 20–30% of the RCl consumption, thus decreasing reagent efficiency in etherification. Alkyl-celluloses are white-to-yellowish, nontoxic solids that exhibit, depending on substituent and DS, varying solubility in various media. The DS of MC is typically in the range of 1.5–2.0; in this case, it is water soluble.

The alkylene oxide derivatives of MC are (in parentheses typical values of DS and MS): **hydroxyethylmethylcellulose** (HEMC) or methylhydroxyethylcellulose (MHEC) (DS_{methyl} 1.3–2.2 and $MS_{hydroxyethyl}$ 0.06–0.5), **hydroxypropylmethylcellulose** (HPMC) or methylhydroxypropylcellulose (MHPC) (DS_{methyl} 1.1–2.0 and $MS_{hydroxypropyl}$ 0.1–1.0), as well as **hydroxybutylmethylcellulose** (HBMC) (DS_{methyl} ≥1.9 and $MS_{hydroxybutyl}$ ≥0.04). These products are used like hydroxyalkylcelluloses in construction materials (e.g., cements and mortars), latex and wall paper paints, agricultural and food products (e.g., mayonnaise and dressings), cosmetics, and pharmaceuticals (e.g., tablets and formulations). The most common DS range of EC products is 2.2–2.7; they are soluble in many organic solvents. For example, **hydroxylethylethylcellulose** (HEEC) or ethylhydroxyethylcellulose (EHEC) is used in lacquers and inks. In addition, **propylcellulose** (PC) and **benzylcellulose** (BC) are produced to some extent.

Cyanoethylcellulose Cell-O-CH_2CH_2-C≡N is the most common cellulose ether that is formed in the presence of NaOH via the reaction of alkali cellulose and an α,β-unsaturated compound that contains a strongly electron-attracting group (reaction time is some hours at 30–50°C). The cellulose anion attacks the carbon atom with a small positive charge in acrylonitrile (CH_2=CH-C≡N) to form a resonance-stabilized intermediate anion that then adds a proton from water, resulting in the formation of cyanoethylcellulose under the simultaneous liberation of an $HO^{\ominus}$ ion. All reaction steps are reversible and because of the regeneration of $HO^{\ominus}$ ions no alkali is consumed. The process is accompanied with the consumption of acrylonitrile in many side reactions, such as the formation of di(cyanoethyl)ether or 3,3′-oxydipropionitrile ($O(CH_2CH_2CN)_2$). The solubility of cyanoethylcelluloses depends on their DS. For solubility in alkali, a DS of 0.25–0.5 and a uniform distribution of the substituents are necessary, while products with a DS of about 2.5 are soluble in polar organic solvents. Due to an unusually high dielectric constant and low dissipation factor, they can be used as insulating materials. Cyanoethylated paper has also a good thermal and dimensional stability.

Although its reactivity is lower than that of acrylonitrile, acrylamide (CH_2=CH-$CONH_2$) can be added to alkali cellulose in a similar manner, leading to the formation of **carbamoylethylcellulose** Cell-O-CH_2CH_2-$CONH_2$ or the carbamoylethyl ether of cellulose. Cyanoethylcellulose and carbamoylethylcellulose both decompose in an aqueous medium of higher alkalinity (NaOH) at elevated temperatures to give by saponification the sodium salt of **carboxyethylcellulose** Cell-O-CH_2CH_2-CO_2Na (CEC) as the stable end product. Here, the amide group (-$CONH_2$) is more easily saponified to a carboxylic acid group (-CO_2H) than the nitrile group (-C≡N). Besides the nitrile and amide groups, several other substituents are also able to activate the C=C bond to the addition reaction with alkali cellulose. Typical examples of other reagents include methacrylonitrile (H_2C=C(CH_3)-C≡N), α-methyleneglutaronitrile (HO_2CC(=CH_2)(CH_2)$_3$-C≡N), α-chloroacrylonitrile (H_2C=CCl-C≡N), *trans*-crotonitrile (H_3CCH=CH-C≡N), and allyl cyanide (H_2C=$CHCH_2$-C≡N). There are also other common functionalized cellulose ethyl ethers, such as **aminoethylcellulose** Cell-O-CH_2CH_2-NH_2 and **sulfoethylcellulose** Cell-O-CH_2CH_2-SO_3H.

9.4.4. *Other cellulose products*

In addition to straightforward manufacture of cellulose derivatives, other ways to tailor the properties of cellulose and its derivatives are *crosslinking* and *graft copolymerization* with certain monomers. In the latter case, the main aim has been to find routes to combine the advantages of cellulose with those of synthetic polymers.

The formation of covalent crosslinks between the cellulose chains in heterogeneous systems is the most important route to modify the macromolecular skeleton of cellulose. This commercial modification improves the performance of cellulose textiles ("fabrics finishing" — e.g., crease and wrinkle resistance, wash-and-wear performance, and durable-press properties), although the rigid three-dimensional network formed may be rather brittle as well. Thus, one purpose of covalent crosslinking is to avoid undesirable changes of cellulose goods in the wet state. Reagents capable of forming crosslinks through ether bonds with the hydroxyl groups of cellulose are preferred, since ester crosslinks have a low stability against alkalis. The traditional method of cellulose crosslinking dates back to the beginning of the 20th century, when the action of formaldehyde on cellulose was studied and its positive effect on the strength of the fibers was noted by Xavier Eschalier in 1906. In principle, this is a two-step reaction via a cellulose hemiacetal as a methylolcellulose intermediate:

$$\text{Cell–OH} + \text{HCHO} \rightleftharpoons \text{Cell–O–CH}_2\text{–OH}$$

$$\text{Cell–O–CH}_2\text{–OH} + \text{HO–Cell} \rightleftharpoons \text{Cell–O–CH}_2\text{–O–Cell} + \text{H}_2\text{O}$$

As can be seen, both steps proceed as equilibrium reactions. In practice, the total reaction is usually performed as a wet process in the presence of acids; the crosslinking takes place within a few minutes during the subsequent curing at 100–130°C. This process has been applied to improve the dimensional stability of rayon fibers. However, it has been replaced by other crosslinking agents, primarily methylolated or alkoxymethylated derivatives of different *N*-containing compounds, such as ureas, cyclic ureas, carbamates, and triazines or acid amides. For example, after impregnation of the cellulose substrate with an aqueous solution containing urea-formaldehyde precondensates, crosslinking takes place during a subsequent short heating at 130–160°C in the presence of acid.

Examples of other crosslinking routes (or crosslinking agents) are as follows:

- — recombination of cellulose macroradicals formed chemically or by irradiation,
- — reaction of anionic cellulose derivatives with divalent metal cations,
- — formation of disulfide bridges (-S-S-) from mercapto groups (-SH) attached to cellulose,

— formation of urethane bridges (R-NHCO$_2$-R) in a reaction of cellulose hydroxyl groups with isocyanates (R-N=C=O),
— formation of ester groups (R-O-CO-R) in a reaction of cellulose hydroxyl groups with polycarboxylic acids (e.g., five-membered anhydrides), and
— formation of ether bonds (R-O-R) with bifunctional etherifying agents (e.g., alkyl halides, epoxides or divinyl sulfone).

Graft copolymerization was presented in 1943 by the Russian scientist S.N. Ushakov, who synthetized vinyl and allyl ethers of cellulose and then copolymerized these cellulose derivatives with the esters of maleic acid. Since that time a large number of scientific reports on grafting polymer side-chains onto cellulose and other polymers have appeared. However, the vast majority of cellulose grafting methods involve polymerization of vinyl monomers of different types of CH$_2$=CH-X (X is an inorganic moiety or an organic substituent). The grafting is traditionally performed by applying liquid or gaseous monomers onto solid cellulosic materials; the course of this heterogeneous reaction is strongly dependent on the raw materials. The most commonly used procedure is the so-called "grafting-from approach", where the growth of polymer chains occurs from initiating sites on the cellulose backbone.

Graft copolymerization of various monomers onto cellulose or its derivatives can be classified into three categories: (i) free radical polymerization, (ii) ionic and ring opening polymerization, and (iii) condensation or addition polymerization. The first group has received the greatest amount of attention among all grafting methods; most products available are obtained by this method. Main reasons for this paramount position include the possibility to use a wide range of monomers (e.g., acrylic acid, (meth)acrylic acid esters, (meth)acrylamides, acrylonitrile, vinyl acetate, butadiene, styrene, dimethylaminoethyl methacrylate, hydroxylacrylates, vinylpyridines, and *N*-vinyl-2-pyrrolidone), tolerance of different reaction conditions (generally rather "mild") and the presence of water or other impurities, as well as a simple process technology together with its ability to provide almost an unlimited number of copolymers.

The concept "living polymerization" originates from the early 1980s. This term is used to describe a chain growth process where chain breaking reactions, such as chain transfer or irreversible termination, have been minimized and it is possible to control the composition and molar mass distribution of polymers (e.g., to obtain a narrow range of polydispersity) and tailor the macromolecules with complex architectures. Recent advances include free-radical polymerizations, such as "nitroxide-mediated polymerization" (NMP), "atom transfer radical polymerization" (ATRP), and "reversible addition-frag-mentation chain transfer (RAFT) polymerization". Of these techniques, the last method can be applied to the widest range of radically polymerizable monomers using reaction conditions that are similar to those of free radical polymerization.

As a high-molar-mass biopolymer, cellulose is suitable for many applications that require a rigid matrix with a large surface area. It is also possible to introduce different chemical groups into the backbone of cellulose, typical examples being the preparation of cation and anion exchangers. Although cellulose-based ion exchangers are chemically less stable than most of the synthetic ion exchange resins and their capacities are is relatively low, they are useful for biochemical applications, such as protein separations.

Ion exchange is the most common chromatographic method for separating inorganic and organic ions. Ionogenic groups are classified as being either strong (e.g., $-SO_3^{\ominus}$) or weak (e.g., $-CO_2^{\ominus}$, $-OPO_3^{2\ominus}$, aromatic-$O^{\ominus}$ or aromatic-$S^{\ominus}$) acids, or strong (e.g., $-NR_3^{\oplus}$) or weak (e.g., $-NH_3^{\oplus}$, $-NRH_2^{\oplus}$ or $-NR_2H^{\oplus}$) bases. Strong acid and base groups are highly dissociated and the exchangers containing these groups resemble insoluble strong electrolytes that possess a permanent positive or negative charge. For example, quaternary ammonium ($-CH_2N(CH_3)_3^{\oplus}$) resins contain strong anion-exchange groups that are effective throughout the pH range 2–12, like strong cation-exchange groups. Similarly, common weak acid and base groups resemble insoluble weak electrolytes and their ability to participate in ion exchange as cation and anion exchangers, respectively, depends on the dissociation of these groups (i.e., on their ionization constants) and the pH of the environment (roughly, effective in pH ranges of 6–10 and 2–9, respectively). In general, the exchanger is packed into a column through which the sample solution flows. In this arrangement the ion-exchange reaction is intrinsically reversible and goes to completion in the desired manner. Although ion-exchange columns are easy to use, the theory behind them is rather complicated.

Table 9.5. Examples of cellulose ion exchangers

Cation exchangers			
	(oxidized)	$-CO_2^{\ominus}$	Weak
CM	(carboxymethyl)	$-OCH_2CO_2^{\ominus}$	Weak
P	(phosphate)	$-OPO_3^{2\ominus}$	Intermediate
SM	(sulfomethyl)	$-OCH_2SO_3^{\ominus}$	Strong
SE	(sulfoethyl)	$-OCH_2CH_2SO_3^{\ominus}$	Strong
SP	(sulfopropyl)	$-OCH_2CH_2CH_2SO_3^{\ominus}$	Strong
Anion exchangers			
AE	(aminoethyl)	$-OCH_2CH_2NH_3^{\oplus}$	Weak
DEAE	(diethylaminoethyl)	$-OCH_2CH_2N^{\oplus}(CH_2CH_3)_2$	Weak
TEAE	(triethylaminoethyl)*	$-OCH_2CH_2N^{\oplus}(CH_2CH_3)_3$	Strong
QAE	(quaternary aminoethyl)	$-OCH_2CH_2N^{\oplus}(CH_2CH_3)_2CH(OH)CH_3$	Strong

*Only a part of the functional groups are of this type.

A series of cellulose-based ion exchangers has been prepared for commercial use (Table 9.5). Their chemical structure is a hydrophilic cellulose network that contains acidic or basic groups. The most common type of carboxylic acid-containing weak ion exchanger is preswollen, microgranular CM-cellulose (the counter ion is usually $Na^{\oplus}$ and the effective pH range is above 4) with a low DS, resulting in a cation exchange capacity of 0.4–0.7 meq/g. CM-celluloses of higher capacity cannot be used in chromatography unless they are cross-linked to prevent extensive swelling. By partially oxidizing the primary hydroxyl groups of cellulose into carboxylic groups, for example, by nitrogen oxide, another carboxylic acid cellulose cation exchanger can be obtained. Dry fibrous P-cellulose represents an intermediate divalent cation exchanger based on orthophosphoric acid groups. It can be prepared by reacting disodium chloromethylphosphonate and cellulose. SE-cellulose, like SM- and SP-celluloses, is a strong cation exchanger with an approximative capacity of 0.5 meg/g. It can be prepared by reacting α-chloroethanesulfonic acid with cellulose in the presence of sodium hydroxide. PEI-cellulose is not chemically modified cellulose, but a complex of cellulose with polyethyleneimine.

DEAE-cellulose is the most common weakly basic anion exchanger (the counterion is usually $Cl^{\ominus}$ and the effective pH range is below 9); it can be made from cotton linters, previously cross-linked with formaldehyde or 1,3-dichloropropanol, using 2-chlorotriethylamine as a reagent. Preswollen, microgranular AE-cellulose is also a weakly basic anion exchanger that can be obtained from a reaction of cellulose with α-aminoethylsulfuric acid in the presence of sodium hydroxide. Its capacity is about 0.2 meq/g, but it is possible to increase it by cross-linking to about 0.7 meq/g. Strongly basic quaternary cellulose anion exchangers have been prepared by reacting DEAE-cellulose with alkyl halides under anhydrous conditions to result, for example, in TEAE-cellulose. The commercially available TEAE-cellulose resembles the weakly basic DEAE-cellulose, showing only slight conversion to a quarternary anion exchanger. A product with an ion exchange capacity of 0.3–0.4 meq/g has also been made in a reaction between cellulose, sodium hydroxide, triethanolamine, and epichlorohydrin and is normally referred to as ECTEOLA-cellulose.

Since cellulose ion exchangers are available, besides as powders, also in the form of fiber and paper, they can be used in paper chromatography and thin-layer chromategraphy (TLC) as well. In the former case, papers are impregnated with ion exchange reagents or finely distributed ion exchange powder is embedded into the paper, or the paper itself is prepared from cellulose ion exchanger. For TLC, the sheets are usually prepared from finely ground ion exchange powder fixed on cellulose fibers with an inert adhesive material. Today, only limited numbers of ion exchange layers (e.g., PEI- and DEAE-cellulose) are available. It should be noted that other carbohydrate supports, such as cross-linked dextran gels (the products are known as trade name "Sephadex") are very hydrophilic and like cellulose, easily derivatized with ionogenic functional groups. They have been used for analysis and purification of biological molecules, such as various proteins. The term "ion exclusion" describes the mechanism of using ion exchangers in the fractionation of neutral and ionic species. This chromatography technique is a mode of high-performance liquid chromatography (HPLC); normal HPLC equipment can be used with the proper eluent, column, and detection-techniques. In HPLC and in principle also in TLC, biopolymer-based stationary phases, such as certain cellulose derivatives (e.g., cyclodextrins) have proved effective in resolving a wide class of racemic forms encompassing a variety of structures. In these cases, the stationary phases interact with a particular enantiomer through hydrogen bonding, charge transfer, and inclusion.

Qualitative tests in medicine provide results that are either positive or negative for the substance to be detected. They can be performed in a simple way and the results can be read within a few minutes from color change, instead of exact numerical values from more sensitive and expensive quantitative tests. For example, in the detection of human chorionic gonadotropin (hCG) from a urine sample, a positive test result indicates that the patient is most likely pregnant. In this case, an antibody specific to the whole hCG molecule or its polypeptide chain is chemically bound to a nitrocellulose filter paper and reacts with the urine sample hCG. Another example is an enzymatic glucose test based on the activity of an enzyme, glucose oxidase. In this case, a firm plastic strip with a stiff absorbent cellulose area is impregnated with a buffered mixture of glucose oxidase together with other reagents. The positive result can be seen as a visible color change. There are also a great number of similar indicative spot tests utilizing filter paper as a carrier matrix.

Among other specific applications, many cellulose-based filters (i.e., cellulose or its derivatives, such as cellulose nitrates and organic esters) are used as a collection technique (total particulate/aerosol sampling) for many analytical purposes. In a typical case, metal-containing particles in air-dust or water samples are examined by drawing air or water through a filter and subsequently analyzing the filter on which the particles are trapped. A similar approach is "column chromatography", where a column is packed, for example, with cellulose adsorbent and the sample solution passes through this affinity column. Electrophoresis is defined as the motion of dispersed particles relative to a fluid under the influence of a spatially uniform electric field. In this method, the charged molecules move through the pores of the support medium toward the opposite charge. In an ideal case, the support medium does not interact with the charged molecules, but acts as a filter to retard the movement of particles with different sizes and shapes. Commercially available cellulose acetate sheets have a homogeneous micropore structure; paper electrophoresis based on this cellulose derivative plays an important role, for example, in the analysis of hemoglobin as well as in the separation of other blood proteins, enzymes, mucopolysaccharides, and urine.

9.4.5. *Other polysaccharides*

The carbohydrates of wood and plant biomass mainly comprise cellulose and various *hemicelluloses*. E. Schulze introduced the term "hemicelluloses" in 1891 when it was noted that specific polysaccharides extracted from plant tissues with diluted alkali could be more readily hydrolyzed with acids than cellulose. This term still has no "unique definition" in spite of its long use. First, these cellulose-resembling polysaccharides (i.e., they differ from (1→4)- and (1→3)-glucans and their derivatives) were misleadingly believed to represent intermediates of the biosynthesis of cellulose (note the prefix hemi-) (cf., Chapt. "8.2. Polysaccharides"). Essential data were obtained by the 1970s about the main hemicelluloses from many important origins. In this section, only the chemistry of the most common hemicelluloses is briefly discussed. Today, due to many biorefinery concepts suggested, the interest towards the commercial utilization of hemicelluloses is increasing.

Hemicelluloses are heteropolysaccharides that are less well-defined than cellulose. Their chemical and thermal stability (they have linear frameworks with certain side-groups) is clearly lower than those of cellulose, presumably due to their lack of crystallinity and lower DP (100–200). Moreover, hemicelluloses differ from cellulose with respect to their solubility in alkalis. This characteristic property is most commonly utilized when fractionating various polysaccharides in lignin-free samples. Some hemicelluloses, such as fragments of hardwood xylan and arabinogalactan, primarily from larch species, are partly or even totally water-soluble. Therefore, in these cases, the distinction between water-soluble hemicelluloses, sugars (mainly mono- and disaccharides), and extractive-derived compounds is sometimes difficult. The polysaccharide chain of hemicelluloses is usually linear, but it is often branched and contains side groups or short side chains, whether basically linear or branched.

The building units of hemicelluloses are hexoses (D-glucose, D-mannose, and D-galactose), pentoses (D-xylose, L-arabinose, and D-arabinose) or deoxyhexoses (L-rhamnose or 6-deoxy-L-mannose and rare L-fucose or 6-deoxy-L-galactose). Small amounts of uronic acids (4-*O*-methyl-D-glucuronic acid, D-galacturonic acid, and D-glucuronic acid) are also present. These units mainly exist as six-membered

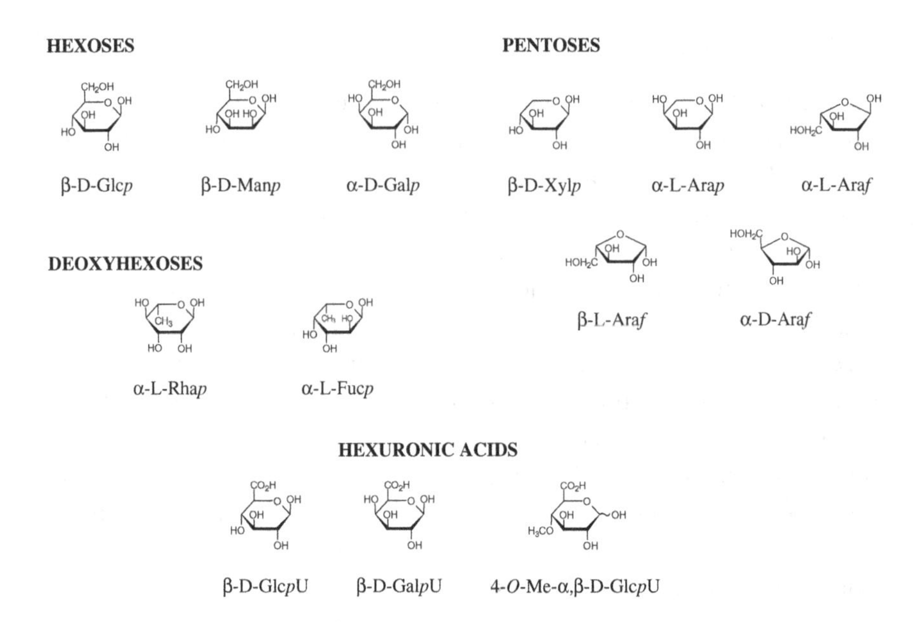

Figure 9.5. Sugar moieties of wood hemicelluloses (Alén 2011a). Note that U is used for uronic acid (e.g., GlcU) instead of A (e.g., GlcA), also commonly used in the literature.

(pyranose) structures in α or β forms (Fig. 9.5). Softwoods and hardwoods differ not only in the content of total hemicelluloses but also in the percentages of individual hemicellulose constituents (mainly glucomannan and xylan), and in the detailed composition of these constituents. Softwood hemicelluloses have more mannose and galactose units and less xylose units and acetylated hydroxyl groups than those from hardwoods.

As mentioned above, chemical structures of hemicelluloses from different origins vary; thus, for example, hardwood xylan(s) and softwood glucomannan(s) are normally used terms. In practice, this means that these shorter names are used instead of very specific "correct names"; thus, *O*-acetyl-(4-*O*-methyl-α-D-glucurono)-β-D-xylan is called "glucuronoxylan" or "hardwood xylan", or simply "xylan". The average relative portions of monosaccharide moieties as anhydrosugar units (percentage of component) in Scots pine (*Pinus sylvestris*) and silver birch (*Betula pendula*) wood are (data contain monosaccharides from hemicelluloses and part of D-glucose from cellulose, which is separately shown in parentheses) as follows:

Component	Pine	Birch
L-Arabinose	<5	+
D-Xylose	10	25
D-Galactose	<5	+
D-Glucose	60 (55)	55 (52)
D-Mannose	15	<5
Uronic acids	5	5
Others (other sugar units and acetyl groups)	<5	10

Softwoods contain (6–11% of the dry wood mass) **L-arabino-(4-*O*-methylglucurono)xylan** or arabinoglucuronoxylan (or xylan). It is composed of a practically linear frame work of (1→4)-linked β-D-xylopyranose (β-D-Xyl*p*) units containing branches of both (1→2)-linked pyranoid 4-*O*-methyl-α-D-glucuronic acid (4-*O*-Me-α-D-Glc*p*U) and (1→3)-linked α-L-arabinofuranose (α-L-Ara*f*) groups. The molar ratios of arabinose:glucuronic acid:xylose may vary considerably, typical values being 1:2:8. In contrast to hardwood xylan, no acetyl groups are present in softwood xylan. The partial chemical structure of xylan from softwoods is as follows:

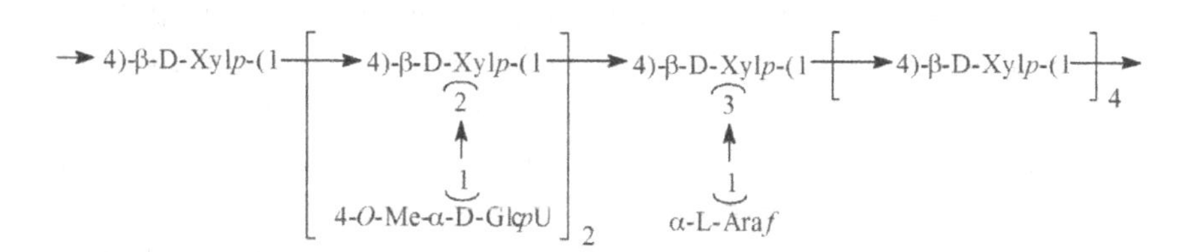

In hardwoods, the content of ***O*-acetyl-(4-*O*-methylglucurono)xylan** is 15–30% of the dry wood mass. It is composed of the same framework (i.e., (1→4)-linked β-D-Xyl*p* units) as the softwood xylan but it contains much fewer (1→2)-linked pyranoid uronic acid substituents that are not evenly distributed within the xylan chain. A typical molar ratio of glucuronic acid:xylose is 0.1:1. The framework moieties are also partly acetylated at C_2-OH and C_3-OH. The acetyl group content varies in the range of 8% to 17% of the total xylan, corresponding to 3.5–7.0 acetyl groups per 10 xylose units. In addition to these structural units, hardwood xylan has been reported to contain small amounts of L-rhamnose (α-L-Rha*p*) and galacturonic acid (α-D-Gal*p*U) in the structural sequence at the reducing end of the xylan molecule. The partial chemical structure of xylan from hardwoods is as follows:

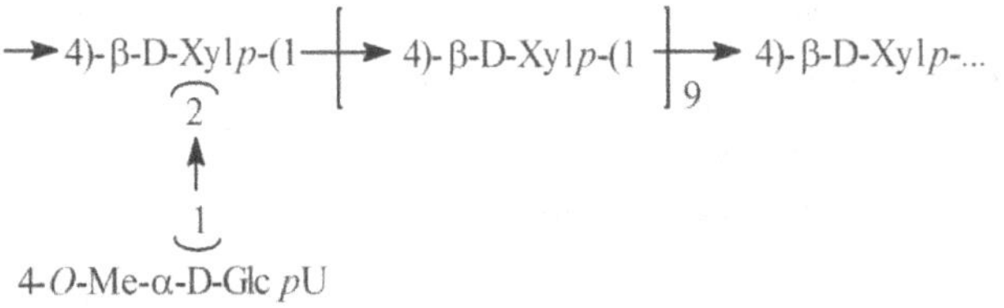

→4)-β-D-Xyl*p*-(1[→4)-β-D-Xyl*p*-(1]$_9$→4)-β-D-Xyl*p*-...
2
↑
1
4-*O*-Me-α-D-Glc *p*U

...→4)-β-D-Xyl*p*-(1→3)-α-L-Rha *p*-(1→2)-α-D-Gal *p*U-(1→4)-β-D-Xyl*p*

In softwoods, the primary hemicellulose component (15–20% of the dry wood mass) is ***O*-acetylgalactoglucomannan** or galactoglucomannan (or glucomannan). Its framework is built of a mainly linear backbone of (1→4)-linked β-D-glucopyranose (β-D-Glc*p*) and β-D-mannopyranose (β-D-Man*p*) units. The framework moieties are partly acetylated at C_2-OH and C_3-OH; the acetyl group content is about 6% of the total glucomannan corresponding to, on the average, one acetyl group per 3–4 hexose units. In addition, it is substituted by (1→6)-linked α-D-galactopyranose (α-D-Gal*p*) units. Glucomannan(s) can be roughly classified into two fractions with different galactose contents. In the galactose-poor fraction (two thirds of the total glucomannan) the molar ratios galactose:glucose:mannose are 0.1–0.2:1:3–4 (it is normally termed "glucomannan"), while in the galactose-rich fraction (one third of the total glucomannan) the corresponding ratios are 1:1:3 (it is normally termed "galactoglucomannan"). The typical overall molar ratios galactose:glucose:mannose are thus 0.5:1:3.5. The partial chemical structure of glucomannan from softwoods is as follows:

→4)-β-D-Glc*p*-(1→4)-β-D-Man*p*-(1→[4)-β-D-Man*p*-(1→]$_2$
6
↑
1
α-D-Gal*p*

The content of glucomannan in hardwoods is clearly lower (1–4% of the dry wood mass) than that in softwoods and, unlike softwood glucomannan, hardwood glucomannan is not acetylated. It has a practically linear backbone with a ratio of glucose:mannose of 1:1–2. The partial chemical structure of glucomannan from hardwoods is as follows:

→4)-β-D-Glc*p*-(1→4)-β-D-Man*p*-(1→4)-β-D-Man*p*-(1→

Arabinogalactan may occur in significant proportions (10–20% of the dry wood mass) in the heartwood of larches (*Larix sibirica*/*L. decidua*), while its content in other softwoods is generally less than 1% of the dry wood mass. It consists of a backbone of (1→3)-linked β-D-galactopyranose (β-D-Gal*p*) residues, most of which carry a side group or chain attached to their C_6 position. The side chains consist of (1→6)-linked β-D-Gal*p*) chains of variable length and arabinose substituents (α-L-Ara*f* and β-L-Ara*p*). In whole larch wood, the molar ratio of arabinose to galactose is typically 1:5–6; this ratio varies considerably in other softwoods. Unlike all the other wood hemicelluloses (matrix substances), larch arabinogalactan is extracellular and it can be extracted almost quantitatively from the heartwood with water.

In addition, different **acidic galactans** (about 10% of the dry wood mass) are mainly present in reaction wood (in case of softwoods, so-called "compression wood" and in hardwoods, "tension wood"). For example, an acidic galactan, built up of (1→4)-linked β-D-Gal*p* units substituted at C_6 mainly with a single α-D-Gal*p*U unit (α-D-Glc*p*U units are also present in small amounts), is a major hemicellulose in compression wood:

→4)-β-D-Gal*p*-(1→4)-β-D-Gal*p*-(1→[4)-β-D-Gal*p*-(1→]$_{40}$
6 6
↑ ↑
1 1
β-D-Gal*p*U β-D-Glc*p*U

A small amount of **rhamnoarabinogalactan** is present in hardwoods; it consists of a slightly branched backbone of (1→3)-linked β-D-Gal*p* units. For example, in sugar maple (*Acer saccharum*) the molar ratios of galactose:arabinose:rhamnose are 1.7:1:0.2. The arabinose and rhamnose components are respectively α-L-Ara*f* and α-L-Rha*p*.

Other miscellaneous polysaccharides in woods and plants are various (1→3)- and (1→4)-glucans (however, they are not generally classified as hemicelluloses) present in small amounts; for example, besides starch (see (1→4)-α-glucan, below), **callose** ((1→3)-β-glucan), **laricinan** ((1→3)-β-glucan), **xyloglucan** (with a backbone of β-(1→4)-linked glucose residues), **fucoxyloglucan**, and **rhamnoarabinogalactan**.

Pectic substances (in Ancient Greek, "pektikós" means "congealed" or "curdled") form a heterogeneous group of carbohydrates; in wood chemistry, they are considered to include polysaccharides that contain acidic groups, such as galacturonans and galactans (see "acidic galactans", above), but also non-acidic arabinans. While there is no agreed-upon definition of these substances, they are traditionally connected only with pectic acids; these substances are galacturonoglycans or poly(α-D-galactopyranosyluronic acids). They consist of linear backbones of (1→4)-linked α-D-Gal*p*U residues that are normally in the form of methyl esters or calcium salts (pectates). Pectins are commercially produced as white powders, mainly extracted from young citrus fruits and berries. They are used in food as gelling agents in jams and jellies, but they also have many other uses in the food manufacturing industry. Pectins are present in most primary cell walls and in the non-woody parts of terrestrial plants. Henri Braconnot first isolated and described pectin-like substances in 1825. They may contain many other monosaccharide residues, such as arabinose and xylose. For example, **rhamnogalacturonan II** (RG-II) is a structurally complex pectic polysaccharide present in the walls of growing plant cells, and it was first identified in 1978. *Amidated pectins* are modified forms of pectins; some of their galacturonic acid groups are converted with ammonia to carboxylic acid amide groups. They are, for example, more tolerant of varying calcium concentrations in their use.

Starch (in German "Stärke" and in Latin "amylum" — in Greek, "amylon" means "not ground at a mill" and in Old English, "stercan" means "to stiffen") is composed of linear amylose and branched amylopectin parts:

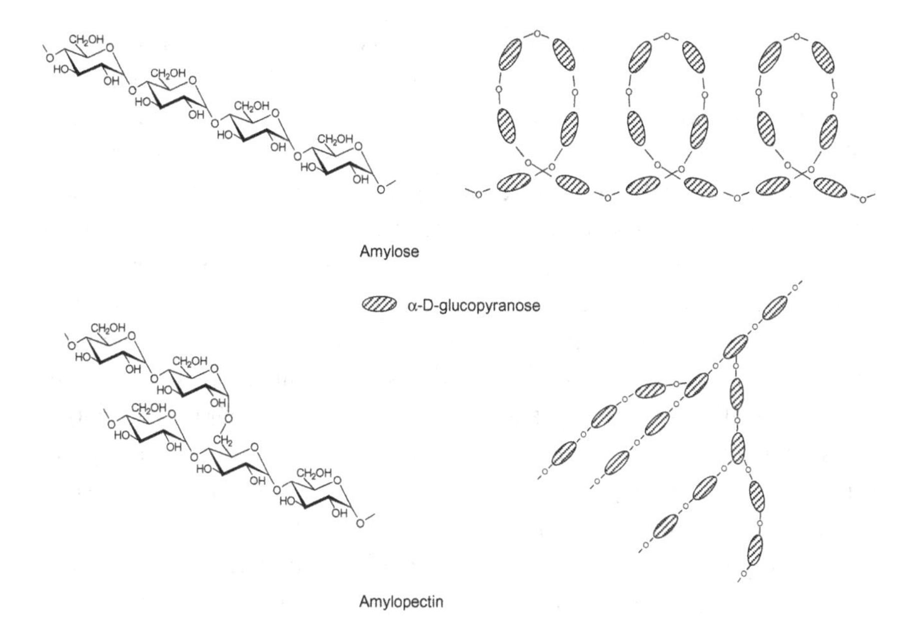

Starch is manufactured in the leaf cell chloroplasts or green plastids through polymerization of excess glucose produced in photosynthesis (cf., Chapt. "8.2. Polysaccharides"). This type of starch consists of small granules and is called "transient starch" or "primary starch". On the other hand, amyloplasts of plant storage tissue cells, for example, in the root tubers of the potato plant (*Solanum tuberosum*), synthezise and store larger, often elliptic starch granules. This type of starch is called "storage starch". Starch granules first appear in the plant cell as minute points, which grow rapidly to fill the cell. Amyloplasts also convert storage starch back into glucose when the plants need energy; i.e., they serve the plant as a reserve food supply.

If starch granules from higher plants are observed through a light microscope, stratified layers of starch (called "growth rings") around a nucleus (called a "hilum") are observed. Growth rings contain alternating regions of semicrystalline and amorphous material. The granule is systematically structured with the starch molecules oriented in specific spherocrystalline patterns. It seems that the crystalline regions are predominantly made up of amylopectin polymers (see below), where the outer branches are hydrogen bonded to each other to form crystallites;

i.e., the segments of the amylopectin chains intertwine to form parallel arrays of double helices responsible for the crystallinity of starch. The amorphous regions of granules are mainly composed of amylose (see below) and amylopectin branch points. The shape, size, and structure of the granules vary substantially between plant organs and between species. By observing microscopically the shape and size of the granule, the plant source of a starch can be identified even in mixtures of dry starch. This is of practical importance when evaluating the quality of flour and the starch content of food. Examples of average diameters of starch granules are (the percentage of starch in dry solids in parentheses): rice 3–8 μm (90%), wheat 50 μm (70%), and potato 15–150 μm (75%). The relative mass percentages of amylose and amylopectin in cereal endosperm starches typically range between 20–30% and 70–80%, respectively. Some mutant genotypes of maize, barley, and rice may contain as much as 70% amylose, whereas some cultivated plant varieties have pure amylopectin starch without amylose (these starches are known as "waxy starches").

Amylose is defined as a linear homoglycan ((1→4)-α-D-glucan), where α-D-glucopyranose units are linked to each other by (1→4)-glycosidic bonds (see above). There are also indications that some molecules are slightly branched by α-(1→6)-linkages. The main structure is a spiral assembly of repeating left-handed helixes (each with six α-D-Glc*p* units, see above). The molar masses of amylose typically vary in the range of 3×10^5–9×10^6 Da. The classic test for the presence of starch is its reaction with iodine (I_2). If starch molecules are present in a substance, the reaction yields a deep blue color. The color results from iodine being trapped inside the helix structure (the total content of I_2 in it is 19–20%).

In contrast to amylose, **amylopectin** is the highly branched component of starch. It is a branched polysaccharide composed of hundreds of short (1→4)-α-D-glucan chains that are interlinked (about 5%) by (1→6)-α-D-glucan chains with an average DP of 20–30. Therefore, amylopectin forms a cluster structure and several diagrams for this cluster-type architecture have been proposed. Typical molar masses of amylopectin vary in the range of 10^7–10^9 Da. Its reaction with iodine yields a reddish brown color; the total content of I_2 in it is <1%. In plants, a wide range of minor components are associated with starches; typical examples are lipids and glycolipids, proteins, amino and nucleic acids, triglycerides, mineral compounds, and water.

Starch is a white, tasteless, and odorless powder that is practically insoluble in cold water and ethanol. When starch is heated in water, amylose becomes soluble and amylopectin forms a slimy colloid. Upon continuous heating, granules gradually swell as water is absorbed by amorphous regions within the granules. This eventually leads to a complete separation of amylose and amylopectin. Starch polysaccharides are broken down by enzymes known as amylases (glycogenases). The saliva of humans and some other mammals contains α-amylase or dextrinogenic amylase ((1→4)-α-D-glucan glucanohydrolase), present also in pancreatic juice and malt, as well as in some bacteria and molds. It rapidly breaks down starch, first into α-dextrin (cf., Chapt. "9.3.2. Other oligosaccharides") with some tens of glucose units and ultimately yielding mainly maltose. However, α-amylase cannot affect glycosidic (1→6)-bonds. Another form of amylase, β-amylase or saccharogen amylase ((1→4)-α-D-glucan maltohydrolase) synthesized by bacteria, fungi, and plants, first catalyzes the hydrolysis of the second α-(1→4)-glycosidic bond from the non-reducing end of (1→4)-α-D-glucans (this process is important in brewing). In the case of amylose, the main end product is glucose, but in the case of amylopectin, besides maltose and glucose, some short chained β-dextrin or "limit dextrin" is also obtained. The main reason for the formation of β-dextrin is that unlike α-amylase, β-amylase cannot break down (1→6)-bonds at amylopectin branch points. The oldest criteria for the linearity of amylose were based on its complete hydrolysis by β-amylase. The "modern history" of amylases began in 1833, when Anselme Payen and Jean-François Persoz (1805–68) isolated an amylase complex from germinating barley and named it "diastase".

Most commercial starch is made from varities of maize, wheat, and rice or from tubers of potato and cassava root (*Manihot esculenta*) (known as "tapioca starch"). In the refining process, purified corns or tubers are first crushed or ground, and the pulp obtained is mixed with water. Large particles are then separated by filtration, starch is separated from the suspension by centrifugation, and the final product is washed in several stages with water. Due to the higher amounts of starch-binding proteins in corns, the manufacture of starch from potato tubers is normally easier. Starch is hygroscopic, binding moisture from the air and the water content of the "dried product" may vary in the range of 10–20%.

Starch in foods has always been the most common item and a source of energy in the human diet. Historically, the use of starch to stiffen linen was known in Egypt over 5000 years ago. This skill spread to Europe in the 14th century (cf., Chapt. "2.2 The Era Before the 1800s"). In addition, papyrus had traditionally been sized with modified wheat starch to produce a smooth surface and resistance to ink penetration. Nowadays, starch has been largely replaced by many other products, but its acid-modified products and certain derivatives are important in food, pharmaceutical, and cosmetic industries. In many applications, starch gelatinization is one of its important and unique properties. Aside from its basic nutritional use, starch is used in brewing and as a thickening agent in baked goods and confections. Modified starch is widely used in the surface sizing of paper and board and as wallpaper paste. Large quantities of starch are also used in the textile industry as warp sizing to strengthen the thread during weaving.

Selective periodate oxidation of starch introduces water-solubility-increasing and highly reactive carbonyl groups to its structure. Oxidation cleaves the C_2-C_3-bonds of the anhydroglucose units of the starch chains to form dialdehyde groups (i.e., the formation of dialdehyde starch, DAS). These kinds of **oxidized starches** produce softer and cleaner gels, which are used in the paper industry, for example, as wet-strength additives, but they have many other applications as well. The water solubility of starches can also be increased by their partial acid hydrolysis with dilute mineral acid. The products, **acid-modified starches**, have much less viscosity, and they form gels, like oxidized starches, with improved clarity and increased strength. **Dextrins** (see above) can be produced from starch using enzymes such as amylases, or by dry heating with or without an acidic or alkaline catalyst. They are white, yellow, or brown powders that are partially or fully water soluble. Dextrins have widely varying properties and a wide range of utilization in mining, paper, foundry, leather, and textile industries.

Starch syrups are starch sugars; they represent the most common starch-based food ingredient and they are used as sweeteners in many drinks and foods. They can be prepared from starch by acids, enzymes, or a combination of the two. They are often known as "dextrins". Their extent of conversion is typically quantified by "dextrose equivalent" (DE), which roughly indicates the fraction of glycosidic bonds in starch that have been broken. In general, depending on the grade, the content of maltose in these products is relatively high (50–70% of the carbohydrates), and they do not contain significant amounts of polysaccharides. Typical products include maltose starch syrups (DE>38), high-saccharified starch syrups (DE>45), confectionary starch syrups (DE 36–44), and low-saccharified starch syrups. Starch syrups or corn syrups are distinct from **high-fructose corn syrups** (HFCSs), which are manufactured from starch syrups by enzymatically converting a large proportion of their glucose into fructose, thus producing sweetener products with high levels of fructose (cf., Chapts. "9.1.1. Aldose monosaccharides" and "9.3.1. Disaccharides").

Starch ethers are produced to some extent. Commercially most significant products are **hydroxyethyl starch**, **hydroxypropyl starch,** and different **cationic starches**. The latter contain quaternary and tertiary ammonium groups. Examples of these products are **diethylaminoethyl starch** and **2-hydroxy-3-(trimethylammonium)propyl starch**. On today's high-speed paper machines, the retention of anionic native starch to anionic fibers is poor. The way to improve this retention is the use of cationic starches. In contrast, **anionic starches** can be used in extremely acid, cationic pulp systems and in dual-retention systems as an anionic component. Anionic starch is typically used to neutralize the ionic nature of an over-cationic system and to improve starch retention and overall retention. A large number of *starch esters* have also been prepared. The most significant such products are **starch acetates** (used in food and textile industries) and **starch phosphates** (mainly used as food additives and thickeners). In the manufacture of *cross-linked starches*, bifunctional reagents that form diethers or diesters are used. Such starches are used in the food industry, where heat-resistance, high viscosity, and/or low pH are needed. It is also possible to cross-link starch ethers and esters. In this case, typical uses include canned foods and salad dressings.

Glycogen is a more branched version of amylopectin; it functions as a secondary long-term energy storage in animals, the primary energy stores being fats.

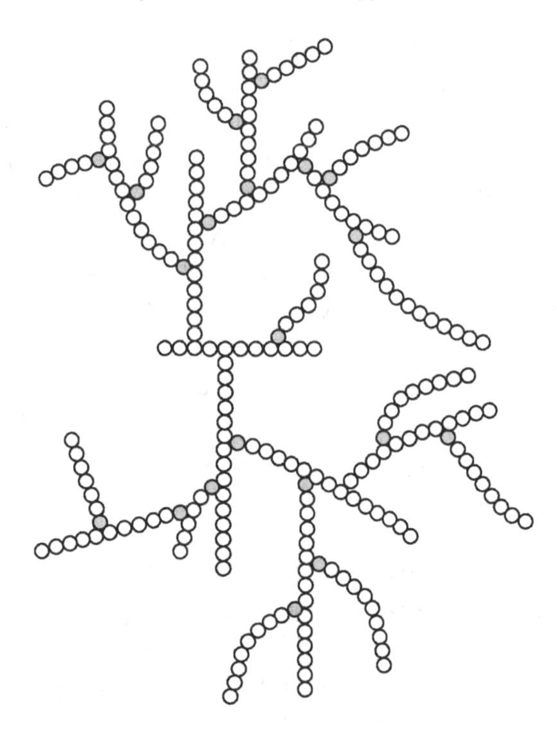

In humans, glycogen is produced and stored in a hydrated form primarily in the liver (it may contain 5–6% of this polysaccharide in its fresh mass), fat cells, and muscles (1–2% of muscle mass) hydrated with three or four parts of water. Small amounts of it can also be found in the kidneys. The amount of glycogen stored in the body depends on physical training, basal metabolic rate, and eating habits. Glycogen is a multibranched polysaccharide that consists of linear (1→4)-α-D-glucan chains with further (1→6)-α-D-glucan chains (DP 20–30) branching off at about every 10 glucose units. The entire glycogen granule may contain about 30,000 α-D-Glc*p* units; its molar mass is similar to or even higher than that of amylopectin. The lower-molar-mass fractions of glycogen are water soluble, but its higher-molar-mass fractions are water insoluble. The separated glycogen preparations usually contain small amounts of proteins.

Chitin, (1→4)-2-acetamido-2-deoxy-β-D-glucan, is the most widespread amino polysaccharide in living organisms. It is a characteristic cellulose-type supporting component in the cell walls of fungi, as well as in certain yeasts and algae. It is mainly found in the exoskeletal elements of arthropods (crustaceans, such as crabs, lobsters, and shrimps) and in worms, insects, molluscs, and cephalopods.

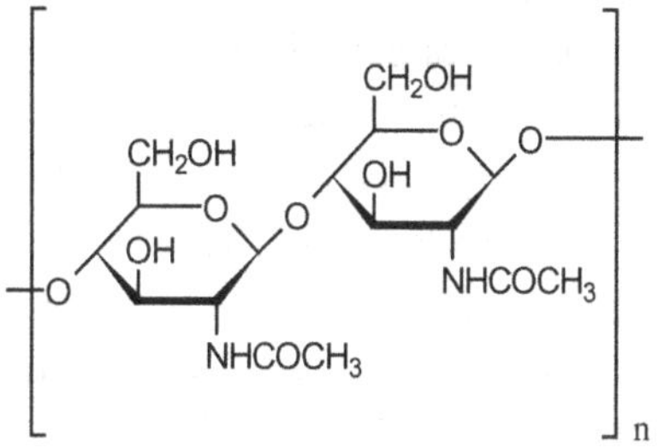

Chitin is a homoglycan that contains 2-acetamido-2-deoxy-D-glucopyranose units joined via β-(1→4)-glycosidic bonds. This polysaccharide probably with a very high DP is useful in several medical and industrial applications. Chitinase enzymes are found in organisms (e.g., in the gastric juice of snails) that either need to reshape their own chitin or dissolve and digest the chitin of fungi or animals. These are hydrolytic enzymes that break down glycosidic bonds in chitin resulting in the formation of *N*-acetylglucosamine. This amine has been prepared from chitin by hydrolysing its repeating structural unit, ***N,N'*-diacetylchitobiose** (molar mass 424.40 ($C_{16}H_{28}O_{11}N_2$), melting point 247°C, $[\alpha]_D$ +18 (*c* 1, H_2O)):

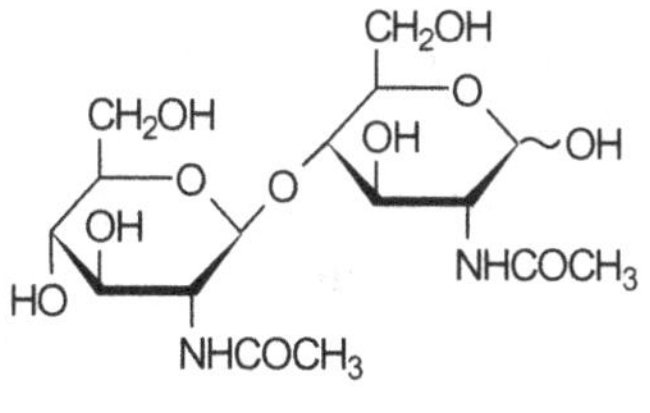

Chitosan is mainly composed of β-(1→4)-linked 2-amino-2-deoxy-D-glucose units; it is prepared from chitin in a partial hydrolysis of its *N*-acetyl groups. It is obtained by treating shrimp and other crustacean shells with sodium hydroxide. This polysaccharide not only gives dry strength to paper, but it is also an efficient wet-strengthening agent. However, its production is time-consuming and expensive and it has not been exploited to any major extent in the paper industry. In spite of this, chitosan has a number of agricultural, horticultural, and biomedical uses, as well as uses in water filtration, wine-making, and food preservation.

Dextrans are homoglycans with molar masses normally higher than 1000 Da. Their linear backbone typically consists of α-(1→6)-linked D-glucopyranose units, which contain branches of smaller chains of D-glucopyranose moieties linked to the backbone by α-(1→2)-, α-(1→3)-, or α-(1→4)-glycosidic bonds. There are also other types of dextrans with different structural features; their backbone contains alternating α-(1→3)- and α-(1→6)-linked D-glucopyranose units or α-(1→3)-linked D-glucopyranose units with α-(1→6)-linked branches. Dextrans have long been recognized as contamination in sugar processing and other food production. Louis Pasteur discovered them as microbial products in wine in 1861. In 1874, Carl Scheibler (1827–99) created the name "dextran" following the finding that this polysaccharide complex had both a "typical molecular formula of sugars" ($C_6H_{10}O_6$) and a positive optical rotation. Dextrans are obtained from sucrose by certain lactic acid bacteria (e.g., *Leuconostoc mesenteroides* and *Streptococcus mutans*) and, for example, dental plaque is rich in these polysaccharides. They are used medically as antithrombotic or antiplatelet drugs (i.e., to prevent and treat thrombosis by reducing blood viscosity) and volume expanders in hypovolaemia.

Fructans are polysaccharides that contain fructose residues, normally with sucrose units and they are also called "**levans**" or "levan polysaccharides" (levans contain repeating β-(2→6)-linked D-fructofuranose units). They occur in foods, typically in artichokes, asparagus, garlic, onion, and rye, and in grasses. The most typical group of fructans is, besides levans, **inulins**, which belong to a class of dietary fibers and are produced by many plants. Their chemical structure mainly consists of repeating β-(2→1)-linked D-fructofuranose units (DP about 100). Inulins are used in processed foods, and they also have some medical applications.

In addition to common monosaccharide moieties, many polysaccharides contain uronic acid (see pectins, above) or amino sugar residues. In the latter case, it is often spoken about "**glycosaminoglycans**" (GAGs) or, in physiological issues, about "**mucopolysaccharides**". They are long unbranched polysaccharides, which contain a repeating dimeric unit typically consisting of an amino sugar (*N*-acetylglucosamine or *N*-acetyl-galactosamine) along with an uronic acid (glucuronic acid or iduronic acid). For example, **hyaluronic acid** or hyaluronan (HA), with a DP value of about 25,000, is widely distributed among connective, epithelial, and neural tissues. It is a non-sulfated polysaccharide, which contains repeating units of β-D-glucopyranosyluronic acid and 2-acetamido-2-deoxy-β-D-glucopyranose:

$$\left[\rightarrow 4)\text{-}\beta\text{-D-Glc}p\text{U-}(1\rightarrow 3)\text{-}\beta\text{-D-Glc}p\text{NAc-}(1\right]_n\rightarrow$$

Chondroitin-4-sulfate and **chondroitin-6-sulfate** are examples of sulfated GAGs. Their basic repeating units are β-D-glucopyranosyluronic acid and 2-acetamido-2-deoxy-β-D-galactopyranose. Chondroitin sulfates are usually found attached to proteins as part of glycoproteins (proteoglycans) (cf., Chapt. "10.5.1. Hormones"). They are significant structural components of cartilage and provide much of its resistance to compression. Chondroitin sulfates are used for the treatment of osteoarthritis.

The term "industrial gums" is often connected to water-soluble heteroglycans that, in an appropriate solvent or swelling agent, form highly viscous dispersions or gels even at a low content of dry solids. Examples of similar gum polysaccharides (e.g., certain hemicelluloses, pectins, and amylopectins were previously described. Further examples, primarily classified as *vegetable gums*, are given in this section. They can be extracted from a variety of non-wood plant materials, such as fruits, seeds, roots, and tubers. These gums are tasteless, odorless, colorless, and nontoxic and they can be deploymerized by acid- or enzyme-catalyzed hydrolysis to low-molar-mass products. The most important example is **gum arabic** or acacia gum, which is collected by hand as the hard sap exuded from two species of *Acacia* (*A. senegal* and *A. seyal*) growing in Africa throughout south of Sahara, from Senegal and Sudan to Somalia. Its structure is highly branched (DP about 3500) and the main unit in the backbone is β-D-galactopyranose; its side chains contain, depending on the origin, L-arabinofuranose, L-rhamnopyranose, and D-galactopyranose, as well as D-glucopyranosyluronic acid and 4-*O*-methyl-D-galactopyranosyluronic acid units. Gum arabic's desirable properties are unique among gums. It is readily soluble in water and the solutions have a low viscosity; the viscosity of its 20% solution corresponds to that of 1–2% solutions of other gums and resembles thin sugar syrup in its flow properties. Gum arabic is primarily used in the food industry as a stabilizer. Its other uses include printing, paint production, glue, and cosmetics, but it is also used as a viscosity control agent in inks and in the textile industry. Examples of other gums are **gum tragacanth** or tragacanth with D-xylopyranose, L-fucopyranose, D-galactopyranose, and D-galactopyranosyluronic acid units, **gum karaya** with D-rhamnopyranose, D-galactopyranose, D-galactopyranosyluronic acid and D-glucopyranosyluronic acid units, **gum ghatti**, and **okra gum**.

The main backbone of **xanthan** or xanthan gum contains, like cellulose, repeating β-D-glucopyranose moieties linked by (1→4)-glycosidic (or glucosidic) bonds. However, to every second glucose unit in xanthan is attached a trisaccharide side chain β-D-Man*p*-(1→4)-β-D-Glc*p*U-(1→2)-6-*O*-Ac-α-Man*p*-(1→3)- at carbon atom C_3. About half of these side chains also carry a pyruvic acid (2-oxopropanoic acid) group as a 4,6-di-*O*-acetal. Xanthan is the extracellular polysaccharide produced by *Xanthomonas campestris* and its molar mass is at least 2×10^6 Da. It has the desirable property of stabilizing aqueous dispersions, suspensions, and emulsions. When heated, the viscosity of xanthan solutions containing a small amount of salt (0.1%) changes only slightly over the temperature range of 0–95°C, while that of other polysaccharide solutions normally decreases upon heating.

Guaran is obtained from guar gum by purification. Its main backbone contains repeating β-D-mannopyranose units linked together by (1→4)-glycosidic (or mannosidic) bonds. On the average, one of every 1.8 mannopyranose units is replaced by a (1→6)-linked α-D-galactopyranose unit. The molar mass of guaran is about 200,000 Da. Examples of similar polysaccharides are particularly **algins**. They can be extracted from brown algae (*Phaeophyceae*) with an aqueous solution of sodium carbonate. Commercial products are generally sodium, ammonium, or potassium salts and esters of alginic acid (alginates). "Alginic acid" is a generic term for a linear polymer (it contains only (1→4)-glycosidic bonds) with D-mannuronic acid- or L-glucuronic acid-containing homogeneous segments and corresponding uronic acid-containing heterogeneous segments. The most important property of alginates is their ability to form gels by a reaction with calcium ions.

Carrageenans or carrageenins, **agar(s)** or agaragar, and **furcellaran(s)** or Danish agar also normally form corresponding gels. Typical for all these polysaccharides is a linear galactan structure with molar masses of about 250,000 Da. Carrageenans are sulfated polysaccharides, which can be extracted from red edible seaweeds, usually from *Eucheuma cottonii* and *E. spinosum*. Commercial products are primarily composed of three types of polymers (i.e., kappa-, iota-, and lambda-carrageenans); they are widely used in the food industry for their gelling, thickening, and stabilizing properties. Agar consists of two components (cf., starch): the linear polysaccharide **agarose** and a heterogeneous mixture of **agaropectin**. It is found in the cell walls of certain species of red algae, primarily in the genera *Gelidium* and *Gracilaria*. Agar has many uses, but perhaps the most known is its use in an agar plate or Petri dish as a growth medium containing a mixture of agar and several nutrients, where microorganisms (bacteria and fungi) can be cultured and observed under the microscope. Furcellaran is a sulfated polysaccharide, which can also be extracted from red algae (*Furcellaria* species, mainly *F. fastigiata*). It is mainly used as a gelling agent.

10. Carbohydrate Residue-containing Substance Groups

10.1. Glycosides

By acid or enzymatic hydrolysis it is possible to obtain from certain chemical constituents of several plants D-glucose and another organic compound that contains a hydroxyl group, i.e., a non-sugar component, referred to as an "aglycone" or a "genin". These native plant compounds are traditionally termed *glucosides*. Since the carbohydrate portion is in many cases other than D-glucose (i.e., generally "glycose"), the compounds are typically known as *glycosides*. If the carbohydrate portion is exactly known, the compounds can also be named on the basis on this carbohydrate moiety (e.g., as *fructosides*, *rhamnosides*, and *galactosides*).

The carbohydrate portion in plants (sometimes referred to as a "glycone") can be mono-, oligo-, or polysaccharides. An aglycone, often with a complicated chemical structure, is then bonded to the free anomeric hydroxyl group (cf., hemiacetal or hemiketal structures shown in Chapt. "1. Introduction") of reducing carbohydrates. Thus, the formed *glycosidic bond* is either the direct link between an anomeric carbon atom and the nitrogen or carbon atom of an aglycone (i.e., via *N*-glycosidic or *C*-glycosidic bonds, respectively); alternatively, in analogy, the link formed through an oxygen (or a sulfur) atom (i.e., C-O(S)-C). In the latter case (*O*-glycosidic or *S*-glycosidic bonds), for example, the hemiacetal structure is simultaneously changed to an acetal structure. A monosaccharide (mainly aldohexose or aldopentose) has either the form of pyranose (a six-ring structure) or furanose (a five-ring structure), and the "simplified structures" of glycosides are (R refers to an aglycone) as follows:

These Haworth projections show simply the general principles of the structures and only indicate the chiral center at C_1 with the orientation of glycosidic bonds. In practice, this also means that the glycosidic bond can be designated as either α- or β-bond (cf., Chapt. "6.1. Mutarotation"); these glycosides are then classified as α-glycosides or β-glycosides. Furthermore, it should be noted that oligo- and polysaccharides contain partial structures where the aglycone portion is another carbohydrate, i.e., a single monosaccharide, or an oligosaccharide or a polysaccharide group. In general, glycosides can be classified by their aglycone, glycone, or glycosidic bond.

In this section, no detailed data on the properties of "basic simple glycosides", such as a common monosaccharide with an open-chain hydrocarbon-derived aglycone, are described. Examples of such compounds are methyl α-D-glucopyranoside (I), methyl α-D-gulofuranoside (II), and ethyl β-D-fructopyranoside (III):

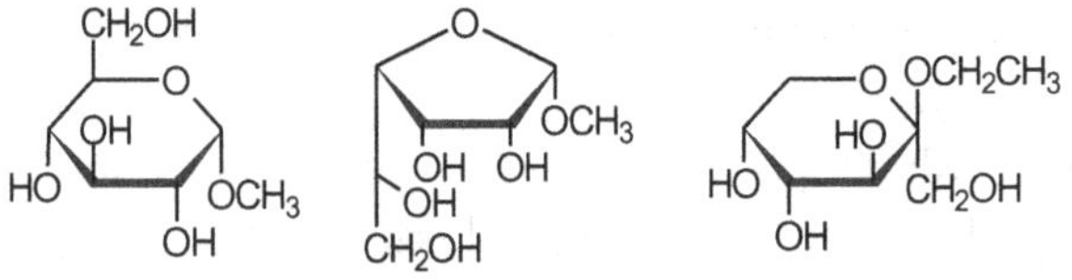

I II III

Glycosides often contain several different monosaccharides, reaching from D-glucose to the relatively rare monosaccharides typically present only in certain glycosides, as well as monosaccharide derivatives, such as uronic acids. The chemical structures of aglycones also show great variety; certain aglycones can bind covalently, depending on the number of hydroxyl groups, many monosaccharide constituents (e.g., *diglycosides*). As already indicated, for example, in *O-glycosides* the hydroxyl group attached to an aliphatic ring (or a phenolic hydroxyl group) reacts with an anomeric hydroxyl group; for example, arbutin or 4-hydroxyphenyl β-D-glucopyranoside is represented as follows:

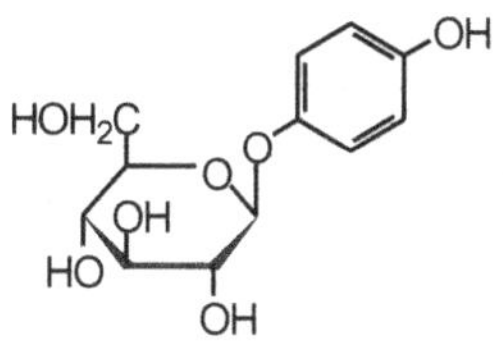

One example of *C-glycosides* or D-glycosyl compounds is pseudouridine or 5-(β-D-ribofuranosyl)uracil):

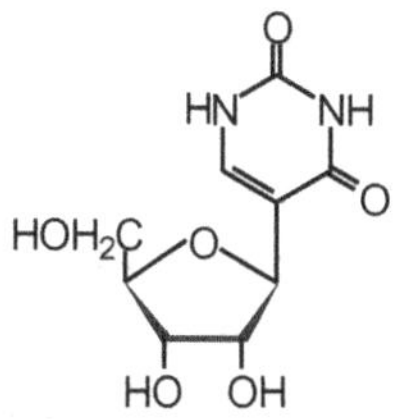

The aglycone portion often belongs to a substance group with a higher chemical priority than carbohydrates. Consequently, sometimes the straightforward classification of glycosides may also be difficult and the categorization is based on the chemical structure of the aglycone. For example, adenosine or 6-amino-9-(β-D-ribofuranosyl)-9*H*-purin is generally not considered an *N-glycoside* or a *glycosylamine*:

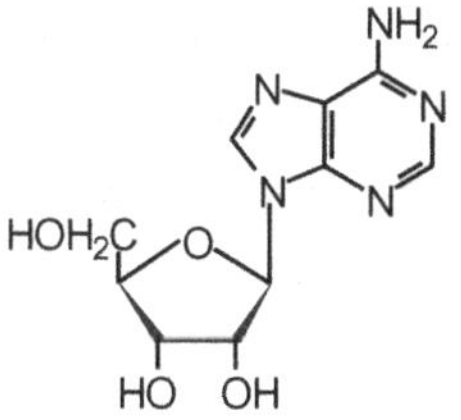

but is seen as a purin-containing nucleoside typically included in nucleic acids (cf., Chapt. "10.2. Nucleic Acids"). Examples of glycosides containing a carbohydrate-derived aglycone are a non-reducing sugar, sucrose (β-D-fructofuranosyl-α-D-glucopyranose, I) and a reducing sugar, α-lactose (4-*O*-β-D-galactopyranosyl-α-D-glucopyranose, II) (cf., Chapt. "9.3.1. Disaccharides"):

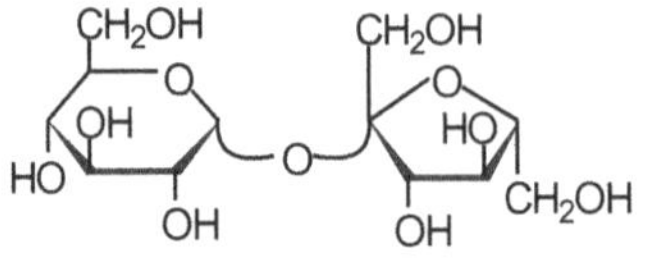

I 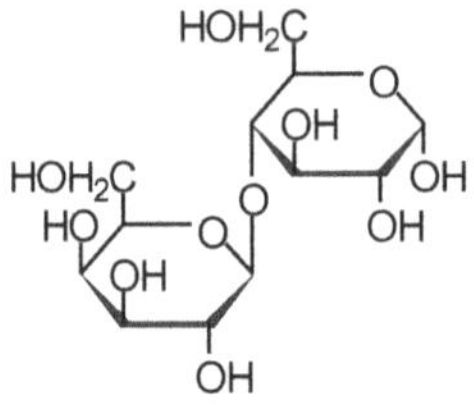 II

In this section, the compounds where the aglycone is a simple acyclic hydrocarbon (see above) are considered glycosides as well as *glycosans* belonging to anhydrosugars (cf., Chapt. "9.2.1. Anhydrosugars"). Likewise, the acetals where the carbonyl group belongs to an initial aglycone can be classified as glycosides. Examples of these structures are methyl β-D-glucopyranoside (I), 1,6-anhydro-β-D-glucopyranose or levoglucosan (II), and 1,2-*O*-isopropylidene-D-glucofuranose (III):

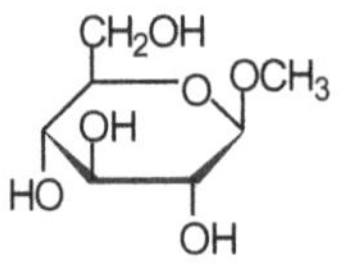

I 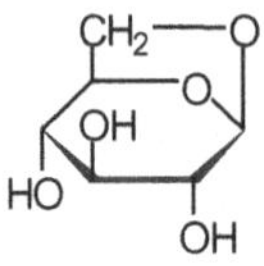 II 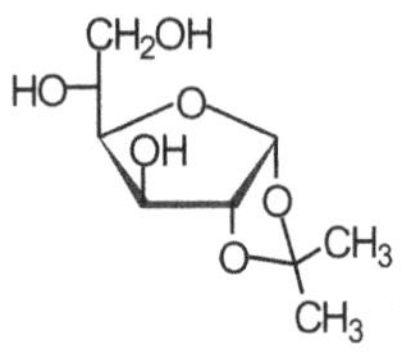 III

Glucosinolates (see sinigrin, below) are typical *S-glycosides* or *thioglycosides*. In addition, there are *selenoglycosides* and *glycosyl halides*.

Glycosides occur widely in nature in varying concentrations as chemical components of seeds, fruits, flowers, roots, cortexes, and leaves of several plants. In certain cases, compounds that belong to different subgroups (e.g., *cardiac glycosides* and *saponins*) of glycosides can be found even in the same plant. However, the role of glycosides is not yet fully understood. They are often considered only residues of metabolic pathways although glycosides are now shown to play numerous important roles in living organisms. Many plants store medicinally important agents in the form of inactive glycosides. In addition, there are research data available on the formation of glycosides in different parts of plants.

It is also typical that along with glycosides various enzymes are present and participate in the synthesis and hydrolysis of glycosides. Hence, it is difficult to isolate the native glycosides with a complex carbohydrate portion. In general, glycosides are readily hydrolysable by acids but relatively stable under alkaline conditions. In addition, their solubility may vary considerably. However, due to a hydrophilic carbohydrate portion, many glycosides dissolve in water and aqueous ethanol solutions. They also have a bitter taste, and some of them are toxic; for example, *cyanogenic glycoside*-rich plants are harmful in cattle feed and cause symptoms of poisoning.

10.1.1. *Cardiac glycosides*

Cardiac glycosides or *steroidal glycosides* represent a family of compounds that occur in a number of plants, especially in those belonging to the genus *Digitalis*. The most common species are common foxglove (*Digitalis purpurea*, native and widespread throughout most of temperate Europe) and woolly foxglove (*Digitalis lanata*, found mostly in Eastern Europe, particularly in Hungary). Both plants are also widely cultivated in Europe. In addition, certain cardiac glycosides are found, for example, in sea onion or squill (*Scilla maritima*) and in woody lianas (*Strophanthus gratus* and *S. kombe*) growing naturally in the tropical regions of southeast Africa. These glycosidic compounds from different plant sources are used as drugs, and they are also generally known, regardless of their origin, as *digitalis* or *digitalis glycosides.*

From ancient times, humans have used crude extracts from cardiac glycoside-containing plants as arrow coatings, homicidal and suicidal aids, and rat poisons. There are also "systematic observations" about the therapeutic benefits of sea onion before 1000 B.C. Ancient Egyptians and Romans used this plant as an emetic and an edema-diminishing agent although cardiac glycoside-rich drugs only entered medical practice at the end of the 18th century. At that time, it was noted that the product extracted from the green parts of common foxglove has a therapeutic effect on edema ("dropsy"). In the classic book "An Account of the Foxglove and Some of its Medical Uses with Practical Remarks on Dropsy and Other Diseases" published in 1785, William Withering (1741–99) described in detail, based on his many years of observations, the main therapeutic uses and toxicity of digitalis, which has formed the basis of its clinical use ever since as a medication for heart failure. Nowadays, it is known that over 200 cardiac glycosides act as heart stimuli by influencing the contractile force of the heart muscle; they can also be used to treat heart conditions, such as atrial fibrillation. However, only a few of them are still in clinical use. In addition, due to the characteristic actions on the heart, cardiac glycosides increase urine flow as a result of the intensified circulation in the kidneys. In general, the advantages of properly using these medicines are their low side effects although in cases of overdose, most of them are extremely toxic and can be fatal. The typical toxic effects include anorexia, nausea, vomiting, diarrhea, confusion, visual disturbances, and cardiac arrhythmia (irregular heartbeat).

The chemical structure of a cardiac glycoside contains, besides a carbohydrate component, a genin part having a steroid structure with a lactone ring. These genin parts are collectively known as *cardenolides* (I) or *bufadienolides* (II):

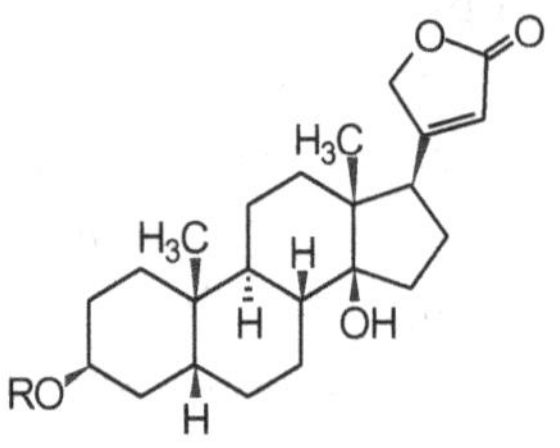

I

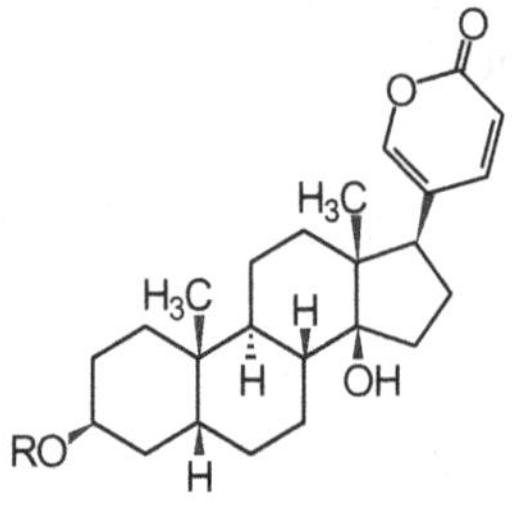

II

An example of the basic structure of the most common ones (I) is digitoxigenin or 14-hydroxy-5β,14β-card-20(22)-enolide, and of the latter ones (II) bufalin or 14-hydroxybufa-20,22-dienolide. The active therapeutic effect of the cardiac glycoside primarily depends on the genin part, whereas the pharmacokinetic properties of the product are caused by the carbohydrate part (R). The structures of monosaccharide components obtained from acid hydrolysis of cardiac glycosides vary significantly. They belong to (i) *aldoses or 6-deoxyaldoses and their methyl ether derivatives* (D-xylose, D-galactose, D-glucose, 6-deoxy-D-allose, 6-deoxy-D-altrose, 6-deoxy-D-galactose or D-fucose, 6-deoxy-D-glucose, 6-deoxy-D-gulose, 6-deoxy-L-mannose or

L-rhamnose, 6-deoxy-L-talose, 3-*O*-methyl-D-glucose, 2,3-di-*O*-methyl-D-glucose, 6-deoxy-2-*O*-methyl-D-allose or D-javose, 6-deoxy-3-*O*-methyl-D-altrose or D-vallarose, 6-deoxy-3-*O*-methyl-L-altrose or L-vallarose, 6-deoxy-2-*O*-methyl-D-galactose or 2-*O*-methyl-D-fucose, 6-deoxy-3-*O*-methyl-D-galactose or D-digitalose, 6-deoxy-3-*O*-methyl-D-glucose or D-thevetose, 6-deoxy-3-*O*-methyl-L-glucose or L-thevetose, 6-deoxy-3-*O*-methyl-L-mannose or L-acofriose, 6-deoxy-3-*O*-methyl-L-talose or L-acovenose, 6-deoxy-2,3-di-*O*-methyl-D-galactose or 2,3-di-*O*-methyl-D-fucose), (ii) *2,6-dideoxy- and 2-deoxy-aldo-hexoses and their methyl ether derivatives* (2,6-dideoxy-D-*arabino*-hexose or D-canarose, 2,6-dideoxy-D-*xylo*-hexose or D-boivinose, 2,6-dideoxy-D-*lyxo*-hexose or D-oliose, 2,6-dideoxy-D-*ribo*-hexose or D-digitoxose, 2-deoxy-D-*arabino*-hexose or 2-deoxy-glucose, 2-deoxy-D-*xylo*-hexose, 2,6-dideoxy-3-*O*-methyl-D-*arabino*-hexose or D-oleandrose, 2,6-dideoxy-3-*O*-methyl-L-*arabino*-hexose or L-oleandrose, 2,6-dideoxy-3-*O*-methyl-D-*xylo*-hexose or D-sarmentose, 2,6-dideoxy-3-*O*-methyl-D-*lyxo*-hexose or D-diginose, 2,6-dideoxy-3-*O*-methyl-L-*lyxo*-hexose or L-diginose, 2,6-dideoxy-3-*O*-methyl-D-*ribo*-hexose or D-cymarose), and iii) *other monosaccharide derivatives* (4,6-dideoxy-hexosone, 2-*O*-acetyl-6-deoxy-3-*O*-methyl-L-altrose or 2-*O*-acetyl-L-vallarose, 2-*O*-acetyl-6-deoxy-3-*O*-methyl-L-glucose or 2-*O*-acetyl-L-thevetose, 2-*O*-acetyl-6-deoxy-3-*O*-methyl-L-mannose or 2-*O*-acetyl-L-acofriose).

Due to the complex structures of cardiac glycosides, their systematic names are also rather long and inconvenient and they are generally known by their trivial names.

Digitoxin, 3β-[(2,6-dideoxy-*O*-β-D-ribohexopyranosyl-(1→4)-2,6-dideoxy-*O*-β-D-ribo-hexopyranosyl-(1→4)-2,6-dideoxy-*O*-β-D-ribohexopyranosyl)oxy]-14-hydroxy-5β,14β-card-20(22)-enolide

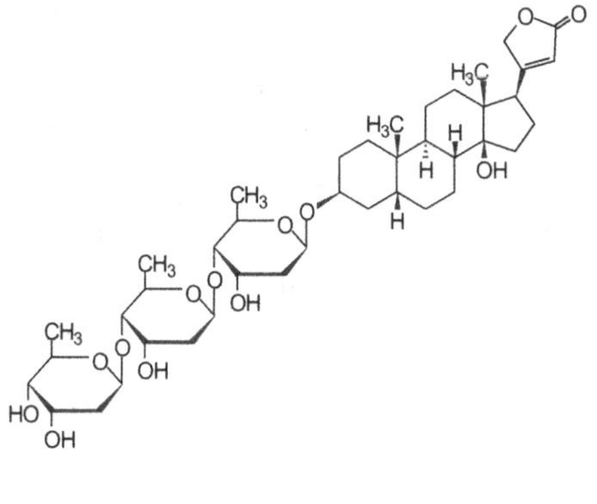

Molar mass 764.95 ($C_{41}H_{64}O_{13}$), melting point 255–7°C (anhydrous).

Digitoxin is obtained (yield 0.2–0.4%) by extraction of the dried leaves of *Digitalis purpurea* and *D. lanata*, as well as many other species of *Digitalis* with a solution of water:ethanol (1:1 v/v). Claude-Adolphe Nativelle (1812–89) first prepared it in crystalline form in 1869 making possible its modern therapeutic use. Digitoxin crystallizes from ethanol in the form of hydrate (contains either 0.5 or 1 mol H_2O) as elongated, small leaflets or it is a crystalline powder. Digitoxin is an odorless, bitter-tasting, and white solid that dissolves at 20°C in ethanol (1 g/60 mL), diethyl ether, and chloroform (1 g/40 mL) and slightly in water (1 g/100 L). It has the most powerful physiologic effects of all the cardiac glycosides occurring in common foxglove (*D. purpurea*) and its effects are similar to those of digoxin, although they are longer lasting. In addition, unlike digoxin (mainly excreted unchanged from the body via the kidneys), digitoxin is mainly eliminated via the liver; it can thus be used in patients with a poor or erratic kidney function. The oral absorption of this glycoside is 90–100% with a biologic half-life of four to six days. However, digitoxin is now rarely used in the current Western medical practice.

Digoxin

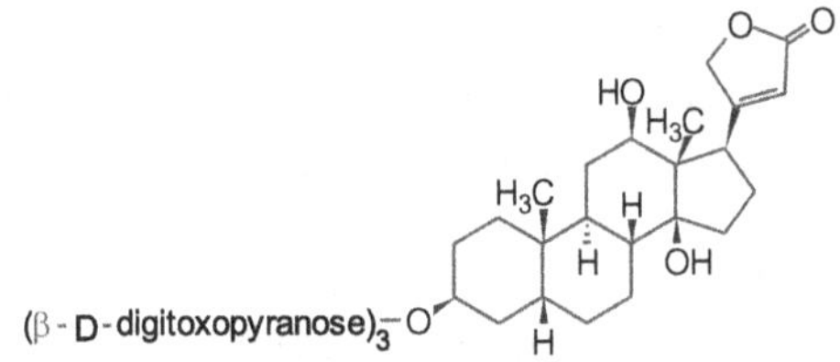

Molar mass 780.95 ($C_{41}H_{64}O_{14}$), melting point 235–65°C (decomposes).

Digoxin (see also digitoxin) was separated for first time in 1930. It can primarily be obtained by extraction of the dried leaves of woolly foxglove (*Digitalis lanata*) or oriental foxglove (*D. orientalis*). It crystallizes as leaflets or it is a white, odorless, and crystalline powder that dissolves in dilute ethanol and pyridine, but not in water, diethyl ether, acetone or chloroform. The oral absorption of this glycoside is 55–75% with a biologic half-life of 36 hours. Nowadays, digoxin is only occasionally used in the treatment of various heart conditions (e.g., atrial fibrillation, atrial flutter, and sometimes heart failure) that cannot be controlled by other medication.

Ouabain, g-strophanthin

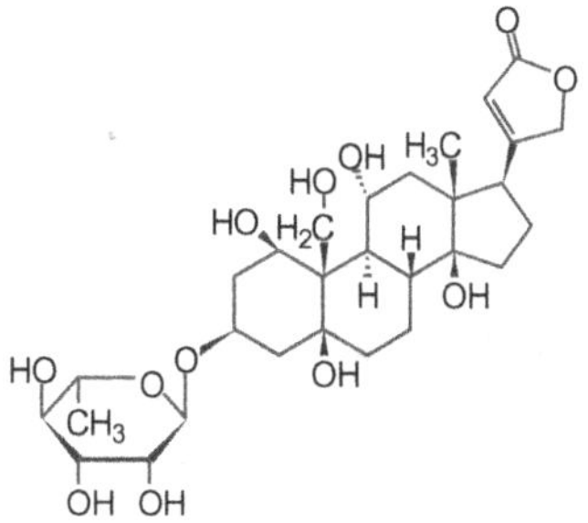

Molar mass 584.66 ($C_{29}H_{44}O_{12}$), melting point 190–200°C (decomposes).

Ouabain was traditionally used as an arrow poison (in Somali, "waabaayo" means "arrow poison" and in French "ouabaïo") in eastern Africa for both hunting and warfare. It belongs to *strophanthosides* or *strophanthines* that can be extracted from the ripe and dried seeds of woody lianas (*Strophanthus gratus* and *S. kombe*) native to southeast Africa and from the root, stem, leaves, and seeds of arrow-poison tree or kivai *Acokanthera schimperi*, native to eastern and central Africa as well as to southern Yemen. In this compound, L-rhamnose forms the glycosidic bond with ouabagenin or g-strophanthidin. It crystallizes from water in the form of octahydrate as lustrous platelets that decompose on standing in light and dissolve in water (1 g/75 mL) and ethanol (1 g/100 mL) and slightly in diethyl ether, chloroform, and ethyl acetate. Ouabain has some medical uses in treating hypotension and cardiac arrhythmias although it is too poorly and unreliably absorbed to be used orally. However, due to its extremely fast onset of action (5–10 min), it is useful for rapid digitalization in emergencies (e.g., atrial flutter, nodal tachycardia, and acute congestive heart failure). On the other hand, an overdose of ouabain causes, for example, respiratory distress, increased and irregular heartbeat, rise in blood pressure, rapid twitching of the neck and chest musculature, and convulsions. Its biologic half-life is 21 hours, and it is mainly excreted unchanged from the body in the urine.

Strophanthin, k-strophanthin, k-strophanthoside

Strophanthin can be mainly found in *Strophanthus kombe*. Native African tribes used this compound among other toxins as arrow poison. Its basic structural units are α-D-glucopyranose (19%), β-D-glucopyranose (19%), 2,6-dideoxy-3-*O*-methyl-β-D-*ribo*-hexopyranose or β-D-cymarose

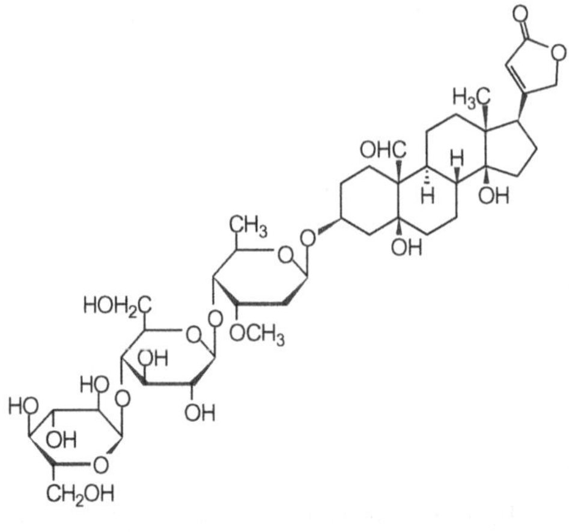

Molar mass 872.97 ($C_{42}H_{64}O_{19}$).

(15%), and strophanthidin (47%). However, the corresponding product is a mixture of glycosides containing, besides the former compound, k-strophanthin-β, in which α-D-glucopyranose in the carbohydrate part is replaced by β-D-glucopyranose. Strophanthin is a white or yellowish, bitter-tasting powder that decomposes on standing in light and can readily bind water (even 10% of its initial mass). It dissolves in water and dilute ethanol, but only slightly in diethyl ether, benzene, and chloroform.

Cymarin, k-strophanthin-α

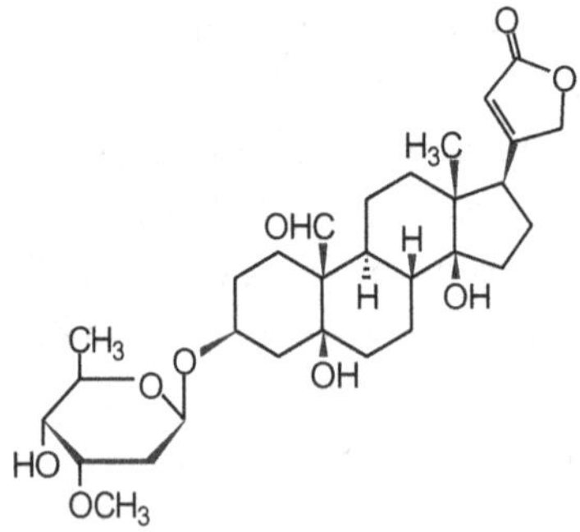

Molar mass 548.67 ($C_{30}H_{44}O_9$), melting point 148°C.

Like strophanthin, cymarin can mainly be found in *Strophanthus kombe*. It also exists in *Apocynum cannabinum* used as a source of fiber by Native Americans and *A. venetum* used as herbal tea in China. In this compound, 2,6-dideoxy-3-*O*-methyl-D-*ribo*-hexose or D-cymarose forms the glycosidic bond with strophanthidin. It crystallizes from methanol as needles that dissolve in methanol and chloroform, but not in water.

Convallatoxin

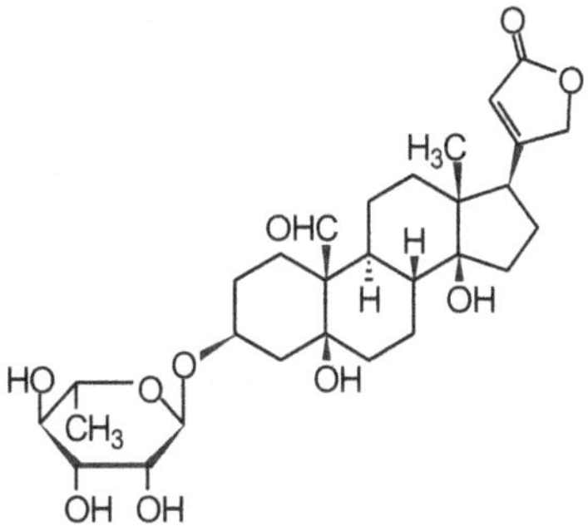

Molar mass 550.64 ($C_{29}H_{42}O_{10}$), melting point 235–42°C.

Convallatoxin occurs in the flowers of lily-of-the valley (*Convallaria majalis*). Its structure is similar to that of cymarin; the carbohydrate part D-cymarose in cymarin is replaced by L-rhamnose. It forms prisms that dissolve in ethanol and acetone, but only slightly in water and chloroform. Convallatoxin is mainly used for acute and chronic heart failure.

Sarmentocymarin

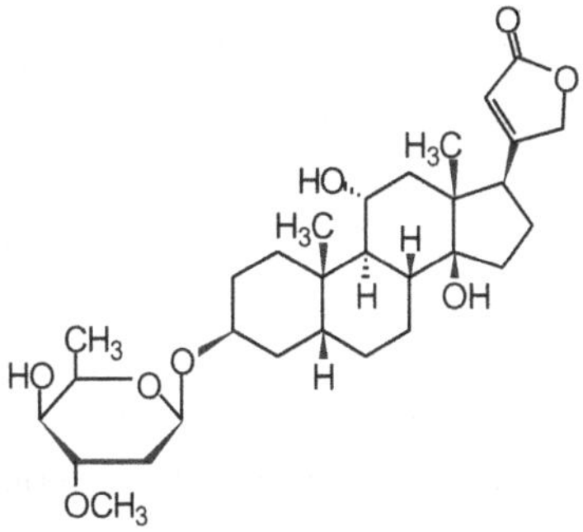

Molar mass 534.68 ($C_{29}H_{46}O_{8}$).

Sarmentocymarin can be extracted from the seeds of *Strophanthus sarmentosus* var. *senegambiae*. Its acid hydrolysis leads to an isomer of D-cymarose, 2,6-dideoxy-3-*O*-methyl-D-*xylo*-hexose or D-sarmentose, and sarmentogenin (molar mass 390.52; $C_{23}H_{34}O_{5}$), from which cortisone can be prepared.

Lanatoside C is found in the leaves of woolly foxglove (*Digitalis lanata*). The oral absorption of this glycoside is relatively low (20–60% with a biologic half-life of 33 hours) although it then metabolizes to the active digoxin and acetyldigoxin. It can be used in the treatment of congestive heart failure and cardiac arrhythmia (irregular heartbeat). In addition, **lanatoside A**, **B**, and **D** can be separated from certain woolly foxgloves.

Digitonin, digitin. Molar mass 1229.33 ($C_{56}H_{92}O_{29}$), melting point 235–40°C.

Digitonin (a steroidal saponin or saraponin) can be obtained by extraction of the seeds of common foxglove (*Digitalis purpurea*). Its genin part is digitogenin (a spirostan steroid) containing six aliphatic rings. The backbone of the carbohydrate part consists of D-galactose, D-glucose, and D-xylose, to whose D-glucose unit is linked a disaccharide side chain that contains D-galactose and D-glucose moieties. It forms with water a saponifiable suspension and it dissolves slightly in ethanol (1 g/220 mL). Digitonin has several potential membrane-related applications in biochemistry and is used for determination of cholesterol in the plasma and serum, as well as in the quantitative precipitation of steroid alcohols.

Examples of other cardiac glycosides are **diginatin** (molar mass 796.95; $C_{41}H_{64}O_{15}$), **diginin** (molar mass 488.62; $C_{28}H_{40}O_{7}$), and **digitalin** (molar mass 712.83; $C_{36}H_{56}O_{14}$). In addition, **proscillarin** can be found in sea onion (*Scilla maritima*). It is readily hydrolyzed by gastric acid and the products are only partly digested in the stomach and intestines, thus quickly excreted almost entirely.

10.1.2. *Cyanogenic glycosides*

The distribution of *cyanogenic glycosides* in nature is relatively wide; these phytotoxins occur (stored in inactive forms in the vacuole) in at least about 2500 plant species that primarily belong to the families *Fabaceae*, *Rosaceae*, *Linaceae*, and *Compositae*. There are about 25 known cyanogenic glycosides. They are present particularly in the seeds of apples, apricots, cherries, raspberries, peaches, plums, crabapples, and quinces, but are also found in bitter almonds, sorghum, lima beans, cassava, corn, yams, chickpeas, cashews, and kirsch. These compounds are the products of secondary metabolism, which are classified as phytoanticipins.

Cyanogenic glycosides are simply *O*-glycosides where the aglycone (α-hydroxynitrile) contains a cyanide group (-C≡N). Thus, poisonous hydrogen cyanide (HCN) can be released from the glycoside if acted upon by an appropriate degrading enzyme (e.g., when chewed or digested). For this reason, cyanogenic glycosides are regarded, on the one hand, as having an important role in plant defense against herbivores, but, on the other hand, as also having additional roles as storage compounds of reduced nitrogen and carbohydrate that can be mobilized when needed in primary metabolism. Plants show variety in the amount of the produced HCN; for example, cassava (dried root cortex) about 2500, cassava (leaves) about 400, sorghum (whole immature plant) about 2500, and lima beans about 3000 mg HCN/kg. The potential toxicity of the edible parts of plants used for human or animal consumption thus depends primarily on the production ability of HCN of these plants and the concentration of HCN that is toxic when humans or animals are exposed to it. In general, HCN is readily absorbed after oral administration and rapidly distributed in the body through the blood. HCN is also readily absorbed after inhalation exposure and through the skin. However, human poisoning by cyanogenic glycosides is rare. It was once thought that these compounds have even anti-cancer properties, but this theory was disproven (see amygdalin, below).

In most cases, the sugar unit is D-glucose and the compounds are termed as "cyanogenic glucosides". In general, there are mainly three types of cyanogenic glycosides: (i) amygdalin-type compounds resulting from hydrolysis of mandelonitrile or its homologues, (ii) linamarin-type compounds resulting from hydrolysis in acetone, and (iii) gynocardin-type compounds resulting from hydrolysis of trioxy ketones.

Amygdalin, [(6-*O*-β-D-glucopyranosyl-β-D-glucopyranosyl)oxy](phenyl)-acetonitrile, mandelonitrile-β-gentiobioside, amygdaloside

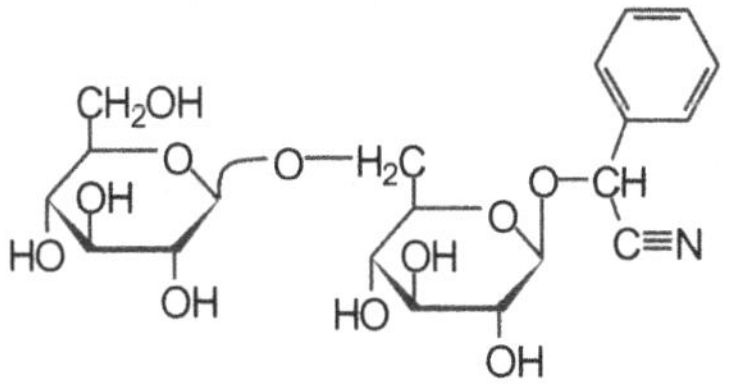

Molar mass 457.44 ($C_{20}H_{27}O_{11}N$), melting point 223–6 (trihydrate) and 214–6 (anhydrous) °C.

Amygdalin was the first glycoside identified in 1830 by Pierre Jean Robiquet (1780–1840) and Antoine-François Boutron-Charlard

(1796–1879). It can be found in many plants, most notably (2–8%) in the kernels of the rose family (*Rosaceae*, particularly in the *Prunus* genus). Prominent examples are bitter almond (*P. amygdalus* or *Amygdalus communis*), as well as many other drupes, such as apricot (*P. armeniaca*), plum (*P. domestica*), peach (*P. persica*), apple (*P. malus* or *Malus domestica*), and wild cherry (*P. avium*). Amygdalin is a colorless and crystalline powder with a bitter taste. It dissolves in water (1 g/12 mL), slightly in ethanol (1 g/900 mL), and is insoluble in diethyl ether. The melting point of the melted and resolidified compound decreases to 125–130°C. The enzyme emulsin (also enzymes β-glucosidase and amygdalase) hydrolyzes in the presence of water the amygdalin in crushed kernels to D-glucose and mandelonitrile, which is then hydrolyzed to benzaldehyde (synthetic oil of bitter almond) and hydrogen cyanide or hydrocyanic acid:

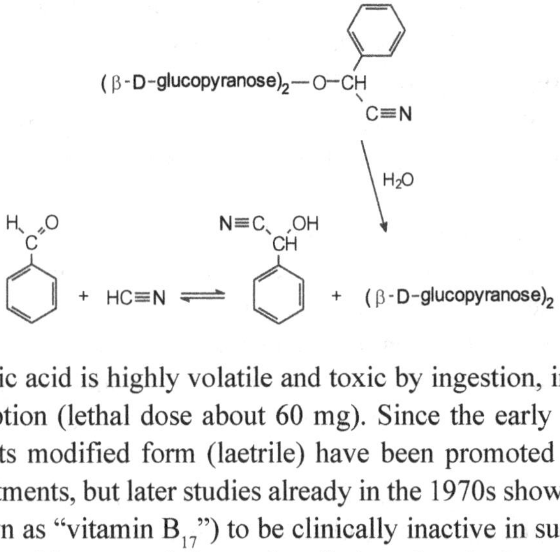

(β-D-glucopyranose)$_2$—O—CH(C≡N)

H_2O ↓

HC(=O) + HC≡N ⇌ N≡C–CH–OH + (β-D-glucopyranose)$_2$

Hydrocyanic acid is highly volatile and toxic by ingestion, inhalation, and skin absorption (lethal dose about 60 mg). Since the early 1950s, amygdalin and its modified form (laetrile) have been promoted as alternative cancer treatments, but later studies already in the 1970s showed them (both often known as "vitamin B_{17}") to be clinically inactive in such treatments. In addition, neither amygdalin nor laetrile is a vitamin. In small quantities, amygdalin exhibits expectorant, sedative, and digestive properties.

Examples of other common cyanogenic glycosides are as follows:

Prunasin, D-mandelonitrile-β-D-glucopyranoside

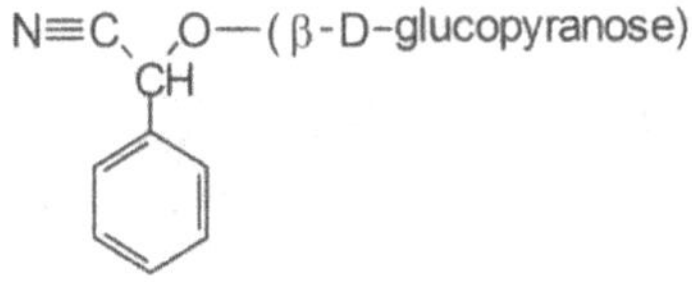

N≡C–CH–O—(β-D-glucopyranose)

Molar mass 295.29 ($C_{14}H_{17}O_6N$).

Dhurrin, *p*-hydroxymandelonitrile-β-D-glucopyranoside

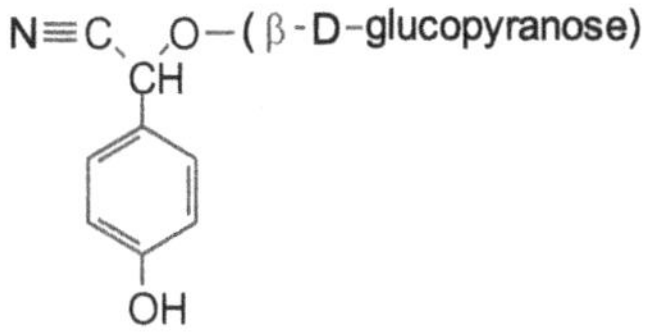

N≡C O–(β-D-glucopyranose)
CH
OH

Molar mass 311.29 ($C_{14}H_{17}O_7N$).

Zierin, *m*-hydroxymandelonitrile-β-D-glucopyranoside

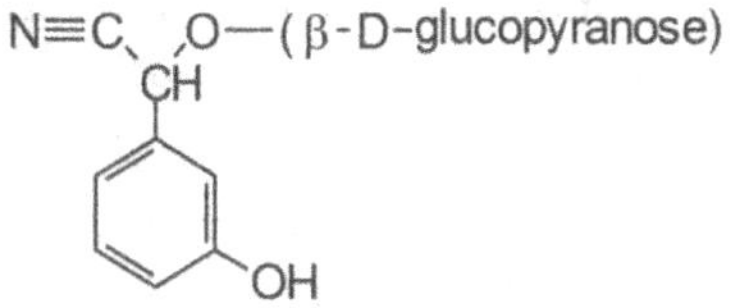

N≡C O–(β-D-glucopyranose)
CH
OH

Molar mass 311.29 ($C_{14}H_{17}O_7N$).

Linamarin, acetonecyanohydrin-β-D-glucopyranoside

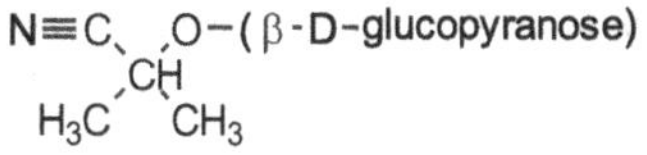

N≡C O–(β-D-glucopyranose)
CH
H_3C CH_3

Molar mass 247.25 ($C_{10}H_{17}O_6N$).

Lotaustralin

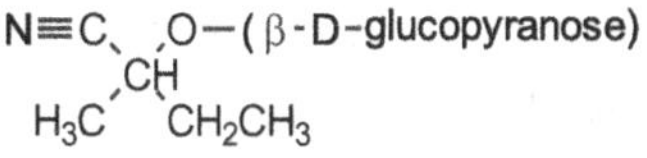

N≡C O–(β-D-glucopyranose)
CH
H_3C CH_2CH_3

Molar mass 261.27 ($C_{11}H_{19}O_6N$).

Acacipetalin

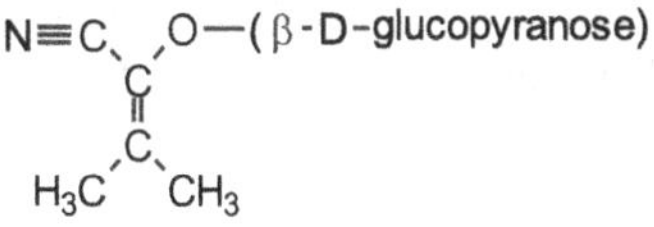

N≡C O–(β-D-glucopyranose)
C
C
H_3C CH_3

Molar mass 259.26 ($C_{11}H_{17}O_6N$).

Sambunigrin, L-mandelonitrile-β-D-glucopyranoside
Prulaurasin, D,L-mandelonitrile-β-D-glucopyranoside
Godiaglycoside, *p*-oxymandelonitrile-β-D-glucopyranoside

10.1.3. *Glucosinolates*

Glucosinolates or mustard oil glucosides are a class of sulfur- and nitrogen-containing compounds that belong to *S*-glucosides, where

D-glucopyranose is β-glycosidically bound via a sulfur atom to the thiocyanate derivative:

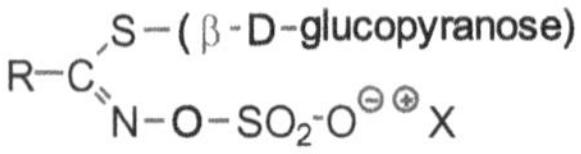

$$R{-}C{\lesssim}^{S-(\beta\text{-D-glucopyranose})}_{N-O-SO_2\text{-}O^{\ominus}\,{}^{\oplus}X}$$

In most cases, X refers to potassium, but it may also be the quaternary ammonium base (see sinalbin). Almost all these glucosides can be found in crucifers or mustard flowers (the family name *Brassicaceae*, formerly *Cruciferae*), but are present in many other plants as well. They are traditionally named by using the prefix gluco- in the Latin-derived general name originating from the family of flowering plants or cruciferous vegetables, thus also indicating their origin. Thus, only a few glucosinolates are known by their "normal trivial names" (see sinigrin, sinalbin, and progoitrin). In general, glucosinolates are secondary metabolites of plants that use these compounds as natural pesticides and as defense against herbivores. They are also responsible for the bitter or sharp taste of several common foods, such as mustard, cabbage, horseradish, and Brussels sprout; their pungency is due to mustard oil produced from glucosinolates when the plant material is chewed, cut, or otherwise damaged. In spite of this, they are enjoyed in small amounts by humans and are believed to partly contribute to the health-promoting properties of cruciferous vegetables.

Sinigrin, sinigroside, potassium myronate

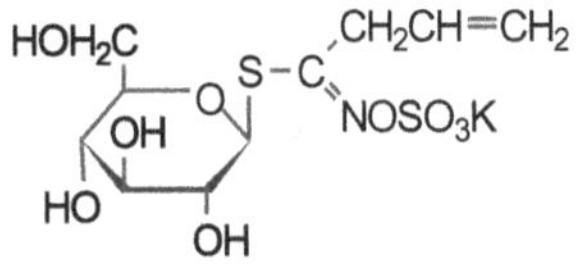

HOH$_2$C, OH, HO, OH, S–C(CH$_2$CH=CH$_2$)(NOSO$_3$K)

Molar mass 397.46 ($C_{10}H_{16}O_9NKS_2$), melting point 127–9 (monohydrate) and 179 (anhydrous) °C.

After the isolation of sinigrin in 1840 by Antoine-Alexandre Brutus Bussy (1794–1882), it was described in the literature in 1863 by William Körner (1839–1925). This glucosinolate can be extracted from the ripe and dried seeds of black mustard (*Brassica nigra*), but it can also be obtained, for example, from the fresh root of horseradish (*Armoracia rusticana*). Sinigrin is a crystalline solid that readily dissolves in water and hot ethanol. It is hydrolyzed by water or the specific enzyme myrosinase to D-glucose, allyl isothiocyanate (mustard oil), and potassium hydrogen sulfate. Sinigrin is used in pharmaceutical preparations for promoting the action of the stomach and intestines in the same way as the natural horseradish.

Examples of other glucosinolates are as follows:

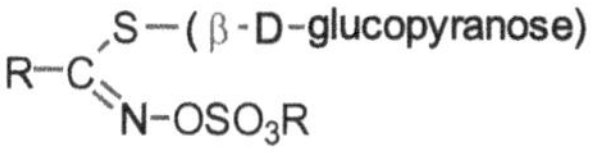

$$R-C(-S-(\beta\text{-D-glucopyranose}))=N-OSO_3R$$

Glucocapparin, methyl glucosinolate

$R = CH_3$ Molar mass 332.32 ($C_8H_{14}O_9NS_2$).

Glucoputranjivin

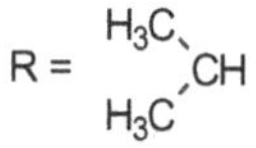

$R = (H_3C)_2CH$ Molar mass 360.37 ($C_{10}H_{18}O_9NS_2$).

Gluconapin

Molar mass 372.38 ($C_{11}H_{18}O_9NS_2$).

$R = H_2C{=}CHCH_2CH_2$

Glucocochlearin

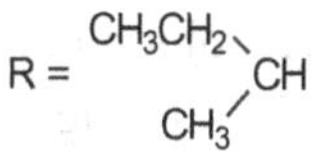

$R = (CH_3CH_2)(CH_3)CH$ Molar mass 374.40 ($C_{11}H_{20}O_9NS_2$).

Glucorapiferin, progoitrin

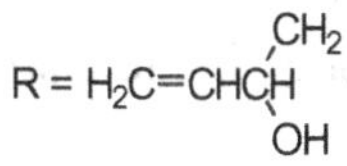

$R = H_2C{=}CHCH(OH)CH_2$ Molar mass 388.38 ($C_{11}H_{18}O_{10}NS_2$).

Glucoberteroin

$R = CH_3SCH_2(CH_2)_3CH_2$ Molar mass 434.51 ($C_{13}H_{24}O_9NS_3$).

Glucoalyssin

$R = CH_3SOCH_2(CH_2)_3CH_2$ Molar mass 450.51 ($C_{13}H_{24}O_{10}NS_3$).

Glucotropaeolin

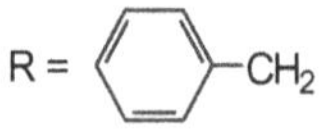

Molar mass 408.42 ($C_{14}H_{18}O_9NS_2$).

Sinalbin, glucosinalbate

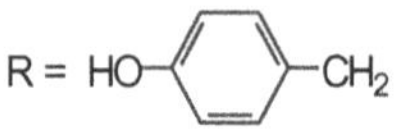

Molar mass 424.42 ($C_{14}H_{18}O_{10}NS_2$).

Gluconasturtin

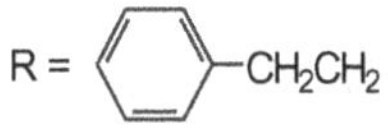

Molar mass 422.44 ($C_{15}H_{20}O_9NS_2$).

10.1.4. *Other glycosides*

Besides the glycosides described above, in nature there are many other glycosides as well, without any significant chemical utilization. They also have a wide distribution within different plant species. In some cases, their straightforward chemical classification is difficult.

Saponins are a class of compounds widely distributed in nature (especially in licorice). These compounds characteristically produce a permanent froth (soap-like foaming) when shaken in aqueous solutions. They are powerful emulsifiers and can cause hemolysis of red blood cells (and therefore, are toxic), which may be related to their binding to cholesterol in cell membranes. In saponins, the carbohydrate component is bound to the aglycone portion termed "sapogenin", which usually is either a C_{30} triterpene or a C_{27} steroid. In the latter case (i.e., a steroidal saponin, saraponin), the structure contains a typical spiroketal side chain also found, for example, in sarsasapogenin:

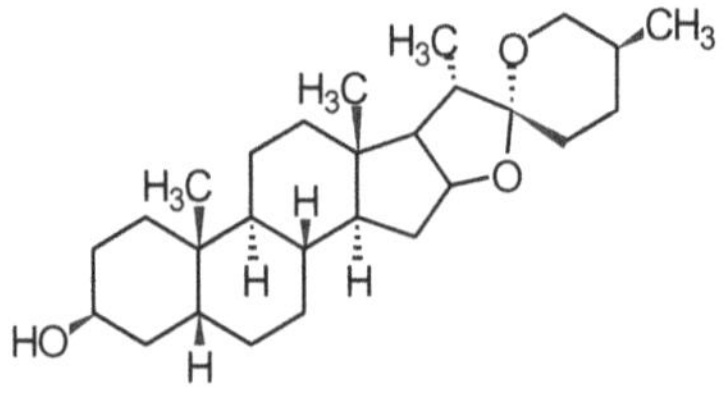

According to its pharmacological properties, digitonin can be classified among cardiac glycosides. On the other hand, based on its chemical structure and physical properties, it belongs to steroidal saponins. However, other saponins described in this section contain the triterpene aglycone portion.

***Quillaja* saponins** can be obtained from the dried and light inner bark of the soap bark tree or soapbark (*Quillaja saponaria*), where their content is about 10%. A heterogeneous mixture of these saponins can be extracted from this bark material by hot water and the extract contains, besides these saponins, tannin, quillaic acid, and carbohydrates. The isolated product is a light yellow, hygroscopic powder with a bitter taste that strongly irritates mucous membranes of respiratory organs causing sneezing. Like soap, it dissolves into water and when shaken it readily forms a long-lasting foam. In *Quillaja* saponins, different aldohexoses (including glucose) are bound via a glycosidic bond to carbon atom C_3 of quillaic acid (R is the carbohydrate portion):

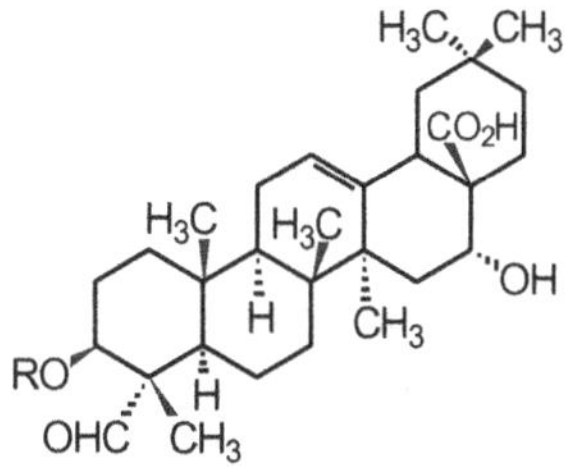

H_3C CH_3 CO_2H H_3C H_3C OH H CH_3 RO H OHC CH_3

Quillaja saponins (sometimes distinguished simply as *Quillaja* saponin) are comercially the important saponin products normally handled as aqueous extracts. They have a wide range of uses (within the recommended dosages), including the use as foaming agents and emulsifiers in the food and beverages industry, foaming agents in fire-fighting foams, and ingredients in pharmaceuticals and personal care products.

Glycyrrhizin

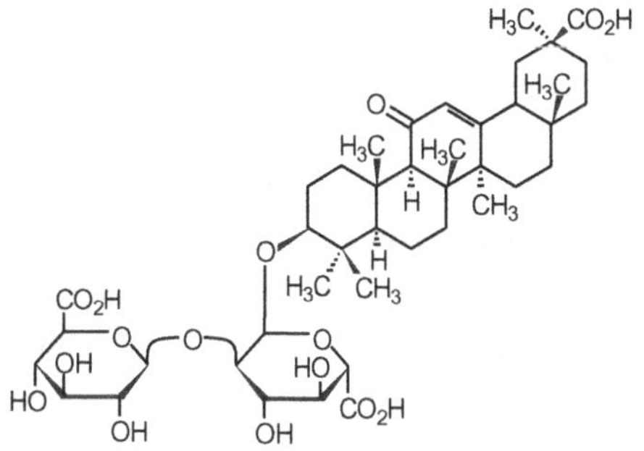

H_3C CO_2H H_3C O H_3C H_3C H CH_3 O H H_3C CH_3 CO_2H O O O OH HO HO CO_2H OH OH

Molar mass 822.94 ($C_{42}H_{62}O_{16}$).

Glycyrrhizin can be obtained from the root of the genus *Glycyrrhiza*, in particular from liquorice or licorice (*G. glabra*), Russian liquorice (*G. echinata*), and Chinese liquorice (*G. uralensis*). It forms colorless needles with a strong sweet taste that dissolve into hot water and ethanol. Its aglycone portion is glycyrrhizic acid or glycyrrhizinic acid and the carbohydrate component is uronic acid. It is used as an emulsifier and gelforming agent in foodstuff and cosmetics as well as a medicine; it is given intravenously for the treatment of chronic viral hepatitis and cirrhosis.

Glycyrrhizin forms salts or glycyrrhizinates. One example is **ammonium glycyrrhizinnate** that crystallizes as a pentahydrate (molar mass 930.05 ($C_{42}H_{75}O_{21}N$), melting point 212–7°C (decomposes)). The commercial product is a white to yellowish crystalline powder used in a wide range of medical and cosmetic products as well as in beverages.

α-Hederin, helixin

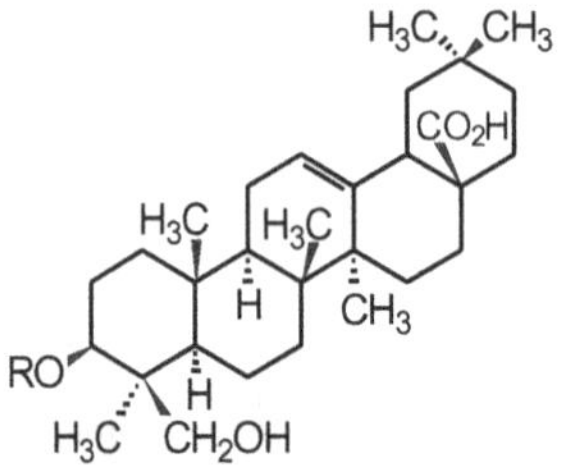

Molar mass 734.97 ($C_{41}H_{66}O_{11}$), melting point 256–9°C, density 1.32 kg/dm^3.

α-Hederin can be found in common ivy or English ivy (*Hedera helix*) that, besides being a decorative plant, is also historically well known for being a medicinal plant in Ancient Greece. Its aglycone portion is a compound chemically similar to that in *Quillaja* saponins and the carbohydrate component (R in the formula) is a disaccharide that contains L-rhamnose and L-arabinose moieties. α-Hederin is used, for example, in the treatment of catarrhs of the respiratory track (as an expectorant) and the symptoms of chronic bronchitis, as well as for reducing cough frequency (see wheeping-cough).

The root of common soapwort (*Saponaria officinalis*) contains significant amounts of saponins, and **saponarin** (molar mass 594.53; $C_{27}H_{30}O_{15}$) can be found in its leaves. This *flavonoid glycoside* crystallizes as a monohydrate. It is a light-yellow, crystalline solid (melting point

228°C) that slightly dissolves into hot water and ethanol. Saponarin produces a strong yellow coloring under slightly alkaline conditions and causes a blue fluorescence in strong sulfuric acid. Its chemical structure is a trihydroxyflavone diglucoside (both *O*- and *C*-glucoside) and its aglycone portion is similar to that found in rutin.

Rutin, rutoside, 3-[6-*O*-(6-deoxy-α-L-mannopyranosyl)-β-D-glucopyranosyloxy]-2-(3,4-dihydroxyphenyl)-5,7-dihydroxy-4*H*-1-benzopyran-4-one, 2-(3,4-dihydroxyphenyl)-5,7-dihydroxy-3-[α-L-rhamnopyranosyl-(1→6)-β-D-glucopyranosyloxy]-4*H*-chromen-4-one, 3′,4′,5,7-tetrahydroxy-3-[α-L-rhamnopyranosyl-(1→6)-β-D-gluco-pyranosyloxy]-flavone, 3,3′,4′,5,7-pentahydroxyflavone-3-rutinoside, quercetin-3-*O*-rutinoside, sophorin

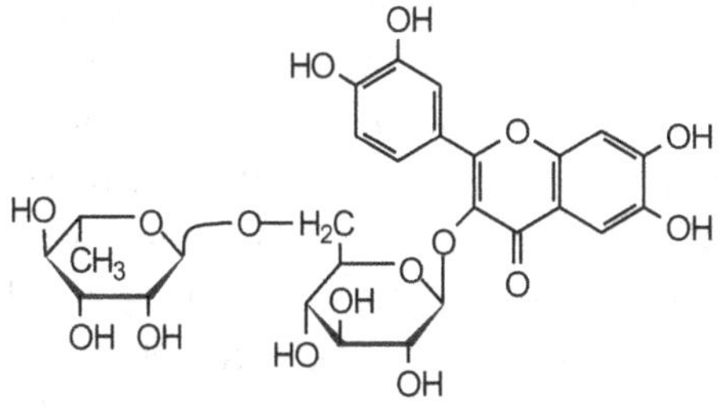

Molar mass 610.52 ($C_{27}H_{30}O_{16}$).

Rutin can be extracted from many plant sources including, for example, buckwheat (*Fagopyrum esculentum*), cultivated tobacco (*Nicotiana tabacum*), weeping forsythia (*Forsythia suspensa*), hortensia or hydrangea (*Hydrangea hortensia*), and the leaves of eucalypt (the family *Myrtaceae*). It crystallizes from water (as a trihydrate) as yellow needles that gradually turn brown on standing in light. When heated, rutin loses water of crystallization at 110°C for 12 h under vacuum (10 mmHg). The anhydrous product is a hygroscopic powder that darkens at 125°C, is plastic at 195–197°C, and decomposes with foaming at 214–215°C. It dissolves in pyridine and slightly in ethanol, acetone, ethyl acetate, and hot water (1 g/200 mL). Rutin inhibits platelet aggregation and decreases capillary permeability (making the blood thinner and improving circulation). It is used in vitamin products.

Phenolic glycosides are common glycosides in nature; their occurrence varies depending on the plant part. In addition, many glycosides containing hydroxyl groups can also contain carbonyl groups or quinoid structures (especially, an anthraquinone structure in *anthraquinone glycosides*). For this reason, the mutual chemical classification of phenolic glycosides and quinone glycosides (see also flavonoid glycosides) is often difficult.

Arbutin, 4-hydroxyphenyl-β-D-glucopyranoside, hydroquinone β-D-glucopyranoside

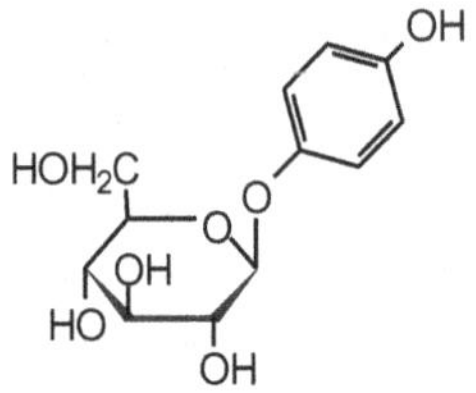

Molar mass 272.25 ($C_{12}H_{16}O_7$), melting point 199.5°C.

Arbutin occurs (often together with methylarbutin or 4-methoxyphenyl β-D-glucopyranoside) in the leaves of common bearberry (*Arctostaphylos uvaursi*) and in mountain cranberry, blueberry, and cranberry, all belonging to the same family of flowering plants (the heath family, *Ericaceae*). However, this glucoside is widely distributed in nature; significant amounts of it also occur in the leaves of bergenia or elephant's ears in the family *Saxifragaceae* (in the genus *Bergenia*) and, for example, in Siberian tea or winter-blooming bergenia (*Saxifraga crassifolia* or *Bergenia crassifolia*). Arbutin can be extracted (its solubility in water is 5 g/100 mL) from the dried leaves of these plants, but its synthetic preparation from acetobromoglucose and hydroquinone is also possible. The product crystallizes from ethyl acetate as hygroscopic needles. Arbutin has traditionally been used in the treatment of urinary tract infections (it also has a diuretic effect) and as a skin-lightening agent. In the human body, it decomposes hydrolytically, resulting in the formation of antiseptic hydroquinone. In addition to the pure product, various extracts, especially from common bearberry, are normally utilized.

Aloin

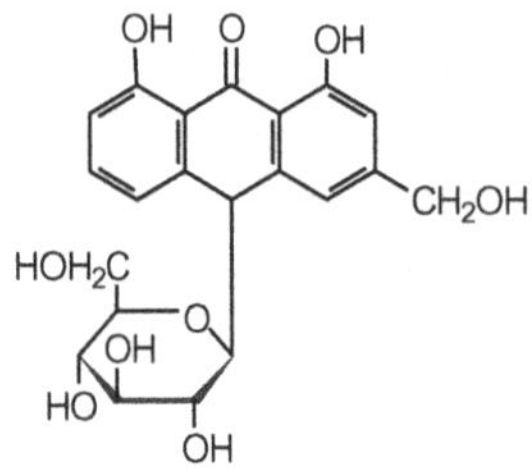

Molar mass 418.40 ($C_{21}H_{22}O_9$), melting point 70–80 (mono-hydrate) or 148–9 (anhydrous) °C.

Aloin (*C*-glucoside and anthraquinone glucoside) is usually separated by extraction from the bitter yellow exudate (commonly referred to as the aloe latex) that seeps out from just underneath the skin of many aloe leaves of plants in the lily family (*Liliaceae*). The natural product is a mixture of two diastereomers; **aloin A** or barbaloin and **aloin B** or isobarbaloin. The purified aloin forms lemon yellow, bitter needles with a characteristic smell that dissolve in pyridine, slightly in methanol, acetone, and acetic acid, and sparingly in diethyl ether, chloroform, and carbon disulfide. This compound and its dried extracts are used as laxatives for treating constipation by inducing bowel movements. In addition, they have been used in quite small quantities as natural bittering agents in alcoholic beverages. Similar pharmaceutical glycoside-containing drugs are **senna** (the effective compounds are **sennoside A** and **sennoside B**) from the leaves of Alexandrian senna (*Senna alexandrina*, *Cassia acutifolia*, *C. alexandrina*, or *C. acutifolia*) and the extracts from the dried bark of alder buckthorn (*Rhamnus frangula*, where the effective compounds are **glucofrangulin A** and **glucofrangulin B**) and from cascara or cascara buckthorn (*Cascara sagra* or *Rhamnus purshiana*). These drugs represent traditional herbal medication to treat constipation and empty the large intestine before surgery.

Other glycosides are, for example, *lactone glycosides* that contain a structural unit of 2-pyrone, coumarin, or isocoumarin. In spite of this, with respect to physiological effects and structural similarities, these compounds should be distinguished from the cardiac glycosides, whose steroid genin portion also contains a five- or six-ring lactone. Lactone glycosides are widely distributed in nature; these compounds, often with a pleasant smell, occur in different parts of plants. In a wide variety of plants there are also a wide range of other glycosides that contain uniform structural units (e.g., iridoids, stilbenes, lignans, and xantons). So far, these compounds are not of significant commercial importance.

10.2. Nucleic Acids

Nucleic acids or *polynucleotides* are among the most important biological macromolecules and they are present in abundance in all living things. These biopolymers function in encoding, transmitting, and expressing genetic information for the biosynthesis of proteins. Johan Friedrich Miescher (1844–95) discovered them over 150 years ago in 1868, and they form a foundation for genome and forensic science as well as for modern biological and medical research. Nucleic acids consist of *nucleotides*, each of them having three chemically bound components: a nitrogenous base, an aldopentose, and a phosphate group (related to phosphoric acid). In nucleotides, the base components are derivatives of either pyrimidine or purine, and the carbohydrate components are either β-D-ribose (I, in ribonucleic acid, RNA) or 2-deoxy-β-D-ribose (II, in deoxyribonucleic acid, DNA):

I II

The base and carbohydrate components are chemically bound to form a *nucleoside*:

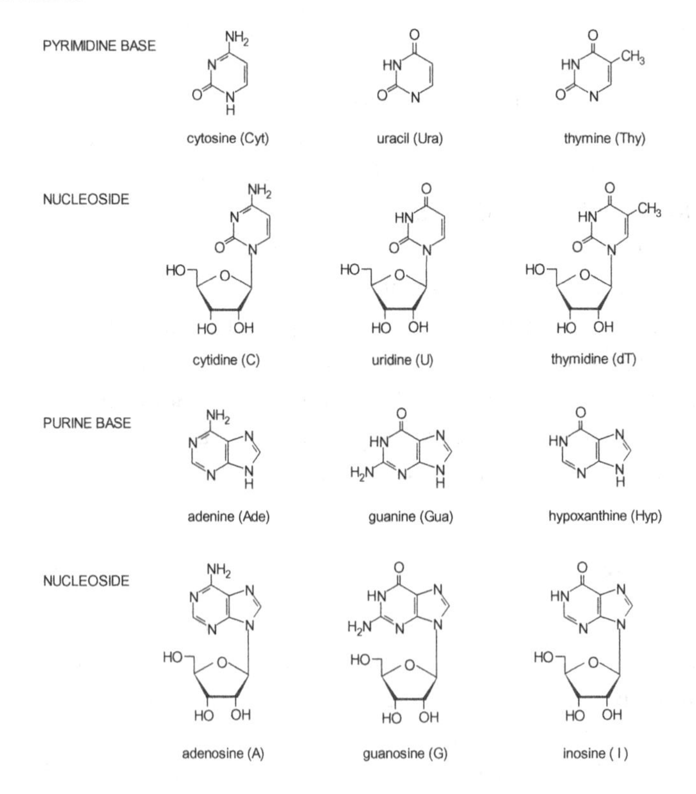

Nucleosides are glycosylamines that can be considered as nucleotides without a phosphate group. Thus, a nucleotide is a phosphorous acid ester of a nucleoside; carbon atom C_5 in the carbohydrate part (corresponding to carbon atom 5′ in the nucleotide) is involved in the esterification. The nucleotides are linked together so that each phosphorous acid component also esterifies with a hydroxyl group at carbon atom C_3 (corresponding to carbon atom 3′ in the nucleotide) of another carbohydrate part.

Most DNA of each living organism is found in chromosomes of the nucleus that contain the genetic information or the genes. This information appears to depend on the arrangement of the bases (thymine, adenine, cytosine, and guanine) along the phosphate-carbohydrate backbone; thus, the arrangement of these four nucleotides in different sequences specifies DNA. It should be noted that DNA contains thousands of nucleotide units existing in the structure formed by double-stranded molecules of nucleic acids (known as a "double helix") held together by hydrogen bonds between the base units. However, in the DNA double helix only certain bases can form base pairs (e.g., thymine ↔ adenine and cytosine ↔ guanine). For example, the sequence -G-T-C-A-T- is as follows:

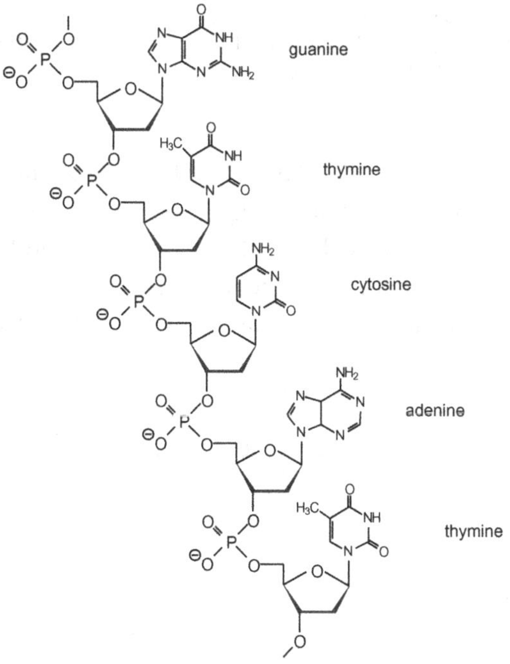

The double-helix model of DNA structure (the previous model suggested was a triple-stranded DNA) was first reported in 1953 by James Dewey Watson (1928–) and Francis Crick (1916–2004); it was based on the crucial X-ray diffraction image of DNA obtained in 1952 by Rosalind Franklin (1920–58). Maurice Wilkins (1916–2004), among others, further clarified this image. Crick, Wilkins, and Watson received the 1962 Nobel Prize in Physiology or Medicine for their contributions to the original discovery. Franklin died in 1958, making him ineligible for a Nobel Prize.

In RNA, the nucleotides are chemically bound in a way similar to DNA. RNA functions in converting genetic information from genes into amino acid sequences of proteins. There are three universal types of RNA: transfer RNA (tRNA), messenger RNA (mRNA), and ribosomal RNA (rRNA), all having their characteristic functions. Nowadays, many other classes of RNAs are known.

Cytosine, 4-amino-1*H*-pyrimidin-2-one, 2-oxo-4-aminopyrimidine, 4-amino-2-hydroxy-pyrimidine (enol form)

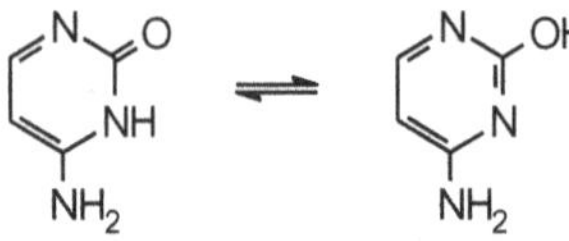

Molar mass 111.40 ($C_4H_5ON_3$), melting point 320–5°C (decomposes).

Albrecht Kossel (1853–1927, the 1910 Nobel prize in Physiology or Medicine) first prepared cytosine by hydrolysis of calf thymus tissues in 1894. It occurs as a part of DNA, RNA, and nucleotides. Besides by hydrolysis of nucleic acids, it can be synthesized by treating pyrimidine with phosphorus pentachloride followed by a treatment of the formed 2,4-dichloropyrimidine by ammonia:

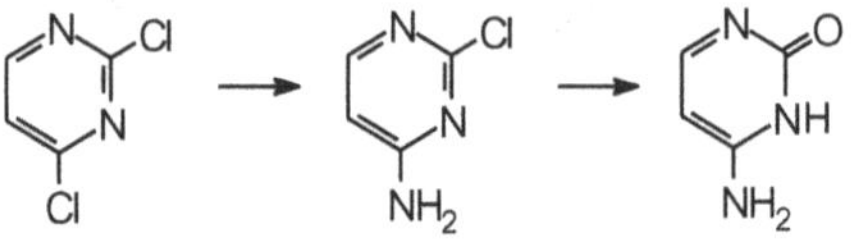

Cytosine forms lustrous platelets that dissolve in water and ethanol, but not in diethyl ether. It is used for biochemical research.

5-Methylcytosine, 4-amino-5-methyl-3*H*-pyrimidin-2-one, 4-amino-2-hydroxy-5-methylpyrimidine (enol form)

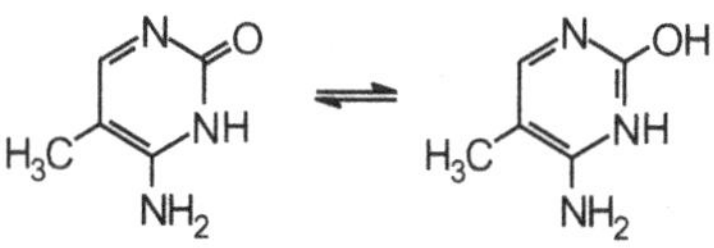

Molar mass 125.13 ($C_5H_7ON_3$), melting point 270°C (decomposes).

5-Methylcytosine is a methylated form of cytosine. The methylation reaction is caused by an enzyme called "DNA methyltransferase". This compound can be prepared by hydrolysis of the methylated DNA.

Uracil, pyrimidine-2,4(1*H*,3*H*)-dione, 2,4-dioxypyrimidine

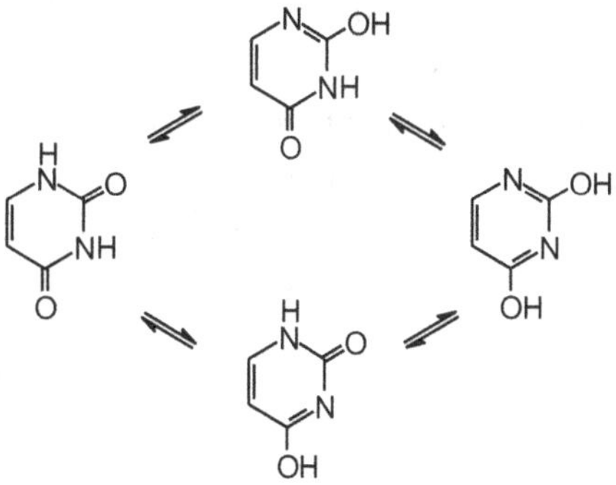

Molar mass 112.10 ($C_4H_4O_2N_2$), melting point 338°C (decomposes).

Abramo Alberto Ascoli (1877–1957) first prepared uracil by hydrolysis of yeast nuclein in 1900. It was also found in bovine thymus and spleen, herring sperm, and wheat germ. Besides hydrolysis of RNA, it can be synthesized by a condensation reaction of maleic acid with urea in fuming sulfuric acid:

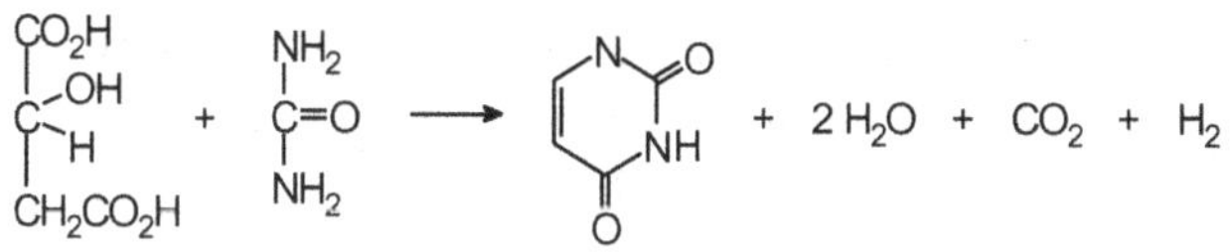

$$+ \ 2\,H_2O + CO_2 + H_2$$

It can also be prepared from urea and ethyl formylacetate. Uracil undergoes amide-imide tautomeric shifts (i.e., amide or lactam structure; 2,4-dioxypyrimidine vs. imide or lactim structure; see 2,4-dihydroxypyrimidine, above). It forms crystalline needles that dissolve in hot water and alkalis, but not in ethanol and diethyl ether. Uracil is used in biochemical research.

Thymine, 5-methyluracil, 5-methylpyrimidine-2,4(1*H*,3*H*)-dione, 2,4-dioxo-5-methyl-pyrimidine

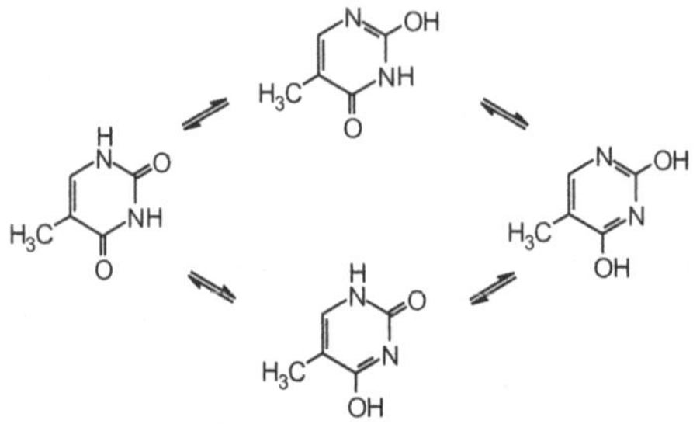

Molar mass 126.12 ($C_5H_6O_2N_2$), melting point 326°C (decomposes), boiling point (sublimes).

Thymine can be obtained by hydrolysis of DNA, but it can also be synthesized from *N*-methyl cyanoacetyl urea by catalytic reduction. Like uracil, it undergoes amide-imide tautomeric shifts (i.e., amide or lactam structure; 2,4-dioxo-5-methylpyrimidine vs. imide or lactim structure; see 2,4-dihydroxy-5-methylpyrimidine, above). It forms crystalline needles that dissolve readily in alkalis, slightly in hot water and diethyl ether, but not in cold water or ethanol. Thymine is used in biochemical research.

10.3. Vitamins

Vitamins are biologically active compounds that are in limited, small amounts necessary in human metabolism or generally, for the normal functions of an organism. The human body stores most vitamins for varying times. In some cases, it takes a considerable amount of time before a deficiency condition is developed. A vitamin deficiency leads either to a disease or to a chronic or long-term condition (avitaminosis) caused by a temporarily increased need or an absorption disorder. Vitamins are among those nutrients that the organism cannot synthesize either at all or in sufficient quantities, and they or their precursors, *provitamins* (e.g., carotene, a provitamin of vitamin A), must be obtained through the diet. Thus, provitamins are substances with little or no vitamin activity that can be converted into active vitamin forms by normal metabolic processes. The most important difference between vitamins and hormones (cf., Chapt. "10.5.1. Hormones") is that the latter ones are produced by glands in multicellular organisms, and they mainly regulate physiological and behavioral activities (e.g., metabolism, tissue function, stress, growth, and mood). However, recent investigations have indicated that certain vitamins and hormones cannot be easily distinguished from each other merely based on this general definition.

The term "vitamin" (in Latin, "vita" means "life" — thus, the term actually means "amine of life" or practically "vital amines") was based on the theory described in 1911 by Casimir Funk (also originally known as Kazimierz Funk 1884–1967) that the organic micronutrient food factors that prevent beriberi in humans and polyneuritis in birds and perhaps other similar dietary-deficiency diseases might be amines. At that time, he thought that the key compound would be thiamine (vitamin B_1, "anti-beriberi-factor"), which chemically is an amine. However, most vitamins are not amines but a rather heterogeneous class of compounds consisting of *fat-* or *lipid-soluble vitamins* and *water-soluble vitamins*. The trivial naming of vitamins was established so the first "vitamin bioactive" ever isolated was initially called "vitamin A", followed by "vitamin B", etc.

The fat-soluble vitamins belong either to partly cyclized isoprenoid (vitamins A, E, and K) or sterol derivatives (vitamins D). These vitamins are usually associated with the dietary lipids of food and are absorbed from the intestine with them. They are stored in moderate amounts in the tissues of the body and are essential to maintaining normal metabolism and biochemical functions; depending on their chemical structure, they can act as oxidation-reduction agents (vitamins A, E, and K), coenzymes or enzyme activators (vitamins A, D, and K), or an enzyme inhibitor (vitamin E). The deficiency of fat-soluble vitamins may primarily result in malabsorption; this phenomenon refers to a number of disorders in which the intestine cannot adequately absorb certain nutrients into the bloodstream. Water-soluble vitamins (vitamins B_1, B_2, B_3, B_5, B_6, B_9, B_{12}, C, and H) are structurally diverse, often consisting of nitrogen-containing heterocyclic moieties. The storage of these vitamins in the tissues of the body is usually not significant. They act as coenzymes or enzyme activators (vitamins B_1, B_3, B_5, B_6, B_9, B_{12}, and H) and oxidation-reduction agents in enzyme reactions (vitamins B_2, B_3, B_9, B_{12}, and C). In addition, water-soluble vitamins can participate in the synthesis of nucleic acids (vitamins B_9, B_{12}, C, and H) and they probably have an influence on mitochondria (vitamins B_2, B_3, and C).

The organisms are able to synthesize certain compounds that act like vitamins at least in some quantities. However, the vitamin character of these vitamin-like substances is not fully clarified in all cases. In addition, there are specific compounds that are generally called "vitamins" but without vitamin-like effects, or their vitamin activities are of minor importance compared to their main functions. Examples of these compounds are vitamin F ("essential fatty acids", such as α-linolenic, linoleic, and arachidic acids), *meso*-inositol (*myo*-inositol), vitamin H_1 or B_{10} (*p*-aminobenzoic acid), vitamin L (*o*-aminobenzoic acid), choline or bilineurine, vitamin B_4 (adenine or 6-aminopurine), vitamin C_2 (dehydroascorbic acid), vitamin K_3 (menadione or 2-methyl-1,4-naphthoquinone), vitamin K_6 (1,4-diamino-2-methylnaphthalene), vitamin K_5 (4-amino-2-methyl-1-naphthol hydrochloride), cysteine (a substitute for vitamin B), phtiocol (2-hydroxy-3-methyl-1,4-naphthoquinone or a substitute for vitamin K), vitamin U (*S*-methyl-L-methionine), carnitine (vitamin B_T or vitamin B_{11}), orotic acid (vitamin B_{13}), xanthopterin (vitamin B_{14}), pangamic acid (vitamin B_{15}), laetrile (vitamin B_{17}), ubiquinone (coenzyme Q_{10}), and lecithin.

The production of vitamins used as food or feed additives, medical or therapeutic agents, or health and cosmetic aids is generally based on synthetic pathways or biotechnological processes (i.e., fermentation and microbial/enzymatic transformation). In some cases, vitamins can be separated from plant or animal sources. Vitamins are administered orally in tablets (or capsules) or by injection into muscle as an aqueous solution. Commercial vitamin tablets may contain many additives; a filler (e.g., microcrystalline cellulose, lactose, or maltodextrin), a lubricant (e.g., magnesium stearate), a flow agent (e.g., silicon dioxide), and a disintegration agent (e.g., carboxymethylcellulose). In addition, vitamin tablets are usually coated, for example, with materials made from a cellulose base. Recommended daily doses of vitamins depend on many factors, and the values normally given are only for guidance. These recommended amounts are generally given as milligrams (mg) or micrograms (μg). For fat-soluble vitamins, the international units (IU) are also used; they facilitate a reliable comparison of the actual biological effect of different forms and preparations.

In this section, only vitamins that are either derivatives of carbohydrates (vitamin C) or have carbohydrate-containing structural units (vitamins B_2, B_5, and B_{12}) are described. They all belong to water-soluble vitamins.

Riboflavin, vitamin B_2, lactoflavin, 7,8-dimethyl-10-(1-D-ribityl)isoalloxazine, 7,8-dimethyl-10-(D-*ribo*-2,3,4,5-tetrahydroxypentyl)isoalloxazine, 7,8-dimethyl-10-(D-*ribo*-2,3,4,5-tetrahydroxypentyl)-3*H*,10*H*-benzo[*g*]-pteridine-2,4-dione, ovoflavin, hepatoflavin, verdoflavin, uroflavin, lyochrome, vitamin G

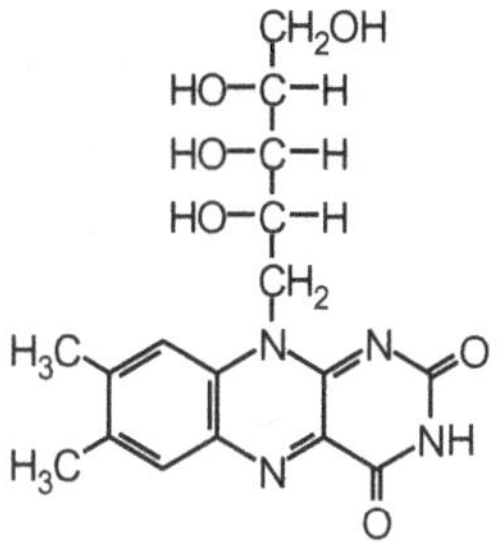

Molar mass 376.37 ($C_{17}H_{20}O_6N_4$), melting point 280°C (decomposes, darkens at about 240°C).

Riboflavin (in Latin, “flavus” means “yellow”) or vitamin B_2 occurs in the free form in the retina of the eye, in urine, and in whey, which is the liquid remaining after milk has been curdled and strained. It also exists, often together with thiamine (vitamin B_1), in many natural products and nutritive preparations, such as dairy products (0.1–0.5 mg/100 g), eggs (0.5 mg/100 g), meat (0.1–0.3 mg/100 g), liver (3.0–3.6 mg/100 g), fish (0.1–0.3 mg/100 g), some vegetables (0.1–0.3 mg/100 g), yeast, and mushrooms. Since the beginning of the 20th century it was known that certain foodstuffs contain small amounts of a yellow, organic substance which was named “flavin”. Richard Kuhn (1900–67, the 1938 Nobel Prize in Chemistry) and his co-workers (Paul György 1893–1976 and Theodor Wagner-Jauregg 1903–92) first isolated riboflavin from eggs in 1933 in a pure, crystalline state (it was named “ovoflavin”) having a biological activity characteristic for vitamins. At the same time, an impure crystalline preparation of riboflavin was isolated from whey and named “lyochrome” (and later renamed “lactoflavin”). Soon thereafter, Paul Karrer (1889–1971, the 1937 Nobel Prize in Chemistry) and his co-workers (Kuhn and Friedrich Weygand 1911–69) separated this compound from a wide variety of animal organs and vegetable sources and they named it “hepatoflavin”. The chemical structure of this substance was finally confirmed in 1935 by the synthesis of riboflavin by Karrer, Kuhn, and Weygand. However, in nature this vitamin generally fulfills its metabolic function in a complex form (see below); it is the central component of the cofactors flavin mononucleotide (FMN) and flavin adenine dinucleotide (FAD). These coenzymes take part in many redox reactions and they are chemically bound to specific proteins (apoenzymes) via the cations $Fe^{2\oplus}$, $Cu^{2\oplus}$, and $Mo^{2\oplus}$ to form flavoenzymes. Riboflavin is normally produced for therapeutic use by chemical synthesis. One example is the condensation of 6-D-ribitylamino-3,4-xylidine with alloxane or a reaction of 1-D-ribitylamino-2-*p*-nitrophenylazo-4,5-dimethylbenzene with barbituric acid in acetic acid:

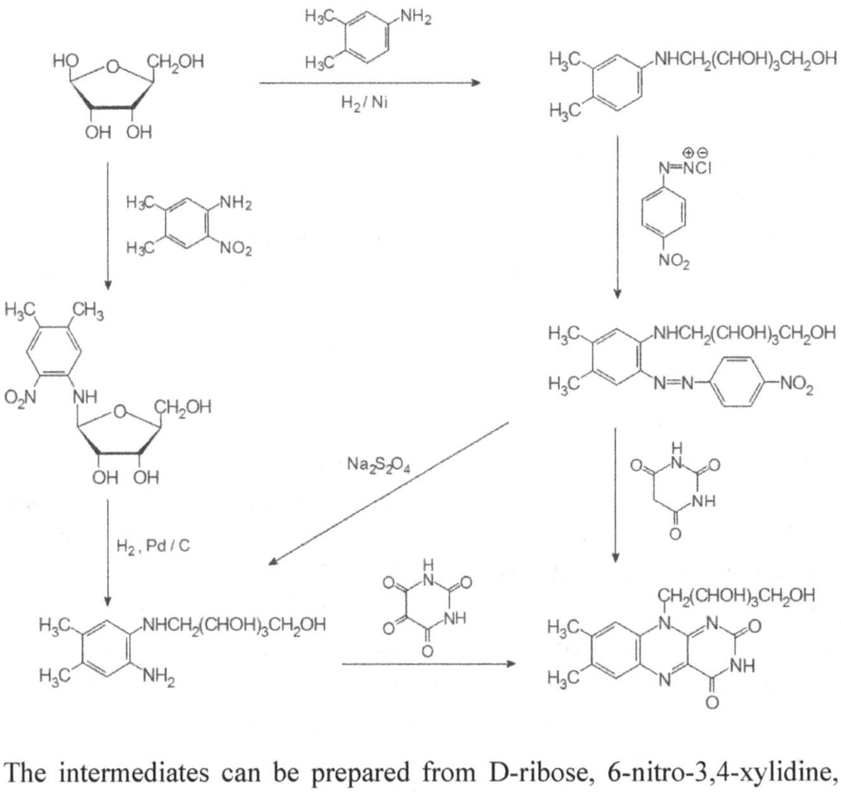

The intermediates can be prepared from D-ribose, 6-nitro-3,4-xylidine, 3,4-xylidine, and *p*-nitrophenyldiazonium salt. For poultry and livestock feeds, various concentrates of riboflavin are typically manufactured from brewer's yeast using microorganisms, such as *Ashbya gossypii* and *Eremothecium ashbyii*, as well as the species *Clostridium* and *Asperigillus*. Riboflavin forms yellow to orange yellow needles with a bitter taste that dissolve slightly in water (10–13 mg/100 mL, 25.0–27.5°C) and ethanol (4.5 mg/100 mL), but not in diethyl ether, acetone, benzene, or chloroform. It is an optically active compound ($[\alpha]_D$ at 20°C is +56.5–59.5, 0.5% soln. in dil. HCl), which decomposes rapidly on standing in light in neutral or acid solutions to 7,8-dimethylalloxazine (lumichrome):

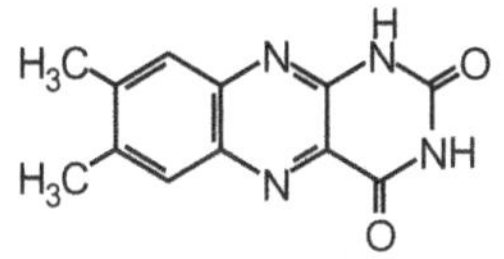

In contrast, in alkaline solutions, the irradiation product is 7,8,10-trimethylisoalloxazine (lumiflavin):

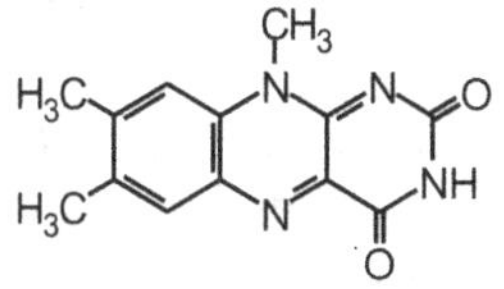

Riboflavin is generally stable against common oxidizing agents, but upon reduction by conventional agents it readily forms 1,5-dihydroriboflavin (molar mass 378.38; $C_{17}H_{22}O_6N_4$). In higher mammals, this vitamin is readily absorbed by the intestines and is distributed to all tissues. It is an important vitamin, especially for the function of ectodermal tissues (the skin and cornea) and myelinated nerve cells. In addition, a sufficient dietary intake is necessary for the development of the fetus. Nowadays, due to the versatile food, the symptoms of riboflavin deficiency (ariboflavinosis), such as stomatitis including painful red tongue (magenta tongue) with sore throat, chapped and fissured lips (cheilosis), and inflammation of the corners of the mouth (angular stomatitis), are relatively rare. In addition, the eyes can become itchy, watery, and sensitive to light. However, the deficiency can be effectively treated during few weeks by a dietary intake of 10 mg/day. Although riboflavin is practically a nontoxic substance for humans, its overdoses (25–200 mg/day) may result in some symptoms, such as itching and effects on the peripheral nerves. The plasma of adult men and women normally contains 2.0–3.5 mg/100 mL of this vitamin. The recommended intake for essentially healthy people depends on age, varying in the range of 0.4–1.8 mg/day (during pregnancy and for the lactating woman 1.8–2.0 mg/day).

Riboflavin-5′-phosphate, vitamin B_2 phosphate, flavin mononucleotide, FMN

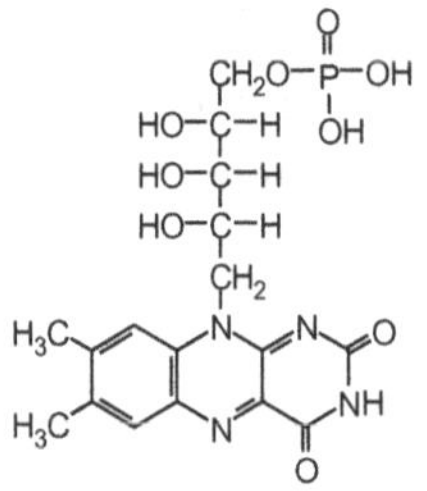

Molar mass 456.35 ($C_{17}H_{21}O_9N_4P$), melting point 195°C.

Riboflavin-5′-phosphate or flavin mononucleotide occurs in nature in flavoenzymes (see riboflavin). This compound is generally classified as a nucleotide, but actually it does not belong to this category. The chemical reason for this is that its carbohydrate component is an alditol (D-ribitol) and not the corresponding aldose (D-ribose). Otto Warburg (1883–1970) and W. Christian first isolated riboflavin-5′-phosphate in 1932 as a crystalline calcium salt from the yellow enzyme in yeast. It can be synthesized by phosphorylation of riboflavin with chlorophosphoric acid, pyrophosphoric acid, or metaphosphoric acid. Riboflavin-5′-phosphate is a yellow, microcrystalline solid and its specific rotation $[\alpha]_D$ at 28°C is +44.5 (2% soln. in conc. HCl). The commercial product is the monosodium salt dihydrate (molar mass 514.36; $C_{17}H_{25}O_{11}N_4NaP$) with a water solubility (3 g/100 mL at 25°C, but tends to gel), more than 200 times that of riboflavin. Its monodiethanolamine salt also crystallizes as a dihydrate (molar mass 598.52; $C_{21}H_{37}O_{13}N_5P$). These products are used in multivitamin mixtures and as food color additives.

Riboflavin-5′-adenosine diphosphate, flavin adenine dinucleotide, FAD

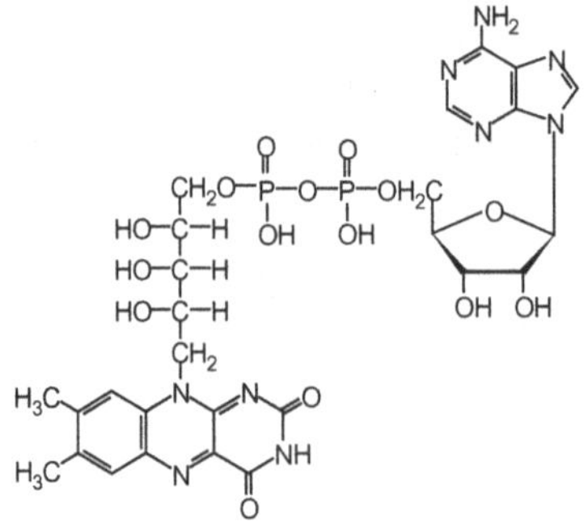

Molar mass 785.56 ($C_{27}H_{33}O_{15}N_9P_2$).

Riboflavin-5′-adenosine diphosphate occurs naturally (see riboflavin). Otto Warburg and W. Christian first isolated it in 1938 from D-amino acid oxidase as its prosthetic group. It can be synthesized from riboflavin-5′-phosphate and adenosine 5′-monophosphate (AMP) with di-*p*-tolylcarbodiimide or trifluoroacetic acid anhydride. This compound can exist in many different redox states (it is converted between these states by accepting or donating electrons): the quinone, semiquinone, and hydroquinone. For this reason, it is a prosthetic group (a redox cofactor) that is involved in many significant reactions in metabolism.

Methylol riboflavin, (hydroxymethyl)riboflavin

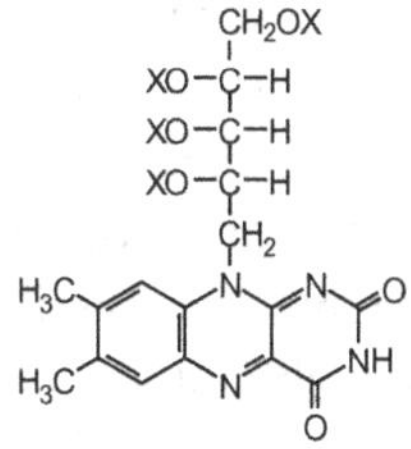

X = H tai CH_2OH

A mixture of methylol riboflavin derivatives is formed by the action of formaldehyde on riboflavin in weakly alkaline solutions and it exhibits the same biological activity as riboflavin. The product is a nearly odorless (it may have a faint, formaldehyde-like smell), orange to yellow, hygroscopic powder that dissolves in water and is nearly insoluble in ethanol, diethyl ether, benzene, and chloroform. As dry powder, it is unstable and on standing it gradually loses its biological activity by liberation of formaldehyde. Methylol riboflavin is used for nutrition as a vitamin source.

Pantothenic acid, vitamin B_5, *N*-(2,4-dihydroxy-3,3-dimethylbutyryl)-β-alanine (optically active D-(+)- and L-(−)-forms and an optically inactive, racemic D,L-form)

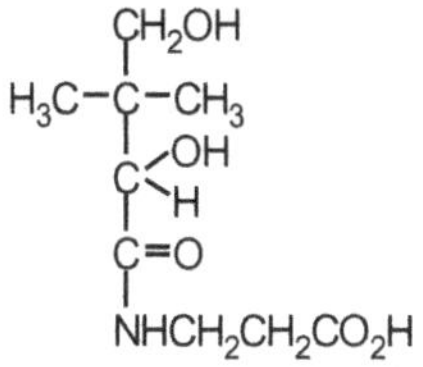

Molar mass 219.24 ($C_9H_{17}O_5N$).

Pantothenic acid (in Greek, "pantothen" means "from everywhere") or vitamin B_5 and its derivatives are readily absorbed and widely distributed in all living cells and tissues, and they are essential nutrients for many animals; their highest concentrations in the human body are found in the liver, kidneys, adrenal glands, and heart. Animals require this vitamin to synthesize coenzyme A (CoA), which is important in the metabolism and the biosynthesis of many significant compounds. Pantothenic acid is also found in nearly every food, in high amounts in meat, milk products, avocados, broccoli, whole-grain cereals, legumes, and eggs, and also in brewer's yeast and sugar molasses. Roger John Williams (1893–1988) discovered it in 1933. The natural pantothenic acid is its D-(+)-form, being the only stereoisomer with a vitamin activity and generally called "vitamin B_5". It can be synthesized from 2,4-dihydroxy-3,3-dimethylbutanoic acid and β-alanine. The product is a yellowish, viscous, and hygroscopic oil that dissolves in water, diethyl ether, ethyl acetate, and glacial acetic acid, but is insoluble in benzene and is unstable against heat. Because of its ubiquitous nature deficiency states are exceptionally rare they can be reversed with pantothenic acid. Symptoms of deficiency are similar to those of other vitamins B typically including fatigue, irritability, gastrointestinal problems, cardiovascular instability, and neurological symptoms (e.g., numbness, paresthesia, and muscle cramps). Since pantothenic acid is nontoxic, it is used, besides the treatment of vitamin deficiency, for other medical purposes. The plasma of adult men and women normally contains 10–40 mg/100 mL of this vitamin. The recommended intake for essentially healthy people varies in the range of 10–15 mg/day.

Sodium pantothenate

$HOCH_2C(CH_3)_2CH(OH)CONHCH_2CH_2CO_2Na$

Molar mass 241.22 ($C_9H_{16}O_5NNa$), melting point 122–4°C.

Sodium pantothenate forms hygroscopic needles that dissolve in water.

Calcium pantothenate

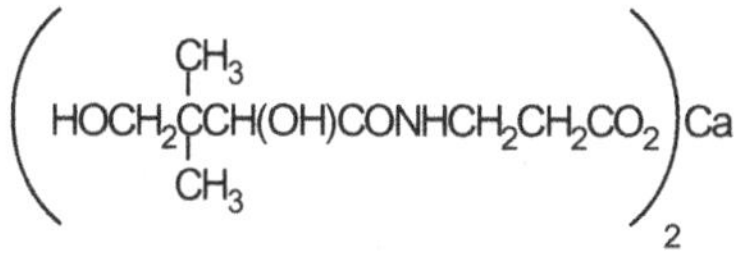

$$\left(\begin{array}{c} CH_3 \\ HOCH_2\overset{|}{\underset{|}{C}}CH(OH)CONHCH_2CH_2CO_2 \\ CH_3 \end{array}\right)_2 Ca$$

Molar mass 476.55 ($C_{18}H_{32}O_{10}NCa$), melting point 170–2°C, boiling point 195–6°C (decomposes).

Because of the instability of the free pantothenic acid, this compound is commercially used as the calcium salt or as its alcohol analog, pantothenol (see below). Calcium pantothenate is a white, odorless, and slightly hygroscopic powder that is stable in air and light as well as upon heating, and it has a sweetish taste. It is soluble in water (its aqueous solutions have a pH about 9) and glycerol, but is insoluble in ethanol, acetone, diethyl ether, chloroform, or benzene. Only its D-(+)-form with a specific rotation $[\alpha]_D$ of +28.2 (5% soln. in H_2O) at 25°C has vitamin activity. This compound is also normally the most common form of vitamin B_5 although the racemic calcium pantothenate provides a more economic source of this vitamin. Calcium pantothenate is used as a supplement in food and animal feeds and as a medicine for treatment of dermatitides.

Pantothenyl alcohol, pantothenol, panthenol, 2,4-dihydroxy-3,3-dimethyl-*N*-(3-hydroxypropyl)butyramide

$$\begin{array}{c} CH_3 \\ HOCH_2\overset{|}{\underset{|}{C}}CH(OH)CONHCH_2CH_2CH_2OH \\ CH_3 \end{array}$$

Molar mass 205.26 ($C_9H_{19}O_4N$), boiling point 118–20°C (0.02 mmHg), density D_{20} 1.2 kg/dm^3 (20°C), refractive index n_D 1.497 (20°C).

Pantothenyl alcohol (provitamin B_5) is quickly oxidized to pantothenic acid in organisms. It is a highly viscous, slightly hygroscopic oil that dissolves in water and ethanol and slightly in diethyl ether and chloroform. Only its D-(+)-form with a specific rotation $[\alpha]_D$ of +28.4 to 30.7 (5% soln. in H_2O) has vitamin activity and this compound is used, like calcium pantothenate, as a dietary supplement and as a humectant and emollient in cosmetics. Also in this case, the racemic pantothenyl alcohol is used for the same purposes.

Vitamin B_{12}, cyanocobalamin, 5,6-dimethylbenzimidazoyl cyanocobamide

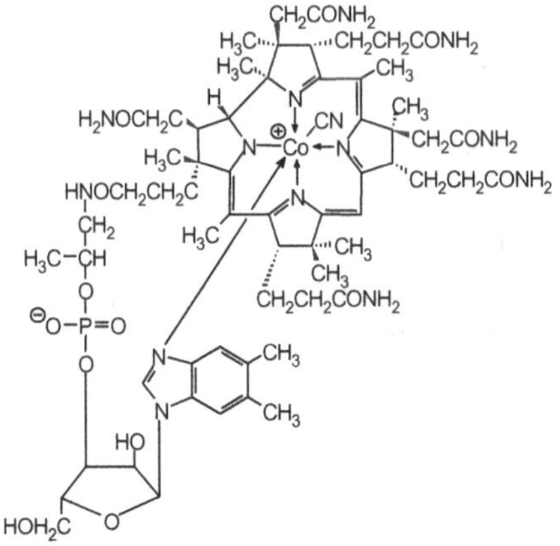

Molar mass 1355.38 ($C_{63}H_{88}O_{14}N_{14}CoP$), melting point >300°C.

Vitamin B_{12} or cobalamins are of a class of chemically related compounds with vitamin activity. There are several active forms: for example, cyanocobalamin, hydroxocobalamin, methylcobalamin, and adenosylcobalamin. They all belong to a group of organo-cobalt complexes that are generally called "*corrinoids*" based on the skeleton of corrin; this cyclic system contains four reduced pyrrole rings joined into a macrocyclic ring similar to porphyrins. A common form of vitamin B_{12}, cyanocobalamin, has besides this corrin structure a cyano group attached on the upper side (β-side) of the ring structure; a nucleotide part exists on the opposite side (α-side) that is further chemically bonded to one side group of the ring structure. These compounds are generally termed "complete corrinoids" (i.e., they contain a nucleotide as the cobalt α-(lower)ligand), differing from "incomplete corrinoids", which contain simple cobalt α-ligands, such as cyanide. Although incomplete corrinoids occur in nature, their biological significance is less than that of complete corrinoids. Corresponding natural ring structures exist, for example, in hemoglobin and chlorophyll.

In 1926, George Richards Minot (1885–1950) and William Perry Murphy (1892–1987) reported that pernicious anemia (also known as vitamin B_{12} deficiency anemia, Addison's anemia, Biermer's anemia, or Addison-Biermer anemia) in humans could be treated by ingesting large amounts of liver juice (or liver). The loss of ability to absorb vitamin B_{12} is the common cause of this anemic disease that was usually fatal before the discovery of its proper medication. It was found that pernicious anemia is mainly due to a lack of an intrinsic factor, a protein essential for absorption of vitamin B_{12} in the ileum. The anemia results in hematological changes in the formation of blood cells (i.e., red blood cells or erythrocytes, white blood cells or leukocytes, and platelets or thrombocytes) besides several other symptoms and signs, for example, neurological complications, a delay in physical growth in children, disturbances in the carbohydrate metabolism (e.g., the excretion of methylmalonic acid), and changes in the digestive tract. The "everyday signs" typically consist of fatigue, depression, diarrhea, dyspepsia, sore tongue, angular cheilitis, dark circles around the eyes, and brittle nails, as well as difficulties in proprioception and walking (muscle weakness). The main symptoms were accurately described already in the 19th century by James Scarth Combe (1796–1883), Thomas Addison (1793–1860), and Anton Bierner (1827–92). In 1928, Edwin Joseph Cohn (1882–1953) prepared an effective liver extract ("the purified liver extract" — 50 to 100 times more potent than the natural liver products), and its use as the first workable treatment (liver therapy) for pernicious anemia was then systematically studied in the early 1930s by George Hoyt Whipple (1878–1976), Minot, and Murphy, also sharing the 1934 Nobel Prize in Physiology or Medicine. In 1948, Karl August Folkers (1906–97) and Ernest Lester Smith (1904–92) finally isolated the active substance, and it was named "vitamin B_{12}". However, it was until in 1956 that Dorothy Mary Crowfoot Hodgkin (1910–94, the 1964 Nobel Prize in Chemistry) determined the complicated chemical structure of vitamin B_{12}, utilizing the technique of X-ray crystallography.

Vitamin B_{12} occurs as dark red (darkens further upon heating to 210–220°C), odorless, tasteless, and very hygroscopic (may absorb about 12% water) crystals or as an amorphous or crystalline powder. It loses its activity by the influence of strong ultraviolet or visible light or reducing agents (e.g., ascorbic acid). Vitamin B_{12} dissolves slightly in water (1.2%), in higher amounts in ethanol, dimethylformamide, dimethyl sulfoxide, low-molar-mass aliphatic carboxylic acids, and liquid ammonia, but not in diethyl ether, acetone, or benzene. It is found in most animal-derived foods, such as liver (60 mg/100 g), meat (2–3 mg/100 g), fish (0.5–10 mg/100 g), milk products (0.1–3 mg/100 g), and eggs (0.4–1.2 mg/100 g). In contrast, plants contain it only in very low concentrations. Due to the complicated structure of vitamin B_{12}, this vitamin is industrially produced only by fermentation from various carbohydrate sources (see riboflavin, vitamin B_2) and the products are known under many trade names. Neither plants nor animals (including humans) are independently capable of producing vitamin B_{12}, and it is synthesized only by bacteria and archaea that have the enzymes required for this biosynthesis. However, many foods are natural sources of vitamin B_{12} because of bacterial symbiosis. Specific derivatives of vitamin B_{12} (coenzymes) perform vital functions in the methylmalonate-succinate isomerization as well as in the methylation of homocysteine to methionine. In addition, the reactions of vitamin B_{12} and its derivatives in the cell participate, for example, in the syntheses of DNA, RNA, proteins, porphyrins, and choline, as well as the metabolism of carbohydrates and lipids. Thus, it can be concluded that this vitamin has a key role in the normal functioning of the brain and nervous system and the formation of blood. The initial treatment of pernicious anemia usually involves injections of vitamin B_{12} with a significant daily dose (1 mg) into muscle; the treatment is repeated every 2–3 months. Vitamin B_{15} is considered safe when used orally in amounts that do not exceed the recommended dietary allowance (RDA). The plasma of adult men and women normally contains 20–90 ng/100 mL of this vitamin. The recommended in-take for essentially healthy people depends on age, varying in the range of 0.3–4.0 mg/day.

In the derivatives of cyanocobalamin the cyano group chemically attached to the cobalt atom is replaced by other groups:

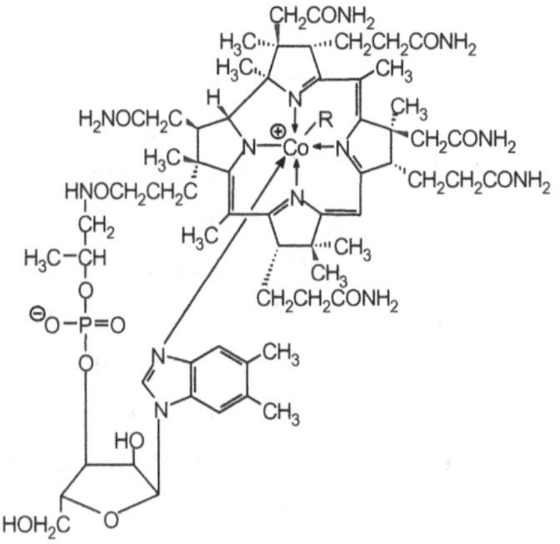

Hydroxocobalamin, vitamin B_{12a}

R = OH

Molar mass 1346.37 ($C_{62}H_{89}O_{15}N_{13}CoP$).

Aquacobalamin, vitamin B_{12b}

R = H_2O

Molar mass 1347.38 ($C_{62}H_{90}O_{15}N_{13}CoP$).

Nitritocobalamin, vitamin B_{12c}

R = NO_2

Molar mass 1375.37 ($C_{62}H_{88}O_{16}N_{14}CoP$).

Methylcobalamin, methyl B_{12}

R = CH_3

Molar mass 1344.40 ($C_{63}H_{91}O_{14}N_{13}CoP$).

5′-Deoxy-adenocyclobalamin, adenocyclobalamin, coenzyme B_{12}

R = HO OH H$_2$C O N N N N NH$_2$

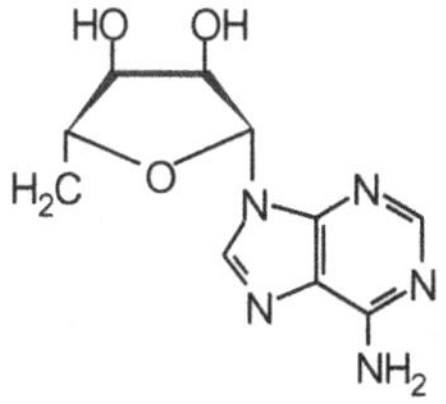

Molar mass 1579.60 ($C_{72}H_{100}O_{17}N_{18}CoP$).

Vitamin C, L-ascorbic acid, ascorbic acid, L-*xylo*-ascorbic acid, L-*threo*-hex-2-enonic acid γ-lactone, 3-oxo-L-gulonic acid γ-lactone

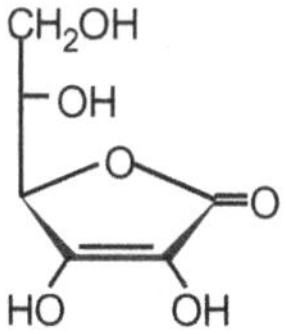

CH$_2$OH OH O O HO OH

Molar mass 176.13 ($C_6H_8O_6$), melting point 192°C (decomposes), density 1.65 kg/dm^3, pK$_1$ value 4.2 and pK$_2$ value 11.6.

L-Ascorbic acid ("a" means "no" and "scorbutus" "scurvy", see below) or vitamin C occurs widely in the vegetable and animal kingdoms. Although many animals can synthesize this vitamin, humans and other primates are not able to produce it, due to the lack of the enzyme L-gulono-γ-lactone oxidase, which catalyzes the terminal step of the biosynthetic process. For this reason, humans require it as a dietary micronutrient.

L-Ascorbic acid is found in fruits (10–1000 mg/100 g) (especially, in rose hips (1000 mg/100 g), hawthorn berries (150–300 mg/100 g), and common guava fruits (300 mg/100 g)), in vegetables (10–200 mg/100 g) (especially, in cabbage (30–180 mg/100 g), parsley (170 mg/100 g), and peppers (125–200 mg/100 g)), as well as in internal organs (10–40 mg/ 100 g), meat (<2 mg/100 g), and milk (1–2 mg/100 g). It was isolated from lemon in 1921 and in 1933, Walter Norman Haworth and Tadeus Reichstein (1897–1996) first synthesized it from L-xylose. L-Ascorbic acid was also the first industrially produced vitamin. Nowadays, it is manufactured for commercial purposes from D-glucose through the conventional multistage Reichstein-Grüssner synthesis:

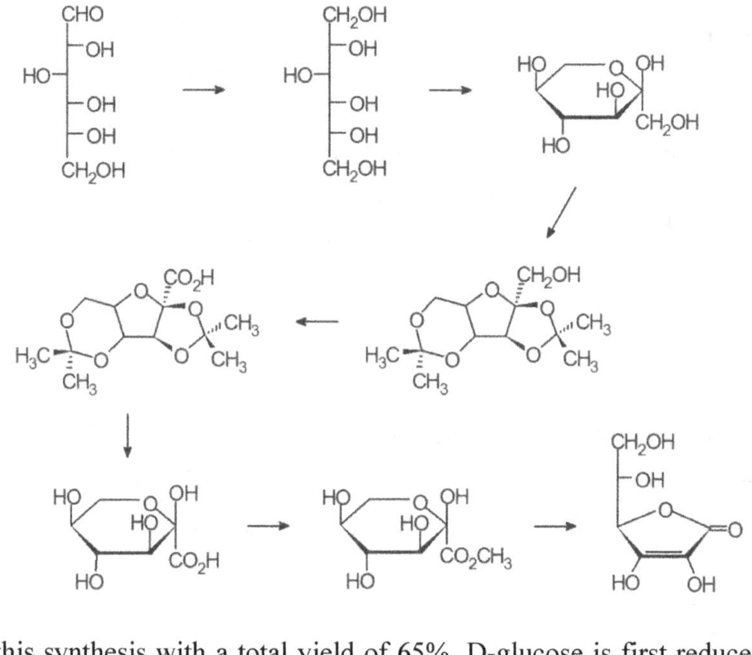

In this synthesis with a total yield of 65%, D-glucose is first reduced to D-glucitol (D-sorbitol) which is then microbiologically (*Acetobacter suboxydans*) oxidized to L-sorbose. In the next phase, L-sorbose is treated with acetone and sulfuric acid and the formed intermediate, 2,3:4,6-bis-*O*-isopropylidene-α-sorbose, is oxidized with potassium permanganate under alkaline conditions to 2,3:4,6-bis-*O*-isopropylidene-2-oxo-L-gulonic acid. This compound is hydrolyzed to 2-oxo-L-gulonic acid (L-*xylo*-2-hexulos-onic acid), which is converted into the final product by heating in water. L-Ascorbic acid forms white or yellowish crystals (plates or needles) that dissolve in water (0.3 g/mL), slightly in ethanol (0.03 g/mL), propylene glycol (0.05 g/mL), and glycerol (0.01 g/mL), but not in diethyl ether, benzene, or chloroform. It is optically active (its specific rotation $[\alpha]_D$ at 20°C is +23) with antioxidant properties and its chemical structure is readily changed on storing in air or light. In addition, it decomposes upon heating and in the presence of acids and bases.

Based on the very old knowledge on herbal cures of scurvy, in the 16th century it was already known that certain types of food may have a curative influence (although inconsistent) on this disease. However, systematic experiments performed during long-distance voyages in the 1750s by James Lind (1716–94) indicated that the prevention and treatment of this deficiency disease is possible by taking citrus fruits, apple, and some other fruits. In 1932, Albert Szent-Györgyi (1893–1986), Charles Glen King (1896–1988), and W.A. Waugh showed that L-ascorbic acid is the active compound behind scurvy. Typical deficiency symptoms of scurvy include fatigue, malaise, lethargy, and lack of appetite followed by the formation of spots on the skin, spongy gums, and bleeding from the mucous membranes. As the disease advances, there can be loss of teeth, yellow skin, fever, emotional changes, open, suppurating wounds, and finally death from bleeding. L-Ascorbic acid performs important metabolic functions. It readily oxidizes to dehydro-L-ascorbic acid with identical vitamin ability:

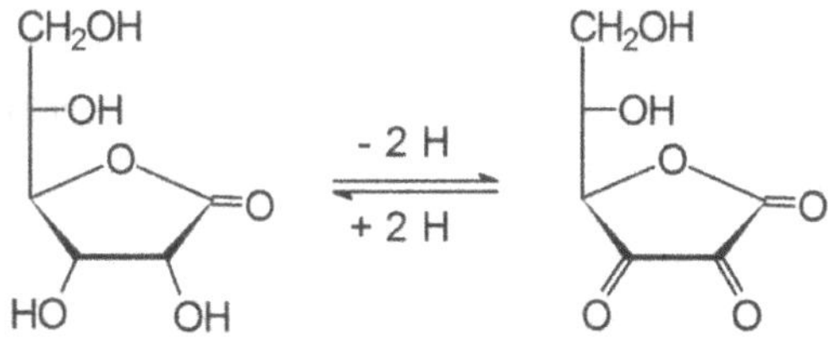

These compounds form a redox pair that participates as a coenzyme in hydroxylation reactions (proline and deoxycorticosterone) and in many enzymatic oxidation-reduction reactions (peroxidases, catalases, amidases, esterases, and glucosidases). The cells need L-ascorbic acid for the synthesis of polysaccharides, steroids, and collagen. Furthermore, it is an important antioxidant that maintains peroxidase reactions and controls the cell functions in the immune system (phagocytosis) against viruses, bacteria, and harmful cells. In general, this vitamin has several important functions in the body and its role, especially as an antioxidant, has given rise to a lively debate. L-Ascorbic acid is considered safe and any excess in the body is rapidly excreted in the urine. However, relatively large doses of it may cause indigestion, particularly when taken on an empty stomach. The clear deficiency can be effectively treated by a dietary intake of 1–3 g/day. The plasma of adult men and women normally contains 0.8–1.4 mg/100 mL of this vitamin. The recommended intake for essentially healthy people depends on age, varying in the range of 36–60 mg/day. In addition to its vitamin use, L-ascorbic acid is, for example, used as a flavoring and preservative additive in meats and other foods, an oxidant in bread doughs, and an analytical reagent, as well as for the abscission of citrus fruit in harvesting.

Dehydro-L-ascorbic acid, dehydroascorbic acid, vitamin C_2, L-*threo*-2,3-hexodiulos-onic acid γ-lactone

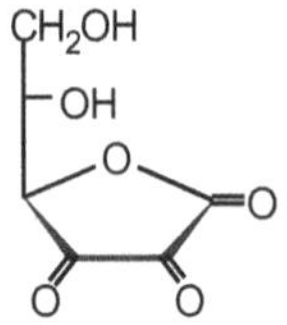

Molar mass 174.11 ($C_6H_6O_6$).

Sodium ascorbate

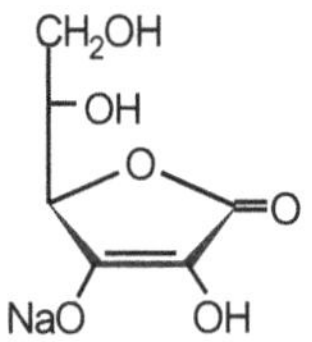

Molar mass 198.11 ($C_6H_7O_6Na$), melting point 218°C (decomposes).

L-Ascorbic acid exists as a lactone (i.e., an intermolecular ester) and its acid nature is due to an enol structure of this lactone ring. Sodium ascorbate ("mineral ascorbate") is formed by adding to its water solution an equivalent amount of sodium hydrogen carbonate and the product can be precipitated by an addition of isopropanol. It forms small crystals that dissolve in water (62 g/100 mL H_2O, 25°C). Its water solutions are readily oxidized in air. Sodium ascorbate is used as a food additive (an antioxidant and an acidity regulator) and for producing vitamin C.

Examples of other salts of L-ascorbic acids include **calcium** (molar mass 390.32; $C_{12}H_{14}O_{12}Ca$), **magnesium** (molar mass 374.54; $C_{12}H_{14}O_{12}Mg$), and **strontium ascorbate** (molar mass 437.86; $C_{12}H_{14}O_{12}Sr$). **Ascorbyl palmitate** (molar mass 414.54 ($C_{22}H_{38}O_7$), melting point 116–7°C) is a fat-soluble form of vitamin C. It is a white to yellowish powder with a lemon-like odor that dissolves in ethanol and vegetable oils and slightly in water. Ascorbyl palmitate is used as a source of vitamin C and an antioxidant food additive, especially in fats and oils. In addition, many other natural derivatives of L-ascorbic acid are known without being of any notable commercial importance.

D-*erythro*-Ascorbic acid, D-ascorbic acid, isoascorbic acid, erythorbic acid

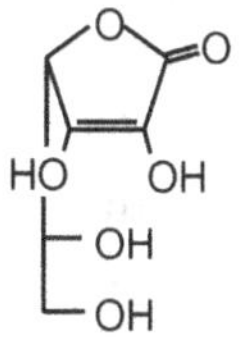

Molar mass 176.13 ($C_6H_8O_6$), melting point 192°C (decomposes).

D-*erythro*-Ascorbic acid forms faintly yellow, shiny crystals that dissolve in water (0.1 g/mL) and ethanol, but not in diethyl ether, acetone, and chloroform. Its specific rotation $[\alpha]_D$ is –17.3 (10% soln in H_2O) and it has no vitamin C-like activity. D-*erythro*-Ascorbic acid is used as an antioxidant, especially in the brewing industry.

10.4. Antibiotics

According to the conventional definition the term "antibiotics" means a group of chemical compounds produced by various species of microorganisms that are useful in small concentrations in inhibiting the growth of, and even in destroying other microorganisms. This initial definition was brought into general use in 1942 by Selman Abraham Waksman (1888–1973, the 1952 Nobel Prize in Physiology or Medicine) although the term "antibiosis" (in Greek, "anti" means "against" and "bios" "life") had already been introduced in 1889 by Jean Paul Vuillemin (1861–1932).

The use of antibiotics has a rich and colorful history. In 1877, Louis Pasteur noted the antagonism of some growing organisms for other groups during his investigations on the rate of growth of different species of bacteria. The therapeutic application of certain harmless bacteria started in the late 19th century to treat some infectious diseases. However, the use of substances with antimicrobial properties is known to have been a common practice as early as 500 to 600 B.C., when molded curd of soybean was utilized in Chinese medicine to treat boils and carbuncles. In ca. 350 B.C., the Sudanese-Nubian civilization used corresponding treatment based on some type of tetracycline antibiotics (i.e., based on the recent identification of tetracycline residues in the ancient bones). Additionally, moldy cheese had been employed for centuries by Chinese and Ukrainian peasants to treat infected wounds. The actual breakthrough in the development of antibiotics took place in 1928, when Alexander Fleming (1881–1955) accidentally discovered the first antibiotic, penicillin, from molds of *Penicillium notatum* for which he shared the 1945 Nobel Prize in Physiology or Medicine with Howard Walter Florey (1898–1968) and Ernst Boris Chain (1906–79). Fleming noticed that one culture of staphylococci on an agar plate was contaminated with a mold, and the colonies of staphylococci immediately surrounding the mold had been destroyed, whereas other staphylococci colonies farther away were normal. In 1939, Florey and Chain finally isolated the active substance in its pure form. Since the mold was a strain of *Penicillium*, they named this antibacterial substance "penicillin" and also started its mass production after the bombing of Pearl Harbor (on December 7th, 1941). In fact, by the Normandy landing operations (on June 6th, 1944), enough penicillin had been already produced to treat all the wounded within the Allied forces.

Based on the historical background of antibiotics, it is easy to understand that they were first considered to represent only one subgroup of chemotherapeutic agents. Thus, antibiotics were thought to be agents that could be only manufactured biologically by microorganisms. However, along with many advances achieved by medicinal chemistry to modify naturally occurring antibiotics and to prepare synthetic analogues, it became possible to also include semisynthetic and synthetic derivatives in the classic definition of antibiotics. On the other hand, the term "chemotherapy" is nowadays not fully clear. Hence, the term "antimicrobial drugs" is preferably used for all agents that are administered to patients for killing microbes. In addition, often "antibiotics" are still only considered as anti-infectives from bacterial sources and the interchangeable term "antibacterial" is used for those drugs which may work against bacteria but are synthetically derived.

The theory behind the traditional concept of chemotherapy (in Greek, "chemo" means "chemical" and "therapie" "treatment"), introduced by Paul Ehrlich (1854–1915, the 1908 Nobel Prize in Physiology and Medicine), was the knowledge about the distribution of coloring agents in tissues (especially, in hematological staining); the general observation was that some drugs were specifically enriched in certain cells. The discovery of arsphenamine (salvarsan) in 1909 was the first notable step in the development of chemotherapeutic agents with tolerable side effects. This chemotherapeutic agent was the first effective drug for syphilis. However, the real breakthrough of chemotherapy only occurred in the 1930s, when sulfonamides were developed.

Nowadays, the number of known antibiotics may approach 20,000, although most of them are too toxic and only a minor portion can be utilized in medicine. Since some serious, bacteria-causing diseases are gradually becoming resistant (i.e., "antimicrobial resistance", AMR; this is a growing danger, especially in hospitals) to most commonly available antibiotics, new antibiotics have to be continuously developed by discovering the total new ones or modifying the old ones. Antibiotics for human use are generally characterized by a "selectivity spectrum of activity", which indicates the ability of an antibiotic to penetrate into bacterial and fungal cells. However, designations of spectrum of activity are of somewhat limited use, unless they are based on clinical effectiveness of the antibiotic against a specific microorganism.

Bacterial species can be differentiated into two large groups to make them more visible under a microscope: Gram-positive and Gram-negative bacteria. In the Gram staining or Gram's method, Gram-positive bacteria are colored purple by crystal violet dye, while a counterstain (normally safranin or fuchsine) added after the crystal violet gives all Gram-negative bacteria a red or pink coloring. Hans Christian Gram (1853–1938) developed this technique, and it classifies bacteria according to the chemical and physical properties of their cell walls by detecting either lipids, i.e., peptidoglycan present in a thick layer (50–90% of cell envelope) in Gram-positive bacteria, or carbohydrates present in a thinner layer (10% of cell envelope) in Gram-negative bacteria. The term "broad-spectrum antibiotics" refers to antibiotics that act against a wide range of disease-causing bacteria (i.e., both Gram-positive and Gram-negative bacteria). In contrast, "narrow-spectrum antibiotics" are only effective against specific families of bacteria.

The chemical structures and molar masses (typically in the range of 140–1900 Da) of antibiotics vary significantly. In spite of this, it is possible to classify most antibiotics according to their structures into characteristic subgroups: for example, β-lactams, aminoglycosides, tetracyclines, macrolides, lincosaminides, peptides, chloramphenicol and analogues, and others. However, a useful classification based on the actual effects of different antibiotics is difficult to formulate because the selectivity spectra of activities even of relatively common antibiotics are not fully known. In addition, besides the traditional pharmaceutical preparations (also for veterinary applications), antibiotics are used as food and feed additives. For this reason, in many cases, a straightforward and comprehensive determination of the effects of varying antibiotics and their mixtures is not an easy task.

An interesting area of carbohydrate chemistry was opened up when it was noted that many natural carbohydrate-containing compounds possessed antibiotic properties. In this section, only those antibiotic groups that contain carbohydrates or their derivatives as well as some typical examples of their compounds, are described. The main antibiotic groups include aminoglycosides, macrolides, lincosaminides, and others.

10.4.1. *Aminoglycosides*

The term "*aminoglycosides*" or "*aminocyclitol antibiotics*" (or aminoglycoside-aminocyclitol antibiotics) is commonly used for bacteria-killing antibiotics whose structures are based either on **D-streptamine** (I, see below, R = OH, molar mass 178.19; $C_6H_{14}O_4N_2$) or **2-deoxy-D-streptamine** (I, see below, R = H, molar mass 162.19; $C_6H_{14}O_3N_2$). Of these aminocyclitols (1,3-diaminohexane derivatives), the derivatives of 2-deoxy-D-streptamine are therapeutically more significant than those of D-streptamine. One or more amino sugars are attached to an aminocyclitol residue comprising at least one amino hexose unit and in some cases, a pentose unit with an amino group. With respect to carbohydrate chemistry, aminoglycosides represent an interesting family of compounds, since these compounds consist completely of carbohydrate-based material with typical characteristic properties of antibiotics.

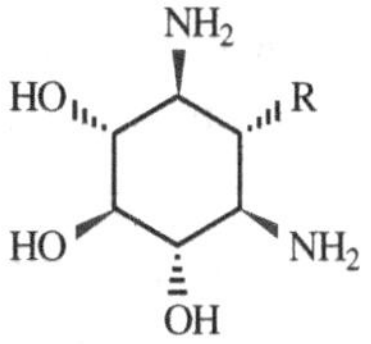

I

Aminoglycosides are among the oldest known antibiotics; they were discovered by Selman Abraham Waksman and his co-workers who systematically studied soil microorganisms and various antibiotics produced by these organisms in 1939. The first member of the series, streptomycin, was isolated in 1943 from the species *Streptomyces griseus*. It was shown in 1944 to inhibit the growth of many Gram-positive and Gram-negative bacteria and especially, the tubercle bacillus (*Mycobacterium tuberculosis*). The discovery of this antibiotic and its chemotherapeutical use further inspired many researchers, and quite soon after, several corresponding compounds were found, besides in the *Streptomyces* genus, also in the *Micromonospora* species. In the former case, compound names with the characteristic ending *-mycin* is used, whereas in the latter case, this ending is *-micin*.

All aminoglycosides are polar, slightly fat-soluble cationic compounds, which typically have a low rate of diffusion through lipid bilayers. They are strong bases and exist at physiological pH values as polycations. Their inorganic salts are readily soluble in water, but dissolve only slightly in organic solvents. The corresponding compounds and their water solutions are chemically relatively stable although they may decompose during treatments in an autoclave.

Aminoglycosides are broad-spectrum antibiotics that have varying activities against different bacteria. They are primarily effective against aerobic Gram-negative bacteria, for example, *Pseudomonas aeruginosa* and *Serrata marcescens*, and Gram-positive staphylococci. They are generally not effective when used alone against streptococci, pneumococci, and enterococci. Moreover, most anaerobic bacteria (e.g., *Bacteroides fragilis*, clostridia, and anaerobic cocci) are resistant against aminoglycosides because the transportation of these antibiotics to the cell requires oxidative phosphorylation.

The effect of aminoglycosides involves their uptake followed by a binding to bacterial ribosomes (to one ribosomal subunit), resulting in the inhibition of protein synthesis by interfering with the translation of the genetic code. Resistance against aminoglycosides is frequently associated with being in the presence of the bacterium of antibiotic-inactivating enzymes or the possibility that the antibiotic may be too poorly transported to the bacterial cell to cause a sufficient antibiotic concentration inside the bacterium. In general, aminoglycosides are not well absorbed (only about 1% of the dose) from the alimentary track; they are usually given parenterally by intramuscular injection. In this case, they are also rapidly absorbed and the maximum concentration levels are attained in 30–90 min. With a common dose, sufficient concentration (optimum 5–10 mg/mL) in the plasma is normally maintained for about eight hours.

Examples of the most significant aminoglycosides used in chemotherapeutical treatments are as follows:

Streptomycin A

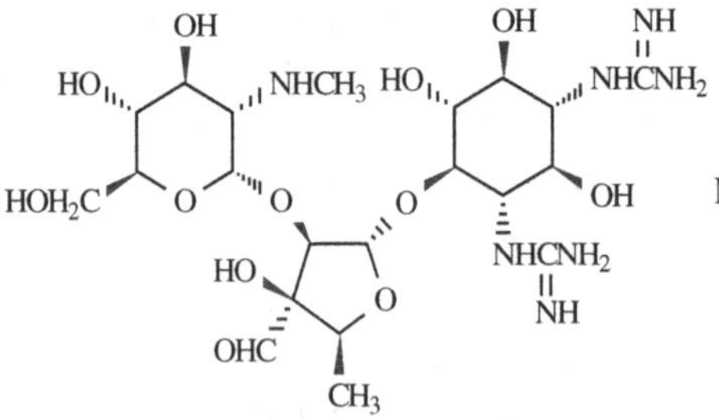

Molar mass 581.58 ($C_{21}H_{39}O_{12}N_7$).

Streptomycin A is produced together with other specific antibiotics (e.g., hydroxystreptomycin, streptomycin B, and cycloheximide) by *Streptomyces griseus*. In principle, the production takes place in the same way as that of penicillins although in the former case, an effective isolation and purification stage is needed. In the chemical structure of streptomycin A, streptidine (1,3-dideoxy-1,3-diguanido-*scyllo*-inositol) binds through a glycosidic bond streptobiosamine that is a combination of L-streptose (5-deoxy-3-*C*-formyl-α-L-lyxofuranose) and *N*-methyl-L-glucosamine. The product is normally isolated as a salt, such as sulfate, hydrochloride, or phosphate. The typical salt in medical usage, **streptomycin sulfate**, is a white, odorless, and hygroscopic powder that is stable on storage towards air and light. It readily dissolves in water and slightly in ethanol, but practically not in most other organic solvents. The antibiotic activity of aqueous solutions (in the optimum pH range 4.5–7.0) of streptomycin sulfate remains at room temperature for one week. As a chemotheraupetic agent, this antibiotic is active against numerous Gram-positive and Gram-negative bacteria, and it is very effective against the tubercle bacillus. However, the greatest drawback in the use of streptomycin sulfate is the relatively rapid development of resistant strains of microorganisms. Due to this drawback together with other notable disadvantages (e.g., not absorbed when given orally, possible development of damage to the otic nerve, and allergy), its medical usage as such or combined with another antibiotics has gradually led to a decreasing use of streptomycin A products. For this reason, for example, the use of pyrazinamide for the treatment of tuberculosis has correspondingly increased. Finally, it should be noted that the term "streptomycin" is normally used instead of "streptomycin A".

Streptomycin B, mannosidostreptomycin

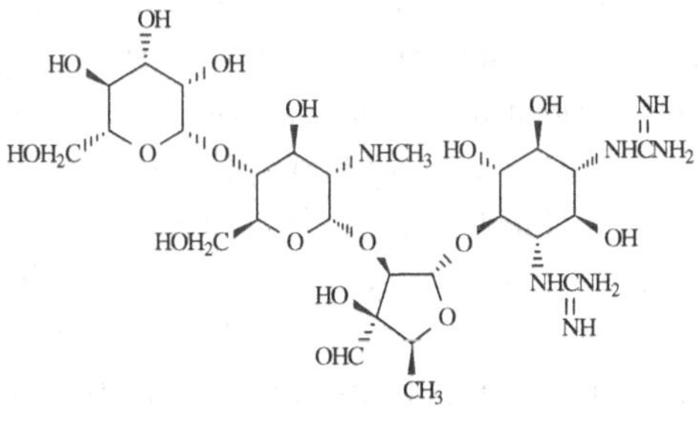

Molar mass 743.72 ($C_{27}H_{49}O_{17}N_7$).

Streptomycin B is a derivative of streptomycin A, where L-mannose is glycosidically bound to the *N*-methyl-L-glucosamine component of streptomycin A. It is produced by *Streptomyces griseus* and can be isolated as one antibiotic component from the concentrate formed. Streptomycin B exists as a trihydrochloride monohydrate (molar mass 871.12 ($C_{27}H_{54}O_{18}N_7Cl_3$), melting point 190–200°C (anhydrous)). It is a white, amorphous powder and its antibiotic activity is only one fourth of that of Streptomycin A.

Spectinomycin or decahydro-4α,7,9-trihydroxy-2-methyl-6,8-bis(methylamino)-4*H*-pyrano[2,3-*b*][1,4]benzodioxin-4-one (molar mass 332.35; $C_{14}H_{24}O_7N_2$) does not structurally belong to aminoglycosides (i.e., it does not exactly contain a carbohydrate part) although it is biogenetically related to streptomycins. It is produced by *Streptomyces spectabilis* and *Streptomyces flavopersicus*. Its chemical structure is a combination of 1,3-di-(*N*-methylamino)-*myo*-inositol and 4,6-dideoxy-2,3-hexodiulopyranose. This antibiotic is, as a dihydrochloride pentahydrate, useful for the treatment of gonorrhea infections (*Neisseria gonorrhoeae*), especially in cases, where the patient has penicillin allergy or shows resistance to penicillin. Spectinomycin is given parenterally by intramuscular injection.

Neomycin was found in 1949 when searching for less toxic antibiotics than streptomycin. It is produced by *Streptomyces fradiae* as a mixture of closely related substances ("neomycin complex"); besides the main component **neomycin B**, the antibiotics **neamine** (originally designated "neomycin A"), **neomycin C**, and **fradisine** are present. The purified neomycin product does not contain fradisine, which has some antifungal properties but no antibacterial activity. The use of neomycin has increased steadily since its discovery, and it is today one of the most useful antibiotics in the treatment of gastrointestinal and dermatologie infections as well as acute bacterial peritonitis. In addition, neomycin is employed in abdominal surgery to reduce or avoid complications caused by infections from bacterial flora of the bowel. It is a broad-spectrum antibiotic and has activity against a wide range of organisms without developing easily resistant strains of microorganisms. It shows a low incidence of toxic and hypersensitivity reactions, and it is also very slightly absorbed from the digestive tract. Hence, the oral use of this antibiotic does not ordinarily produce any systemic effect. Neomycin is normally used as **neomycin sulfate**, which is a white to slightly yellow, hygroscopic, and crystalline powder that readily dissolves in water. It is photosensitive, but its water solutions are chemically stable in a wide pH range and in autoclave treatments.

Neomycin C consists of 2-deoxy-D-streptamine, to which at the carbon atoms C_4 and C_5, respectively, neosamine C (2,6-diamino-2,6-dideoxy-β-D-glucopyranose) and a disaccharide (a combination of neosamine C and D-ribose) are glycosidically linked. Neomycin B (molar mass 614.65; $C_{23}H_{46}O_{13}N_6$) is an isomer of neomycin C; the configuration of the carbon atom C_5 in neosamine C is epimerized, thus corresponding to the structure of 2,6-diamino-2,6-dideoxy-α-L-iodopyranose (neosamine B). The structure of neamine differs from those structures above containing 2-deoxy-D-streptamine, to which neosamine C is glycosidically attached at the carbon atom C_6.

Neomycin C

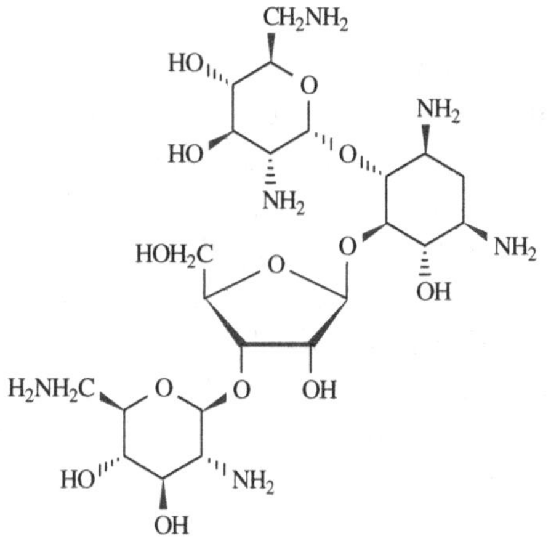

Molar mass 614.65 ($C_{23}H_{46}O_{13}N_6$).

Neamine, neomycin A

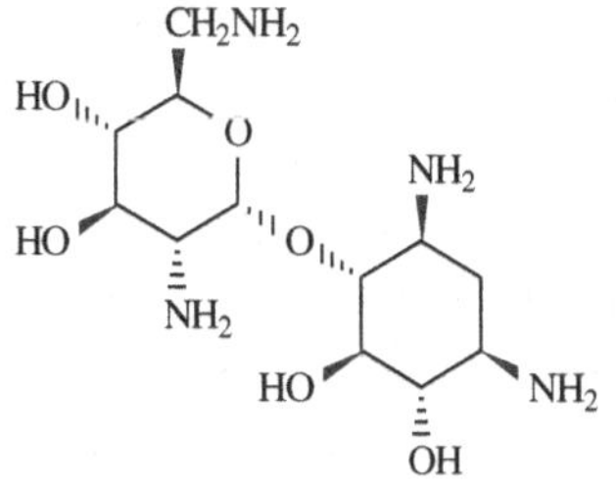

Molar mass 322.36 ($C_{12}H_{26}O_6N_4$).

Kanamycin was discovered in Japan in 1957 and produced by *Streptomyces kanamyceticus*. It was quite soon noted that this antibiotic has activity against mycobacteria and several intestinal bacteria, as well as a number of pathogens that show resistance to other antibiotics. However, the present usage is mainly restricted to infections of the intestinal tract and to systematic infections arising from Gram-negative bacilli (e.g., *Klebsiella*, *Proteus*, *Enterobacter*, and *Serratia*). The main reason for this reduction in use is its poor absorption from the intestinal tract. It should be given parenterally by intramuscular injection, which is rather painful and the concomitant use of a local anesthetic is needed. In addition, similarly to streptomycin A, the use of this antibiotic may cause a decrease in, or complete loss of, hearing. Kanamycin is chemically a base and it forms salts of acids through its amino groups. As a free base it dissolves readily in water although in therapy it is used as **kanamycin sulfate** which is even more water soluble.

The detailed structural determination of the kanamycin(s) has indicated that it actually consists of three closely related compounds that are designated **kanamycins A**, **B**, and **C**. However, the commercial product is almost pure kanamycin A, the least toxic of the three forms. In all these forms, carbon atom C_4 of 2-deoxy-D-streptamine is linked to 3-amino-3-deoxy-D-glucose or 3-D-glucosamine, but the forms differ from each other with respect to a monosaccharide glycosidically bound to carbon atom C_6; in kanamycin A it is 6-amino-6-deoxy-D-glucopyranose, in kanamycin B 2,6-diamino-2,6-dideoxy-D-glucopyranose, and in kanamycin C 2-amino-2-deoxy-D-glucopyranose.

Tobramycin was separated in 1976 as the most active compound from the aminoglycoside-type components in the “nebramycin complex” produced by *Streptomyces tenebrarius*. This antibiotic is given parenterally by intramuscular injection and its most important property is a high activity (i.e., two- to fourfold when compared to gentamicin) against the Gram-negative bacillus (*Pseudomonas aeruginosa*). The chemical structure of tobramycin is close to that of kanamycin B and both these compounds are assumed to be liberated from the 6-*O*″-carbamoyl derivative by hydrolysis during its isolation from the fermentation broth. The commercial product is **tobramycin sulfate**.

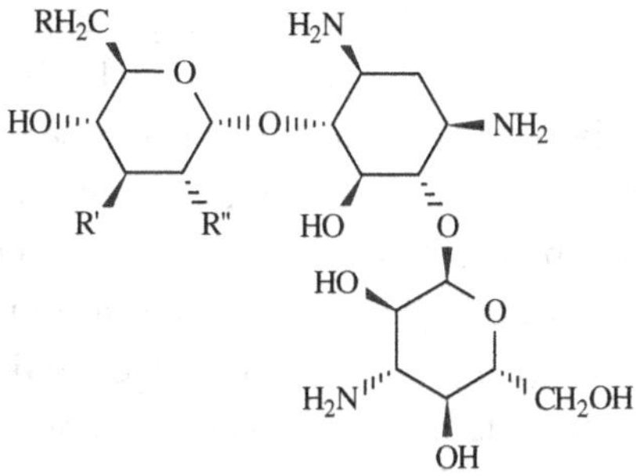

	R	R'	R''
I	NH_2	OH	OH
II	NH_2	OH	NH_2
III	OH	OH	NH_2
IV	NH_2	H	NH_2

Kanamycin A (I, molar mass 484.50; $C_{18}H_{36}O_{11}N_4$)
Kanamycin B (II, molar mass 483.52; $C_{18}H_{37}O_{10}N_5$)
Kanamycin C (III, molar mass 484.50; $C_{18}H_{36}O_{11}N_4$)
Tobramycin (IV, molar mass 467.52; $C_{18}H_{37}O_9N_5$)

Amikacin, 3-*N*-(4-amino-2-hydroxybutanoyl)kanamycin A

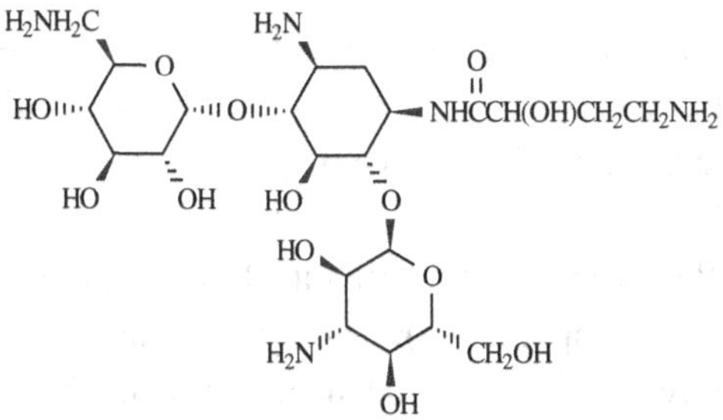

Molar mass 585.61 ($C_{22}H_{43}O_{13}N_5$).

Amikacin is a semisynthetic aminoglycoside first prepared from kanamycin A in Japan in 1972. In the synthesis, the amino group at the carbon atom C_3 in the 2-deoxy-D-streptamine component of kanamycin A is acylated with L-amino-α-hydroxybutanoic acid (L-AHBA). As a result of this reaction, the activity of kanamycin A decreases about 50% against Gram-positive bacilli. However, the remarkable new feature obtained is that amikacin resists attack by most bacterial inactivating enzymes; it is effective against strains of bacteria that are resistant to other aminoglycosides and only a few enzymes (e.g., tobramycin and gentamicin have at least six such enzymes) can inactivate it. It is furthermore noteworthy that the optimum concentration of amikacin in the plasma is 20–30 mg/mL (for other aminoglycosides 5–10 mg/mL). This higher value is due to the fact that, compared to other aminoglycosides, amikacin has a three- to fivefold higher MIC value; i.e., “minimum inhibitory concentration”, the lowest concentration of an antimicrobial that will inhibit the visible growth of a microorganism after overnight incubation. Because of the clear advantages, this antibiotic is usually recommended to be given parenterally for the treatment of serious infections caused by bacterial strains resistant to other aminoglycosides.

Dibecacin (molar mass 451.52; $C_{18}H_{37}O_8N_5$) is a semisynthetic derivative of kanamycin B.

Paromomycin was discovered in 1956 and it is produced by specific *Streptomyces* species. The product consists of two compounds: **paromomycin I** and **paromomycin II**. The chemical structures of the paromomycins are similar to those of neomycins B and C. The most significant difference is the replacement of the glycidically linked neosamine C (2,6-diamino-2,6-dideoxy-β-D-glucopyranose, at the carbon atom C_4 of the 2-deoxy-D-streptamine component of neomycins) in the neomycins by D-glucosamine in the paromomycins. Hence, paromomycin II corresponds, with respect to a disaccharide side chain (it consists of D-ribose and either neosamine C or neosamine B), to the structure of neomycin C with neosamine C (2,6-diamino-2,6-dideoxy-β-D-glucopyranose). On the other hand, paromomycin I corresponds to the structure of neomycin B with neosamine B (2,6-diamino-2,6-dideoxy-α-L-iodopyranose). Paromomycin has a broad-spectrum antibacterial activity and it has been mainly used for the treatment of gastrointestinal infections caused by *Salmonella*, *Shigella*, and *Escherichia coli*. However, its use is currently restricted. The commercial product, **paromomycin sulfate**, dissolves in water and its water solutions are chemically stable in a wide pH range.

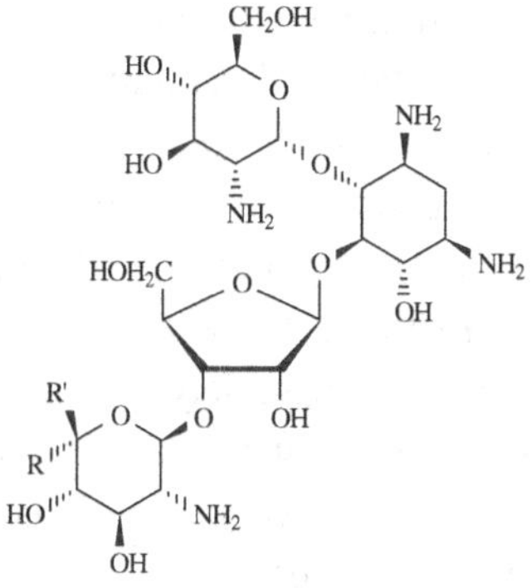

	R	R'
I	CH_2NH_2	H
II	H	CH_2NH_2

Paromomycin I (I, molar mass 615.64; $C_{23}H_{45}O_{14}N_5$)
Paromomycin II (I, molar mass 615.64; $C_{23}H_{45}O_{14}N_5$)

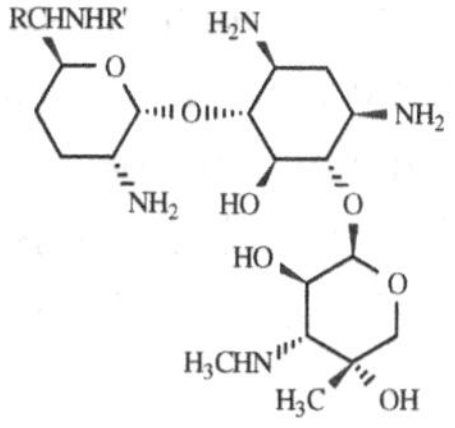

	R	R'
I	CH_3	CH_3
II	CH_3	H
III	H	H

Gentamicin C$_1$ (I, molar mass 477.60; $C_{21}H_{43}O_7N_5$)
Gentamicin C$_2$ (II, molar mass 463.58; $C_{20}H_{41}O_7N_5$)
Gentamicin C$_{1a}$ (III, molar mass 449.55; $C_{19}H_{39}O_7N_5$)

Gentamicin was isolated in 1958 as the "gentamisine complex" and reported in 1963 to belong to the aminoglycosides. It is commercially produced by the microorganism *Micromonospora purpurea.* Gentamicin has a broad spectrum of activity (see tobramycin and amikacin) against many pathogens of both Gram-positive and Gram-negative types. This antibiotic is employed particularly in the treatment of hospital-acquired infections caused by Gram-negative bacteria (*Pseudomonas*, *Enterobacter*, and *Serratia*). However, the most significant property of gentamicin is its high degree of activity against *Pseudomonas aeruginosa* and other Gram-negative enteric bacilli. It is used parenterally as **gentamicin sulfate**, but also in the form of a topical cream or ointment for the treatment of a variety of skin infections. Gentamicin sulfate is a white to brown-yellow substance that dissolves in water, but not in ethanol, acetone, and benzene. Its water solutions are chemically stable in a wide pH range and in autoclave treatments.

The *Micromonospora* species coproduce, besides **gentamicins** $\mathbf{C_1}$, $\mathbf{C_2}$, and $\mathbf{C_{1a}}$, also the compounds **gentamicin A** (molar mass 468.50; $C_{18}H_{36}O_{10}N_4$) and **gentamicin B** (molar mass 482.53; $C_{19}H_{38}O_{10}N_4$), which are not present in the commercial product. **Isepamicin** (molar mass 569.61; $C_{22}H_{43}O_{12}N_5$) is a semisynthetic product of gentamicin B.

The properties and structure of **sisomicin** are similar to those of gentamicin. It can be isolated from the fermentation broth of *Micromonosporum inyoensis*. It is a broad-spectrum aminoglycoside and is highly active, especially against Gram-positive bacteria. A semisynthetic aminoglycoside, **netilmicin** or 3-*N*-ethylsisomicin, can be prepared from sisomicin by reductive ethylation. Netilmicin is active against many gentamicin-resistant strains, especially among *Escherichia coli*, *Enterobacter*, *Klebsiella*, and *Citrobacter*. In both cases, the commercial product is a sulfate derivate, thus corresponding to **sisomicin sulfate** and **netilmicin sulfate**.

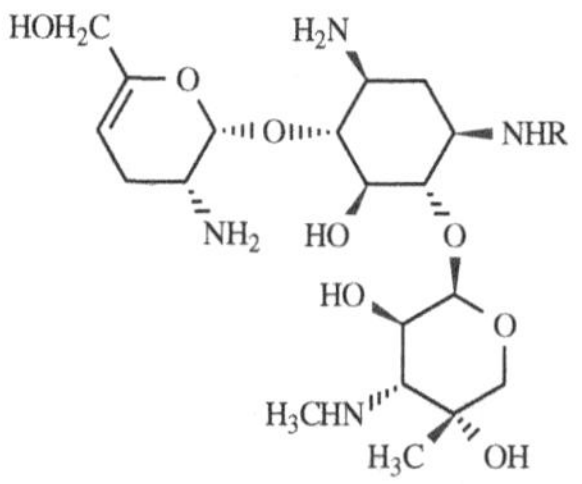

	R
I	H
II	CH_2CH_3

Sisomicin (I, molar mass 448.52; $C_{19}H_{36}O_8N_4$)
Netilmicin (II, molar mass 476.57; $C_{21}H_{40}O_8N_4$)

Examples of other compounds with aminoglycoside-type structures are **apramycin** (molar mass 539.58; $C_{21}H_{41}O_{11}N_5$), **ribostamycin** (molar mass 454.48; $C_{17}H_{34}O_{10}N_4$), **hygromicin B** (molar mass 527.53; $C_{20}H_{37}O_{13}N_3$), and **butirosin B** (molar mass 555.58; $C_{21}H_{41}O_{12}N_5$).

10.4.2. *Macrolides*

Macrolide antibiotics are structurally complex, well established antimicrobial agents produced as secondary metabolites of soil microorganisms, mainly by various strains of *Streptomyces*. The products are often chemically and microbiologically further modified to semisynthetic antibiotics that are therapeutically more active than the parent antibiotics. The carbon chain structures of macrolides are partly branched and have three common chemical characteristics: (i) a large lactone ring (i.e., an aglycone), (ii) a ketone group, and (iii) a glycosidically linked amino sugar, and normally also a glycosidically linked other carbohydrate-based unit. The amino sugar component (normally present as a *N*-dimethylated amino sugar) can be a compound, such as D-angolosamine, D-desosamine, D-forosamine, L-megosamine, or D-mycaminose, and the sugar component represents the monosaccharide or its derivative, such as D-aldgarose, L-amiketose, L-arcanose, L-cinerulose, L-chalcose, L-cladinose, 6-deoxy-D-allose, D-javose, L-mycarose, D-mycinose, L-oleandrose, L-olivose, or L-rhodinose. Because of the presence of the dimethylamino group on the moiety, the macrolides are bases that form clinically useful salts (pK_a value 6–9). The free bases are only slightly soluble in water, but they dissolve in somewhat polar organic solvents. The macrolides are chemically stable at room temperature, but are inactivated by bases, acids, and heat.

Therapeutically used macrolide antibiotics can be divided into three main groups: (i) *low-molar-mass macrolide antibiotics*, (ii) *polyene antibiotics or high-molar-mass antibiotics*, and (iii) *ansamacrolides* or *ansamycins*. The last compounds do not contain any carbohydrates or their derivatives, and this class of compounds is not included in this section. The low-molar-mass macrolide antibiotics (over 100 natural members known) are classified into three families according to the size of the lactone ring: the 12-, 14-, or 16-membered macrolides. In contrast, polyene antibiotics (about 200 members known) fall into two subgroups based on the size of the macrolide ring: the 26- and 38-membered macrolides. However, often the polyene macrocyclic lactones as well as the ansamacrolides are not included among the typical macrolide antibiotics.

The low-molar-mass macrolides are considered to be medium- or narrow-spectrum antibiotics. Some possess primarily activity against Gram-positive bacteria, both cocci and bacilli, but may also exhibit useful effectiveness against Gram-negative cocci, in particular *Neisseria* species. These macrolides are frequently active against bacterial strains that are resistant to penicillin and are effective, for example, in the treatment of infections of the respiratory tract. Furthermore, the low-molar-mass macrolide antibiotics are generally effective against *Mycoplasma*, *Chlamydia*, *Campylobacter*, and *Legionella* species. These antibiotics are mainly administered enterally via digestion, but also parenterally by intramuscular or intravenous injection. The rare side effects include dysfunction, irritation, or malaise in the intestines and the stomach as well as allergies. The polyene antibiotics are broad-spectrum antifungical agents with potent activity against pathogenic yeasts, dermatophytes, and molds. In contrast, only a few ansamacrolides are of therapeutical importance.

There are only some low-molar-mass macrolide antibiotics in the subgroup of the 12-membered macrolides that are therapeutically used.

Methymycin

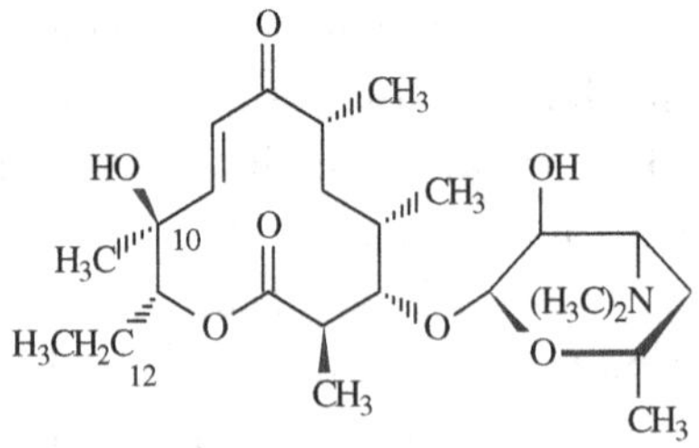

Molar mass 469.62 ($C_{25}H_{43}O_7N$), melting point 195.5–7.0°C.

Methymycin is produced by specific *Streptomyces* species and it was the first macrolide whose structure was determined. It was also the first conventional macrolide prepared by total synthesis. Its aglycone is **methynolide** (molar mass 312.41; $C_{17}H_{28}O_5$) and the amino sugar component is β-D-desosamine. Methymycin crystallizes as prisms that dissolve in ethanol, acetone, and mild acids, but only slightly in water. The isomeric (the hydroxyl group is attached at the carbon atom C_{12} instead of C_{10}) **neomethymycin** (molar mass 469.62; $C_{25}H_{43}O_7N$) is also coproduced with methymycin and its aglycone is **neomethynolide** (molar mass 312.41; $C_{17}H_{28}O_5$) and the amino sugar component β-D-desosamine.

The low-molar-mass macrolides in the subgroup of the 14-membered macrolides form the most significant class of these compounds. Examples of these macrolides used therapeutically are as follows:

Erythromycin, erythrocin

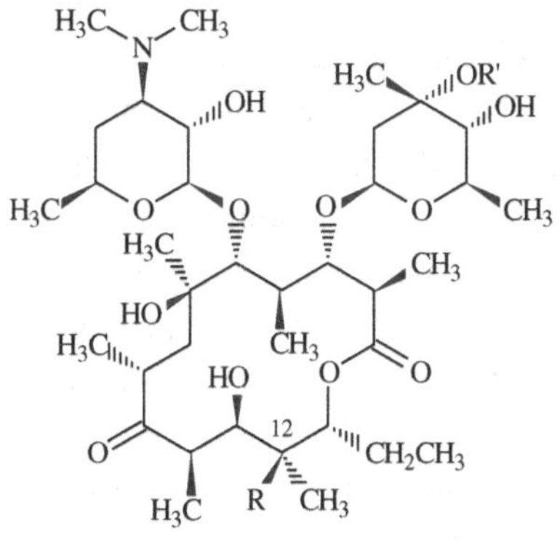

	R	R'
I	OH	CH_3
II	H	CH_3
III	OH	H

The commercial erythromycin corresponds to erythromycin A (formula I, molar mass 733.94 ($C_{37}H_{67}O_{13}N$), melting point 135–40°C), which is produced by *Streptomyces erythreus*, nowadays reclassified as *Saccharopolyspora erythrea*. It was isolated first in 1952 from a Philippine soil sample and its chemical structure was determined in 1957 (the complete stereochemical structure in 1965). The aglycone part is **erythronolide A** (molar mass 418.53; $C_{21}H_{38}O_8$) and the carbohydrate derivatives are β-D-desosamine (3-*N*,*N*-dimethylamino-3,4,6-trideoxy-β-D-*xylo*-hexopyranose) and β-L-cladinose (2,6-di-deoxy-3-*C*-methyl-3-*O*-methyl-β-L-*ribo*-hexopyranose). Erythromycin A is a white or yellowish white, crystalline powder with a bitter taste that dissolves in ethanol, acetone, diethyl ether, chloroform, acetonitrile, and ethyl acetate, and in lesser amounts in water (about 2 g/L, pH of the saturated solution is 8.0–10.5). Its specific rotation $[\alpha]_D$ at 25°C in ethanol is −78. Erythromycin A is the most active low-molar-mass macrolide and this bacteriostatic agent is primarily active against Gram-positive bacteria, especially those causing infections in the respiratory tract and soft tissues. Erythromycin A and its derivatives are mainly used as stopgap antibiotics, for example, for the treatment of scarlet fever, erysipelas, syphilis, gonorrhea, and diphtheria in the cases, where penicillins and cephalosporines cannot be used due to allergies or resistance factors. Tetracyclines can also be replaced by them in the treatment of chlamydia infection, while they are used as primary antibiotics in mycoplasma and legionella pneumonia. Erythromycin A can be employed as a free base for topical administration or in oral dosage forms although it is rapidly decomposed in the stomach if the pH decreases below 4 (the food in the stomach also delays its absorption). For this reason, a variety of enteric-coated and delayed-release dose forms of it (e.g., relatively water-insoluble, acid stable-salts, esters, and/or formulations) have been developed and the maximum concentration can be obtained within 4 hours. In general, this antibiotic is considered a safe medicine and a major part of it metabolizes in the human body.

The organic salts of erythromycin A are more stable in the stomach and can be better absorbed than the parent antibiotic. Examples of these salts are **erythromycin stearate**, **erythromycin ethylsuccinate**, **erythromycin estolate** (the laurylsulfate salt of the 2′-propionate ester of erythromycin), **erythromycin gluceptate** (erythromycin glucoheptonate), and **erythromycin lactobionate**. These salt derivatives are crystalline compounds that are sparingly soluble in water, but soluble in common organic solvents. Erythromycin estolate that crystallizes as long needles may cause cholestatic hepatitis and its use is gradually reduced in many countries.

Several minor antibiotics are coproduced with erythromycin A. The most significant ones are **erythromycin B** (formula II above, molar mass 717.94; $C_{37}H_{67}O_{12}N$) and **erythromycin C** (formula III above, molar mass

719.91; $C_{36}H_{65}O_{13}N$). Compared to erythromycin A, erythromycin B lacks a hydroxyl group at the carbon atom C_{12} and erythromycin C contains the neutral sugar L-mycarose instead of L-cladinose. The B analogue is more acid-stable, but has only 80% of the activity of erythromycin A. In contrast, the C analogue appears to be as active as erythromycin A, but only very small amounts of it are present in fermentation broths. In addition, **erythromycin D** (molar mass 703.91; $C_{36}H_{65}O_{12}N$), **erythromycin E** (molar mass 747.92; $C_{37}H_{65}O_{14}N$), and **erythromycin F** (molar mass 749.94; $C_{37}H_{67}O_{14}N$) have been isolated from culture broths. Erythromycin F is also considered the biosynthetic precursor of erythromycin E.

Oleandomycin

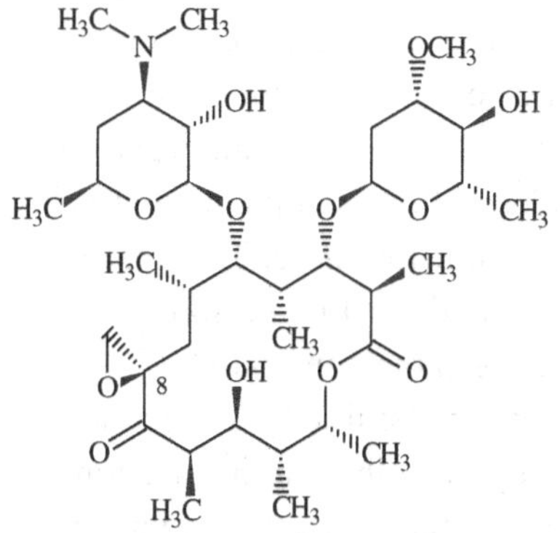

Molar mass 687.87 ($C_{35}H_{61}O_{12}N$).

Oleandomycin is produced by *Streptomyces antibioticus*. Its aglycone, **oleandolide** (molar mass 386.49; $C_{20}H_{34}O_7$), differs chemically from other erythronolides at several centers; most notable is the epoxide structure at the carbon atom C_8 and it contains, besides the amino sugar β-D-desosamine, the neutral sugar 2,6-dideoxy-3-*O*-methyl-β-L-*arabino*-hexopyranose. It and its triacetyl derivative, **triacetyloleandomycin** or troleandomycin (TAO) (molar mass 813.98; $C_{41}H_{67}O_{15}N$), are used as an alternative to oral forms of erythromycin A. Triacetyloleandomycin is a white, crystalline solid that is nearly insoluble in water.

Examples of other 14-membered macrolides are **pikromycin** (molar mass 525.68; $C_{28}H_{47}O_8N$), which was the first macrolide separated in 1950, and **narbomycin** (molar mass 509.68; $C_{28}H_{47}O_7N$). Their aglycones are, respectively, **pikronolide** (molar mass 368.47; $C_{20}H_{32}O_6$) and

narbonolide (molar mass 352.47; $C_{20}H_{32}O_5$). In addition, the megalomines obtained from *Micromonospora* species and the lankamycins or kujimycins isolated from various broths are known. The latter antibiotics lack amino sugars, but they contain the neutral sugars D-chalcose and L-arcanose or their derivatives.

Several semisynthetic derivatives are also prepared from erythromycin A and oleandomycin, some of which are commercial products. One example of them is **azithromycin** (molar mass 557.00; $C_{38}H_{72}O_{12}N_2$), which is more acid-stable than erythromycin A. It is mainly used for the treatment of venereal diseases and infections of the throat, the skin, and soft tissues. **Clarithromycin** (molar mass 747.96; $C_{38}H_{69}O_{13}N$) is primarily used to treat a number of bacterial infections including strep throat, pneumonia, and skin infections. In particular, it is effective against the bacterium *Helicobacter pylori* and thus, it is generally recommended as therapy for gastric ulcer (i.e., to heal it or prevent it from coming back).

The natural low-molar-mass macrolides in the subgroup of the 16-membered macrolides can be divided into leucomycin- and tylosin-related groups, which have some structural differences. They are usually produced by several different organisms and for this reason, alternative trivial names have often been given for the same compound.

The leucomycin compounds are further classified into characteristic subgroups, but they are not systematically described in this section. Only some common examples of therapeutically used 16-membered macrolides are given.

Carbomycin A, deltamycin A_4, magnamycin

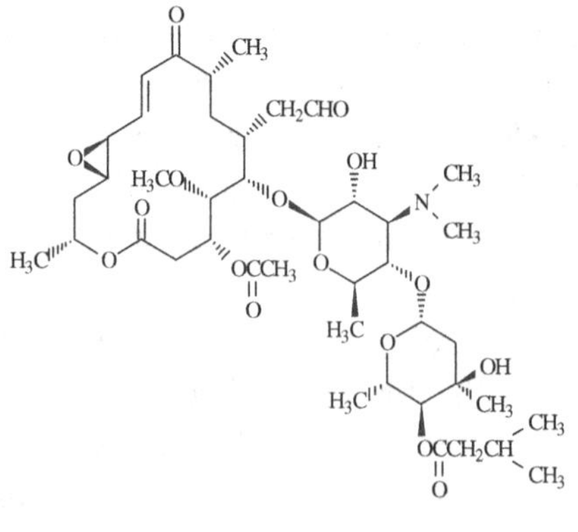

O CH3 CH2CHO O OH CH3 H3CO N O CH3 H3C O OCCH3 O O O H3C O OH H3C CH3 CH3 OCCH2CH O CH3

Molar mass 841.99 ($C_{42}H_{67}O_{16}N$), melting point 214°C.

Carbomycin A was the first 16-membered macrolide discovered and it is produced by *Streptomyces halstedii* and also *Streptomyces thermotolerans*. A disaccharide derivative, where the amino sugar is β-D-mycaminose or 3,6-dideoxy-3-*N*,*N*-dimethylamino-β-D-*gluco*-hexopyranose is linked to its aglycone part, and the neutral sugar component is β-L-mycarose or 2,6-dideoxy-3-*C*-methyl-β-L-*ribo*-hexopyranose. Carbomycin A crystallizes as thick needles.

Carbomycin B

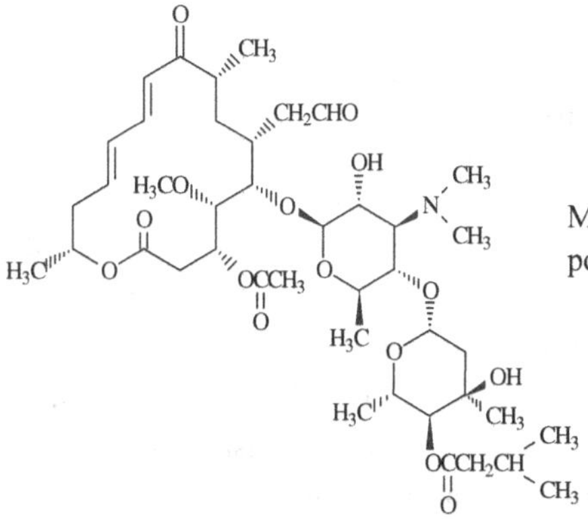

Molar mass 825.99 ($C_{42}H_{67}O_{15}N$), melting point 141–4°C (decomposes).

Carbomycin B is a minor component in the fermentation broths of *Streptomyces halstedii*. It crystallizes as platelets.

Leucomycin A$_3$ or josamycin (molar mass 828.01; $C_{42}H_{69}O_{15}N$) is obtained from culture broths of *Streptomyces narbonensis* var. *josamyceticus*. It can be used for the treatment of throat infections and venereal diseases and it is effective against Gram-positive bacteria. **Rosamicin** or rosaramicin is produced by *Micromonospora rosaria*. Its properties are similar to those of erythromycin A, but with a better activity against Gram-negative bacteria. This antibiotic is mainly used for the treatment of venereal diseases. In addition, there are several spiramycins and platenomycins although their therapeutical effects are lower than those of erythromycin A.

The most significant difference between the leucomycin- and tylosin-related compounds is dealing with the substitution patterns of their aglycones. One difference is a hydroxylmethyl group at the carbon atom

C_{14}. This functional group may be glycosidically substituted by a neutral sugar, such as D-mycinose.

Tylosin

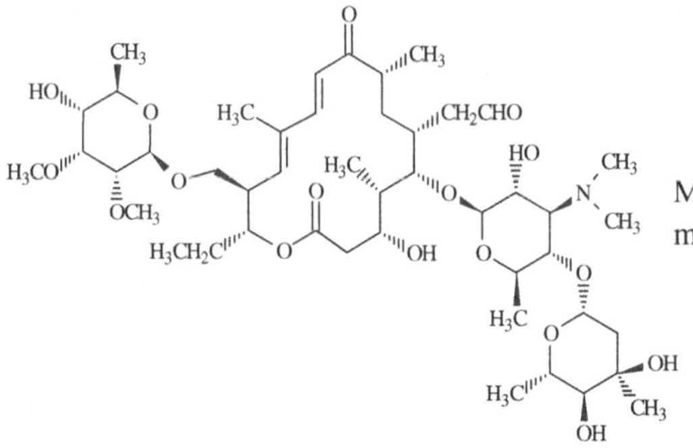

Molar mass 916.11 ($C_{46}H_{77}O_{17}N$), melting point 128–32°C.

Tylosin, the most important macrolide in this group, is produced by *Streptomyces fradie*. It crystallizes from water and dissolves in water (5 g/L, 25°C), ethanol, benzene, diethyl ether, and chloroform. Its specific rotation $[\alpha]_D$ at 25°C in methanol (2 g/100 mL) is −46. To its aglycone part is linked, besides the neutral sugar β-D-mycinose, a disaccharide derivative (as also in carbomycin A), where the amino sugar is β-D-mycaminose or 3,6-dideoxy-3-*N*,*N*-dimethylamino-β-D-*gluco*-hexopyranose and the neutral sugar component is β-L-mycarose or 2,6-dideoxy-3-*C*-methyl-β-L-*ribo*-hexopyranose. If both neutral sugars are liberated from tylosin by acid hydrolysis, 5-*O*-mycaminosyltylonide (OMT) (molar mass 597.75; $C_{31}H_{51}O_{10}N$) is obtained. However, under acidic conditions only the carbohydrate component β-L-mycarose is normally hydrolyzed resulting in the formation of **desmycosine** (molar mass 771.94; $C_{39}H_{65}O_{14}N$). Tylosin is a bactericide and is mainly used as a veterinary antibiotic.

Other corresponding compounds belonging to this substance group are **relomycin** or 20-dihydrotylosin (molar mass 918.13; $C_{46}H_{79}O_{17}N$) and **macrosin** (molar mass 902.09; $C_{45}H_{75}O_{17}N$). Many compounds within this substance group have a tylosin-type aglycone with only one amino sugar that is either D-mycaminose or D-desosamine. In addition, there are a wide range of mycinamicins and algamycins.

Polyene antibiotics or polyene antimycotics are antifungal agents and only a few of them have therapeutical uses. These compounds are broad-spectrum antibiotics with potent activity against pathogenic yeast, molds, and dermatophytes. They are typically obtained from some species of *Streptomyces* soil bacteria. Polyene antibiotics differ from the low-molar-mass macrolide antibiotics in the size of the lactone ring and the presence of the conjugated ene system (i.e., referred to as the "polyenes"). They can be classified into two families according to the size of the lactone ring: (i) the 26-membered ring polyenes (e.g., natamycin) and (ii) the 38-membered ring polyenes (e.g., amphotericin B, nystatin, and candicidin D). A common feature of the currently available polyene antibiotics is also a glycosidically linked deoxyaminohexose, β-D-mycosamine or 3-amino-3,6-dideoxy-β-D-mannopyranose. Moreover, they differ in the number of double bonds (four to seven and all of them in the *trans* configuration) present in the lactone ring; thus, they are often called "tetraenes" (e.g., natamycin), "pentaenes" (e.g., filipin), "hexaenes" (e.g., nystatin), and "heptaenes" (e.g., amphotericin B and candicidin D).

Natamycin, pimaricin

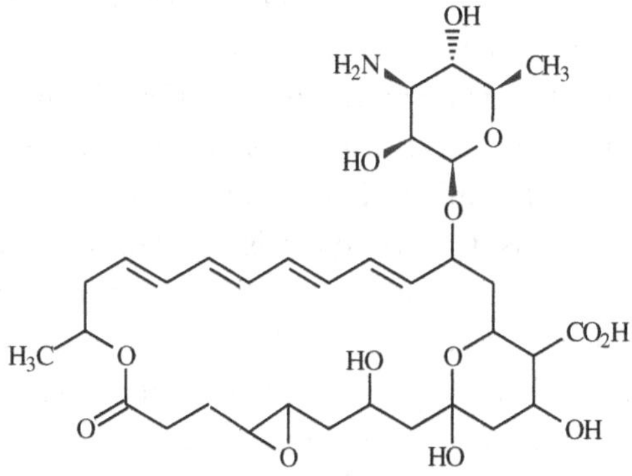

Molar mass 665.74 ($C_{33}H_{47}O_{13}N$).

Natamycin is produced by *Streptomyces natalensis*. It is an amphoteric compound that dissolves in common organic solvents, but only slightly in water. Its structure consists of a 26-membered lactone ring. When taken orally, almost none is absorbed from the gastrointestinal tract, making it inappropriate for systemic infections. For this reason, it is applied topically to treat fungal infections as a cream, in eye drops (to treat fungal keratitis), or in a lozenge (to treat oral infections). It is particularly effective against *Aspergillus* and *Fusarium* corneal infections and *Trichomonas vaginalis* that is an anaerobic, flagellated protozoan parasite and the causative agent of trichomoniasis. Natamycin is also used in the food industry as a natural preservative.

Amphotericin B

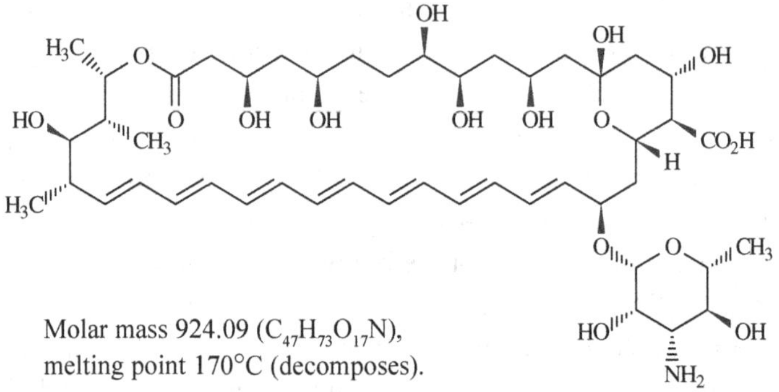

Molar mass 924.09 ($C_{47}H_{73}O_{17}N$), melting point 170°C (decomposes).

Amphotericin B is produced by *Streptomyces nodosus*, and it was first isolated in 1955 from a soil sample collected in the Orinoco River region of Venezuela. The isolated antifungal material was a mixture of two structurally similar compounds, amphotericins A and B, the latter of which was shown to be a more preeminent drug that is still in use for therapeutical purposes. The chemical structure of amphotericin B was not determined until the early 1970s. It crystallizes as deep yellow prisms or needles that are sparingly soluble in polar organic solvents, but not in water. It forms salts with both acids and alkalis and these salts slightly dissolve, respectively, in acidic (pH 2) or alkaline (pH 11) water solutions (about 0.1 mg/mL); its name also originates from the amphoteric nature of this macrolide. Amphotericin B is also a light- and heat-sensitive compound. One of the main uses of this macrolide is to treat a wide range of serious systemic fungal infections, being sometimes the only effective treatment. However, its use may cause common side effects, such as fever, headache, shaking chills, gastrointestinal distress, anorexia, low blood pressure, muscle and joint pain, and malaise, as well as kidney problems. It is not suitable for oral administration, and is injected intravenously, but must not be administered intramuscularly. A variety of topical forms with a typical concentration of 3%, including creams and lotions, are also produced.

Streptomyces noursei produces **nystatin** (originally named "fungicidin"), which was first isolated in 1951. The product consists of three active compounds; nystatins A_1, A_2, and A_3. It occurs as a yellow to light tan powder with a cereal-like odor that is only very slightly soluble in water and sparingly soluble in organic solvents. It is also decomposed by moisture, light, and heat. Nystatin is traditionally used for the treatment of local and gastrointestinal infections involving the skin, mouth, oesophagus, and vagina. For this reason, many mold and yeast infections, mostly caused by notable *Candida*, are sensitive to it. Combinations of this macrolide have been long employed to prevent monilial overgrowth due to destruction of the bacterial microflora of the intestine. Nystatin is not well absorbed from the digestive tract and its large oral doses may produce a number of adverse effects including diarrhea, abdominal pain, malaise, and vomiting. However, its injectable formulations are not generally used. In contrast, due to its minimal absorption through mucous membranes, it can be safely given topically. The commercial products include creams, lotions, and powders.

Nystatin A_1 is the most common form of nystatins. It is a hexaene, although all its six ring double bonds are not conjugated.

Nystatiini A_1

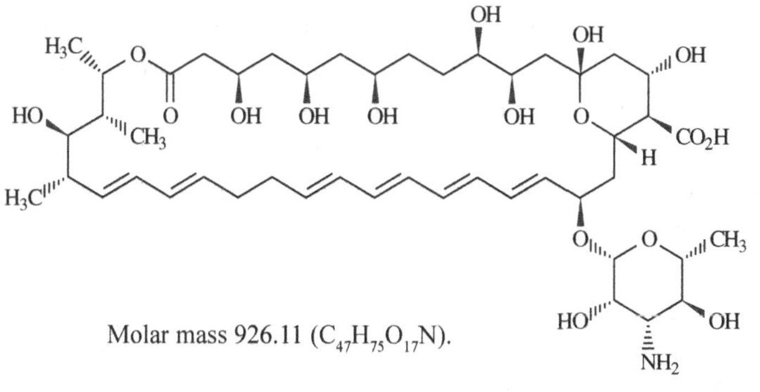

Molar mass 926.11 ($C_{47}H_{75}O_{17}N$).

Candicidin D is produced by *Streptomyces griseus* and was first isolated in 1953. In practice, the product is a mixture of closely related heptaene macrolides (candicidins A, B, C, and D), of which candicidin D is the principal component. Although its antifungal properties had

Candicidin D

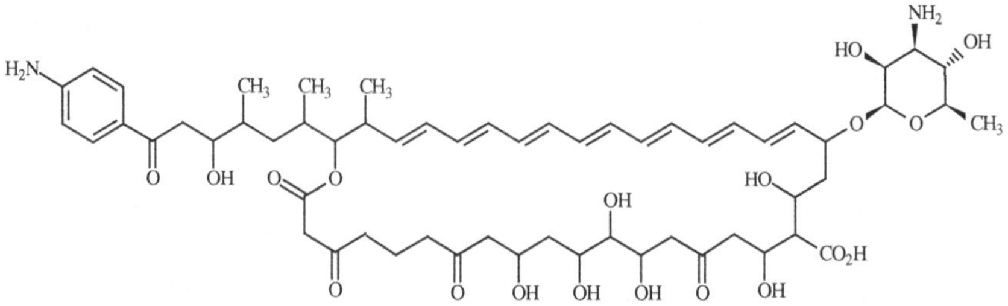

Molar mass 1109.32 ($C_{59}H_{84}O_{18}N_2$).

been recognized for some time already after its discovery, it was not until in the middle 1960s that it became available for medical use. The chemical structure of candicidin D was determined as late as in 1979. This macrolide is recommended for use in the treatment of vaginal candidiasis.

10.4.3. *Lincomycins*

Lincomycins or *lincosaminides* are sulfur-containing antibiotics that can be isolated from *Streptomyces lincolnensis*. This class of compounds resembles the macrolides in antibacterial spectrum and biochemical mechanisms of action. These antibiotics are mainly active against Gram-positive organisms, particularly the cocci. In addition, lincomycins are effective against nonspore-forming anaerobic bacteria, actinomycetes, mycoplasma, and some species of *Plasmodium*.

Lincomycin, methyl 6,8-dideoxy-6-(1-methyl-L-*trans*-4-*n*-propylpyrrolidine-2-carboxamino)-1-thio-D-*erythro*-D-*galacto*-octopyranoside

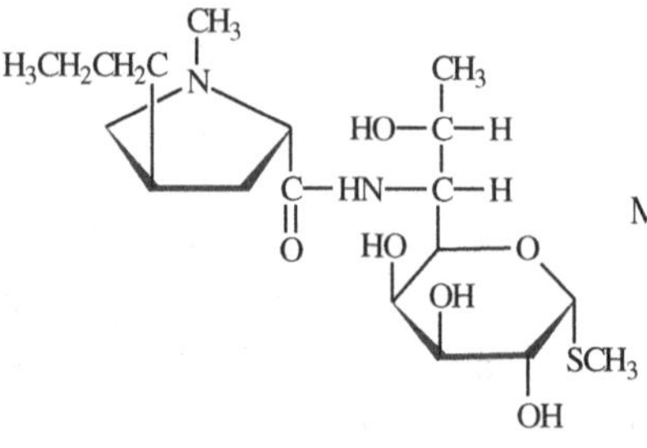

Molar mass 406.54 ($C_{18}H_{34}O_6N_2S$), pK_a 7.5.

The discovery and biological properties of lincomycin were described as the first member of this group of drugs in 1962, and its total synthesis was accomplished in 1970. This antibiotic consists of a carbohydrate moiety, methyl 6-amino-6,8-dideoxy-1-thio-D-*erythro*-α-D-*galacto*-octopyranoside or methyl 1-thio-α-lincosaminide, bound to an amino acid, L-*trans*-4-*n*-propylhygric acid, by an amide linkage. Lincomycin is a white, crystalline powder that is produced by several species of *Streptomyces*, and it is generally the most active and medically useful of the compounds obtained from such fermentation. This drug is primarily employed as a hydrochloride for the treatment of infections caused by Gram-positive organisms, such as streptococci, pneumococci, and staphylococci; in the treatment of diseases of the ear, throat, nose, skin, respiratory, soft tissues, bones, joints, and septicemic infections. Only 15–20% of it is absorbed by oral administration (it is widely distributed in the tissues) and the absorption by the gastrointestinal tract is retarded in the presence of food. Most of it (60–90%) is metabolized in the body. Traditionally, lincomycin has been considered a nontoxic compound, with a low incidence of allergy (skin rashes) and occasional gastrointestinal complaints. However, it may cause severe diarrhea (in about 10% of patients) and the development of pseudomembraneous colitis (similar to *Colitis ulcerosa*). This condition is usually reversible when the drug is withdrawn. Because of these adverse effects, it is rarely used today and reserved for patients allergic to penicillin or cases where bacteria have developed resistance. In general, clindamycin (see below) is superior to lincomycin for the treatment of most infections for which these antibiotics are indicated.

Lincomycin hydrochloride semihydrate. Molar mass 452.00 ($C_{18}H_{36}O_{6.5}N_2ClS$).

Lincomycin hydrochloride is a white, crystalline powder that is chemically stable in the dry state and in aqueous solutions at room temperature, but decomposes gradually in acidic solutions. It is readily soluble in water and ethanol. Its specific rotation $[\alpha]_D$ in water (1 mg/mL) at 25°C is +137.

Clindamycin, 7*S*-7-chloro-7-deoxy-lincomycin

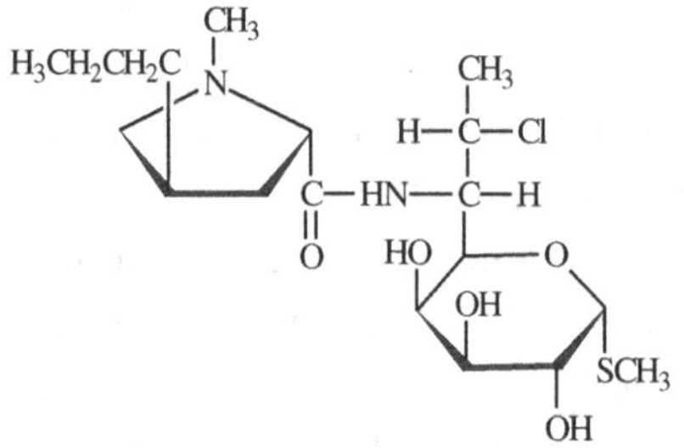

Molar mass 424.99 ($C_{18}H_{33}O_5N_2ClS$).

Clindamycin is a semisynthetic product of lincomycin that was first prepared in 1966 by replacing the 7*R*-hydroxy group of lincomycin by chlorine with thionyl chloride (note the inversion of configuration at C_7). Nowadays, the starting materials are normally lincomycin and triphenylphosphine dichloride or triphenylphosphine in carbon tetrachloride. Compared to lincomycin, this antibiotic has a more significant bacteriostatic effect and better pharmacokinetic properties. Clindamycin is rapidly absorbed by the gastrointestinal tract, even in the presence of food. Also in this case, most of it (60–90%) is metabolized in the body and about 10% of the dose is excreted in the urine. The therapeutical and side effects of clindamycin are the same as those of lincomycin; in pharmaceutical use lincomycin is gradually being replaced by clindamycin that is a more effective antibiotic. Both these antibiotics are useful for patients allergic to penicillin and for treating bacterial infections caused by anaerobic bacteria (particularly *Bacteroides fragilis*). However, a number of reports describe clindamycin(and lincomycin)-associated gastrointestinal toxicity, which ranges in severity from diarrhea to an occasionally serious pseudomembranous colitis. The commercial pharmaceutical products are **clindamycin palmitate hydrochloride** (molar mass 731.86; $C_{34}H_{64}O_8N_2Cl_2S$) and **clindamycin phosphate** (molar mass 504.97; $C_{18}H_{34}O_8N_2ClPS$). The palmitic acid ester is a tasteless prodrug without any bacteriostatic effects. When administered orally, it is readily hydrolyzed by the enzyme esterase to active clindamycin and palmitic acid. When given intramuscularly, **clindamycin hydrochloride** (molar mass 461.45; $C_{18}H_{34}O_5N_2Cl_2S$) causes pain at the injection site, whereas the inactive clindamycin phosphate can be administered in this way. The latter compound is also rapidly hydrolyzed by the enzyme phosphatase to active clindamycin.

Examples of other antibiotics belonging to lincomycins are **celesticetin** (molar mass 528.62; $C_{24}H_{36}O_9N_2S$) and **desalicetin** (molar mass 408.51; $C_{17}H_{32}O_7N_2S$).

10.4.4. *Other antibiotics*

Among the many hundreds of natural or semisynthetic antibiotics that have been tested for activity or have already gained significant clinical attention, some antibiotics with varying chemical structures cannot be classified into the main groups described above. Some of these antibiotics have quite specific activities against a narrow spectrum of microorganisms. On the other hand, some are useful in therapy as substituents for other antibiotics to which patients have developed resistance. In this section, only a few examples of antibiotics that are either derivatives of carbohydrates or have carbohydrate-containing structural units are described.

The *vancomycin-ristocetin compound group* is a subclass of glycopeptides (cf., Chapt. "10.5.1. Hormones"). They are composed of a linear peptide chain containing seven amino acids (the configurations of amino acids are 1*R*, 2*R*, 3*S*, 4*R*, 5*R*, 6*S*, and 7*S*) with a carbohydrate substituent that is cross-linked to generate a characteristic stereochemical configuration. This configuration forms the basis of a complexation with the D-alanine-D-alanine end group of uridine diphosphate-*N*-acetylmuramylpeptide that catalyzes mucopeptide polymerization in the bacterial cell wall structure. Because the mechanism of action is the specific feature of these peptides, they are also called "*dalbaheptides*" ("D-al(anine-D-alanine) b(inding) a(ntibiotics) (having) hept(apept)-ide (structure) → dalbaheptides") to distinguish them within the larger and diverse groups of glycopeptide antibiotics. Almost 50 naturally produced typical dalbaheptides have been isolated since 1953. Many of these antibiotics are groups of strictly related factors called "complexes" and their molar masses vary in the range of 1150–2200 Da. Dalbaheptides are traditionally produced by extraction with acetone or methanol either from the fermentation broth or the mycelial mass. Pure compounds are colorless or whitish, amorphous, and normally water-soluble powders, the commercial products of which are (except teicoplanins, see below) hydrochlorides or sulfates. They are levorotatory, thus representing the (–)-forms.

Vancomycin

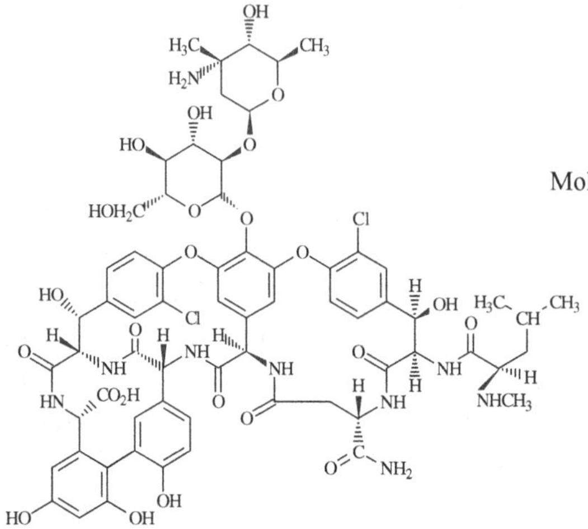

Molar mass 137.35 ($C_{66}H_{73}O_{24}N_9$).

Vancomycin is produced by *Streptomyces orientalis* and it was originally obtained from cultures of an Indonesian soil sample in 1955 and subsequently from samples of Indian soil. It was introduced into clinical practice already in 1958 before its structure determination in 1983. The glycosidically linked carbohydrate component of vancomycin consists of two carbohydrate moieties, β-D-glucose and β-L-vancosamine (3-amino-3-*C*-methyl-2,3,6-trideoxy-β-L-*lyxo*-hexopyranose) that form the glycosidic bond with the carbon atom C_2 of β-D-glucose. This antibiotic decomposes slightly already at 37°C upon heating. To obtain reasonably stability, it should be used as such in the presence of a stabilizator (glycine) at <5°C and the pH of its aqueous solutions should be between 3 and 5. However, it is clinically used as **vancomycin dihydrochloride** (molar mass 1449.27; $C_{66}H_{75}O_{24}N_9Cl_2$). The hydrochloride derivative is a tan to brown powder that is relatively stable in the dry state and is very soluble in water and insoluble in acetone or chloroform. Vancomycin dihydrochloride is active against Gram-positive cocci, particularly streptococci, staphylococci, and pneumococci, but is not active against Gram-negative bacteria. Nowadays, it is mainly used when infections have not responded to treatment with more common antibiotics or when the infection is known to be caused by a resistant organism (e.g., multiresistant *Staphylococcus aureus*). Only small amounts of vancomycin dihydrochloride are rapidly absorbed from the gastrointestinal tract and it should be administered intravenously (never intramuscularly, because of strong irritation) by slow injection or by continuous infusion, for the treatment of systemic infections. In short-term therapy, the toxic side reactions are normally slight and the allergy may emerge as a fever. In contrast, continued use may lead to the impairment of auditory acuity, renal damage, phlebitis (thrombophlebitis), and skin rashes. Only about 10% of it is metabolized and it is mainly excreted in the urine. On the other hand, because vancomycin is not well absorbed by the gastrointestinal tract, it may be administered orally with clindamycin (cf., lincomycins) therapy for the treatment of intestinal infections.

Teicoplanin refers to a complex of related natural products by *Actinoplanes teichomyceticus* and it was first reported in 1975. Teicoplanin is a ristocetin-type lipodalbaheptide with a spectrum of activity and chemical structure similar to vancomycin. This antibiotic is used in the prophylaxis and treatment of serious infections caused by Gram-positive bacteria (e.g., *Staphylococcus aureus* and *Enterococcus faecalis*). Teicoplanin is obtained as an internal salt or as a partial monoalkaline (sodium) salt depending on the pH of the aqueous solution in the final purification step. It is five to ten times more hydrophilic than vancomycin and chemically stable in neutral, aqueous solutions (the optimum pH is 7.4).

Examples of other glycopeptides (and dalbaheptides) are (in parentheses are shown the discovery year and the producing organism) as follows: **actaplanin complex** (1971, *Actinoplanes missouriensis*), **actinoidin A** and **B** (1956, *Proactinomyces actinoides*), **actinoidin A_2** (1987, *Nocardia* sp.), **ardacin complex** (1983, *Kibdelosporangium aridum*), **avoparcin complex** (1966, *Streptomyces candidus*), **decaplanin** (1990, *Actinomyces* sp.), **eremomycin** (1987, *Actinomyces* sp.), **izupeptin A** and **B** (1986, *Nocardia* sp.), **chloropolysporin A** (1983, *Faenia interjecta*), **orienticin** (1987, *Nocardia orientalis*), **parvodicin** (1986, *Actinomadura parvosata*), **ristomycin A** and **B** (1962, *Proactinomyces fructiferi* var. *ristomycini*), **ristocetin A** and **B** (1953, *Nocardia lurida*), and **synmonicin A**, **B**, and **C** (1986, *Synnemomyces mamnoorii*).

Novobiocin, albamycin, cathomycin, streptonivicin, 7-[6,6-dimethyl-3-hydroxy-4-carbamoyloxy-5-metoxytetrahydropyran-2-yloxy]-4-hydroxy-3-[4-hydroxy-3-(3-methyl-2-butenyl)benzamido]-8-methylcoumarin, 4-hydroxy-3-[4-hydroxy-3-(3-methyl-2-butenyl)benzamido]-7-(3-*O*-carbamyl-α-L-noviopyranosyl)-8-methylcoumarin

Molar mass 612.63 ($C_{31}H_{36}O_{11}N_2$).

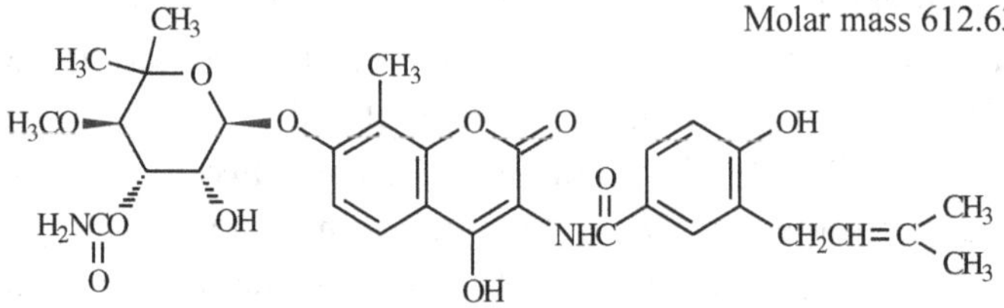

Novobiocin is an *aminocoumarin antibiotic* that was discovered in 1955 and is produced by *Streptomyces spheroides* (or *Streptomyces niveus*). Its chemical structure (an amide) consists of three entities: the derivative of α-L-noviose (3-*O*-carbamyl-α-L-noviose) is glycosidically linked to novobiocinic acid (an aglycone) containing 3-amino-4,7-dihydroxy-8-methylcoumarin and benzoic acid derivative moities. Novobiocin is a pale yellow, somewhat photosensitive solid that crystallizes in two chemically identical forms with different melting points. It dissolves in ethanol and acetone, but is quite insoluble in less polar solvents. Its solubility in water is affected by pH; it deteriorates in basic aqueous solutions and is precipitated from acidic aqueous solutions. Novobiocin reacts as a diacid; the enolic hydroxyl group on the coumarin moiety behaves as a rather strong acid by which the salts are normally formed, whereas the phenolic hydroxyl group on the benzoic acid derivative moiety behaves as a weaker acid. This antibiotic is clinically used as a sodium salt, **sodium novobiocinate** (molar mass 633.61; $C_{31}H_{34}O_{11}N_2Na$), but also as a quite water-insoluble calcium salt, **calcium novobiosinate** (molar mass 1059.32; $C_{62}H_{70}O_{11}N_2Ca$), which is used in aqueous oral suspensions. The sodium salt is stable in dry air but decreases in activity in the presence of moisture. Due to its acidic characteristics, novobiocin can also form salt complexes with basic antibiotics with a combined antibiotic effect although these products are not of practical importance. The action of novobiocin is mainly bacteriostatic, probably inhibiting bacterial protein and nucleic acid synthesis. Its medical use is not wide and only includes the treatment of staphylococcal infections resistant to other antibiotics and sulfas and for patients allergic to these drugs. The usefulness of novobiocin is also limited by the relatively high frequency of adverse reactions, such as urticaria, allergic skin rashes, hepatotoxicity, and blood dyscrasias. About 50% of it is absorbed by oral administration, while most of it decomposes in the body (its biologic half-life is three to eight hours), the remainder being mainly excreted in the bile.

Other aminocoumarin antibiotics include **clorobiocin** or chlorobiocin (molar mass 697.13; $C_{35}H_{37}O_{11}N_2Cl$) and **coumermycin A1** (molar mass 1110.08; $C_{55}H_{59}O_{20}N_5$).

Anthracyclines are a large class of antibiotics. In these compounds, a glycosidic linkage exists between the hydroxyl group at the carbon atom C_7 of anthracyclinone (i.e., an aglycone containing the anthraquinone chromophore) and a monosaccharide derivate with L-configuration, thus resembling the tetracyclines. The anthracyclines differ from each other in the number (two or three) and location (at the carbon atoms C_4, C_6, or C_{11}) of phenolic hydroxyl groups as well as the degree of the two-carbon side chain at C_9 and the possible presence of a carboxylic acid ester at C_{10}. These antibiotics are generally used to treat many types of cancers.

Daunorubicin, daunomycin

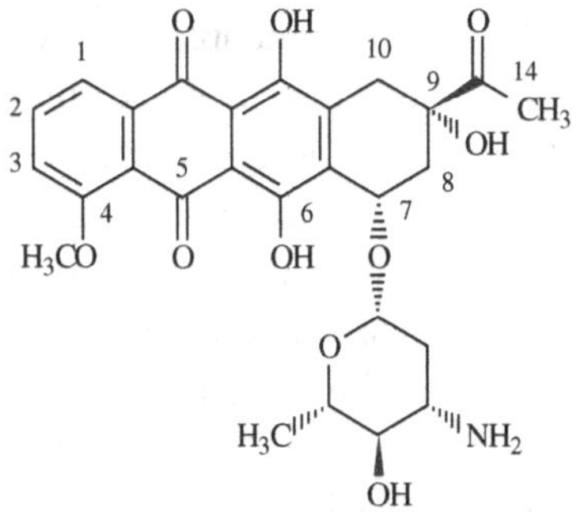

Molar mass 527.52 ($C_{27}H_{29}O_{10}N$).

Daunorubicin was discovered in 1963 and it is produced by *Streptomyces peucetius*. The chemical structure of this antibiotic comprises daunomycinone and β-L-daunosamine (3-amino-2,3,6-trideoxy-β-L-*lyxo*-hexopyranose). It is primarily used (often together with other chemotherapy drugs) to treat specific types of leukemia; acute myeloid leukemia and acute lymphocytic leukemia. This drug is administered only intravenously and the administration depends on the type of tumor and the degree of response. The usual daily dose is 30 mg/m^2 to 60 mg/m^2 of body surface given on days one, two, and three every three to four weeks. Daunorubicin slows or stops the growth of cancer cells in the body; it probably interacts with DNA by intercalation and inhibition of macromolecular biosynthesis (cf., actinomycins). Toxic effects include stomatitis, hair loss, gastrointestinal disturbances, and bone marrow depression, and at higher doses cardiac toxicity may develop. The commercial product **daunorubicin hydrochloride** (molar mass 563.98; $C_{27}H_{30}O_{10}NCl$) is a red, crystalline powder that dissolves in water and ethanol. The pink color of its aqueous solution turns blue at alkaline pH values. It is relatively stable at room temperature, but after reconstitution with sterile water it should be clinically used within six hours.

The discovery of daunorubicin led in 1969 to the isolation of **doxorubicin**, which is produced by *Streptomyces peucetius* var. *caesius*. It is a 14-hydroxy analogue of daunorubicin, thus having a hydroxyl group at the carbon atom C_{14} ($-CH_3 \rightarrow -CH_2OH$). Doxorubicin is commonly used (often in combination chemotherapy with various other agents) in the treatment of a wide range of cancers, including blood cancers (like leukemia and lymphoma), many types of solid tumors (carcinoma), and soft tissue sarcomas. It is one of the most widely used antineoplastic agents. Like all anthracyclines, it seems to work by intercalating DNA. Common adverse effects of this antibiotic include (practically the same as those for daunorubicin) hair loss, myelosuppression, nausea and vomiting, oral mucositis, diarrhea, and skin reactions together with less common, yet serious reactions such as hypersensitivity reactions, radiation recall, heart damage, and liver dysfunction. Doxorubicin is administered intravenously and the usual daily dose is 60 mg/m^2 to 75 mg/m^2 of body surface at 21-day intervals (or 30 mg/m^2 of body surface on each of three successive days, repeated every four week). The commercial product **doxorubicin hydrochloride** (molar mass 579.98; $C_{27}H_{30}O_{11}NCl$) forms orange-red needles which are soluble in water and ethanol. Above pH 9, its orange aqueous solutions turn blue-violet.

Doxorubicin, 14-hydroxydaunorubicin

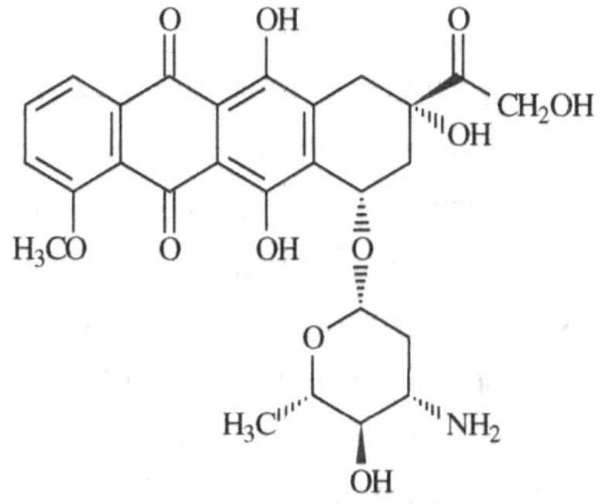

Molar mass 543.52 ($C_{27}H_{29}O_{11}N$).

In addition, several derivatives of daunorubicin and doxorubicin have been produced. Examples of them are **daunomycinol**, **adriamycinol**, **4′-deoxydoxorubicin**, and **4′-epidoxorubicin**. In **aclarubicin** or aclasinomycin A the β-L-daunosamine side group is replaced by a trisaccharide derivative.

Nogalamycin

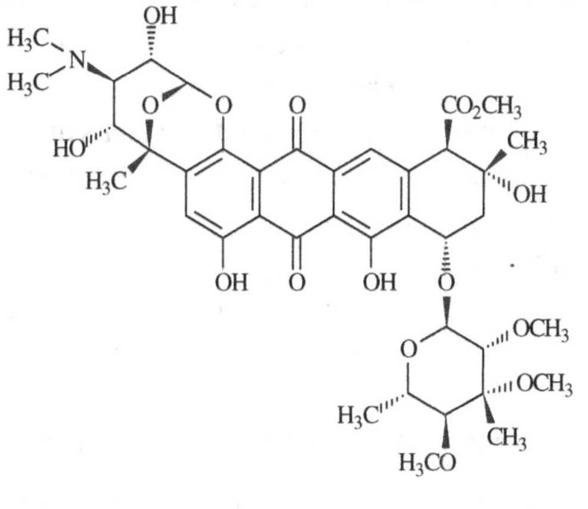

Molar mass 787.82 ($C_{39}H_{49}O_{16}N$).

Nogalamycin is produced by *Streptomyces nogalater* and it has a significant antitumor activity. It differs from other anthracyclines in that the amino sugar is joined to the basic anthracycline structure with a carbomethoxy group at C_{10} by a carbon-carbon bond and a cyclic acetal structure. In addition, there is a nonamino monosaccharide derivative, α-L-nogalose (6-deoxy-3-*C*-methyl-2,3,4-*O*-trimethyl-α-L-mannopyranose), glycosidically linked at the usual C_7 position. This antibiotic is highly cardiotoxic and it is currently not clinically used.

Plicamycin is produced by *Streptomyces plicatus* and *S. argillaceus* and it belongs to the *aureolic acid group of anti-tumor antibiotics*. Its aglycone is a tetrahydroanthracene derivative (the carbon atom C_1 belongs to a ketone group and a methyl group is linked to C_7) with phenolic hydroxyl groups at C_6, C_8, and C_9, an aliphatic hydroxyl group at C_2, and a highly-oxygenated pentanyl side group (-CH(OCH_3)-CO-CH(OH)CH(OH)CH_3) at C_3. In addition, the hydroxyl groups at C_2 and C_6 bind glycosidically oligosaccharide chains with 2,6-dideoxy sugar units (a C-2 chain contains three and a C-6 chain two units), such as D-oliose (2,6-dideoxy-D-*lyxo*-hexose), D-olivose (D-canarose or D-chromose C or 2,6-dideoxy-D-*arabino*-hexose), and D-mycarose (2,6-dideoxy-3-*C*-methyl-D-*ribo*-hexose). Plicamycin is a yellow powder that dissolves in polar organic solvents and aqueous basic solutions (in which it is gradually oxidized by air). It readily forms complexes with magnesium and other divalent metals and the specific rotations of these complexes differ significantly from each other. This antibiotic is an RNA synthesis inhibitor and it has been traditionally used in the treatment of advanced embryonal tumors of the testes, but its use has been largely superseded by newer agents (e.g., bleomycin and cisplatin). The other main uses have included Paget's disease of bone, in which it reduces alkaline phosphatase activity and relieves bone pain, and, rarely, the management of hypercalcemia or hypercalciuria (i.e., high levels of calcium in the blood) resulting from advanced metastatic cancer involving bones. The treatment with plicamycin may produce severe hemorrhaging, but bone marrow, liver, and kidney toxicity can also occur. However, in case of hypercalcemia, a lower total dose results in less toxicity. The usual daily dose is 25–30 μg/kg of body weight for eight to ten days.

Plicamycin, mithramycin, aureolic acid. Molar mass 1085.15 ($C_{52}H_{76}O_{24}$), melting point 180–3°C.

The antibiotics **chromomycin A_3** and **olivomycin A** are also similar to plicamycin. **Bleomycin** was discovered in 1966 and it is produced by *Streptomyces verticillus*. This antibiotic is a mixture of closely related compounds (cytotoxic glycopeptides) that is partly resolved before formulation for clinical use. The main compound in the commercial products (mainly sulfates) is **bleomycin A_2** and a low amount of **bleomycin B_2** is also present. Bleomycin is a white, highly water-soluble powder that, like its analogues, occurs naturally as a blue copper chelate. Copper-free bleomycin products are preferred for chemotheraby because of their decreased toxicity. The complexation of bleomycin with metal ions readily occurs, being also a key factor in its mode of action; inside the cell, bleomycin forms a chelate with Fe(II). This antibiotic is used for the palliative treatment of squamous cell carcinomas of the head and neck, esophagus, skin, and genitourinary tract as well as for the treatment of Hodgkin lymphoma or Hodgkin's disease. It is also used against testicular carcinoma, normally in combination with other agents (e.g., cisplatin and vinblastine). Toxic effects include fever, nausea, ulcers, and hair loss. However, the principal toxicities are for skin and lungs; they do not contain an intracellular aminopeptidase enzyme bleomycin hydrolase that rapidly inactivates bleomycin. In addition, it is inactivated under mild alkaline conditions.

Streptozocin, 2-deoxy-2-(3-methyl-3-nitrosoureido)-D-glucopyranose

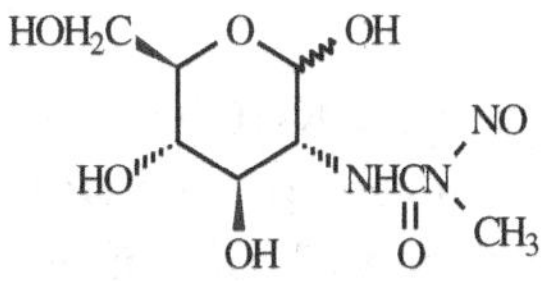

Molar mass 265.22 ($C_8H_{15}O_7N_3$).

Streptozocin was discovered in 1960 and it is obtained from cultures of *Streptomyces achromogenus* subspecies *streptozoticus*. It can also be synthesized from D-glucosamine. Streptozocin readily dissolves in water and it is used only for metastatic cancer of the pancreatic islet cells. Therapy is limited to patients with symptomatic or progresssive disease or whose cancer cannot be removed by surgery. The main reason for this limited use is its inherent renal toxicity (about 70% of patients) and its use may also lead to liver dysfunction. However, the principal toxicities are nausea and vomiting (over 90% of patients). After intravenous injection, the drug is rapidly cleared from the plasma (its biologic half-life is 35 minutes). A typical daily dose is 500 mg/m^2 of body surface for five consecutive days (repeated every six weeks).

Oligosaccharide antibiotics belong to narrow-spectrum antibiotics and they have complex and unique structural features. These compounds are also sometimes referred to as "orthosomycins" because characteristically they possess two acid-sensitive *ortho*-ester linkages, a cleavage of which leads to the complete loss of antibiotic activity. Another feature common to all oligosaccharide antibiotics is the presence of a substituted phenol ester. This phenolic group is essential for the antibiotic activity. The compounds are colorless, crystalline solids with defined melting points and specific rotations. This family of antibiotics can be divided into two main groups: everninomicins and avilamycins. General structures of the everninomicins are as follows:

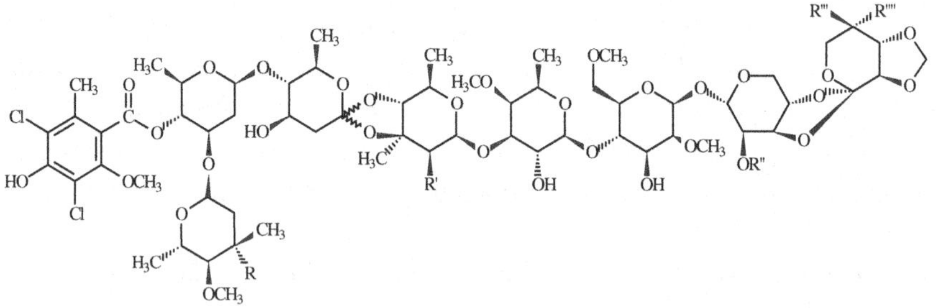

Of these compounds, **everninomicin D** is produced by *Micromonospora carbonacae*. Examples of other members are **everninomicin B** and **everninomicin C** as well as **everninomicin 2** (molar mass 1350.21 ($C_{58}H_{86}O_{31}Cl_2$), melting point 212–6°C), **everninomicin 3** (molar mass 1490.39 ($C_{66}H_{98}O_{33}Cl_2$), melting point 157–8°C), and **everninomicin 7** (molar mass 1508.40 ($C_{66}H_{100}O_{34}Cl_2$), melting point 173–5°C).

Everninomicin D (R = NO_2, R′ = H, R″ = CH_3, R‴ = (*S*)-CH(CH_3)OCH_3, and R⁗ = OH). Molar mass 1537.40 ($C_{66}H_{99}O_{35}NCl_2$), melting point 169–71°C.

Everninomicin B (R = NO_2, R′ = OH, R″ = CH_3, R‴ = (*S*)-CH(CH_3)OCH_3, and R⁗ = OH). Molar mass 1553.40 ($C_{66}H_{99}O_{36}NCl_2$), melting point 184–5°C.

Everninomicin C (R = NO_2, R′ = H, R″ = CH_3, R‴ = OH, and R⁗ = H). Molar mass 1479.32 ($C_{63}H_{93}O_{34}NCl_2$), melting point 181–4°C.

General structures of the avilamycins are as follows:

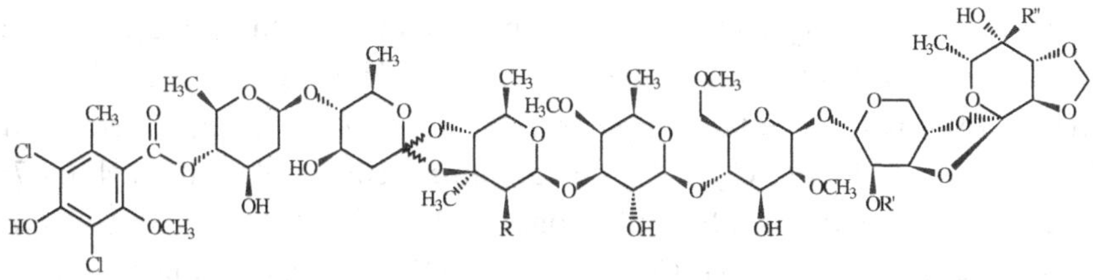

The primary components produced by *Streptomyces viridochromogenes* are **avilamycin A** and **C**. **Curamycin A** is obtained from *S. curacoi* and **flambamycin** from *S. hygroscopicus*. About 20 compounds belonging to this substance group are generally known.

Avilamycin A (R = H, R′ = COCH(CH_3)$_2$, and R″ = $COCH_3$).
Molar mass 1404.26 ($C_{61}H_{88}O_{32}Cl_2$), melting point 181–2°C.

Avilamycin C (R = H, R′ = COCH(CH_3)$_2$, and R″ = CH(OH)CH_3).
Molar mass 1406.27 ($C_{61}H_{90}O_{32}Cl_2$), melting point 188–9°C.

Curamycin A or avilamycin B (R = H, R′ = $COCH_3$, and R″ = $COCH_3$).
Molar mass 1376.20 ($C_{59}H_{84}O_{32}Cl_2$), melting point 192–9°C.

Flambamycin monohydrate (R = OH, R′ = COCH(CH_3)$_2$, and R″ = $COCH_3$).
Molar mass 1438.27 ($C_{61}H_{90}O_{34}Cl_2$), melting point 202–3°C.

The naturally occurring *nucleoside and nucleotide antibiotics* can be divided into the *C*- or *N*-glycosides (cf., Chapt. "10.1. Glycosides"). They contain a variety of purine and pyrimidine rings (aglycones), but also the structures of the carbohydrate portions vary. The nucleoside and nucleotide antibiotics are obtained from predominantly microbial sources, but their antibiotic significance is not very high. Examples of the *C*-nucleosides are **ezomycin A_1** (molar mass 734.70; $C_{26}H_{38}O_{15}N_8S$), **ezomycin A_2** (molar mass 530.45; $C_{19}H_{26}O_{12}N_6$), **formycin** (molar mass 267.24; $C_{10}H_{13}O_4N_5$), **maleimycin** (molar mass 153.14; $C_7H_7O_3N$), **oxazinomycin** (molar mass

245.19; $C_9H_{11}O_7N$), **pyrazomycin** (molar mass 259.22; $C_9H_{13}O_6N_3$), and **showdomycin** (molar mass 229.19; $C_9H_{11}O_6N$). The subgroup of the *N*-nucleosides is very wide and only some examples of the compounds are given: **adenomycin** (molar mass 757.69; $C_{25}H_{39}O_{18}N_7S$), **2′-amino-2′-deoxy-adenosine** (molar mass 266.26; $C_{10}H_{14}O_3N_6$), **amipurimycin** (molar mass 495.49; $C_{20}H_{29}O_8N_7$), **arginomycin** (molar mass 436.47; $C_{18}H_{28}O_5N_8$), **blasticidin S** (molar mass 422.44; $C_{17}H_{26}O_5N_8$), **fosfadecin** (molar mass 467.27; $C_{13}H_{19}O_{10}N_5P_2$), **herbidicin E** (molar mass 536.52; $C_{23}H_{30}O_{10}N_5$), **mureidomycin A** (molar mass 840.91; $C_{38}H_{48}O_{12}N_8S$), **neplanocin A** (molar mass 263.26; $C_{11}H_{13}O_3N_5$), **oxanosine** (molar mass 284.23; $C_{10}H_{12}O_6N_4$), **puromycin** (molar mass 471.52; $C_{22}H_{29}O_5N_7$), **tubercidin** (molar mass 266.26; $C_{11}H_{14}O_4N_4$), and **tunicamycin I(A_0)** (molar mass 802.87; $C_{36}H_{58}O_{16}N_4$).

Streptothricin is a mixture of antibiotics (originally thought to be a single substance) produced by an actinomycete (*Actinomyces lavendulae*). They are active against both Gram-positive and Gram-negative bacteria and some fungi. The *N*-glycoside structures differ only in the number of repeating residues in the peptide side chain; streptothricins F, E, D, C, B, A, and X are known and have, respectively, one to seven β-lysine residues in their structures. Thus, **streptothricin F** (molar mass 492.45; $C_{19}H_{24}O_8N_8$) contains one β-lysine unit. As a hydrochloride, it is a white powder that dissolves in water and dilute mineral acids, but not in diethyl ether or chloroform.

10.5. Other Substance Groups

10.5.1. *Hormones*

Hormones (in Greek, "hormaein" means "to rouse" or "set in motion") are bioactive substances that are produced in the body and are transported by the circulatory system, the tissue fluid, and other body fluids to distant target organs to regulate physiology and behavior. A low secretion of hormones may result in difficult diseases; for example, diabetes mellitus (due to either the pancreas not producing enough insulin or the cells of the body not responding to the insulin produced) or cretinism (due to congenital deficiency of thyroid hormones). On the other hand, their hypersecretion may also be harmful; examples are immune diseases, such as Graves' disease or Basedow's disease (i.e., the thyroid gland produces too much thyroid hormone; this disease is generally called "hyperthyroidism").

Hormones form a relatively heterogeneous class of compounds. They can be divided according to their chemical structures into (i) phenols (part of thyroid gland hormones and adrenal gland hormones), (ii) oligopeptides, polypeptides, proteins, and glycoproteins (anterior and posterior pituitary gland hormones, hypothalamus and pancreas hormones, and part of thyroid gland hormones), and (iii) steroids (corticosteroids, estrogens, progestins, androgens, and anabolic steroids). Oligopeptides contain less than 10, polypeptides 10 to 100, and proteins over 100 amino acid residues. Thus, polymers containing covalently bound carbohydrates and amino acid residues are generally termed "glycopeptides" (or "peptidoglycans") or "glycoproteins". However, it is not possible to give precise distinctions between these terms.

The pituitary gland in adult humans is composed of two distinct lobes: anterior and posterior. The anterior pituitary synthesizes and secretes several hormones that play an important role, for example, in the regulation of physiological processes including stress, growth, carbohydrate metabolism, reproduction, and lactation — in contrast, the posterior pituitary stores and secretes, but does not synthesize hormones. These hormones have been isolated in their pure forms and their structures have also been determined. In addition, most of these hormones have been synthesized. They can be divided according to their chemical structures into the three main subgroups (only examples of hormones are shown): (i) simple peptides (corticotropins, melanotropins, and lipotropins), (ii) simple proteins (prolactin and somatotropin or growth hormone), and (iii) glycoproteins (thyrotropins and gonadotropins, such as follicle-stimulating hormone and lutropin or luteinizing hormone). In these structures, the amino acid residues are linked to each other via a peptide bond (-HN-CO-), and the structural units are indicated by the known symbols: the common amino acid moieties in hormones are Ala alanine, Asn asparagine, Asp aspartate, Arg arginine, Cys cysteine, Gln glutamine, Glu glutamate, Gly glycine, His histidine, Ile isoleucine, Leu leucine, Lys lysine, Met methionine, Phe phenylalanine, Pro proline, Ser serine, Thr threonine, Trp tryptophan, Tyr tyrosine, and Val valine.

Carbohydrates linked to lipid molecules are collectively termed "glycolipids", and they are common components of biological membranes.

Glycoproteins and glycolipids are normally together called "glycoconjugates", which are formed in glycosylation and have a wide variety of cellular functions. However, glycoconjugates can broadly be considered to comprise many subclasses, the major groups being, besides glycoproteins and glycolipids, glycopeptides, peptidoglycans, lipopolysaccharides, and glycosides; i.e., "all types of compounds" consisting of carbohydrates covalently linked with other types of chemical constituents. For example, a peptidoglycan contains repeating units of D-glucosamine and either muramic acid (2-amino-3-*O*-[*R*-1-carboxy-ethyl]-2-deoxy-D-glucose) or L-talosaminuronic acid (2-amino-2-deoxy-L-taluronic acid), which both are usually *N*-acetylated or *N*-glycosylated. In addition, the carboxylic acid group of muramic acid is normally substituted by a peptide with L- and D-amino acid structures, whereas the carboxylic acid group of L-talosaminuronic acid binds only L-amino acid structures. In general, in a glycopeptide, a carbohydrate portion is linked to an oligopeptide containing L- and/or D-amino acid residues (see above). In contrast, in a glycolamino acid, a carbohydrate portion is joined to one amino acid residue via any possible covalent bond, whereas in a glycosyl amino acid, these components are linked together by a glycosidic bond (i.e., they are *O*-, *N*-, or *S*-glycosides).

10.5.2. *Alkaloids*

Alkaloids (from "alkali" + "-oid", in Arabic, "ali-quali" means "ashes of plants" and the Greek suffix "eidos" "like") are a large group of naturally occurring secondary metabolites that contain mostly basic nitrogen atoms; the nitrogen originates either from specific amino acid structures (ornithine, lysine, phenylalanine, tyrosine, tryptophan, histidine, and anthranilic acid) or from other nitrogen-containing simple compounds, such as ammonia and alkyl amines. Furthermore, they can be formed in other biosynthetic reactions. Carl Friedrich Wilhelm Meissner (1792–1853) first proposed the term "alkaloid" (in German, "Alkaloide" means "alkali-like") in 1819. Mankind has used alkaloids, for example, in medicines, poisons, and potions for at least 3000 years.

Alkaloids primarily occur in flowering plants (also in microorganisms, marine organisms, and animals), especially in plant roots, trunk barks, leaves, and seeds. However, the pronounced concentration of alkaloids in a specific plant part does not necessarily indicate the initial formation site of them. It is also usual that one plant contains several different alkaloids or related compounds. In spite of the enormous number of structural types of alkaloids, their distribution in nature is not as wide as that of terpenoids and phenols. In addition, their significance has so far not been fully clarified and with this respect, they can be compared to many extractive-like substance groups.

The classification of natural nitrogen-containing compounds is generally rather difficult; they are not necessarily bases, and most of them contain acidic functional groups. On the other hand, alkaloids have a wide range of pharmacological activities (although some alkaloids do not have this property) and many have found use in traditional or modern medicines or as starting materials for drug discovery. For this reason, the straightforward classification of alkaloids is difficult. They are generally considered low-molar-mass compounds (100–900 Da) with varying chemical structures that characteristically differ from those of simple amines, amides, short-chain peptides, and other primary metabolites.

Alkaloids are often divided into three major groups: (i) "true alkaloids" (e.g., ornitine-derived compounds, such as tropane alkaloids and tropane carboxylic acid alkaloids, lysine-derived compounds, such as piperidine alkaloids and quinolizidine alkaloids, phenylalanine- and tyrosine-derived compounds, such as morphine alkaloids and isoquinoline alkaloids, tryptophan-derived compounds, such as ergot alkaloids, *Rauwolfia* alkaloids, and China alkaloids, histidine-derived compounds, and nicotinic acid-derived compounds), (ii) "protoalkaloids" (e.g., mescaline, adrenaline, and ephedrine), and (iii) "pseudoalkaloids" (e.g., caffeine, theobromine, and theophylline). Likewise, there are polyamide alkaloids as well as peptide and cyclopeptide alkaloids. Normally, alkaloids do not contain carbohydrates, and in this section, only a few examples of these rare alkaloids are presented.

Tomatine, lycopersicin. Molar mass 1034.20 ($C_{50}H_{83}O_{21}N$), melting point 263–8°C.

Tomatine is a *steroid alkaloid* (steroidal glycoalkaloid or glycoalkaloid), which can be found in the stems and leaves of tomato plant (*Lycopersicon esculentum* or *Solanum lycopercium*). It is a steroid glycoside derivative in which the aglycone is **tomatidine**, also present in the roots of tomato plant. The carbohydrate portion is a tetrasaccharide with two D-glucose, one D-xylose, and one D-galactose unit. Tomatine is a white, crystalline solid that dissolves in ethanol and dioxane, but practically not in water or diethyl ether. It is stable against alkalis, but is hydrolyzed by acids; this case, tomatidine can be crystallized from the corresponding hydrolysate. Its specific rotation $[\alpha]_D$ in pyridine at 20°C is –30. Tomatine as well as tomatidine have been shown to have some health benefits and tomatine has antimicrobial properties against certain classes of microbes. In addition, tomatine is used as a reagent in analytical chemistry for precipitating cholesterol.

Solanine. Molar mass 868.07 ($C_{45}H_{73}O_{15}N$), melting point 265°C (decomposes).

Like tomatine, solanine is a steroid alkaloid, which can also be found in the species of the nightshade family (*Solanaceae*), such as the common potato (*Solanum tuberosum*) and other corresponding species including, for example, European black nightshade (*S. nigrum*), bittersweet nightshade (*S. dulcamara*), and tomato (*S. lycopercium* or *Lycopersicon esculentum*). It crystallizes from ethanol as thin needles that turn dark and sinter at about 190°C. Solanine dissolves in hot ethanol, but practically not in water or diethyl ether, chloroform, or benzene. Its specific rotation $[\alpha]_D$ in pyridine at 20°C is –60. Hydrolysis of solanine results in **solanidine** (molar mass 397.64; $C_{27}H_{43}ON$) with a melting point of 218–219°C that sublimes (partly decomposes) upon heating.

Chaconine or α-chaconine (molar mass 852.07; $C_{45}H_{73}O_{14}N$) occurs also in plants of the *Solanaceae* family. This steroidal glycoalkaloid is a natural toxicant (together with solanine) produced in green potatoes and gives the potato a bitter taste (if its concentration is above 14 mg/100 g). It has the same aglycone (soladine), with solanine, but the structures of their carbohydrate side chains differ. There have been many reported cases of human poisoning due to the ingestion of greened or otherwise damaged potatoes. For example, the symptoms of low-grade solanine (and chaconine) poisoning are typical acute gastrointestinal upset with diarrhea, vomiting, and pain. However, in more severe cases, neurological symptoms, such as drowsiness and apathy, confusion, weakness, and vision disturbances, followed by unconsciousness and even death may appear. In general, the production of solanum-type glycoalkaloids is favored by the same conditions that promote the development of chlorophyll. Therefore, their highest concentrations can be found in potato sprouts and green potato skins, as well as in green tomatoes. It should be noted that these alkaloids are not destroyed by cooking and drying at high temperatures. An average content of solanine in the common potato is 8 mg/100 g, but the interior of the potato normally contains it clearly less, and thus, it does not cause any toxic effects (the toxic dose is 20 mg to 25 mg).

11. Characteristic Reactions of Carbohydrates

A number of carbohydrate derivatives and their preparation have already been discussed in Chapts. “9. Natural Carbohydrates and Their Derivatives” — in particular, in Chapt. “9.4. Polysaccharides and Their Derivatives” — and “10. Carbohydrate Residue-containing Substance Groups”. In addition, many basic reactions of carbohydrates, especially in the context of industrial utilization of biomass resources, are presented in Chapt. “12. Utilization of Biomass”.

The objective of this chapter is to present a general view of the dominant reactions of carbohydrates with the help of a few typical examples. While several of the basic reactions of the monosaccharides possess their own historical significance, the behavior of polysaccharides in either alkaline or acid environment is in many cases of major industrial importance.

11.1. Monosaccharides

The reactions of monosaccharides can generally be divided into three main categories: (i) *reactions at the anomeric center*, (ii) *reactions at the non-anomeric carbon atoms*, and (iii) *reactions of the hydroxyl groups*. Significant reactions of the first group include reactions that shorten or lengthen a monosaccharide carbon chain, while those of the other two groups mostly relate to the production of various, often rather specific derivatives. Especially, the protection reactions of the hydroxyl groups of monosaccharides are quite important in the gas chromatographic analysis of monosaccharides in preventing the formation of hydrogen bonds between the molecules and so ensuring the volatility of the monosaccharide derivatives formed that this type of analysis requires.

In the early days of structural studies of monosaccharides, a few specific reactions were used to either increase or decrease the number of carbon atoms in their chain (Fig. 11.1). The most traditional one is the Kiliani cyanohydrin synthesis, which is also known as the "Fischer-Kiliani synthesis". In this reaction, a cyanide ion reacts in a nucleophilic attack with a carbonyl carbon of an aldehyde or a ketone forming cyanohydrin as an addition product. This method was traditionally used to produce from aldoses, for example, aldonic acids, by hydrolyzing the formed intermediate product containing a cyano group in acid or alkaline water leading, for example, from D-arabinose to D-gluconic acid and D-mannonic acid. The corresponding aldoses with a higher number of carbons than that of the initial aldose were then traditionally obtained by reducing the obtained aldonic acids with mercury amalgam (i.e., an alloy of mercury). Starting with a 2-ketose, the final product is an aldonic acid with a hydroxymethyl group attached on its carbon atom C_2.

The Grignard reaction (François Auguste Victor Grignard 1887–1935, the 1912 Nobel Prize in Chemistry) offers an interesting way to lengthen the carbon chain of a monosaccharide (Fig. 11.1). A reagent specific to this method, an $R^{\ominus}$ ion in RMgBr, is allowed to react nucleophilically with a carbonyl carbon in the absence of water (even the hydroxyl groups of the monosaccharide have to be protected). For example, a reaction between ethynylmagnesium bromide and 2,3:4,5-di-*O*-isopropylidene-L-arabinose leads to two heptitol derivatives, from which a partial catalytical hydrogenation (acetylene group → alkene group) followed by a well-known ozonization reaction finally results in L-glucose and L-mannose.

The Ruff degradation (Otto Ruff 1871–1939) is performed by treating the calcium salt of an aldonic acid, obtained by oxidizing an aldose with bromine water, with hydrogen peroxide in the presence of iron(III) ions (Fig. 11.1). In this process, for example, D-arabinose forms D-erythrose. The rather complicated process is based on the quite specific reaction of the hydroxyl radical ($HO^{\bullet}$) with a "complex" formed by the aldonic acid salt, iron(III) ions, and hydrogen peroxide. Used on a large scale, this

(a) $H(C{=}O)R \xrightarrow[\text{(pH 9)}]{NaCN} C{\equiv}N\text{–}CHOH\text{–}R \xrightarrow{H_2O/H^{\oplus}} CO_2H\text{–}CHOH\text{–}R \xrightarrow{[H]} H(C{=}O)\text{–}CHOH\text{–}R$

(b) $H(C{=}O)R \xrightarrow{HC{\equiv}CMgBr/Et_2O} C{\equiv}CH\text{–}CHOH\text{–}R \xrightarrow{\text{i) } H_2/Pd \text{ ii) } O_3} H(C{=}O)\text{–}CHOH\text{–}R$

(c) $H(C{=}O)\text{–}CHOH\text{–}R \xrightarrow{Br_2/H_2O} CO_2H\text{–}CHOH\text{–}R \xrightarrow{\text{i) } Ca^{2\oplus} \text{ ii) } Fe^{3\oplus}/H_2O_2} H(C{=}O)R + CO_3^{2\ominus}$

(d) $H(C{=}O)\text{–}CHOH\text{–}R \xrightarrow{NH_2OH} H(C{=}NOH)\text{–}CHOH\text{–}R \xrightarrow[-H_2O]{Ac_2O} C{\equiv}N\text{–}CHOH\text{–}R \xrightarrow[-HCN]{} H(C{=}O)R$

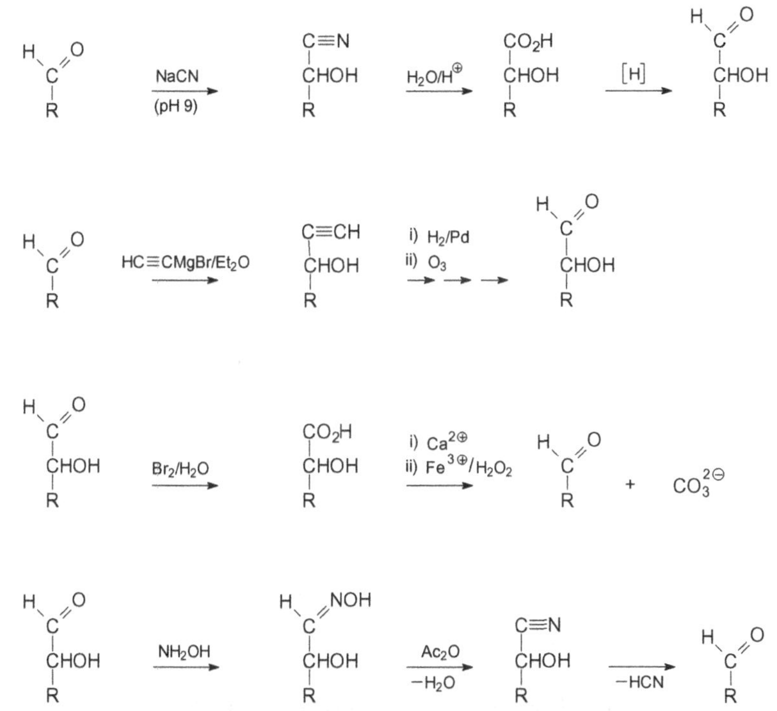

Fig. 11.1. Examples of reactions suitable for shortening or lengthening the carbon chain of monosaccharides (see the text). The Fischer-Kiliani synthesis (a), the Grignard reaction (b), the Ruff degradation (c), and the Wohl degradation (d).

process is nowadays known under the general name of "Fenton's reaction" (Henry John Horstman Fenton 1854–1929), where the reagent is $Fe^{3\oplus}/H_2O_2$ as mentioned above.

Aldoses react typically with hydroxylamine, but the *oximes* formed are not well suited for the characterization of carbohydrates because of their high solubility in water. In addition, these derivatives show mutarotation typical of *glycosylamines* (*N*-glycosides) caused by an equilibrium between an open-chain form and cyclic structures (cf., Chapt. "6.1. Mutarotation"). This reaction is, however, historically important in the shortening of the carbon chain of monosaccharides following the Wohl degradation (Alfred Wohl) (Fig. 11.1). In the formation of glycosylamines, the aldoses condense with either ammonia or primary or secondary amines with the liberation of water.

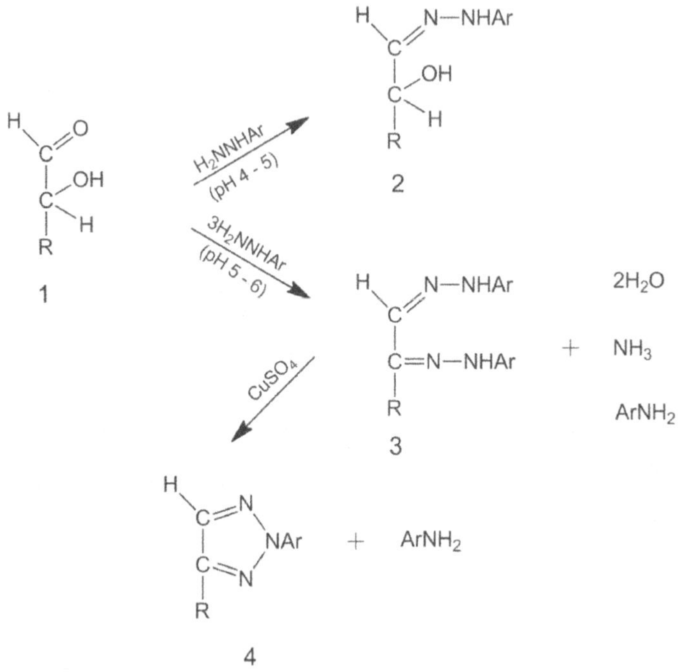

Fig. 11.2. Reaction with aldoses (1) and arylhydrazine leads to arylhydrazones (2). The reaction with an excess of arylhydrazine leads to arylosazones (3), which are readily oxidized under mild conditions to osotriazoles (4).

Aldoses and ketoses form in reaction with water *arylhydrazones*, if the molar amounts of the initial materials are equal (Fig. 11.2). These derivatives were to some extent used to identify monosaccharides, as they are crystalline (e.g., *p*-nitro- or 2,5-dichlorophenylhydrazones) and less soluble in water than oximes. Fischer noticed later in 1884 that free sugars formed very crystalline *phenylosazones*, if phenylhydrazine reagent was added in excess (Fig. 11.2). These derivatives had already gained importance in the clarification of the configurational relationships between aldoses and ketoses. Upon oxidizing phenylosazones with copper(II) sulfate, *osotriazoles* were obtained, which were still more useful for identification because of their typically sharp melting points and their lack of various equilibrium forms in solution.

When aldoses that have a reducing end group are treated with alcohols in the presence of an acid catalyst, *glycosides* (hemiacetal → acetal) are formed; they were examined in detail in Chapt. "10.1. Glycosides". This reaction is also known under the general name the "Fischer glycosidation". The glycosidic bonds can, on the other hand, be dissolved by both acids and bases and also by specific enzymes (cf., Chapts. "11.2. Polysaccharides" and "12.2. Production of Chemicals"). In addition, monosaccharides can form intramolecular glycosides; for example, 1,2-, 1,3-, and 1,4-anhydroaldoses (cf., Chapt. "9.2.1. Anhydrosugars").

In conjunction with these nitrogen nucleophilic reactions, carbon nucleophilic reactions also take place at anomeric carbon atoms; typical examples are the Fischer-Kiliani synthesis and the Grignard reaction discussed above. Similar reactions are obtained, among others, with nitroalkanes, diazomethane, Wittig reagent, and aldol-type condensation reactions (e.g., the Knoevenagel condensation). It should be emphasized that many reagents are able to cause oxidation reactions (aldoses → aldonic acids, see the Ruff degradation) and also reductions (aldoses and ketoses → alditols) at anomeric centers.

Anomeric carbon atoms can also participate in specific sulfur-based nucleophilic reactions leading to *thioacetals* and *thioglycosides*. Here, the hemiacetal structures of the aldoses react with alkanethiols in the presence of acid catalysts, the main products being open-chain dialkyldithioacetals (intermediate products are hemithioacetals and thioglycosides). These reactions thus clearly differ from the Fischer glycosidation mentioned. For instance, 1-thioglucosides (e.g., sinigrin) are found in nature and they can also be synthesized by several routes.

Figure 11.3 shows typical etherification reactions taking place at the hydroxyl groups of carbohydrates and acetylation as an example of esterification reactions. Nowadays, for example, silylation is commonly used in analytical chemistry because it is easy to carry out; it has gradually replaced traditionally used acetylation. In nature, only partially methylated monosaccharides (e.g., 3-*O*-methyl-D-galactose, 6-deoxy-2-*O*-methyl-D-allose (javose), and 6-deoxy-2,3-di-*O*-methyl-D-galactose) are often found, but in analytical work, the objective generally is to methylate all free hydroxyl groups (also including the anomeric one); we then speak of "per(methyl)ated derivatives". Analogous naming practices are used for other derivatives as well (e.g., per(trimethylsilyl)ated monosaccharides).

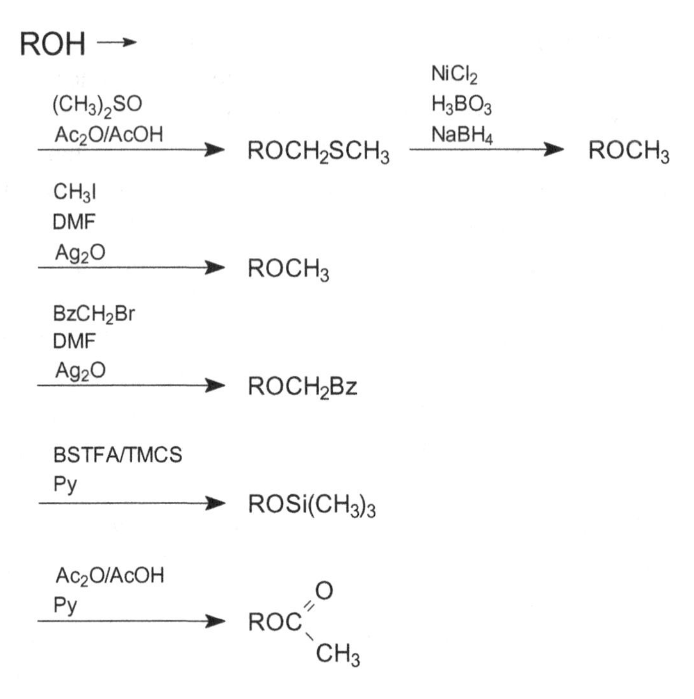

Fig. 11.3. Typical reactions of the hydroxyl groups of carbohydrates. AcOH is acetic acid, Ac_2O acetic anhydride, BSTFA *N,O*-bis(trimethylsilyl)trifluoroacetamide, Bz benzene, DMF *N,N*-dimethylformamide, Py pyridine, and TMCS trimethylchlorosilane.

Classical methylation has been successfully used in elucidating the structures of polysaccharides. The hydroxyl groups of the compound studied are first converted into methyl ethers, and the glycosidic bonds are then cleaved with acid hydrolysis; the formed, partially methylated carbohydrate units are analyzed by gas chromatography (with identification by mass spectrometry). This procedure will yield information about the structural features and bonds of the original polysaccharide. Several methylation procedures have been developed for this purpose. Among them, in the conventional Purdie method (also known as the "Irvine-Purdie methylation" after Thomas Purdie 1843–1916 and James Irvine 1877–1952), the reaction is carried out with methyl iodide in the presence of silver oxide. Richard Kuhn (1900–67) modified the method by using *N,N*-dimethylformamide as solvent, since it was found to be more efficient in this procedure than methyl iodide. Also, silver oxide can be replaced with barium or strontium oxide. In the conventional Haworth method, the methylation is carried out with dimethyl sulfate in an aqueous solution of sodium hydroxide. However, dimethyl sulfate is quite poisonous on the skin, through which it will absorb.

11.2. Polysaccharides

Under alkaline conditions various peeling reactions of polysaccharides with a reducing end group (i.e., a hemiacetal structure) are important particularly in the context of kraft pulping or sulfate pulping and, for example, in the manufacture of dissolving pulp (cf., Chapts. "9.4. Polysaccharides and Their Derivatives" and "12.4.1. Kraft pulping"). In the kraft process, the cellulose-containing fibers of the raw material (wood or non-wood materials) are liberated by dissolving the lignin (i.e., delignification) that binds the fibers together under strongly alkaline conditions in an aqueous solution of sodium hydroxide and sodium sulfide ("white liquor"). The substantial carbohydrate losses (especially losses in hemicelluloses) arise primarily from the *peeling reaction*, called the "primary peeling reaction" that takes place already at low temperatures directly with the polysaccharides in the feedstock materials (Fig. 11.4). The *stopping reaction* (a process 50 to 65 times slower) competes with this peeling reaction, which prevents further cleavage of the chains arising from the primary peeling process. In a conventional cooking process, the temperature is raised to its maximum value, 160–170°C, in about one and half hours, so that toward the end of this heating up period, the random *alkaline hydrolysis* of glycosidic bonds becomes energetically possible. As this alkali-catalyzed hydrolysis happens new reducing end groups are also simultaneously formed to the partially degraded carbohydrate chains or those already stabilized by the stopping reaction, thus making possible the "secondary peeling reaction". In addition, already in the beginning phase of the cooking the acetyl groups of hemicelluloses are peeled almost completely under the influence of the alkali (deacetylation), forming sodium acetate that dissolves into the cooking liquor.

The fundamental requirement for the peeling reaction is the presence of a reducing end group that acts in an alkaline environment as shown in Fig. 11.5. This phenomenon was noticed in 1885, and it is called after its

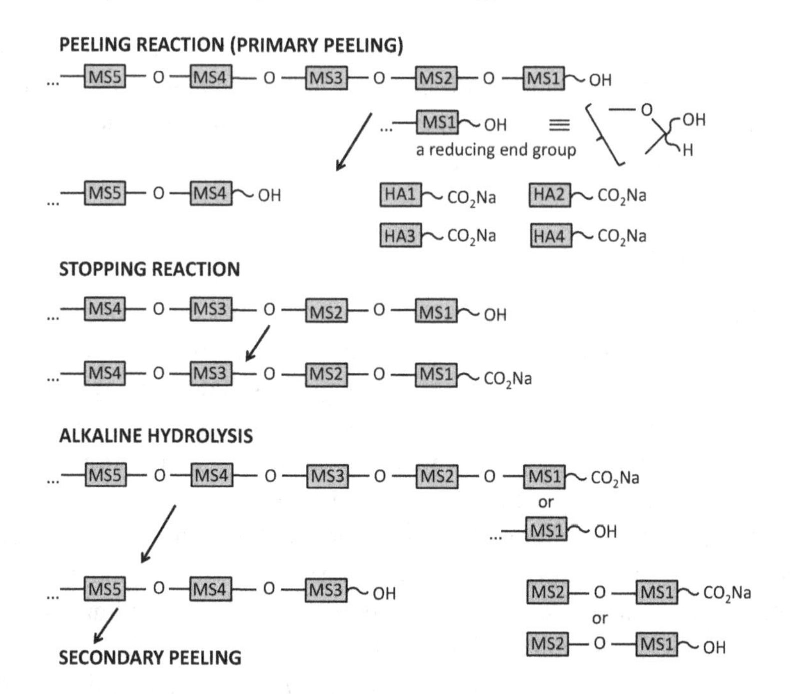

Fig. 11.4. Main reactions of polysaccharides in alkaline pulping. MS is a monosaccharide unit and HA a hydroxy acid component.

Fig. 11.5. Lobry de Bruyn-van Ekenstein transformation of aldoses. If the aldose is D-glucose (1), a 1,2-enediol (2), D-mannose (3), D-fructose (4), a 2,3-enediol (5), and D-allulose (6) are formed in the first stage of this transformation.

observers (Cornelis Adriaan Lobry van Troostenburg de Bruyn 1857–1904 and Willem Alberda van Ekenstein 1858–1937) and referred to as the "Lobry de Bruyn-van Ekenstein transformation" or the "Lobry de Bruyn-Alberda van Ekenstein transformation". The aldose end group is isomerized through a 1,2-enediol intermediate into a 2-ketose end group — two tautomeric forms of this 1,2-enediol are also present, the initial aldose end group and its 2-epimer, besides a 2-ketose end group, which transforms further into 3-ketose end group and so on. The total reactions thus include isomerism, tautomerism, and epimerism. If the initial carbohydrate is D-glucose, this "first phase" forms a mixture of D-glucose, D-fructose, and D-mannose.

After the isomerization of the aldose end group of the carbohydrate chain happens β-*alkoxy elimination* (Fig. 11.6), where the "end structural group"

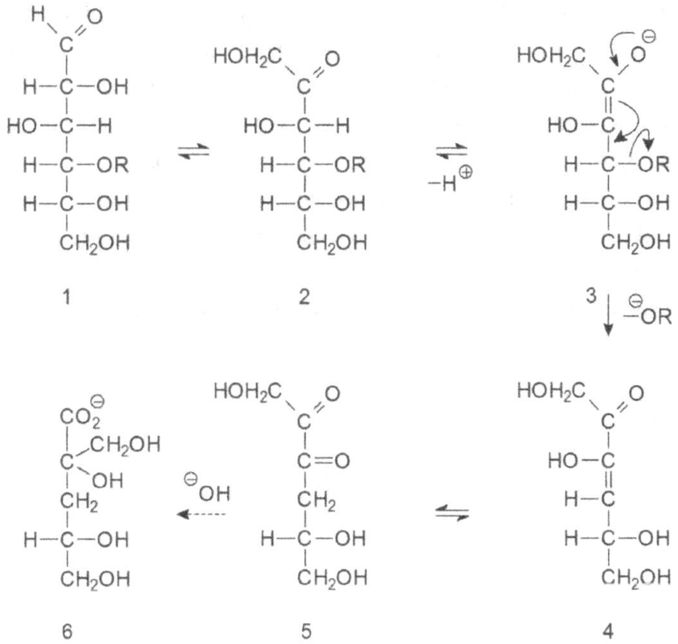

Fig. 11.6. Peeling reaction of cellulose (R is an alkoxy group consisting of a glucan chain). Reaction steps: Isomerization (1→2), formation of a 2,3-enediol structure (2→3), β-alkoxy elimination (3→4), tautomerization (4→5), and benzilic acid rearrangement type reaction (5→6), during which an epimeric glucoisosaccharinic acid (3-deoxy-2-*C*-hydroxymethylpentonic acid) is formed.

dissolves into the cooking liquor (the final kraft cooking liquor is called "black liquor"), and the remaining carbohydrate chain (an alkoxy group originally in the β position with respect to the carbonyl group of a ketose) is shortened by one monosaccharide unit while exposing a new reducing end group open to further peeling. The peeling reaction thus involves the elimination of monosaccharide units that form various carboxylic acids (about one and half equivalents of acids per one monosaccharide unit are formed). The soluble monomeric compound, for its part, forms a dicarbonyl intermediate (2,3-diulose structure) from which the main reaction route produces in a benzilic acid rearrangement-type reaction either glucoisosaccharinic acid or xyloisosaccharinic acid. Several different peeling reactions produce, besides "volatile" formic and acetic acids, "non-volatile" hydroxy monocarboxylic acids (e.g., glycolic, lactic, 2-hydroxybutanoic, 3-deoxy-tetronic, 3,4-dideoxy-pentonic, 3-deoxy-pentonic, xyloisosaccharinic, and glucoisosaccharinic acids) and minor amounts of hydroxy dicarboxylic acids (cf., Chapt. "12.4.1. Kraft pulping"). Of the main hydroxy acid components, 3,4-dideoxy-pentonic and glucoisosaccharinic acids (α- and β-forms) are formed from hexosans (glucomannans and cellulose), whereas 2-hydroxybutanoic and xyloisosaccharinic acids originate from pentosans (xylans). In contrast, the prominent glycolic, lactic, and 3-deoxy-pentonic acids together with volatile formic acid are formed from all polysaccharide constituents present. In addition, Fig. 11.7 shows an example of a possible retro-aldol reaction that degrades the carbon chain of the monosaccharide structure.

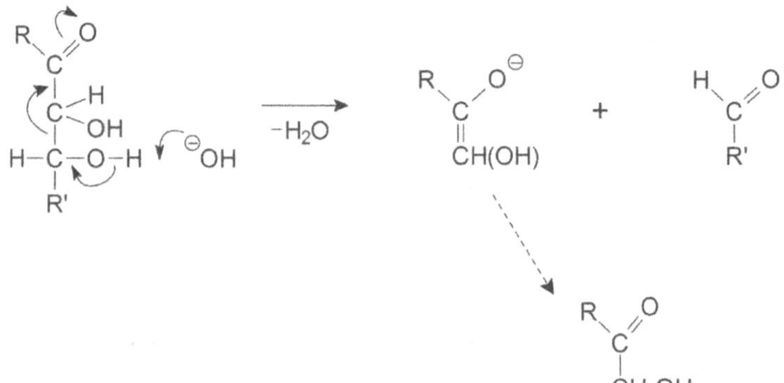

Fig. 11.7. Example of an alkali-catalyzed reaction for cleaving a monosaccharide carbon chain.

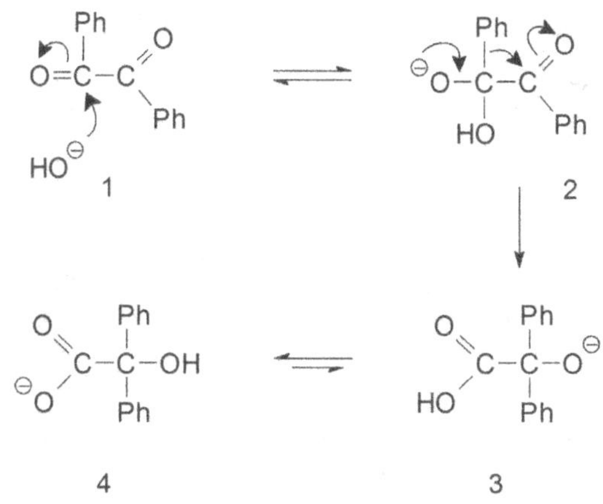

Fig. 11.8. Alkali-catalyzed rearrangement reaction of benzil (1,2-diphenylethane-1,2-dione) (1) via intermediates (2) and (3) to benzilate anion (4). Ph is a phenyl group.

The oxidation of benzoin (PhCH(OH)COPh, Ph is a phenyl group) leads to benzil (PhCOCOPh) with a 1,2-diketone structure that transforms in alkaline environments into an α-hydroxy acid anion ($Ph_2C(OH)CO_2^{\ominus}$) according to Fig. 11.8. When applying to the peeling reaction of carbohydrates (cf., a 2,3-diulose structure), the reaction proceeds as shown in Fig. 11.9. For clarity, the possible formation of the two alternative epimeric structures is not shown; this depends on the direction of approach of the hydroxyl ion at the beginning of the reaction.

The stopping reaction of the carbohydrate chain means *β-hydroxy elimination* directly from the terminal group aldose (Fig. 11.10), so that the peeling of the chain is prevented; with respect to the original carbonyl carbon, the β carbon now has a hydroxyl group instead of an alkoxy group attached to it. Generally, this reaction leads to a terminal group where the functional unit is carboxylic acid, so that isomerization into a 2-ketose is no longer possible. In this case, the dicarbonyl structure (a 1,2-diulose structure) that is also produced undergoes a benzilic acid-type transformation (Fig. 11.11), which primarily leads to the formation of a metasaccharinic acid end group into the polysaccharide chain. Other possibilities are 2-*C*-methylglyceric acid and 2-*C*-methylribonic acid (glucosaccharinic acid) end groups and in small quantities also certain aldonic acid end groups (i.e., mannonic, arabinonic, and erythronic acid groups), indicating

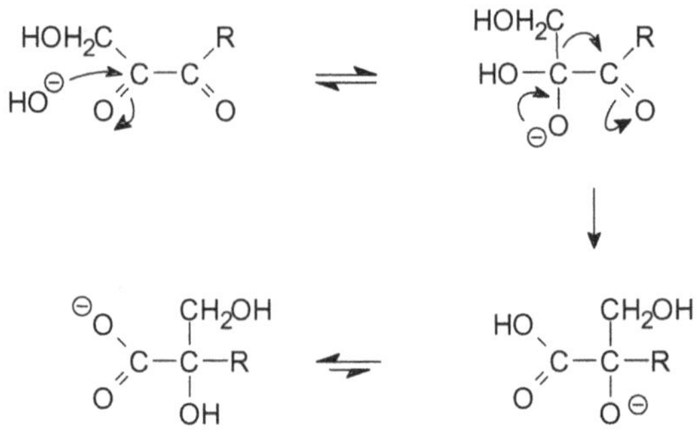

Fig. 11.9. Benzilic acid rearrangement-type reaction of the compound residue dissolved during the peeling reaction (see Fig. 11.8).

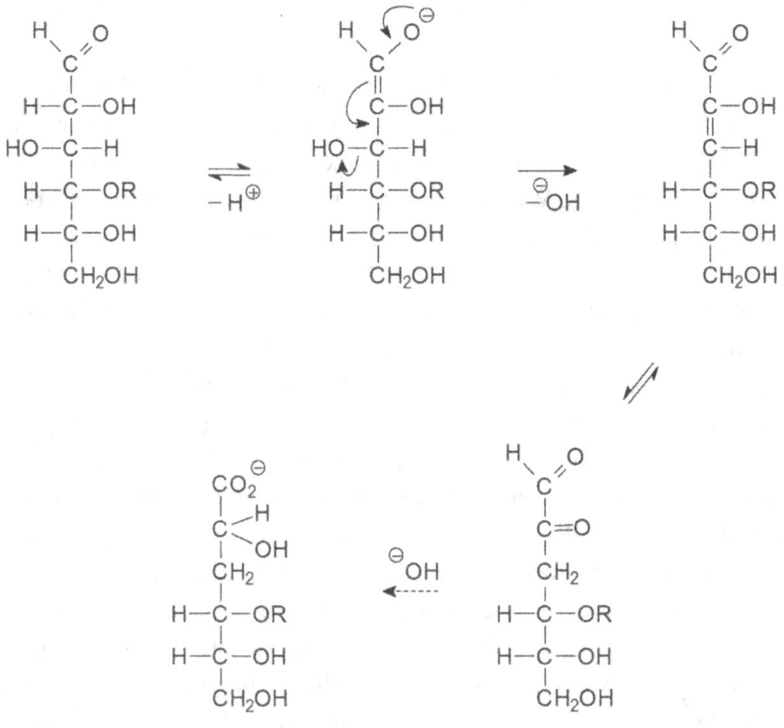

Fig. 11.10. Stabilization of a carbohydrate chain in the stopping reaction (i.e., termination of the peeling reaction). The reaction steps are analogous to those occurring in the peeling reaction (see Fig. 11.6). However, in this case, due to β-hydroxy elimination (i.e., β-alkoxy elimination is not occurring), no cleavage of the carbohydrate chain takes place.

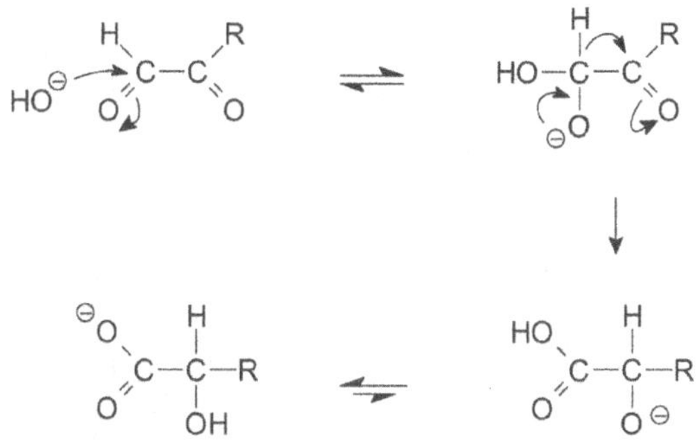

Fig. 11.11. Benzilic acid rearrangement-type reaction of a carbohydrate chain end group during the stopping reaction (see Fig. 11.10).

the presence of oxygen in the reaction. As mentioned above, the stopping reaction is clearly slower than the terminal peeling reaction; therefore, the carbohydrate losses caused for cellulose and the hemicelluloses with a substantially lower degree of polymerization are respectively 10–15% and 50–60% of the initial amounts.

In an acid delignification process, such as conventional acid sulfite cooking, *acid hydrolysis* of the carbohydrate chains into both monosaccharides and the larger-molecule oligo- and polysaccharide fragments takes place. Similar carbohydrate fractions are also obtained in the processing of biomass, and their utilization is possible in many ways (cf., Chapt. "12.2. Production of Chemicals"). Among essential products are acids and alcohols produced by fermentation and furan derivatives.

The *thermochemical reactions* of carbohydrates have been widely studied; one goal has been the production of chemicals (e.g., anhydrosugars). Raw materials based on carbohydrates are, however, primarily used in the production of various types of biofuels through pyrolysis, gasification, or liquefaction (cf., Chapt. "12.3. Production of Biofuels"). In these cases, the chemical bonds in carbohydrates are rather unselectively cleaved on heating. An illustrative example (Fig. 11.12) is the behavior of the organic constituents of black liquor (lignin, aliphatic carboxylic acids, and hemicellulose residues) in the recovery furnace combustion (650–1250°C). In the furnace, the energy of the dissolved organic compounds is recovered while the inorganic matter is converted for reuse after modification (caustization) for the next cooking process. Black liquor

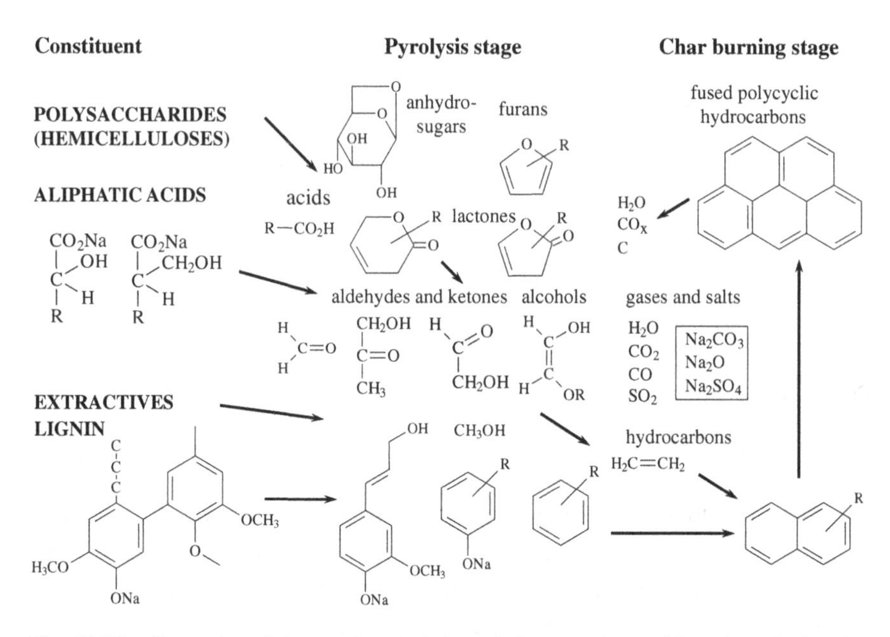

Fig. 11.12. Examples of thermochemical degradation reactions taking place during the combustion of black liquor in a recovery furnace (Alén 2011b).

(65–85% dry matter after evaporation) is sprayed into the recovery furnace (the process is controlled by adjusting the amount of air normally fed by a three level air system of the furnace) as droplets that undergo certain stages (drying, pyrolysis, and char burning) before they land at the char bed that mainly consists of inorganic salts (mainly Na_2CO_3 and Na_2S) in the bottom of the furnace. The char bed is typically conical in shape and contains an active layer with liquid smelt together, with solid carbon causing the reduction of Na_2SO_4 as well as an inactive core with solid smelt. The liquid smelt flows by gravity to the smelt spouts and dissolves into water ("green liquor"). For example, in this process, the polysaccharides first decompose into anhydrosugars; thereafter, a variety of small-molecule compounds (aldehydes, ketones, and alcohols) are obtained, and finally hydrocarbons, such as ethene, which tend to aromatize in the fusion process until char is formed as the final product.

12. Utilization of Biomass

12.1. General Aspects

We produce our energy and chemicals mainly from fossil resources. However, it is an undisputed fact that these reserves are clearly limited. Together with global warming and other environmental issues, this drives us toward replacing fossil-based carbon sources, petroleum, coal, and natural gas, by alternative raw materials. In other words, the production of energy in the form of electricity, heat, and fuels, as well that of organic chemicals and other products, must increasingly be based on *renewable carbon resources* ("biomass").

During most of our history, our living was almost exclusively based on renewable resources; this changed during the first part of the 19th century. Production of organic chemicals and other organic products from fossil resources started as coal-based thermochemistry, coal carbonization, about 150 years ago. Petroleum-based industrial chemistry followed some 60 years ago, leading to an enormous increase in the number of varying products. This chemical industry is now shifting to more efficient utilization of various CO_2-neutral *lignocellulosic feedstocks* and decreasing use of fossil resources. As we move toward a world of greater diversity and balance with the natural cycles of various materials, it is still important to learn the lessons from oil refining and petrochemical industries.

Bioenergy production comprises the production of heat and electric power, or combined heat and power (CHP), by burning solid biomass and the use of bio-oils and synthesis gases thermochemically produced from it (see below). The term "*biofuels*" covers a wide range of fuels including solid biomasses, liquid fuels, and biogases. Many biofuels are more suitable for the production of heat and electricity than, for example, for use as transportation fuels.

Production of solid fuels from biomass is becoming more important. While many raw biomasses already are suitable to be burned at >900°C to provide heat, resources, such as sawdust, shavings, bark, harvesting residues, grasses, and agricultural residues are often upgraded by grinding (i.e., crushing, drying, and milling to an appropriate particle size distribution) and/or densifying to more compact and regular shapes as briquettes and pellets.

"First generation biofuels" produced by well-known technologies are typically ethanol (or "bioethanol") from fermentable sugars and biodiesel from vegetable oils after transesterification. In this process, fatty acid methyl esters (FAMEs) are formed from glycerides (mainly triglycerides, whose three fatty acid residues are esterified with glycerol) by methanol. Pure ethanol is being used as vehicle fuel, but it is normally used as a gasoline additive to increase its octane number and to improve vehicle emissions. Biodiesel is also used as such as vehicle fuel although its main use is as diesel additive. Compared to fossil diesel, it contains a reduced amount of carbon and correspondingly higher amounts of hydrogen and oxygen ("oxygenated fuel"). According to another definition, all first generation biofuels contain products whose manufacture threatens food supplies and biodiversity. In most cases, they are not cost competitive with existing fossil fuels, and their production does not reduce greenhouse gas (GHG) emissions.

The production of "second generation biofuels" or "advanced biofuels" from syngas, i.e., gas-to-liquid (GTL) processes including Fischer-Tropsch diesel, methanol, dimethyl ether (DME, prepared catalytically from methanol), as well that of hydrogen, is under extensive development to solve various technical problems (see below). Diesels obtained by catalytic hydrogenation of various vegetable oils or related materials, including feedstocks for first generation biofuels or fatty acid fractions from tall oil, also belong to this category. These diesels have a high cetane number but weak operating properties at low temperatures; consequently, they are catalytically isomerized to produce branched structures with better properties needed, for example, in the Northern Hemisphere. Other feedstocks for second generation biofuels include animal fats and recycled greases.

The *biorefinery concept* can be defined as a process of fractionating and/or converting biomass, a CO_2-neutral feedstock, in an eco-friendly way through advanced technologies into solid, liquid, and gaseous bioproducts. The main objective is to maximize the value of the product while minimizing the production of waste. This principle is analogous to that of petrorefineries utilizing fossil resources, but biorefineries use a wider range of feedstocks and process technologies.

The production of liquid fuels is the main objective of this *green chemistry* or *green engineering*, but it along with the versatile technological development in progress will lead to the emergence of many other novel bioproducts as well. More effective employment of biomass resources has already generated novel process concepts with many breakthroughs, but a full realization of the biorefinery concept is still a challenge. Huge barriers in developing and commercializing new technologies for full-scale production must be overcome. The term "*bioeconomy*" comprises those parts of the economy that use renewable carbon resources to produce food, energy, and other bioproducts. It is an essential alternative to our current fossil-based economy and can be considered the next wave in our economic development.

The chemical pulp industry is an important branch of global industry based on vast and multidisciplinary technology. Interestingly, the first industrial biorefineries were operated already about 150 years ago in the pulp and paper industry (cf., Chapt. "2.2. The Time After the 1700s"), although "novel" biorefineries were only more recently developed in the agricultural industry. For example, in the kraft process the wood material is primarily fractionated into pulp (mainly cellulose), extractives (turpentine and tall oil), and black liquor containing degraded lignin and mainly hemicellulose-derived aliphatic carboxylic acids (cf., Chapts. "9.4.5. Other polysaccharides", "11.2. Polysaccharides", and "12.4.1. Kraft pulping").

The moisture content of a living tree varies seasonally and even diurnally depending on the weather, average values being in the range of 40–50% of the total wood mass. Approximately two thirds of the dry matter of wood is composed of polysaccharides, i.e., cellulose and various hemicelluloses. However, when studied in more detail, softwoods and hardwoods are found to differ typically from each other in their chemical composition. Figure 12.1 illustrates typical chemical compositions of

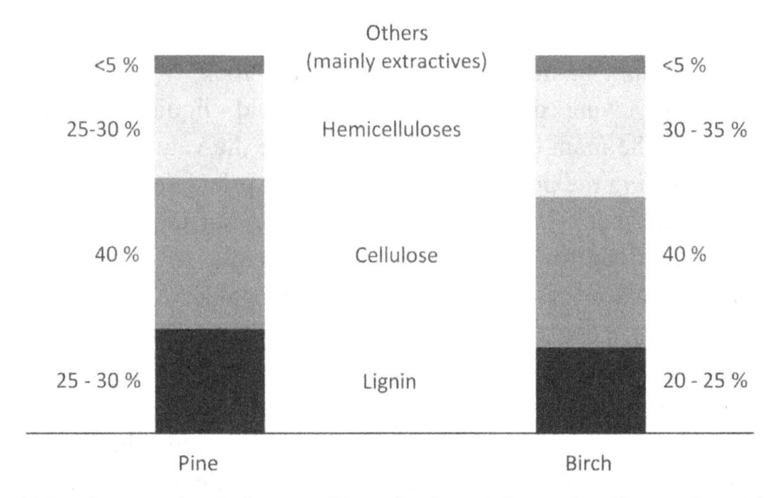

Fig. 12.1. Average chemical composition of softwood Scots pine (*Pinus sylvestris*) and hardwood silver birch (*Betula pendula*). Figures are given as percentages of wood dry solids.

commercial softwood and hardwood, summarizing the differences between Scots pine (*Pinus sylvestris*) and silver birch (*Betula pendula*). The chemical composition of various wood species may naturally differ to some extent from these common examples.

In both cases shown in this figure, the cellulose content is more or less the same (40–45% of the wood dry solids), but softwoods usually contain less hemicelluloses and more lignin; the lignin content of common softwoods is exceeded only by some tropical hardwoods. The typical content of hemicelluloses in softwoods and hardwoods is, respectively, 25–30% and 30–35% of the wood dry solids. On the other hand, the lignin content of softwoods is typically in the range of 25–30% of the wood dry solids, whereas that of temperate-zone hardwoods varies between 20% and 25% of the wood dry solids. Detailed structures of softwood and hardwood hemicelluloses are described in Chapt. "9.4.5. Other polysaccharides" and their structural moieties are presented in Fig. 9.5. The other compounds, mainly extractives, in woods from temperate zones usually make up about 5% of the wood dry solids; tropical species often exceed this value. In woods from temperate zones the macromolecular substances building up the cell walls account for about 95% of the wood material (Fig. 12.2). In contrast, for tropical woods, this can decrease to an average value of 90%.

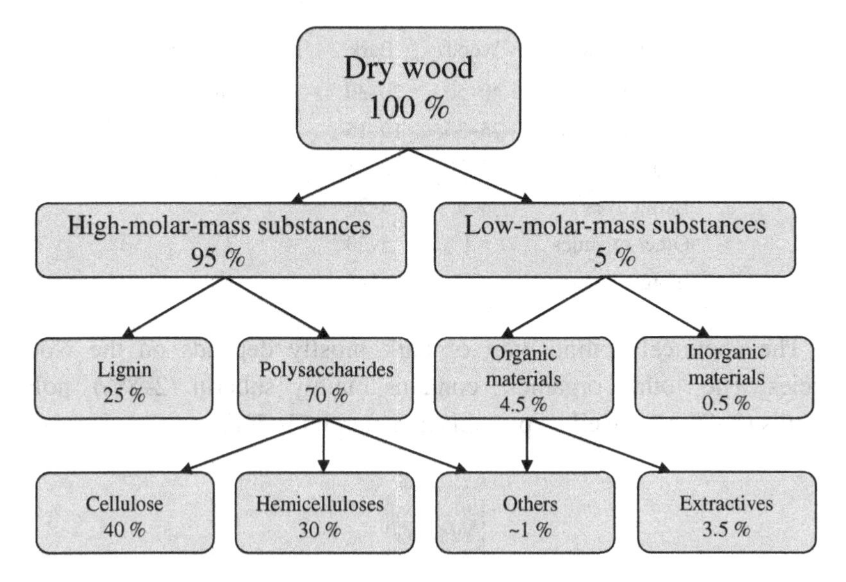

Fig. 12.2. General classification and content of chemical wood components (% of the wood dry solids) (Alén 2011a).

Wood and other cellulosic materials (lignocellulosics) can be traditionally processed in a number of ways using mechanical, chemical, and thermochemical or thermal conversion methods (Fig. 12.3). However, in spite of the huge production volumes, the current fiber-related utilization of wood is still rather limited in relation to the total wood utilization: fuel 50–55%, construction 25–30%, fiber 10–15%, and others 5%. Nowadays, wood obtained from forest trees is the predominant source of cellulosic fiber for pulp and paper manufacture; other lignocellulosic materials, generally collectively referred to as "*non-wood materials*", are a minor feedstock resource of about 10%.

Particularly, in pyrolysis and gasification (see below), only the ultimate composition of the raw material is often of practical importance. For example, the elemental composition (C:H:O:N:others, % of the dry feedstock) of pine (*Pinus sylvestris*) sawdust, pine bark particles, and average Finnish forest residue chips is 51.0:6.0:42.8:0.1:0.1, 52.5:5.7:39.6:0.4:1.8, and 51.3:6.1:40.8:0.4:1.4, respectively. A general comparison between the average chemical compositions of these feedstocks utilized in the production of energy is (% of the feedstock dry solids):

Component	Wood	Bark	Forest residue
Cellulose	40–50	20–30	35–40
Hemicelluloses	25–35	10–15	25–30
Lignin	20–30	10–25	20–25
Extractives	3–4	5–20	~5
Other organics	~1	5–20	~3
Inorganics	<0.5	2–5	~1

The chemical composition of bark mostly depends on the wood species. The "other organics" contains mainly suberin (2–8%), polyphenols (2–7%), as well as proteins and starch (1–5%).

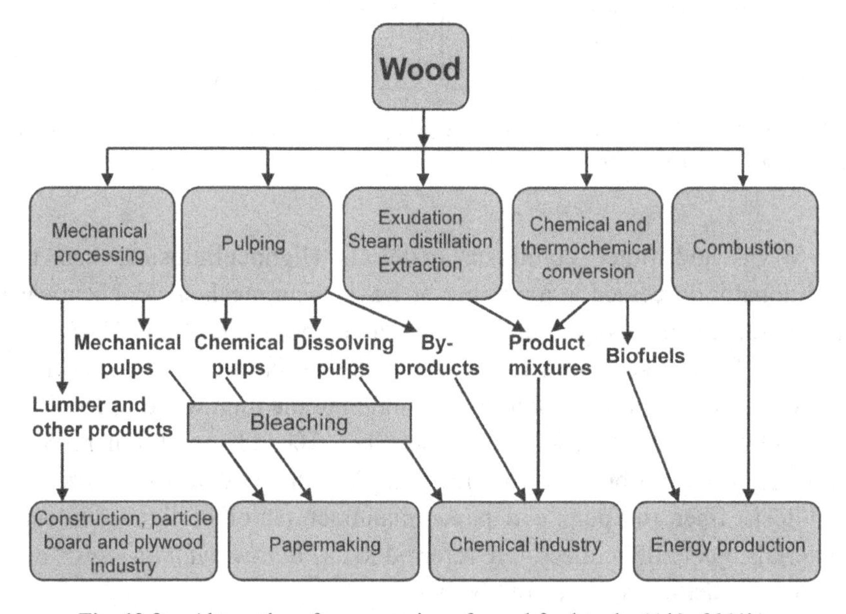

Fig. 12.3. Alternatives for conversion of wood feedstocks (Alén 2011b).

There are several ways to classify non-wood feedstocks, but usually they are divided according to their origin into the following groups:

- Agricultural residues (e.g., sugarcane bagasse, sorghum corn stalks, cotton stalks, rice straw, and cereal straws)
- Natural-growing plants (e.g., bamboo, esparto, sabai, elephant grass, and reeds, such as reed canary grass)

- Non-wood crops grown primarily for their fiber content
  - — Bast (stem) fibers (e.g., jute, ramie, hemp, kenaf, and flax tow)
  - — Leaf fibers (e.g., abaca and sisal)
  - — Seed hair fibers (e.g., cotton linter)

The main chemical components in non-wood feedstocks are the same as those in wood feedstocks; they are found in varying amounts depending on species (genetic differences), presence of special tissues within individual plants, and growing conditions. These differences are important in selective biomass conversion processes, where, for example, main pentose moieties, xylose and arabinose, may be either a desired or an unwanted product. In contrast, these chemical differences are only of minor concern in gasification processses that non-selectively convert biomass raw material into gaseous compounds (synthesis gas). As non-wood feedstocks are available in various forms, their physical and mechanical properties also vary.

Table 12.1 shows a general comparison between the chemical composition of wood and fibrous non-wood feedstocks utilized in pulping. One of the most harmful properties of non-wood lignocelluloses in pulp production is a high content of inorganics, mainly silicon dioxide or silica. Silica, as well as other sparingly soluble materials, promotes scale formation in the evaporators of the recovery cycle of pulping chemicals.

Table 12.1. Typical chemical composition of wood and non-wood feedstocks used for pulping (% of the feedstock dry solids) (Alén 2011a)

Component	Wood feedstock	Non-wood feedstock
Carbohydrates	65–80	50–80
Cellulose	40–45	30–45
Hemicelluloses	25–35	20–35
Lignin	20–30	10–25
Extractives	2–5	5–15
Proteins	<0.5	5–10
Inorganics	0.1–1	0.5–10
SiO_2	<0.1	0.5–7

Agricultural residues, such as wheat straw and especially rice straw contain a large amount of inorganic materials with a high silica content. Natural plants have a somewhat lower silica content and bamboo has the lowest. On the other hand, a high ash content indicates a problematic behavior of the feedstock in gasification and combustion. Difficulties caused by ash fusion and sintering in straw combustion are generally known; for example, the ash of reed canary grass harvested in the spring fuses at a fairly high temperature and should not cause problems under normal gasification conditions. In addition, morphological properties of non-wood fiber plants differ markedly from those of woods, affecting their pulping and papermaking processes and the quality properties of the paper produced.

As a general trend, non-wood feedstocks contain an increased amount of hydrophilic (i.e., water-soluble) extractives, while lipophilic (i.e., organic solvent-soluble) extractives are dominant among wood extractives. The content of proteins is also higher in non-wood materials than in woods. Of the grass family, compared to wood, agricultural residues and natural plants often have less lignin and more hemicelluloses. However, bamboo as a natural plant is an exception with a lignin and hemicellulose content similar to those in wood. Hemicelluloses from different non-wood feedstocks (they contain more xylan, 60–70% of total hemicelluloses, than glucomannan) have many structural features in common. In addition, structural features in xylans from non-wood feedstocks seem rather similar to those detected in xylans from wood materials (cf., Chapt. "9.4.5. Other polysaccharides"). However, the DP of non-wood hemicelluloses is typically lower than that in wood. The variations noted in the hemicellulosic composition of non-wood feedstocks at different stages of growth are mainly reflections of the varying proportions of different types of tissue.

One potential new source of energy is fast growing algae (up to 30 times more energy per area than land crops can be obtained) and currently the production of biodiesel ("algae fuel") from this feedstock ("farming algae") is of increasing interest. In general, photosynthetic algae including microalgae ("single celled algae") as well as cyanobacteria ("blue-green algae") together with seaweeds are effective in converting CO_2 into various lipids and long-chain hydrocarbons by sunlight energy.

Traditional organic chemical industry uses various operations (physicochemical steps) to transform common fossil raw materials into more valuable products that often are more oxidized than the raw materials. This industry is highly competitive and new processes in the utilization of biomass resources are steadily introduced. Although extensive research on the utilization of wood has been going on for several decades, it has so far received little notice in the industry, primarily due to the central position of petroleum as raw material.

The description below shows illustrative examples of the production of chemicals from cellulosic biomass after hydrolysis. An enormous number of chemicals can be manufactured from biomass-derived sugars, not only by traditional chemical conversion but also by biochemical conversion, which seems to gain importance. One advantageous, compared to oil-based feedstocks, is that biomass-based compounds often contain reactive functional groups, such as oxygen-containing alcoholic, aldehyde, ketone, and carboxylic acid groups.

12.2. Production of Chemicals

Wood and other cellulosic biomasses are complex feedstocks. Although their major constituent is cellulose, they also contain substantial amounts of other polysaccharides, hemicelluloses, as well as non-carbohydrate constituents, mainly lignin and extractives. These carbohydrates provide raw materials for future biorefineries and for this reason, current technologies generally focus on the biochemical or chemical utilization of carbohydrates. The most common method is pretreatment of biomass through hydrolysis to glucose and other monosaccharides ("*saccharification*", the cleavage of glycosidic linkages between the monosaccharide units); it and can be accomplished by chemical (with acids like H_2SO_4 or HCl) or biochemical (with enzymes like cellulases or hemicellulases) treatments (Fig. 12.4). Even if the main goal is to obtain fermentable sugars (i.e., soluble carbohydrates) for bioconversion, the monosaccharides liberated are also suitable for other applications, including their chemical modification into a wide range of useful value-added chemicals.

Acid hydrolysis has a long history; it can be accomplished either by concentrated mineral acids at low temperatures or by dilute acid at high

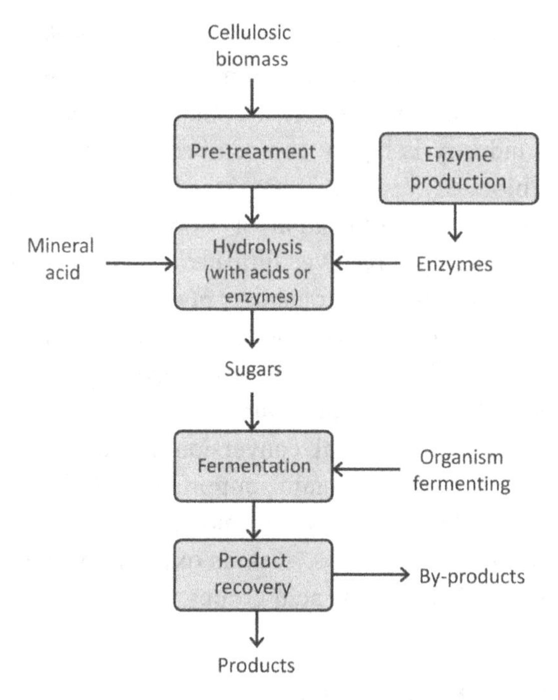

Fig. 12.4. Process scheme for the conversion of biomass-derived carbohydrates into varying products by acid or enzymatic hydrolysis and their subsequent fermentation (Alén 2011b).

temperatures. The first commercial processes called "wood saccharification" were developed in 1901 employing concentrated H_2SO_4 by Alexander Classen, and in 1909 employing dilute H_2SO_4 by Malcom F. Ewen and George H. Tomlinson.

An example of the process conditions in the dilute acid process is the use of 1% H_2SO_4 at higher temperatures (a typical temperature range for biomass is 160–230°C) and pressures of about 1 MPa with a sugar yield about 50% of the initial carbohydrates. In contrast, the concentrated acid hydrolysis process, the Arkenol process, typically works by adding 70–77% H_2SO_4 to biomass dried to 10% moisture content at 50°C, the ratio of acid to biomass being 1.25. Water is then added to dilute the acid to a concentration of 20–30%, and the mixture is heated to 100°C for one hour. A chromatographic column is used to separate H_2SO_4 for recycling from the sugar mixture. In acid hydrolysis, besides H_2SO_4, concentrated

HCl has also attained some commercial significance. Although HCl is more efficient hydrolyzing agent than H_2SO_4, the dilute H_2SO_4 method has been traditionally used in industrial applications.

Many factors influence the reactivity and digestibility of the cellulose fraction of lignocellulosic biomasses. These factors include their lignin and hemicelluloses contents, the crystallinity of cellulose, and the porosity of the materials. Pretreatment of biomasses prior to utilization is necessary, for example, in the biomass-to-ethanol conversion processes. In recent years, acidic treatment of lignocellulosic biomasses with dilute H_2SO_4 has primarily been used as pretreatment for enzymatic hydrolysis of cellulose (see below). A number of commercial-scale wood-based sugar plants producing ethanol and other products, such as furfural and yeast cells, are in operation as a part of the sugar industry in many countries.

Two important carbohydrate reactions take place during the acid hydrolysis process: (i) depolymerization of carbohydrates to their monosaccharide moieties (and in part to their oligosaccharide residues) and (ii) formation of monosaccharide-derived products (e.g., furans) that can inhibit the subsequent fermentation. Hemicelluloses are relatively easily hydrolyzed and the composition of the resulting monosaccharide mixture essentially depends on the raw material. The major hydrolysis products from hemicelluloses are, in addition to glucose, other aldohexoses (mannose and galactose), and aldopentoses (xylose and arabinose). In contrast, cellulose is hydrolyzed more slowly and the glucose concentration increases towards the end of hydrolysis. In order to minimize contamination of glucose, which is in most cases the desired product, a two-stage acid hydrolysis is often recommended. It involves first a mild treatment, where hemicelluloses are preferentially hydrolyzed, followed by a second step to convert the cellulose to glucose. However, a complete hydrolysis of cellulose is difficult.

Acid hydrolysis methods can be costly; they require special equipment due to the corrosion risk and generate large amounts of acid and solid wastes (e.g., sulfur-containing lignin), as well as problems in acid recycling. Due to the presence of lignin and hemicelluloses in feedstock materials, the heterogeneity of the *hydrolysates* is normally a difficult problem. An additional drawback is the further degradation of monosaccharides to a wide range of harmful by-products; i.e., the reaction cannot be limited to hydrolysis only. The problems in wood acid hydrolysis arise from the manufacturing procedure, together with the high energy consumption in recycling the waste acid. New economical "green processes" for the conversion of cellulose into glucose under mild conditions with high selectivity would be essential.

Mild acidic pre-hydrolysis of wood using 0.5% to 1.0% H_2SO_4 at 120°C to 130°C was commercially used first in Germany during the 1940s in two-stage acid-alkaline (kraft) pulping processes. The hydrolysates were rich in monosaccharides in varying proportions depending on the wood raw material, and were thus useful for aerobic fermentation by *Torula* or fodder yeast. Such pre-treatment stages prior to kraft pulping have later been used in several kraft mills and recently restudied as a "potential integrated biorefinery concept" by many researchers (cf., Chapt. "12.5. Integrated Biorefineries"). In addition, pre-hydrolysis with dilute sulfurous acid has been integrated to the sulfite pulping process to produce monosaccharides (cf., Chapt. "12.4.2. Sulfite pulping"). The spent liquors from such processes contain large amounts of sugars (see below); this fraction can be fermented to products as ethanol and protein, or their main monosaccharides, xylose and mannose, can be isolated and used in other applications.

The first detailed study of the kinetics of dilute acid hydrolysis of wood at elevated temperatures was carried out in 1945 by Jerome F. Saeman. Since then the research effort on this topic has continued and a great number of new results, mainly dealing with a broader range of reaction conditions, have been obtained. Recent reactor design and simulation studies suggest that acid hydrolysis technology might again be a viable alternative among effective biomass saccharification processes. In many acid hydrolysis processes, as also in the case of *enzymatic hydrolysis*, economic (capital investment) and even technical obstacles have hampered their utilization. During recent decades biomass hydrolysis processes have been economically attractive only in a few cases, although more economical processes may emerge from the research now in progress.

Hydrolyzing wood and other cellulosic materials by enzymes to fermentable sugars has been intensively studied in recent years. This has mainly been due to the present priority given to efficient conversion of various feedstocks into fuel ethanol. In this method, *cellulases* together with various *hemicellulases*, are chosen as enzymes for preferentially cleaving β-(1→4)-glycosidic linkages in polysaccharide chains, ultimately aiming at almost complete degradation of carbohydrate-containing raw materials to monosaccharides, aldohexoses and aldopentoses.

Lignocellulosic raw materials are generally rather resistant to enzymatic digestion; the rate and extent of their biochemical hydrolysis is influenced not only by the effectiveness of the enzymes, but also by the chemical and morphological characteristics of the heterogeneous substrate. Since many structural and compositional factors hinder enzymatic digestion of carbohydrates in cellulosic biomasses, a dilute acid or alkali pretreatment, as well as a simple thermochemical pre-treatment, is usually needed prior to the enzymatic process.

In addition to substrate characteristics, the nature of the enzyme complex has a significant effect on how effectively a pre-treated cellulosic biomass is hydrolyzed. Furthermore, the use of enzymatic hydrolysis is highly dependent on the cost of the selective enzymes required. Besides the relatively high costs of the enzymes, the low conversion rate due to the heterogeneity of the biomass causes difficulties. In general, enzymatic degradation is slow compared to acid hydrolysis. In spite of this, it should be noted that enzymatic hydrolysis proceeds under non-corrosive and mild conditions, is rather selective, and the formation of harmful by-products is minimized. New and more economical processes are likely to emerge from the versatile research now in progress.

The products of a biorefinery primarily depend on the basic structural composition and integrity of the feedstocks. There is only a limited demand for glucose and other monosaccharides from hydrolysis unless they are further refined. The isolation of individual components often requires complicated separation techniques and thus the most important method for utilization of sugars is fermentation. It enables the use of sugars in the production of food, chemicals, and energy (see the alternative uses of chemicals, such as ethanol and butanol).

There are two types of processes in the production of primary chemicals by fermentation (Fig. 12.5). One type consists of well-established anaerobic or aerobic production of simple chemicals, such as various alcohols and carboxylic acids. The other type of fermentation is

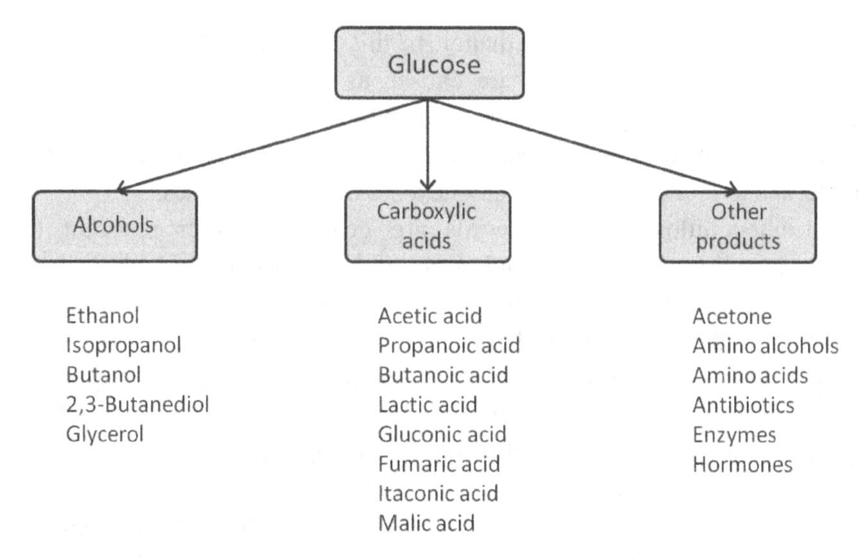

Fig. 12.5. Examples of fermentation products from glucose (Alén 2011b).

used for more complex chemicals, such as antibiotics, enzymes, and hormones. For clarity, glucose is selected here as an example of feedstock sugars, although the same products can be obtained from other sugars as well. An example of typical biochemical modification routes is also enzymatic isomerization in the production of starch-based sweeteners, so-called "high-fructose syrups" (cf., Chapt. "9.4.5. Other polysaccharides"). These sugar mixtures are replacing sucrose in applications where the sweetener can be added in the form of an aqueous solution.

A number of chemicals, representing a substantial volume of today's chemical industry, could be produced from fermentation-derived products of glucose (Fig. 12.6). In principle, the conversion of cellulose to polymers with properties totally different from those of polysaccharides is possible; these platform chemicals are employed, besides as monomers for polymer synthesis, also as precursors for production of other chemicals, for example, by adding oxygen- or nitrogen-containing functional groups. One of the prominent chemicals, gluconic acid, has been excluded because it is utilized as such in the food industry, and also in many other applications as its sodium salt.

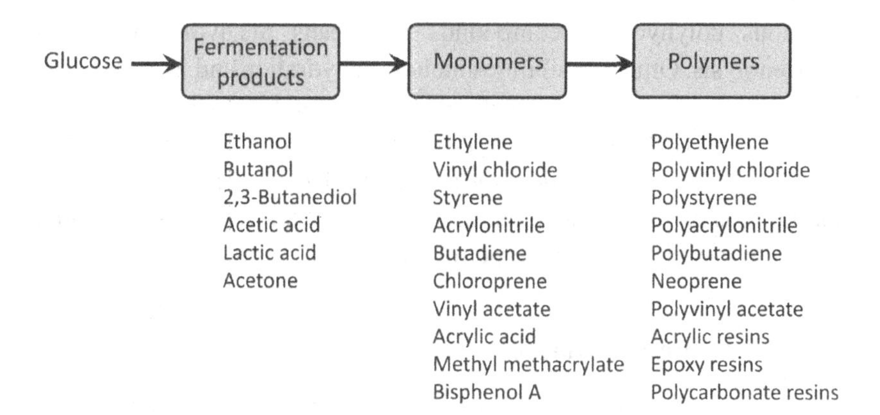

Fig. 12.6. Products from the platform chemicals formed in glucose fermentation (Alén 2011b).

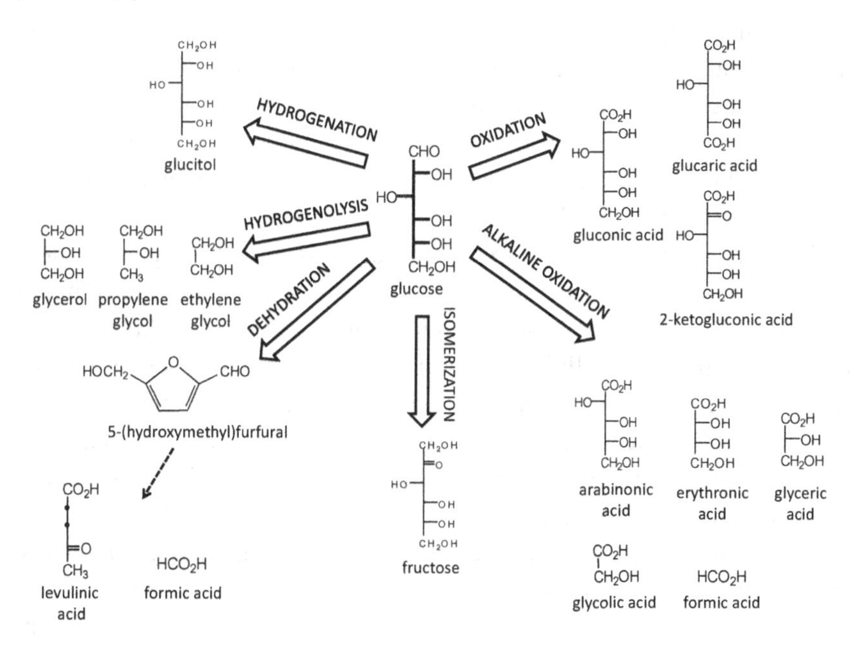

Fig. 12.7. Examples of chemical products from glucose by common chemical treatments (Alén 2011b).

Figure 12.7 illustrates some straightforward examples of possible production routes in making useful chemicals from glucose by well-known chemical processes. Thus, glucose can be oxidized, either under acidic or alkaline conditions, to mono- and dicarboxylic acids, or reduced to various polyhydroxy compounds by means of hydrogenation or hydrogenolysis. Other possibilities include dehydration and isomerization. The production of furans, such as furfural and 5-hydroxymethyl-2-furaldehyde or HMF is important. Levulinic acid, for example, is obtained from HMF via the liberation of formic acid during heating. It is an important platform chemical that can be converted into a variety of other chemicals; it is also a potential petrol additive.

12.3. Production of Biofuels

Thermal or *thermochemical conversion* of cellulosic materials always results in three groups of substances: gases, condensable liquids (tars), and solid char products (in the case of wood, charcoal) (Fig. 12.8). The relative proportions of these products depend on the chosen method and the specific reaction conditions. In comparison with saccharification-based processes, thermal conversion is relatively rapid and it avoids large volumes of water and other external chemicals. The major disadvantage is unselective reactions that yield a large number of products at low individual yields.

The two main thermal conversion methods, *pyrolysis* and *gasification*, differ in that pyrolysis refers to thermal degradation in complete or near complete absence of oxidizing agent, air or oxygen, to provide complex fractions of gases, condensable liquids (tars), and char (solid residue), while in gasification cellulosic materials are converted by heating in the presence of controlled amounts of oxidizing agents, air, oxygen, or steam, primarily to provide a simple gaseous phase. When molecular oxygen (or its molar equivalent) is the oxidizing agent, the amount used is substantially below that required for stoichiometric combustion.

Pyrolysis is generally carried out at lower temperatures (typically about 500°C) than gasification (typically above 800°C). In spite of this distinction, pyrolysis is often more broadly defined as any chemical changes brought about by the application of heat, even with air or other additives. The main products of pyrolysis are char and liquids, which as a result of an incomplete thermal degradation retain much of the structure and complexity of the feedstock undergoing pyrolysis. In gasification, it is always necessary to pass through a pyrolytic stage followed by a total or partial oxidation of the primary products.

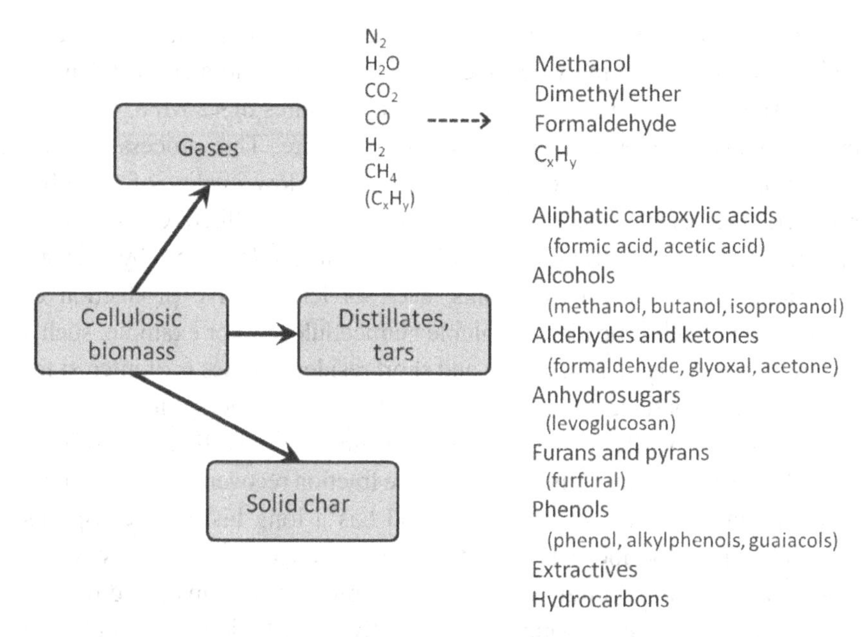

Fig. 12.8. Product groups from thermal conversion of cellulosic biomass (Alén 2011b). Their relative proportions depend on the method and the reaction conditions (see the text). Some examples of individual components and their further products are given.

In addition to pyrolysis and gasification, there are processes where liquid fuels are obtained by *liquefaction* (i.e., direct biomass-to-liquid (BTL) process), or biogases by *anaerobic digestion* (see below). "Direct liquefaction" of biomass (i.e., from a solid to a liquid through "melting") can be accomplished at 300–350°C under high pressure (about 20 MPa) in the presence of CO or H_2, resulting in a main product of a high-viscosity liquid that is insoluble in water. Since "liquefaction" is also generally defined as a change from a gas to a liquid by cooling, "indirect liquefaction" involves successsive production of a volatile product, for example, by pyrolysis and the subsequent conversion of this intermediate to a liquid fuel.

Torrefaction of woody biomass is an upgrading pyrolysis for drying biomass to remove its volatile compounds and partially to decompose its hemicelluloses. It is a mild pyrolysis at 225–300°C for several minutes in the absence of air. This pre-treatment is often used prior to gasification; as it enhances the energy density of wood by increasing its carbon content and net caloric value, it will become more important in the future.

Several *steam explosion* processes are examples of an unusual type of delignification, where the wood feedstock is subjected to a short treatment (<4 min) at temperatures of 200–230°C and pressures of <2 MPa, followed by a fast pressure release to atmospheric pressure. This process has not gained any significant success, mainly due to a low quality of the fiber product. However, steam explosion can be an effective hydrolytic pre-treatment, especially for non-wood feedstocks followed by various enzymatic and chemical treatments, such as fermentative production of ethanol after saccharification of soluble hemicelluloses. For example, such a pre-treatment at high temperatures and short residence times is of interest for biomass fractionation, since high temperature favors the solubilization of hemicelluloses while a short residence time avoids their significant degradation and increases the hemicellulose fraction recovered after washing.

Thermal treatment of resinous wood has a long history. Perhaps its most significant period was from the 17th to the early 20th century, when *wood tar* was commercially produced in tar pits and later in closed retorts. The vapors produced during the destructive distillation of wood were condensed to give wood tar and an aqueous layer (*pyroligneous liquor* or pyroligneous acid). This raw liquor from hardwood contains mainly acetic acid, methanol, and acetone derived from acetic acid. Although the production of chemicals from wood distillates is rarely attractive, there are, in addition to acetic acid and methanol, many other chemicals that could be extracted from raw pyroligneous liquor, including phenols and food flavoring agents. In these liquors and tars as well as in bio-oils, an awesome mixture of various products can be identified, among them phenols, acids, lactones, alcohols, aldehydes, ketones, anhydrosugars, furans, esters, ethers, and hydrocarbons. "Oxidative pyrolysis" can be seen as the first step and an inherent part of combustion processes.

As generally accepted, *carbonization* refers to processes, where char is the principal product of interest, while in *wood distillation* the product is the liquid — in *destructive distillation* both char and liquid.

12.3.1. *Fischer-Tropsch process*

In steam-oxygen gasification cellulosic biomass is converted above 800°C with a controlled amount of oxygen and/or steam into synthesis gas ("syngas") composed mainly of N_2, H_2O, CO, CO_2, H_2, and CH_4. Its composition may vary over a wide range depending upon the basic parameters of the gasification. The product gas can be used either as a fuel or for producing chemicals, such as methanol and ammonia. Methanol can be further processed, for example, to DME or formaldehyde. A wide range of aliphatic hydrocarbons, olefins and paraffins, together with oxygenated products, can be produced from this product gas by catalytic conversion according in the Fischer-Tropsch process.

The process is based on old technology; it was originally developed at the Kaiser-Wilhelm-Institut für Kohlenforschung (now known as the Max-Planck-Institut für Kohlenforschung or the Max Planck Institute for Coal Research) in Mühlheim, Germany in the early 1920s by Franz Fischer (1877–1947) and Hans Tropsch (1889–1935). They used alkalized iron catalysts to produce liquid hydrocarbons rich in oxygenated compounds from coal-derived syngas (the Synthol process). The "preceding phase" of this breakthrough, the synthesis of CH_4 from the mixture of CO and H_2 over transition metal catalysts, was already discovered in 1902 by Paul Sabatier (1854–1941, the 1912 Nobel Prize in Chemistry) and Jean Baptiste Senderens (1856–1937). These reactions were also an important step in the use of metal-catalyzed reactions in large-scale synthesis of organic compounds, as was later demonstrated. In 1928, Walter Julius Reppe (1892–1969) started research on catalytic acetylene reactions under high pressure ("Reppe chemistry") for producing a wide range of chemicals.

During World War II the Third Reich, being petroleum-poor but coal-rich, manufactured replacement liquid fuels with the Fischer-Tropsch process from brown coal. For example, in 1944 about 19.7 million liters of synthetic fuel (about 95,000 barrels) were produced daily in 25 manufacturing plants. In addition, it was produced by this technique for military purposes lubricants and a synthetic rubber copolymer of acrylonitrile and butadiene, nitrile rubber or "Buna N", developed in 1934 by Erich Konrad (1894–1975), Eduard Tschunkur (1874–1946), and Helmut Kleiner (1902–87), and styrene-butadiene rubber ("Buna S"), developed in 1929 by Eduard Tschunkur and Walter Bock (1895–1948).

This conversion process was not considered commercially favorable after the war. However, coal-based liquid fuel production, the Sasol process, was started in South Africa in the 1950s due to a trade embargo, including the import of oil, against South Africa as a result of the apartheid (racial segregation) policy enforced through legislation in 1948. Besides synthetic fuels, a number of chemical products, such as paraffin waxes and other hydrocarbon mixtures, lubricants, and corrugated cardboard coating emulsions are produced. Today, similar manufacturing units can be found, for example, in Qatar, Australia, Malaysia, India, and USA.

In principle, biomass gasification and the Fischer-Tropsch synthesis can be combined to produce renewable biofuels, such as transportation fuels. Gasification of carbonaceous biomaterials is currently being reconsidered because fossil fuels can no longer satisfy our energy needs. For this reason, gasification technology, especially gasification under pressure is under extensive development. However, there are still some technical issues that should be clearly resolved before gasification-based biomass utilization becomes fully competitive with fossil fuels. In syngas-based energy production, the use of gasification instead of the direct combustion of the original fuel is potentially more efficient since syngas can be combusted at higher temperatures or even used in fuel cells. It should be noted that significant energy production from renewable sources requires large quantities of biomass over extensive periods of time.

The original catalytic (iron or cobalt) Fischer-Tropsch synthesis involves a series of chemical reactions that result in a variety of hydrocarbons of the basic formula C_nH_{2n+2}, the overall net reaction being:

$$(2n+1)\ H_2 + nCO \rightarrow C_nH_{2n+2} + nH_2O$$

Starting substances (H_2 and CO) can be produced by other methods, for example, in a partial combustion of hydrocarbons:

$$C_nH_{2n+2} + 1/2\ nO_2 \rightarrow (n+1)H_2 + nCO$$

The production of certain oxo chemicals (e.g., alcohols and aldehydes) from synthesis gas with a chemical composition of 40–45% H_2, 25–30% CO, and 25–30% CO_2 in the Fischer-Tropsch synthesis is under strong development to solve technical problems. Besides Fischer-Tropsch diesel, hydrogen, methanol, and DME prepared catalytically from methanol are manufactured.

In conversion reactions of this type, *catalysts* are one of the key factors — over 80% of the present chemical processes contain a catalytic stage. Catalysts offer an obvious way to reduce energy requirements of chemical reactions. A catalyst is commonly defined as a material that changes the rate of a chemical reaction without being consumed in the process. Catalytic reactions have a lower rate limiting free energy of activation than the corresponding non-catalyzed reactions; this results in a lower overall energy required and thus, a higher reaction rate at a given temperature. In addition to this, use of catalysts generally increases reaction selectivity. The catalyst can be heterogeneous or homogeneous (where catalyst occupies the same phase as the substrate) — various biocatalysts are often considered as a separate group. Besides "positive catalysts" there are "negative catalysts" (inhibitors) and their harmful role may be significant, for example, in fermentation processes. The application of catalytic chemistry to replace conventional stoichiometric processes is one of the key research areas within green chemistry.

12.3.2. *Pyrolysis*

In the slow heat-up of conventional pyrolysis ("*slow pyrolysis*") the most important product is solid char, but volatile products, gases and tar, are also formed. The yield of volatile products can be increased by faster heating ("*fast pyrolysis*"). High yields of liquid fuels are obtained from many cellulosic materials in this way. On the other hand, because of the complicated nature of the liquid product mixtures, this method is not economically very attractive for production of pure chemicals. However, high anhydrosugar yields are possible in fast pyrolysis, although certain pretreatment of the raw materials is necessary in these cases.

In conventional approaches, a fraction of the vapors were non-condensable "wood gas" (that could be burned as fuel) and solid charcoal was left in the retort. The elemental composition of charcoal (it generally contains 80% carbon), as well as its yield and properties, depend on final carbonization temperatures. In the latter part of the 19th century, wood carbonization was a major pyrolysis process that supplied charcoal for iron ore smelting in a rapidly industrializing world. After the 1870s, charcoal was replaced by coal and coke in the production of iron.

Nowadays, wood carbonization to yield charcoal for heating and cooking (it has clearly higher heat content than dry wood) is very common in many developing nations, often utilizing crude carbonization kilns. Charcoal has also found many industrial uses as an additive in paints, inks, and medicines and as an efficient adsorbent for various purposes.

In fast pyrolysis, still in a relatively early stage of development, the heating rate, for example, 300°C/min to 500–700°C is very high and the reaction time is only a few seconds or less. Therefore, chemical reaction kinetics, mass transfer processes, phase transitions, and heat transfer phenomena play important roles and influence the overall outcome. Since the critical issue is to bring the reacting biomass particle to the optimum process temperature, the particle size should be fairly small, 105–250 μm. This also minimizes the formation of charcoal. On the other hand, the gaseous products of fast pyrolysis require rapid cooling to minimize secondary reactions that can result, for example, in harmful carbon deposition. In addition, biomass should normally be dried to about 10% moisture content before fast pyrolysis, since a considerable amount of reaction water is formed in this process.

The liquid product from fast pyrolysis of biomass is known as bio-oil or fast pyrolysis bio-oil; it is a complex mixture of compounds derived from the fragmentation of lignin, cellulose, hemicelluloses, and extractives. Many examples of power generation from bio-oils with or without upgrading have been reported among recent renewable energy processes. The foremost idea behind the use of bio-oils as fuels is that they are much easier to handle and transport than solid biofuels. There are also some negatively bio-oil qualities, mainly low heating value, incompatibility with conventional fuels, high solids content, high viscosity, and chemical instability. For this reason, primary bio-oils need significant modification before use.

Although bio-oils can be readily stored, they still can change during storage due to reactions in pyrolysis failing to reach thermodynamic equilibrium. In addition, water content of bio-oils is typically about 25%, which may lead to the formation of two phases (i.e., lipophilic and hydrophilic ones) during storage. Their heating value can be increased by removing oxygen with unit processes generally utilized in the petroleum refining industry; other undesired characteristics can be improved by physical methods or by addition of proper solvents. All these methods are technically feasible but not always economically attractive. As indicated above, fast pyrolysis has also achieved some commercial success in the production of chemicals.

12.3.3. *Anaerobic digestion*

In naturally occurring ecosystems, organic matter is decomposed by microorganisms under oxygen-limiting or oxygen-free conditions to a methane(CH_4)-rich gas. Industrial biogas generation by *anaerobic digestion* is today an interesting way of producing fuel from various wastes. Raw materials used have traditionally been agricultural, animal, or urban wastes. The production potential of CH_4 strongly depends on plant species; an average value is 200–450 m^3 CH_4/ton organic material, corresponding to an average yield of 2500–6500 m^3 of CH_4 per hectare.

Anaerobic digestion of cellulose-containing material proceeds roughly as a three-stage process, resulting in the formation of CH_4 mixed with a substantial portion of CO_2 (Fig. 12.9). For example, biogas from sewage digesters usually contains from 55% to 65% CH_4, from 35% to 45% CO_2, and <1% N_2, while the same figures for biogas from organic waste digesters and in landfills are, respectively, 60–70% CH_4, 30–40% CO_2, and <1% N_2 and 45–55% CH_4, 30–40% CO_2, and 5–15% N_2. In the first phase of this process, hydrolysis, the feed-stock material is enzymatically decomposed to simple compounds, for example, sugars and "long-chain fatty acids" (LCFA), which are then further broken down in the second phase, acidogenesis, to acetic acid (acetate), H_2, and CO_2. In the last phase, CH_4 is obtained by the action of methanogens that can also be found, for example, in wetlands (i.e., the formation of marsh gas) and in the digestive tracts of animals such as ruminants and humans.

Biogas produced from waste or energy crops can be used as vehicle fuel in natural gas vehicles, since the same engine and vehicle configurations are used for both fuels. When gas is used as vehicle fuel, typical reductions in air emissions compared to diesel fuel emissions have been from 60% to 85% in nitrogen oxides (NO_x), from 60% to 80% in particulates, and from 10% to 70% in CO. Gaseous fuels have a higher ignition temperature and a higher lower flammability limit than those of

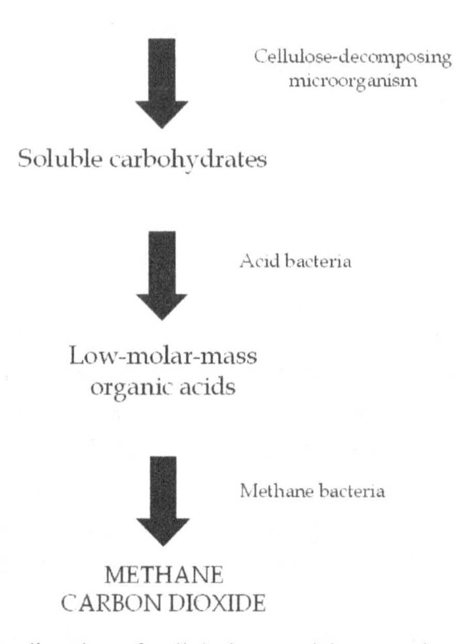

Fig. 12.9. Anaerobic digestion of cellulosic materials to methane (CH_4) and carbon dioxide (CO_2) (Alén 2011b).

liquid fuels; in leakage situations gas fuels rise into the atmosphere unlike liquid fuels. However, when using biogas as vehicle fuel and in fuel cells, a higher CH_4 concentration is required and the raw biogas has to be upgraded. In addition, when biogas is used in vehicles, hydrogen sulfide (H_2S) and halogenated compounds commonly found in biogas can cause corrosion in engines. Primary methods of biogas upgrading include water scrubbing and pressure swing adsorption (PSA), but more sophisticated techniques have been developed.

12.3.4. *Production of bioenergy*

As indicated above, production of bioenergy from a wide range of renewable resources is rapidly increasing. The use of biofuels also often offers advantages as both the production and consumption of energy are normally located in the same area. On the other hand, world energy demand has dramatically increased and new sources of energy are needed

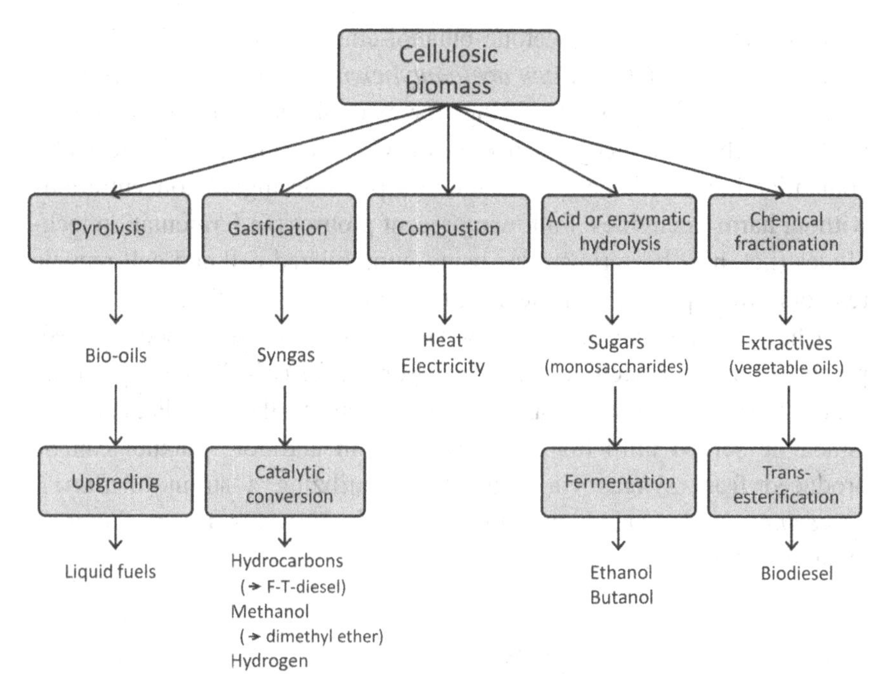

Fig. 12.10. Principal conversion routes for cellulosic biomass to produce various energy sources. (Alén 2011b). F-T refers to Fischer-Tropsch, see the text.

to replace fossil sources. In response to this trend, global strategies are developed to reduce GHG emissions and generally to increase energy efficiency. Meeting these targets requires actions in several different areas, while effectively utilizing the existing infrastructure (Fig. 12.10). The most effective conversion route depends on regional possibilities and the availability of biomass resources. In addition, the water footprint per unit of bioenergy, besides society's carbon footprint, should be kept at a satisfactory level. The production should not be allowed to compete with that of food, if based on the same feedstock.

The figure summarizes thermochemical and chemical routes in the production of bioenergy. Although anaerobic digestion is not included, biochemical production of ethanol and butanol by fermentation of sugars is shown. Along with the established manufacturing of ethanol, production of butanol may offer a promising full-scale way of converting cellulosic biomass into useful transportation fuel. This alcohol can be produced from sugars in "ABE (acetone-butanol-ethanol) fermentation" using anaerobic bacteria *Clostridium acetobutylicum* or *C. beijerinckii.* Butanol is an excellent fuel extender being, for example, superior to ethanol; it contains only 22% oxygen (in ethanol 35%), is able to tolerate water contamination, and is easy to blend with petrol in higher concentrations without harm to engines. Following recent progress in fermentation techniques (i.e., new bacteria strains using immobilized-cell and cell-recycle reactors), this application is nearing final commercialization.

Chaim Azriel Weizmann (1874–1952) originally developed the ABE process to produce acetone that was important for British industry during World War I. Weizmann was elected in 1949 as the first President of Israel; he served until his death in 1952. In addition, butanol can be produced from syngas via fermentation utilizing a strain of *Butyribacterium methylotrophicum* that converts CO directly to butanol, ethanol, butanoic acid, and acetic acid.

12.4. Chemical Delignification

Timber procurement covers wood harvesting (including logging) and all the activities needed to produce timber from the forest for mills. These activities mainly comprise cutting and off-road hauling of timber from the stump to road-side storage points or landings and its transportation from the landings to the mill yard. Pulp mill companies use good forest management practices to ensure a sustainable source of feedstock materials.

According to estimates particular to the Nordic region, 35–40% of the initial dry tree biomass remains as harvesting residues, i.e., stumps, roots, branches, needles, or foliage. These residues, in addition to material from debarking at a mill, can be processed in many ways ("feedstock-based technology") using various mechanical (pulverization and pelletizing), chemical (hydrolysis followed by fermentation), and thermochemical (torrefaction, pyrolysis, gasification, liquefaction, and combustion) methods to produce energy and chemical products. A prerequisite for planning these processes is that the detailed chemical composition of these materials is known and that all the feedstock constituents are taken into account.

In the simplest case, a biorefinery utilizes only one feedstock with a single process that results in a single major product. In more complicated cases, such as modern pulp mills, the biorefinery utilizes one wood or non-wood feedstock, but creates several end products. In the latter case it is important to clarify the chemical reactions to understand the formation of various degradation products (i.e., formation of pulping by products). In order to understand the overall chemical operation of a modern pulp mill, it is essential also to understand the thermochemical behavior of the main constituents of spent liquors as well as of bark, since these components are burned in the recovery and bark-burning furnaces, respectively.

Pulping refers to different processes to convert wood or other fibrous feedstocks into a product mass of liberated fibers by dissolving lignin that binds the cellulose fibers together. This conversion can be accomplished either chemically or mechanically or by combining these two treatments. The term "pulp" is collectively used for chemical, semi-chemical, chemimechanical, and mechanical pulps. Although pulps are mainly used for papermaking, some pulps are processed into cellulose derivatives (cellulose esters and ethers) and regenerated celluloses, for example, viscose or rayon (cf., Chapt. "9.4. Polysaccharides and Their Derivatives").

Table 12.2 gives a broad classification of commercial pulping processes and their yields. The average yield of chemical pulp is in the range of 45% to 55%. Yields of the products in dissolving pulping (i.e., acidic sulfite, multistage sulfite, and pre-hydrolysis kraft methods) are generally 35–40%; these products are utilized in the manufacture of cellulose derivatives and related products. Chemical pulping accounts for 70% of the total worldwide production; about 90% of chemical pulps (about 130 million tons) are currently produced in the dominant kraft (sulfate) process (cf., Chapt. "2.2 The Time After the 1700s"). Sulfite pulping has clearly decreased during recent decades.

Nowadays, wood is the predominant source of cellulosic fiber for pulp and paper manufacture corresponding to about 90% of the total, the rest being various fibrous non-wood feedstocks. The term "high-yield pulp" is often used for different types of lignin-rich pulps (mainly from neutral sulfite pulping) that need mechanical defibration.

Table 12.2. Commercial pulping methods (Alén 2011b)

Method	Yield % of wood
Chemical pulping	35–60
Kraft, polysulfide kraft, and pre-hydrolysis kraft	
Soda anthraquinone (AQ)	
Acid sulfite, bisulfite, and alkaline sulfite-AQ	
Multistage sulfite	
Semichemical pulping	65–85
Neutral sulfite semichemical (NSSC)	
Soda	
Chemimechanical pulping	80–90
Chemithermomechanical (CTMP)	
Chemigroundwood (CGWP)	
Mechanical pulping	91–98
Thermomechanical (TMP)	
Refiner mechanical (RMP)	
Stone groundwood (SGWP)	
Pressure groundwood (PGWP)	

12.4.1. *Kraft pulping*

In conventional kraft pulping, *white liquor* is used for cooking the chips; it is an aqueous solution that contains mainly active cooking chemicals, sodium hydroxide (NaOH) and sodium sulfide (Na_2S). After cooking (or digestion) in a pressurized digester at 160–170°C, the spent cooking liquor (*black liquor*) is separated from the pulp by washing and is concentrated to a 65–80% solids content in multiple-effect evaporators. It is then combusted in the recovery furnace to recover the cooking chemicals and to generate energy. Combustion of black liquor in the recovery furnace produces an inorganic smelt of sodium carbonate (Na_2CO_3) and sodium sulfide (Na_2S, formed from sodium sulfate, Na_2SO_4, by reduction) with a small amount of residual Na_2SO_4. The smelt is dissolved in water to form *green liquor*, which is reacted in the causticizing stage with $Ca(OH)_2$ to convert Na_2CO_3 into NaOH (the formed $CaCO_3$ is insoluble) and to regenerate the original white liquor. Due to incomplete conversion reactions (about 90%) in the recovery cycle, white liquor also contains some Na_2CO_3 and sodium salts of oxidized sulfur-containing anions as primary dead load chemicals.

Roughly one half of the wood substance degrades and dissolves in kraft pulping. Organic matter in the black liquor is composed of degradation products of lignin and polysaccharides in addition to a minor fraction of extractives. Figure 12.11 outlines the basic reactions and

LIGNIN

- Degradation
  ($\bar{M}_w$ decreases)
- Increase in hydrophilicity
  (liberation of phenolic groups)

→ Water /alkali-soluble lignin fragments

HEMICELLULOSES, CELLULOSE

- Cleavage of acetyl groups
- Peeling reaction
- Stopping reaction
- Alkaline hydrolysis

→ Water /alkali-soluble aliphatic carboxylic acids and hemicellulose fragments

EXTRACTIVES

- Hydrolysis of fatty esters
  (fats and waxes)
- Evaporation

→ Tall oil soap and crude turpentine

Fig. 12.11. Main reactions and phenomena in the feedstock constituents during kraft pulping (Alén 2011b).

phenomena between active alkali and the main wood constituents resulting in the formation of various soluble fractions. Of the charged alkali ($NaOH$ + Na_2S, 18–24% of the wood dry solids), 70–75% is required for the neutralization of aliphatic carboxylic acids and about 20% is consumed to neutralize degradation products of lignin. Aliphatic carboxylic acids in black liquor do not contain any sulfur; the sulfur content of kraft lignin is 1–2% of the dry solids, corresponding to 10–20% of the initial charge of sulfur. The approximate content of sodium in soluble aliphatic carboxylic acids and lignin is, respectively, 20% and 8% of the dry solids.

During kraft pulping, or generally during alkaline pulping, due to the lack of selectivity in delignification, a substantial amount of polysaccharides is degraded and converted mainly to low-molar-mass hydroxycarboxylic acids (cf., Chapt. "11.2. Polysaccharides"). The total mass loss of carbohydrates is about 30%, while mass losses of hemicelluloses and cellulose are, respectively, 50–60% and 10–15%. Small amounts of polysaccharides removed are not completely degraded and they can be found as hemicellulose residues in the final black liquor. Dissolution and degradation of lignin during kraft pulping also produces a complex mixture of breakdown products with a wide molar mass distribution, ranging from simple low-molar-mass phenolic compounds to large macromolecules.

Figure 12.12 illustrates a typical material balance of wood organics in the kraft process, including the subsequent oxygen-alkali delignification and the bleaching sequence, which produces kraft pulp with high brightness. Approximately 90% of the lignin is dissolved in the cooking. The total cooking yield of the conventional Scots pine (*Pinus sylvestris*) and silver birch (*Betula pendula*) kraft pulping is, respectively, 47% and 53% of the dry feedstock. If no chlorine-containing chemicals are used in bleaching, bleaching effluents and oxygen delignification effluent can be combined and burned together with black liquor in the recovery furnace. It should be noted that debarked wood yields only about 50% fiber (i.e., about 25% of the initial dry wood) after delignification.

The main by products of softwood kraft pulping are, besides black liquor, *crude sulfate turpentine* and *tall oil soap*. The availability of these

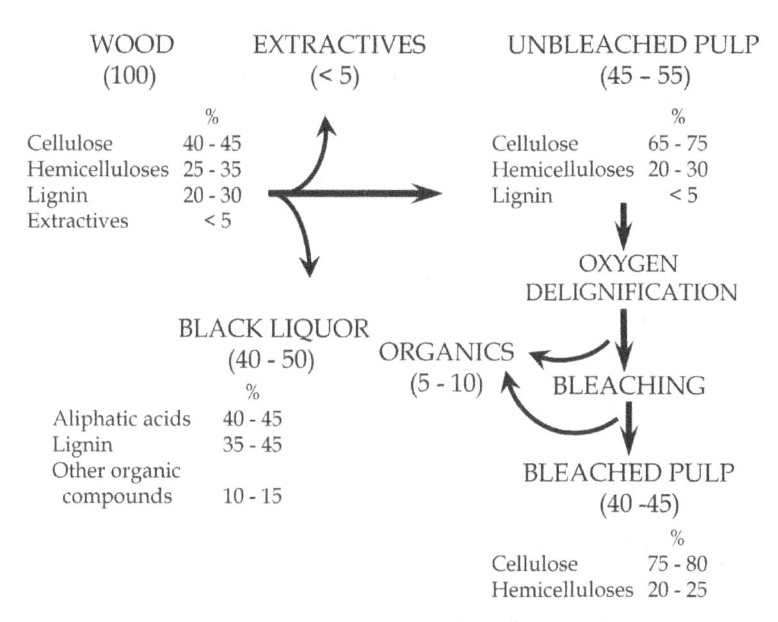

Fig. 12.12. Typical material balance of wood organics in the kraft process for bleached pulp (Alén 2011b).

by-products in a mill strongly depends on the wood species used, the method and time of storing logs and chips, and the growth conditions of the trees; even among extractives-rich pine species there is significant variation in the amount of by-products. After recovery of most extractives-based compounds, the remaining black liquor mainly contains, in addition to inorganic substances, lignin, carbohydrate degradation products (aliphatic carboxylic acids and hemicellulose residues), and residual extractives (Table 12.3).

Although most black liquors have rather similar composition, only a few of them, mainly from conventional processes, have so far been investigated in any detail even in terms of their main compounds. Thus, very little is currently known about the composition of black liquors from alkaline pulping of tropical hardwoods or non-wood materials. Lignin and aliphatic carboxylic acids are today almost exclusively utilized as fuel for production of energy for the pulp and paper mills. Partial utilization of these degradation products as chemical feedstocks may become more attractive.

Table 12.3. Typical composition of dry matter of Scots pine (*Pinus sylvestris*) and silver birch (*Betula pendula*) kraft black liquors (% of the total dry matter) (Alén 2011b)

Component	Pine	Birch
Lignin*	31	25
HMM (>500 Da) fraction	28	22
LMM (<500 Da) fraction	3	3
Aliphatic carboxylic acids	29	33
Formic acid	6	4
Acetic acid	4	8
Glycolic acid	2	2
Lactic acid	3	3
2-Hydroxybutanoic acid	1	5
3,4-Dideoxy-pentonic acid	2	1
3-Deoxy-pentonic acid	1	1
Xyloisosaccharinic acid	1	3
Glucoisosaccharinic acid	6	3
Others	3	3
Other organics	7	9
Extractives	4	3
Carbohydrates**	2	5
Miscellaneous compounds	1	1
Inorganics***	33	33
Sodium bound to organics	11	11
Inorganic compounds	22	22

*HMM and LMM refer to high- and low-molar-mass, respectively.
**Mainly hemicellulose-derived fragments.
***Due to the presence of dead-load inorganics, this mass proportion may be higher.

Black liquor has several features that make its combustion different from that of other fuels. Its high contents of inorganic material (25–40%) and water (15–35%) decrease its heating value to 12–15 MJ/kg dry solids; this can be compared with lignin (22–27 MJ/kg), aliphatic carboxylic acids (5–18 MJ/kg), and extractives (about 35 MJ/kg). Lignin, a significant component of black liquor organics, is mostly responsible for its heating value, which is clearly lower than that of other industrial fuels.

Crude turpentine is recovered from the digester relief condensates. Batch digester recovery systems differ substantially from continuous digester recovery systems in the methods of capturing relief vapors. The average yield of crude turpentine of pine species is 5–10 kg/ton of pulp and somewhat lower for spruce species. Crude turpentine is purified in a distillation process, where impurities, such as methyl mercaptan (MM) and dimethyl sulfide (DMS), as well as higher compounds, are removed. Main fractions obtained are *monoterpene fraction* that primarily consists of α-pinene (50–80% of all compounds), β-pinene, and Δ^3-carene, and *pine oil fraction* that mainly consists of hydroxylated monoterpenes. Monoterpenes have been used as paint thinners, varnishes, and lacquers or as rubber solvents and reclaiming agents. Today, they are used for making different products for chemical industry, for pharmaceutical purposes (liniments), and perfumery. Pine oil is used for a variety of purposes, such as for solvents with good emulsifying and dispersing properties and for flotation of minerals.

Tall oil soap is removed from black liquor during its evaporation process (the optimum dry solids content is 28–32%). Organic acids, fatty and resin acids, in the tall oil soap are liberated by adding H_2SO_4 to yield crude tall oil (CTO). The average yield of CTO is in the range of 30–50 kg/ton of pulp, corresponding to 50–70% of its initial amount in the raw material used for pulping. CTO is normally purified and fractionated in vacuum distillation (3–30 mbar, 170–290°C). The main fractions and their mass proportions are light oil (10–15%), fatty acids (20–40%), rosin (resin acids, 25–35%), and pitch residue (20–30%). Of these fractions, only some parts of the light oil and pitch residue are generally used for combustion, other fractions and parts are utilized for chemical purposes (industrial oils, rust protection, asphalt additive, and oil-well drilling muds). Fatty acids are used for paint vehicles, soaps, printing inks, foam inhibition, lubricants, greases, flotation agents, and industrial oils. Rosin is suitable for alkyd resins, printing inks, adhesives, emulsifiers, paint and lacquer vehicles, and soaps.

Various commercial fatty acid products from tall oil ("*tall oil fatty acids*", TOFAs) are available in various purity and composition. Oleic (monoenoic) and linoleic (dienoic) acids are common in most TOFA products. Predominating resin acids in commercial rosins are abietane (e.g., abietic and dehydroabietic acids) and pimarane (e.g., pimaric, palustric, and levopimaric acids) types and lesser amounts of labdane type can also be present in some rosins. Neutral components in light oil decrease yields of other fractions and in general reduce product quality. However, different extraction methods have been developed for removing neutral components prior to distillation. An example of such processes is the production of sitosterol, which after reduction to sitostanol and esterification can be used in foodstuff fats. This compound has been shown to lower the cholesterol level in human blood.

In addition to conventional burning of black liquor, partial recovery of its organic material (<20%) could be possible without interfering with the recovery of cooking chemicals. A polymeric lignin fraction begins to separate as a fine precipitate when the initial pH (about 13), of black liquor is reduced to about 10 by acidification and it can be precipitated more or less completely by further acidification. Neutralization of the phenolic hydroxyl groups in kraft lignin occurs in the pH range of 9–11, resulting in precipitation of most lignin. Further lowering of pH to about 2 then liberates the carboxylic acid groups (pK_a 3–5) and more lignin precipitates. The liquor is usually heated (80°C) to coagulate lignin and to overcome filtration difficulties at lower temperatures. Approximate yields of 35%, 80%, and 90% of the initial kraft lignin can be obtained at pHs 9.5, 8.0, and 2.0, respectively. In contrast, highly water-soluble lignin (10% of the total kraft lignin), comprising primarily "lignin monomers", cannot be precipitated.

This type of precipitation of spent alkaline cooking liquors has a long history; the general approach has been to use CO_2 from flue gases and from causticization, with possible addition of strong mineral acid, for this purpose. A part of lignin is generally removed if an overloaded recovery furnace is the process bottleneck. It also affords production of solid biofuel with a high energy density and low ash content. Moreover, there are process concepts in non-wood soda pulping, where partial lignin precipitation and sodium recovery are accomplished with carbonation together with electrodialysis (e.g., $Na_2CO_3 \rightarrow NaOH + CO_2$ and $R\text{-}CO_2Na \rightarrow R\text{-}CO_2H + NaOH$). In one possible process concept, black liquor is first evaporated to a solids content of 25–30% and the tall soap skimmings are recovered. Lignin is then precipitated in two-stage carbonation, first with impure CO_2 from flue gases followed by pure CO_2 under a pressure of 1500 kPa, resulting in a yield of about 75%. Lignin particles can also be concentrated by ultrafiltration, but the membrane material must be carefully selected to avoid problems with fouling and plugging.

All alkali lignin materials (see lignosulfonates, below) can be treated under many conditions, from thermochemical methods to chemical processes, and a wide range of products can be obtained: solid and liquid fuels, low cost carbon fiber, activated carbon, and resins ("phenol mixtures") for the plastic industry. Among other products that do not require major modification are binders, surface or dispersing agents, emulsifiers, and sequestrants. Lignin can replace many chemicals currently derived from petroleum-based sources; it is likely to be used as a bio-based additive in polymers. Most obvious use for the bulk of low-cost lignin is still biofuel as powder, pellets, or mixed with other fuels. Lignin recovered from black liquor can be used within the mill instead of fossil fuel oil in the lime kiln, or it can be sold to external customers for use in CHP. As kraft lignins contain some chemically bound sulfur, the combustion plant should be equipped with a suitable flue gas treatment system.

In addition to lignin, large amounts of aliphatic carboxylic acids (cf., Chapt. "11.2. Polysaccharides") are formed in the kraft pulp process and their partial recovery is an interesting alternative to using them as fuel. The basic idea behind this approach is that about two thirds of the total heat produced by the liquor originates from lignin and only one third stems from the remaining constituents, mainly these acids. Their recovery is a complicated separation problem and has, so far, only been solved on a laboratory scale. According to a simplified scheme, after the recovery of tall soap skimmings and carbonated lignin by precipitation (see above), the mother liquor is evaporated to crystallize out sodium bicarbonate/ carbonate ($NaHCO_3/Na_2CO_3$) suitable for causticization. As only a minor part of aliphatic acids can be utilized in the form of their sodium salts, the acids are next liberated (to pH of about 2.5) from sodium by a strong mineral acid, primarily H_2SO_4; simultaneously about half of the remaining lignin precipitates. Sodium sulfate (Na_2SO_4) formed is then crystallized out almost completely by evaporating the liquor under suitable conditions. After recovery of volatile formic and acetic acids by evaporation, a crude hydroxy acid fraction is obtained. Large-scale mutual separation of formic and acetic acids is possible by azeotropic distillation with the aid of ethylene dichloride.

The difficulty in this concept is to find a way of handling the huge amount of Na_2SO_4 formed. Na_2SO_4 is very cheap and has a rather limited use. It is chemically very stable and, for example, can be reduced to sodium sulfide (Na_2S) only at high temperatures. However, electrochemical membrane techniques, such as electrodialysis, offer a possibility of recovering and recycling at least a part of the sulfuric acid (i.e., the reaction $Na_2SO_4 \rightarrow H_2SO_4 + NaOH$). This electrodialysis technique could also be directly applied to carbonated black liquor, resulting in the formation of a crude acid fraction and NaOH.

Purification of this crude fraction of hydroxy acids is also somewhat problematic. A rough fractionation of acids, resulting in a "low-molar-mass acid mixture" of glycolic, lactic, and 2-hydroxybutanoic acids and, on the other hand, a "high-molar-mass acid mixture" of 3,4-dideoxy-pentonic, 3-deoxy-pentonic, xyloisosaccharinic, and glucoisosaccharinic acids, would be possible — they could be purified either by straight-forward distillation under reduced pressure (0.067–0.173 kPa) or by ion-exclusion chromategraphy. Due to difficulties in the distillation of glucoisosaccharinic acid from hydroxy acids and the instability of particularly glycolic acid during heating, overall distillation yields have been about 50% (pine) and 75% (birch). This means that the overall recovery of aliphatic acids is, respectively, about 70% and 85%.

Aliphatic carboxylic acids can be used as single components or as more or less purified mixtures in a number of applications. Of this group, formic, acetic, glycolic, and lactic acids are commercially important today. Most of the uncommon hydroxy acids can be converted into corresponding derivatives by reduction to polyalcohols, oxidation to polycarboxylic acids, or esterification producing emulsifying agents. They are also suitable as starting materials for a great deal of other chemicals and products.

Since the 1950s, the kraft process has become the dominant production method for chemical pulps in the world. The main reasons for this dominance are excellent pulp strength, low demands on wood species and quality, well-established recovery of cooking chemicals, energy surplus, by-products, and short cooking times. The main disadvantages of kraft pulping include relatively low yield, low brightness of unbleached pulps, odor problems, and high investment costs, mainly those of the recovery furnace.

In spite of a long development history, chemical pulping and bleaching procedures will still be continuously modified, resulting in an increasing diversification of processes. A general goal in cooking is to obtain as low a kappa number as possible without deactivating the residual lignin and without a negative influence on product quality. Low kappa numbers in cooking can be achieved by both modified continuous and batch delignification methods followed by oxygen-alkali delignification. Kappa number is used to determine the amount residual lignin in pulp; low numbers indicate low lignin conternts.

The current trend towards close process-water circulation aims at a drastic decrease in the wastewater load. A totally effluent-free (TEF) mill represents the ultimate objective in pulp production. In this respect, to avoid corrosion problems caused by chlorine-containing compounds, the use of oxygen-based chemicals, oxygen, ozone, hydrogen peroxide, and peracids, in a totally chlorine-free (TCF) bleaching offers a potential process alternative. The proportion of elemental chlorine-free (ECF) bleached pulp is gradually increasing in the world's bleached pulp production, being today more than 50%; the corresponding TCF production is about 5%. In addition to the strong development of chemical pulping, different mechanical and high-yield pulping methods are also gaining increasing importance.

12.4.2. *Sulfite pulping*

The sulfite process is still important in certain countries and for some pulp qualities. Its advantages over the kraft process (see above) can be summarized to be higher brightness of unbleached pulps, higher yields at a given kappa number, lower odor problems, and lower investment cost. Unlike the kraft process, the sulfite process covers the whole range of pH (Table 12.4); this enables high flexibility in pulp yields and properties.

Table 12.4. Sulfite pulping processes (Alén 2011b)

Process*	pH range	Base	Active reagent	Pulp type
Acid sulfite	1–2	$Na^{\oplus}$, $Mg^{2\oplus}$, $Ca^{\oplus}$, $H_4N^{\oplus}$	$HSO_3^{\ominus}$, $H^{\oplus}$	Dissolving pulp Chemical pulp
Bisulfite	2–6	$Na^{\oplus}$, $Mg^{2\oplus}$, $H_4N^{\oplus}$	$HSO_3^{\ominus}$, $H^{\oplus}$	Chemical pulp High-yield pulp
Neutral sulfite (NSSC)	6–9	$Na^{\oplus}$, $H_4N^{\oplus}$	$HSO_3^{\ominus}$, $SO_3^{2\ominus}$	High-yield pulp
Alkaline sulfite-AQ	9–13	$Na^{\oplus}$	$HSO_3^{\ominus}$, $HO^{\ominus}$	Chemical pulp

*NSSC refers to neutral sulfite semichemical and AQ anthraquinone.

The pulps extend from dissolving pulps for chemical end uses to high-yield NSSC grades. The latter pulps are produced by cooking the chips with Na_2SO_3/$NaHSO_3$ solutions, followed by the mechanical defibration of the partially delignified wood in disc refiners. The fiber properties of hardwood NSSC pulps make them suitable especially for corrugating mediums. However, most pulps produced by acid sulfite and bisulfite pulping are suitable for different paper grades. Alkaline sulfite-AQ methods generally produce kraft-type pulps.

The active sulfur-containing species in the sulfite process are sulfur dioxide (SO_2), hydrogen sulfite ions ($HSO_3^{\ominus}$), and sulfite ions ($SO_3^{2\ominus}$) together with active ions $H^{\oplus}$ and $HO^{\ominus}$ in proportions that depend on the pH of the cooking liquor. According to the equilibrium, these species are present almost exclusively in the form of $HSO_3^{\ominus}$ in water solutions of SO_2 (the total SO_2 charge is typically 20% of oven dry wood feed-stock) at pH around 4. Below and above this value, the concentrations of SO_2 and $SO_3^{2\ominus}$ ions, respectively, increase. In the old terminology, the total amount of SO_2 ("total sulfur dioxide") present in cooking acid or cooking liquor is divided into so-called "free" and "combined" sulfur dioxide. They are usually expressed as grams of SO_2 per 100 mL of solution.

The active base, normally $Na^{\oplus}$, $NH_3^{\oplus}$, $Mg^{2\oplus}$, or $Ca^{2\oplus}$, is the cation bound to the $HSO_3^{\ominus}$ and $SO_3^{2\ominus}$ and its concentration is usually expressed in grams of Na_2O/liter of solution. The conventional base, calcium, was mainly used due to its low cost (from limestone, $CaCO_3$) and, because of lack of stringent environmental quality regulations, there was no need to recover it. However, due to the limited solubility of calcium sulfite ($CaSO_3$), calcium base can be used only in acid sulfite pulping, where a large excess of SO_2 is available to prevent the formation of $CaSO_3$ from calcium hydrogen sulfite, $Ca(HSO_3)_2$. When using more soluble magnesium as a base, pH can be increased to about 5; above this magnesium sulfite ($MgSO_3$) starts to precipitate, while magnesium precipitates as hydroxide in the alkaline region. In contrast, sodium and ammonium sulfites and hydroxides are easily soluble and the use of these bases places no limitations to the pH of the cooking liquor. In modern sulfite cooking, mainly sodium (and magnesium, at pH <5) are used and inorganic chemicals are recovered and regenerated. Organic solids are burned to generate energy.

In addition to the "basic processes" shown in Table 12.14, sulfite pulping in two or even three stages with different pH regions has been developed as a means of enhancing pulp properties for different applications. One possibility of two-stage (or multistage) applications is to precook the wood chips in a $Na_2SO_3/NaHSO_3$ solution at pH 6–7 and then use acid sulfite pulping in a second cooking stage. In the first stage lignin is sulfonated (see below) to a certain degree, but is mainly retained in the solid wood phase. In the second stage, delignification is accomplished by charging liquid SO_2 to the digester. Compared to conventional acid sulfite pulping, two-stage processes greatly improve the uniformity of lignin sulfonation and can result in an increase in pulp yield (maximally about 8% of dry wood). This yield increase, mainly restricted to softwoods, is associated with increased retention of glucomannan in the pulp. In contrast, two-stage pulping of hardwoods only moderately improves xylan yield compared with conventional pulping and is not used in the pulp industry. An additional advantage of two-stage pulping is that it can process pine heartwood, which is not possible in conventional acid sulfite process. In the first stage at pH 6–7 reactive groups of lignin are protected by their sulfonation, which blocks their condensation reactions with phenolic extractives (see pinosylvin and taxifolin, below).

The term "lignosulfonates" refers to neutralized lignin fragments dissolved in the cooking liquor. A certain amount of base is required for the neutralization of lignosulfonic acids (see below) and other acidic degradation products of the wood substance. If the base concentration is too low in an acid sulfite cook, the pH sharply decreases and the rate of competing lignin condensation reactions increases. This accelerates the decomposition of cooking acid in the interior of the chips and results in dark, hard cores. These harmful reactions cause decreased delignification or completely prevent it. In general, the heat value of lignosulfonates is rather high, roughly one half of that of oil, while the heat value of the carbohydrate fraction is much lower, roughly one third of that of oil.

Two types of reactions, sulfonation and hydrolysis, cause delignification in sulfite pulping (Fig. 12.13). Sulfonation generates hydrophilic sulfonic acid (-SO_3H) groups, while hydrolysis breaks aryl ether linkages between the phenylpropane units, thus lowering the average molar mass and creating new free phenolic hydroxyl groups. Most sulfonic acid groups introduced into lignin replace hydroxyl or etherified substituents at the α-carbon atom of the propane side chain. Both of these reactions, sulfonation and hydrolysis, increase the hydrophilicity of lignin and facilitate its water solubility. Hydrolysis is fast compared to sulfonation under the conditions of acid sulfite pulping although lignin is also sulfonated to a fairly high degree; the molar ratio of sulfonic acid groups to spruce lignin methoxyl groups is about 0.5, promoting extensive dissolution of lignin. In contrast, delignification is slow in neutral and alkaline sulfite pulping, because hydrolysis reactions are very slow compared to sulfonation. The degree of sulfonation of lignin also remains low; the molar ratio of sulfonic acid groups to spruce lignin methoxyl groups is about 0.3. The major portion, 80–90% of the sulfur, is bound in sulfonate groups, but minor amounts of active sulfur are also consumed in the formation of carbohydrate sulfonic acids (i.e., the formation of α-hydroxysulfonic acids). A substantial amount of base is required for the neutralization of all lignosulfonic acids.

Because of the sensitivity of glycosidic linkages toward acid hydrolysis, the most prominent reaction of polysaccharides during acid sulfite and bisulfite pulping is the cleavage of glycosidic bonds in the hemicellulose components, giving rise to monosaccharides and soluble oligo- and polysaccharide fragments. Hemicelluloses also in this case, like in kraft pulping, are more readily attacked than cellulose due to their

LIGNIN

- Degradation (hydrolysis)
  ($\overline{M}_w$ decreases)

- Increase in hydrophilicity
  (sulfonation, liberation of phenolic groups)

→ Water/acid-soluble lignosulfonates

HEMICELLULOSES

- Cleavage of acetyl groups

- Hydrolysis of glycosidic bonds

→ Water/acid-soluble mono-, oligo-, and poly-saccharides, carboxylic acids, and furans

EXTRACTIVES

- Hydrolysis of fats and waxes

- Dehydration

- Sulfonation

- Evaporation

→ Water/acid-soluble fragments and sulfite turpentine (*p*-cymene)

Fig. 12.13. Main reactions and phenomena in the feedstock during acid and bisulfite pulping (Alén 2011b). Note that cellulose is practically stable under these conditions.

amorphous state and relatively low DP. In practice, no cellulose is lost in these acidic delignification processes unless delignification is extended to very low lignin contents and conditions are rather drastic, as in the production of dissolving pulps. When hydrolysis has proceeded far enough, depolymerized hemicellulose fragments are dissolved in the cooking liquor and tend to be hydrolyzed to monosaccharides.

Besides typical depolymerization of glycosidic bonds in polysaccharides, some other reactions occur, including deacetylation (i.e., formation of acetic acid), oxidation of monosaccharides (15–20%) by $HSO_3^{\ominus}$ (i.e., forming aldonic acids), and dehydration (i.e., formation of furan derivatives, such as furfural). Hemicellulose yield losses are generally higher for hardwoods than for softwoods; the total yield in Norway spruce (*Picea abies*) and silver birch (*Betula pendula*) acid sulfite pulping is, respectively, 52% and 49% of the dry feedstock. After bisulfite and neutral sulfite cooking, a large portion of soluble carbohydrates remains as oligo- and polysaccharides. In contrast, polysaccharides are degraded in alkali-catalyzed peeling reactions (see kraft pulping, above) with an excess of alkali present.

During acid sulfite and bisulfite pulping, fatty acid esters are saponified to an extent determined by the cooking conditions. Some resin components can also become sulfonated, resulting in their increased hydrophilicity and better solubility. Dehydrogenation of certain extractives-derived compounds is also possible. The formation of *p*-cymene ("sulfite turpentine") from α-pinene and quercetin from taxifolin are well-known reactions of this type. Due to their unsaturation, diterpenoids, including the resin acids, are probably polymerized to high-molar-mass products causing pitch problems in the subsequent paper production. Lignin can also condense with reactive phenolic extractives; for example, pinosylvin and its monomethyl ethers in pine heartwood can act as nucleophilic agents creating harmful cross links. Similar cross-links between lignin entities can also be generated by thiosulfate in the cooking liquor. This results in retarded delignification and, under exceptional circumstances, in its complete inhibition ("black cook"). The reactions of extractives in alkaline sulfite pulping are probably similar to those generally occurring in alkaline pulping.

A variety of useful products can be obtained from sulfite spent liquors, but today most of their organic solids are burned to generate energy and recover cooking chemicals. Since the dissolved organic solids represent a considerable fuel value, there are rather few industrial applications for the production of chemicals from spent liquor. Typical sulfite spent liquors differ from alkaline ones in many respects (Table 12.5). Because of problems with the separation of extractives from sulfite spent liquors, their utilization has never received the same status as the recovery of extractives from kraft black liquors. Sulfite turpentine has traditionally been the sole extractive-derived by-product from acidic sulfite pulping; *p*-cymene can be separated from the digester gas relief condensates and purified by distillation. The crude product can be used within the mill as a resin cleaning solvent, while the distilled product finds use in the paint and varnish industry.

Table 12.5. Typical composition of Norway spruce (*Picea abies*) and silver birch (*Betula pendula*) spent acid sulfite liquors (kg/ton pulp) (Alén 2011b)

Component	Spruce	Birch
Lignosulfonates	510	435
Carbohydrates	270	380
Monosaccharides	215	305
Arabinose	10	5
Xylose	45	240
Galactose	30	5
Glucose	25	10
Mannose	105	45
Oligo- and polysaccharides	55	75
Carboxylic acids	70	130
Acetic acid	30	75
Aldonic acids	40	55
Extractives	40	40
Others	30	55

High-molar-mass lignosulfonates in sulfite spent liquors can be separated in more or less pure form from low-molar-mass fragments by ultrafiltration. After isolation and purifycation the solution is concentrated by evaporation. The lignosulfonates are usually marketed in a powder form after spray drying. They are useful for a number of applications, especially due to their adhesion and dispersion properties. Their primary utilization includes additives (in oil well drilling muds and Portland cement concretes), dispersing agents and binders (in textiles, products of printing industry, and mineral slurries), and chemical purposes, such as production of vanillin and phenolic resins. Minor uses in fertilizer applications, animal feeds, silages, insecticides, and herbicides also exist.

Fermentation has played a dominating role in the industrial processing of the carbohydrate fraction of sulfite spent liquors. These liquors have been used for producing ethanol and single-cell protein. Production of protein by means of aerobic cultivation using either yeast (*Candida utilis*) or fungi (*Paecilomyces variotii*) has taken place in some sulfite mills. In addition to hexose and pentose sugars, acetic acid and aldonic acids are consumed by these microorganisms. Although fermentation processes were once an effective way of reducing the pollution load from the mill, this kind of by-product utilization is currently not economically attractive. The isolation of single components from spent liquors is often possible only with tedious and complex separation methods. Therefore, the value of the final products must be high to compensate the separation cost.

Various monosaccharides and their degradation products, such as furfural, can also be isolated from sulfite spent liquors and used in other applications. However, so far such processes have been of limited practical interest because of the complexity and expense of separating these substances. A characteristic of the spent liquors from neutral sulfite cooking of hardwood is the high proportion of acetic acid in comparison with other organic compounds. A full-scale process has been developed for extracting acetic acid with an organic solvent from the spent liquor after acidification. The formic acid present in small quantities can be removed by azeotropic distillation if a pure product is needed.

12.4.3. *Non-wood pulping*

The most traditional delignification methods for non-wood feedstocks are soda and soda-AQ processes, although various sulfite and kraft processes are used as well. Pulping of non-wood, as well as wood, raw materials with organic solvents ("*organosolv methods*"), an approach dating back to the beginning of the 1930s (Theodor N. Kleinert and Kurt Von Tayenthal), was not seriously considered for practical application until the 1980s. The idea was to use basically "lignin solvents" for this purpose, but the systems were systematically investigated with a view to develop new industrial pulping systems (Fig. 12.14).

The main driving force for developing organosolv pulping has shifted from original energy-related considerations toward the possibility of sulfur-free, less polluting, and economical (small-size) pulp mills with a simplified chemical recovery system, as well as improved recovery and upgrading of by-product lignin and hemicelluloses. These mainly acidic processes can provide solutions to problems of conventional non-wood

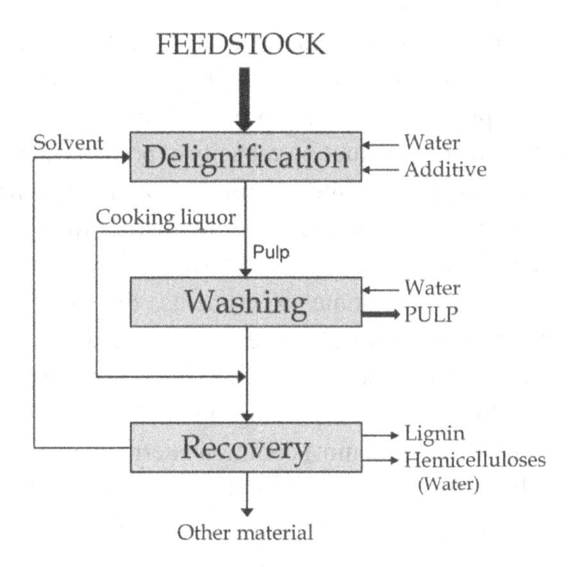

Fig. 12.14. Simplified flowchart of organosolv processes (Alén 2011b). Note that delignification is possible also without water and/or external additives.

pulping, including those related to chemical recovery of silica (it dissolves in alkali); the present alkaline processes already incorporate silica removal systems.

These ambitious goals have given rise to many variations of organosolv pulping with organic solvents, mainly ethanol and methanol, as well as formic and acetic acids. Most of these processes have been intensively studied on a laboratory scale, but only few full-scale applications have emerged. Most of these solvents have been used in the presence of water; the introduction of suitable catalysts (e.g., NaOH, H_2SO_4, HCl, $MgCl_2$, $AlCl_3$, and AQ) as the third component of the cooking liquor has enabled the production of pulps with a satisfactory delignification, yield, and strength. In addition, complicated processes, such as ASAM (alkaline sulfite-AQ-methanol) have been developed. Some organosolv methods, seen as potential examples of biorefinery concepts, may in the future attract increased interest. As the ultimate aim is not only effective production of chemical pulp, but also the utilization of lignin and carbohydrates without their significant degradation, organosolv methods may play a significant role in the future.

Initial trials with simple solvent systems, as well as trials with more complicated cooking liquors, have failed to produce results superior to those of the kraft process. The new processes tend to yield poor pulp strength properties and the production capacity of potential new organosolv mills is too small in relation to the existing raw material supply, making large-scale investments economically unrealistic. In addition, their chemical recovery (i.e., the recovery of solvent and possible alkali added) needs further development. Some experiments, the MD process using methanol and alkali and the Alcell process using ethanol as solvent, have reached pilot and even full-scale production, but no real breakthrough.

In the search for sulfur-free pulping methods, traditional soda and soda-AQ processes are still the most common alternatives. Due to possible health risks, it has been recommended that the use of AQ should be decreased. Hence, delignification of non-woods (and woods as well) with oxygen and alkali may offer a potential sulfur-free method. Numerous studies of one-stage delignification of wood by oxygen in the pH region of 7–9 have been made, especially in the 1950–1970s. The characteristic feature of this process is extensive oxidation degradation of lignin and carbohydrates, resulting in black liquors that have lower heating values than those from the kraft and soda-AQ pulping. The delignification rate in the oxygen-alkali pulping is usually slightly lower than in the usual pulping methods.

12.5. Integrated Biorefineries

Carbohydrates and their derivatives have a large potential as starting materials for useful low- and high-molar-mass products. The primary approach of utilizing wood and other carbohydrate-containing biomasses is their straightforward degradation into fermentable sugars; it can be accomplished by suitable pretreatment followed by acid or enzymatic hydrolysis. The hemicelluloses in wood are more readily hydrolysable by acids than cellulose, and their removal also enhances the reactivity of cellulose in the residual solids. As shown above, the majority of hemicelluloses, along with lignin, are dissolved into cooking liquor during chemical pulping. Thus, one potential approach would be to separate a prominent part of hemicelluloses prior to delignification instead of recovering their degradation products from the spent cooking liquors. This is actually not a recent idea, as industrial pre-hydrolysis of hemicelluloses, especially prior to hardwood kraft pulping, has been used for many years in the production of dissolving pulp. Further, it may be better to recover the hemicelluloses as monosaccharides rather than trying to extract them in large quantities without major degradation.

Especially in the Northern Hemisphere, the pulp and paper industry faces major challenges and needs new products with added value in order to remain competitive. Integration of various pre-treatment stages prior to pulping may be a possibility: new ways of recovering dissolved organic solids during alkaline pulping, as well as from kraft black liquors, need to be considered. In developing concepts for the potential recovery of dissolved organics during alkaline pulping, certain limiting factors, including both technical and economic factors, need to be considered. In all cases, the main product is cellulosic fiber; its strength properties must be maintained without interfering with the recovery of cooking chemicals. The extractives must also be efficiently separated. These prerequisites practically dictate that only partial recovery of the dissolved material is possible. The general aim is to maximize the recovery of carbohydrate-derived material with low heating value, while minimizing the recovery of lignin-derived material with high heating value. It would be also advantageous if sulfur-free by-products could be produced by applying straightforward separation techniques. The typical pre-treatment process is hot-water extraction catalyzed by bases or acids (Fig. 12.15). Depending on the pH used, it is possible to produce a wide range of products that are also present in the cooking liquors after delignification. Such pre-hydrolysis processes of wood chips have been investigated under a variety of conditions from several points of view.

12.5.1. *Alkaline pre-treatments*

Carbohydrate losses are high in the beginning of the cook, when delignification is still slow. This rather low selectivity in softwood kraft pulping means, for example, that at the end of the heating-up period, somewhat after the lignin extraction phase, the mass ratio of the aliphatic acid fraction to the lignin fraction is 1.1–1.2; in the final black liquor, it

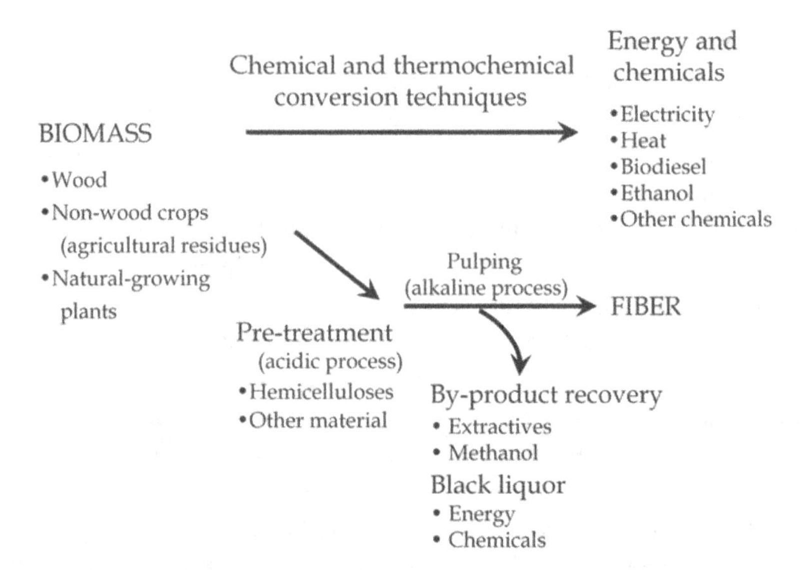

Fig. 12.15. Principles of a biorefinery concept in the forest industry (Alén 2011b).

is 0.8–0.9. The average molar mass of the kraft lignin dissolved during the initial phase, about 1900 Da and 2200 Da, respectively, for pine and birch lignin, is lower than that observed in black liquor, about 3100 Da and 2600 Da, respectively, for pine and birch lignin. Hydrogen sulfide ions react primarily with lignin, whereas carbohydrate reactions are only driven by alkalinity.

In the early stage of delignification under alkaline conditions (i.e., with only aqueous NaOH), a large amount of aliphatic carboxylic acids in relation to lignin, removed mainly by extraction rather than intense degradation, could be obtained. This process, resulting in the formation of sulfur-free fractions of organic material, is also in accordance with limiting factors listed above. In a modern kraft process, when surplus energy from the burning of black liquor is available, partial recovery of dissolved organic solids becomes more attractive, and no major changes in current practices are likely.

Figure 12.16 shows an example of the fractionation process of effluent from alkaline pre-treatment of wood. In practice, about 15% of the dissolved feedstock material can be withdrawn in a process where the chips are treated at 150°C for 90 min with a liquor-to-wood ratio of

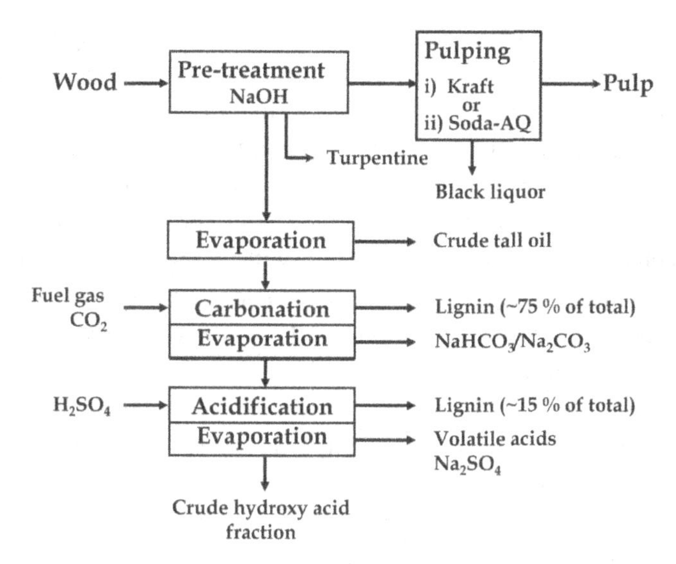

Fig. 12.16. Schematic representation of the fractionation process of black liquor (Alén 2011b).

5 L/kg and an alkali (NaOH) charge of 6–8% on oven-dry feedstock. In the hardwood case, due to the rapid deacetylation of the acetyl groups of xylan, 60–70% of the acids, mainly acetic acid, are volatile and can readily be recovered.

The withdrawal of the pre-treatment spent liquor may also influence the combustion properties of black liquor in the recovery furnace. Besides knowledge of changes in the burning behavior induced by modified methods of cooking, further information is also needed about optimizing the delignification of alkali-pre-treated wood in modern kraft (or soda-AQ) pulping. Depending on the alkali charge, temperature, and time, the target of alkaline pre-treatment can also be the dissolution of hemicelluloses.

12.5.2. *Acidic pre-treatments*

Organic material can be removed, in addition to alkaline pre-treatment, under acidic conditions (Fig. 12.15). Of these methods, "autohydrolysis" is of special interest because, compared to pre-hydrolysis with dilute mineral acids, water is the only reagent, making it environmentally friendly and inexpensive. In this approach, the hydrolysis is catalyzed by hydronium ions ($H_3O^{\oplus}$). The acidic conditions in autohydrolysis are created through the cleavage of acetyl groups from hemicelluloses (i.e., the release of acetic acid) and the formation of formic acid from carbohydrate-degradation products. Effluents from conventional autohydrolysis (hydrolysates, pH of 3–4) contain a mixture of various carbohydrates: in addition to oligo- and polysaccharides, some monosaccharides are present together with minor amounts of other organics including aliphatic carboxylic acids, as well as furans and heterogeneous fractions of lignin- and extractives-derived materials. A few inorganics are also present.

In general, about 15% of the dissolved feedstock material can be withdrawn in a process where the chips are treated at 150°C for 90 min with a liquor-to-wood ratio of 5 L/kg. The hydrolysates typically contain 75–80% of carbohydrates; only one-third is present as monosaccharides. In addition, lignin fragments have been characterized in the hydrolysates. Further hydrolysis, for example by enzymes, is normally needed when aiming at the production of value-added products by fermentation.

Acidic pre-treatments of wood chips with an addition of a mineral acid catalyst have been investigated under varying conditions, mainly to recover hemicelluloses. Chemical utilization of hemicelluloses includes films, surface modifiers, and a wide range of their derivatives. Compared to acidic pre-treatments, alkaline pre-treatments are generally more effective in the solubilization of extractives and lignin. Acidic pre-treatments tend to have a more negative effect on pulp strength properties than alkaline ones; the latter may also enhance the impregnation of cooking alkali in the next cooking stage.

12.6. Concluding Remarks

Petroleum or crude oil ("mineral oil" or "rock oil") (in Greek, "petra" means "rocks" and "elaion" "oil") is fossil fuel formed from ancient fossilized organic materials, microscopic zooplankton, plants, and algae that settle to sea or lake bottoms, mixed with sediments and buried under anaerobic conditions and subjected to intense heat and pressure. This gradually breaks down their organic material: it first turns into a bituminous material known as *kerogen* (in Greek, "keros" means "wax" and "-gen" "birth") and, heated to the right temperatures in the Earth's crust during millions of years, into a mixture of solid (coal), liquid (petroleum; the black, highly viscous variety is called "*bitumen*"), and gaseous hydrocarbons via a process known as "*catagenesis*". The term "kerogen" was introduced in 1912 by Alexander Crum Brown (1838–1922). Hydrocarbon-rich petroleum is mostly recovered from underneath sedimentary deposits via drilling oil wells; natural gas is usually produced along with crude oil. Known natural petroleum springs are very rare. Petroleum is processed and refined into a wide range of products in an oil or petroleum refinery. Kerogen-rich shales that have not undergone decomposition by heating to release their hydrocarbons may form oil shale deposits (see below).

Unrefined crude oil was used for many purposes many thousands of years ago. The modern history of petroleum can be considered to have started in the 19th century, when James Young (1811–83) in 1847 distilled from a natural petroleum seepage a light thin oil suitable for use as lamp oil and, at the same time, prepared a thicker oil suitable as a lubricant. He also developed a method of distilling paraffin from coal and oil shales. The large-scale petroleum industry, the actual "era of oil", started in the United States in the late 1800s.

The First Industrial Revolution was a transition from hand production to machines, new chemical manufacturing, iron production, and improved use of steam energy; briefly, a change from an agrarian handicraft economy to one dominated by industry and machine power. These technological changes included effective use of new energy sources: coal, petroleum, steam, and electricity, as well as new basic materials, chiefly iron and steel, and newly invented machines, such as the spinning jenny for producing textiles. In addition, a new organization of work, known as the "factory system", was introduced. Important developments in transportation, the steamship and steam locomotive, as well as other steam-powered wheeled vehicles, arose with other rapidly increasing applications of science to industry and agriculture. Almost every aspect of daily life was influenced by this revolution; besides technology, it affected socioeconomic and cultural values. Average income, and particularly population, began an unprecedented, sustained growth. This period began in Britain in the 18th century, from about 1760 to sometime between 1820 and 1840; from there, it spread to other parts of the world. In the Second

Industrial Revolution, in the transition years between 1840 and 1870, rapid technological and economic progress led to increasing adoption of steam-powered transport on railways and ships, large-scale manufacture of machine tools, and increasing industrial use of machinery powered by steam.

There are different opinions about how long we can continue to utilize our fossil resources. These reserves are certainly limited and we need, not only to develop new ways of producing energy, but also to find alternative ways of manufacturing important biochemicals. Estimated amounts of economically recoverable oil are mainly limited by technologies of bringing the oil to surface; these estimates vary and are influenced by economic and political pressures. For these recognized reasons and those related to environmental issues, industrial activities are gradually entering a new era, the Green Industrial Revolution, which prompts us to focus on better use of renewable feedstock resources. Refined petroleum contains hydrocarbons of varying molar masses with different physical and chemical properties; this, and its worth as a portable, dense energy source for a vast majority of today's vehicles, makes it one of the world's most important commodities. Petroleum now provides about 40% of world's total energy consumption and about 90% of its vehicular fuel needs. However, the current use of oil reserves is far beyond the estimated "CO_2 tolerance" of the Earth's atmosphere. Therefore, we need new innovations utilizing alternative energy sources and green technologies.

For the near future, fossil fuel resources that partly replace petroleum are, besides natural gas (see also GTL processes), unconventional petroleum deposits: *oil sands* or tar sands (bituminous sands) mainly found in large quantities in Canada and Venezuela, and *oil shale* or kerogen shale (kukersite or "burning rock") found in Estonia and North-West Russia, as well as in China and Brazil. Oil shale from Estonia was named in 1917 by Mikhail Zalessky (1877–1946) after the German name of the Kukruse manor (in German "Kuckers"). Oil sands are either loose sands or partially consolidated sandstone consisting of a mixture of sand, clay, and water that is saturated with bitumen (see above). In the recovery methods of oil from these resources, more water and larger amounts of energy are needed than in conventional oil extraction. Oil shale is a fine-grained sedimentary rock that contains kerogen from which liquid hydrocarbons (*shale oil*) can be recovered although it is more costly than the production of conventional crude oil. Pyrolysis of oil shale under anaerobic conditions at 450–500°C results in the formation of *shale gas,* which upon cooling leads to *liquid shale oil.* The term "shale gas" is normally related to natural gas trapped within shale formations. Apparently, shale gas will greatly expand the worldwide energy supply; it has become an increasingly important source of natural gas, especially in the United States. Also, China has enormous shale gas reserves.

Peat is harvested in many countries as an important source of fuel and several other products. It is formed as an accumulation of partially decayed vegetation in peatlands, bogs or mires. Peat is generally regarded as slowly renewable biomass due to its slow regrowth rate, about one mm per year. The peatland ecosystem is also an efficient carbon sink because its plants capture the CO_2 naturally released from the peat, thus maintaining an equilibrium. Natural carbon sinks, such as forests, soils, oceans, and atmosphere, accumulate and store more carbon than they release, while carbon sources release more carbon than they absorb. As a renewable biomass, forests are considered an effecttive carbon sink although there still are many gaps in our knowledge about the carbon cycle of different forest types. Carbon moves between these sources and sinks in a continuous cycle.

The removal of CO_2 from the atmosphere into carbon sinks is known as *carbon sequestration*. The most important of these sinks are actually fossil fuel deposits buried deep inside the Earth and thus are separated from the carbon cycle in the atmosphere. However, this separation ended when humans started utilizing those fossil resources and returning their carbon as CO_2 into the atmosphere; the result has been that the GHG concentration (the primary component is CO_2) in air is now more than 30% higher than that at the beginning of the First Industrial Revolution. Nowadays, most climate scientists agree that the current average global warming is mainly due to increased concentrations of GHG emissions caused by human activities.

Bibliography

Alén, R., Conversion of cellulose-containing materials into useful products, in: J.F. Kennedy, G.O. Phillips, and P.A. Williams (Eds.), *Cellulose Sources and Exploitation — Industrial Utilization, Biotechnology, and Physico-Chemical Properties*, Ellis Horwood, Chichester, UK, 1990, pp. 453–464.

Alén, R., Structure and chemical composition of wood, in: P. Stenius (Ed.), *Forest Products Chemistry*, Fapet Oy, Helsinki, Finland, 2000, pp. 11–57.

Alén, R., Basic chemistry of wood delignification, in: P. Stenius (Ed.), *Forest Products Chemistry*, Fapet Oy, Helsinki, Finland, 2000, pp. 58–104.

Alén, R. (Ed.), *Papermaking Chemistry*, 2nd edition, Finnish Paper Engineers' Association, Helsinki, Finland, 2007.

Alén, R., *Collection of Organic Compounds — Properties and Uses*, Consalen Consulting, Helsinki, Finland, 2009 (in Finnish).

Alén, R., Structure and chemical composition of biomass feedstocks, in: R. Alén (Ed.), *Biorefining of Forest Resources*, Paper Engineers' Association, Helsinki, Finland, 2011a, pp. 17–54.

Alén, R., Principles of biorefining, in: R. Alén (Ed.), *Biorefining of Forest Resources*, Paper Engineers' Association, Helsinki, Finland, 2011b, pp. 55–114.

Alén, R., Cellulose derivatives, in: R. Alén (Ed.), *Biorefining of Forest Resources*, Paper Engineers' Association, Helsinki, Finland, 2011c, pp. 305–354.

Alén, R., Pulp mills and wood-based biorefineries, in: A. Pandey, R. Höfer, M. Taherzadeh, K.M. Nampoothiri, and C. Larroche (Eds.), *Industrial Biorefineries & White Biotechnology*, Elsevier, Amsterdam, The Netherlands, 2015, pp. 91–126.

Anon., *Walden's Paper Handbook*, 3rd edition, Walden-Mott Corporation, Ramsey, NJ, USA, 1995, pp. 1-1-1-24.

Anon., https://en.wikipedia.org (read in 2015 and 2016).

Asikainen, A., Raw material resources, in: R. Alén (Ed.), *Biorefining of Forest Resources*, Paper Engineers' Association, Helsinki, Finland, 2011, pp. 115–130.

Aspinall, G.O. (Ed.), *The Polysaccharides, Volume 1*, Academic Press, New York, NY, USA, 1982.

Atchison, J.E. and McGovern, J.N., History of paper and the importance of non-wood plant fibers, in: M.J. Kocurek and C.F.B. Stevens (Eds.), *Pulp and Paper Manufacture, Volume 1. Properties of Fibrous Raw Materials and their Preparation for Pulping*, 3rd edition, The Joint Textbook Committee of the Paper Industry (TAPPI & CPPA), 1983, pp. 154–156.

Barton, D., Nakanishi, K., and Meth-Cohn, O. (Eds.), *Comprehensive Natural Products Chemistry, Volumes 1–9*, Elsevier, Oxford, UK, 1999.

BeMiller, J.N., Carbohydrates, in: *Kirk-Othmer — Encyclopedia of Chemical Technology, Volume 4*, 4th edition, John Wiley & Sons, New York, NY, USA, 1992, pp. 911–948.

Binkley, R.W., *Modern Carbohydrate Chemistry*, Marcel Dekker, New York, NY, USA, 1988.

Birch, G.G. (Ed.), *Analysis of Food Carbohydrate*, Elsevier Applied Science Publishers, London, UK, 1985.

Bridgwater, A.V. (Ed.), *Fast Pyrolysis of Biomass: A Handbook, Volume 2*, CPL Press, Newbury, UK, 2002.

Brown, R.C., *Biorenewable Resources — Engineering New Products from Agriculture*, A Blackwell Publishing Company, Ames, IA, USA, 2003.

Budavari, S. (Ed.), *The Merck Index*, 11th edition, Merck & Co., Rahway, NJ, USA, 1989.

Buxton, S.R. and Roberts, S.M., *Guide to Organic Stereochemistry from Methane to Macromolecules*, Addison Wesley Longman Limited, Harlow, UK, 1996.

Carey, F.A. and Sundberg, R.J., *Advanced Organic Chemistry, Parts A and B*, Plenum Press, New York, NY, USA, 1977.

Chaplin, M.F. and Kennedy, J.F. (Eds.), *Carbohydrate Chemistry — A Practical Approach*, IRL Press, Oxford, UK, 1986.

Cheremisinoff, N.P. (Ed.), *Handbook of Polymer Science and Technology, Parts 1 → 4*, Marcel Dekker, New York, NY, USA, 1989.

Collins, P.M. (Ed.), *Carbohydrates*, Chapman and Hall, New York, NY, USA, 1987.

Collins, P.M. (Ed.), *Dictionary of Carbohydrates with CD-ROM*, 2nd edition, Chapman & Hall/CRC, Boca Raton, FL, USA, 2006.

Collins, P. and Ferrier, R., *Monosaccharides - Their Chemistry and Their Roles in Natural Products*, John Wiley & Sons, Chichester, UK, 1998.

Cooke, E.I. and Cooke, R.W.I. (Eds.), Garner's Chemical Synonyms and Trade Names, 8th edition, The Technical Press, Oxford, UK, 1978.

Daintith, J. (Ed.), *A Dictionary of Chemistry*, 3rd edition, Oxford, UK, 1996.

Davidson, E.A., *Carbohydrate Chemistry*, Holt, Rinehart and Winston, New York, NY, USA, 1967.

Davis, B.G. and Fairbanks, A.J., *Carbohydrate Chemistry*, Oxford University Press, Oxford, UK, 2003.

Delgado, J.N. and Remers, W.A. (Eds.), *Wilson and Gisvold's Textbook of Organic Medicinal and Pharmaceutical Chemistry*, J.B. Lippincott Company, Philadelphia, PA, USA, 1991.

Descotes, G. (Ed.), *Carbohydrates as Organic Raw Materials II*, VCH, Weinheim, Germany, 1993.

El Khadem, H.S., *Carbohydrate Chemistry — Monosaccharides and Their Oligomers*, Academic Press, New York, NY, USA, 1988.

Fan, L.T., Gharpuray, M.M., and Lee, Y.-H., *Cellulose Hydrolysis*, Springer-Verlag, Heidelberg, Germany, 1987.

Fengel, D. and Wegener, G., *Wood — Chemistry, Ultrastructure, Reactions*, Walter de Gruyter, Berlin, Germany, 1989.

Fengl, R., Cellulose esters, Inorganic esters, in: *Kirk-Othmer — Encyclopedia of Chemical Technology, Volume 5*, 4th edition, John Wiley & Sons, New York, NY, USA, 1993, pp. 529–540.

Fessenden, R.J. and Fessenden, J.S., *Organic Chemistry*, 3rd edition, Wadsworth, Belmont, CA, USA, 1986.

French, A.D., Bertoniere, N.R., Battista, O.A., Cuculo, J.A., and Gray, D.G., Cellulose, in: *Kirk-Othmer — Encyclopedia of Chemical Technology, Volume 5*, 4th edition, John Wiley & Sons, New York, NY, USA, 1993, pp. 476–496.

Fresenius, P., *Organic Chemical Nomenclature — Introduction to the Basic Principles*, Ellis Horwood Limited, Chichester, UK, 1989.

Gedon, S. and Fengl, R., Cellulose esters, Organic esters, in: *Kirk-Othmer — Encyclopedia of Chemical Technology, Volume 5*, 4th edition, John Wiley & Sons, New York, NY, USA, 1993, pp. 497–529.

Goldstein, I.S. (Ed.), *Organic Chemicals from Biomass*, CRC Press, Boca Raton, FL, USA, 1983.

Grayson, M. (Ed.), *Kirk-Othmer Concise Encyclopedia of Chemical Technology*, John Wiley & Sons, New York, NY, USA, 1985.

Guthrie, R.D. (Ed.), *Guthrie & Honeyman's Introduction to Carbohydrate Chemistry*, 4th edition, Clarendon Press, Oxford, UK, 1974.

Györgydeák, Z. and Pelyvás, I.F., *Monosaccharide Sugars — Chemical Synthesis by Chain Elongation, Degradation, and Epimerization*, Academic Press, San Diego, CA, USA, 1998.

Heinze, T.J. and Glasser, W.G. (Eds.), *Cellulose Derivatives: Modification, Characterization, and Nanostructures*, ACS Symposium Series 688, American Chemical Society, Washington, DC, USA, 1998.

Herrick, F.W. and Hergert, H.L., Utilization of chemicals from wood: Retrospect and prospect, in: F.A. Loewus and V.C. Runecles (Eds.), *The Structure, Biosynthesis, and Degradation of Wood, Recent Advances in Phytochemistry, Volume 11*, Plenium Press, New York, NY, USA, 1977, pp. 443–515.

Hesse, M., *Alkaloids — Nature's Curse or Blessing?*, Wiley-VCH, Weinheim, Germany, 2002.

Hon, D.N.-S., Functional natural polymers: A new dimensional creativity in lignocellulosic chemistry, in: D.N.-S. Hon (Ed.), *Chemical Modification of Lignocellulosic Materials*, Mercel Dekker, New York, NY, USA, pp. 1–10.

Hon, D.N.-S. and Shiraishi, N. (Eds.), *Wood and Cellulosic Chemistry*, 2nd edition, Marcel Dekker, New York, NY, USA, 2001.

Ihde, A.J., *The Development of Modern Chemistry*, Dover Publications, New York, NY, USA, 1984.

Kamide, K., *Cellulose and Cellulose Derivatives*, Elsevier, London, UK, 2005.

Kamm, B., Gruber, P.C., and Kamm, M. (Eds.), *Biorefineries — Industrial Processes and Products. Status Quo and Future Directions, Volumes 1 & 2*, Wiley-VCH Verlag, Weinheim, Germany, 2006.

Kennedy, J.F. (Ed.), *Carbohydrate Chemistry*, Clarendon Press, Oxford, UK, 1990.

Kennedy, J.F. and White, C.A., *Bioactive Carbohydrates: In Chemistry, Biochemistry and Biology*, Ellis Horwood Limited, Chichester, UK, 1983.

Kirk, R.E. and Othmer, D.F. (Eds.), *Encyclopedia of Chemical Technology, Parts 1 → 15*, 1st edition, Interscience Publishers, New York, NY, USA, 1947.

Klemm, D., Philipp, B., Heinze, T., Heinze, U., and Wagenknecht, W., *Comprehensive Cellulose Chemistry — Volume 1, Fundamentals and Analytical Methods*, Wiley-VCH, Weinheim, Germany, 1998.

Klemm, D., Philipp, B., Heinze, T., Heinze, U., and Wagenknecht, W., *Comprehensive Cellulose Chemistry — Volume 2, Functionalization of Cellulose*, Wiley-VCH, Weinheim, Germany, 1998.

Konttinen, J., Reinikainen, M., Oasmaa, A., and Solantausta, Y., Thermo-chemical conversion of forest biomass, in: R. Alén (Ed.), *Biorefining of Forest Resources*, Paper Engineers' Association, Helsinki, Finland, 2011, pp. 262–304.

Lampinen, J. and Mononen, K., Utilisation of solid wood-based materials, in: R. Alén (Ed.), *Biorefining of Forest Resources*, Paper Engineers' Association, Helsinki, Finland, 2011, pp. 151–175.

Lehman, J., *Chemie der Kohlenhydrate*, Georg Thieme Verlag, Stuttgart, Germany, 1976.

Lichtenthaler, F.W. (Ed.), *Carbohydrates as Organic Raw Materials*, VCH, Weinheim, Germany, 1991.

Lindberg, N.J., History of papermaking, in: H. Paulapuro (Ed.), *Papermaking Part 1, Stock Preparation and Wet End*, Fapet Oy, Helsinki, Finland, 2000, pp. 56–71.

Majewicz, T.G. and Podlas, T.J., Cellulos ethers, in: *Kirk-Othmer — Encyclopedia of Chemical Technology, Volume 5*, 4th edition, John Wiley & Sons, New York, NY, USA, 1993, pp. 541–563.

Malm, C.J., Mench, J.W., Kendall, D.L., and Hiatt, G.D., Properties, *J. Ind. Eng. Chem.*, 43(1951), 688–691.

Mann, J., Davidson, R.S., Hobbs, J.B., Banthorpe, D.V., and Harborne, J.B., *Natural Products — Their Chemistry and Biological Significance*, Longman, Essex, London, UK, Englanti, 1994.

Mark, H.F., Othmer, D.F., Overberger, C.G., and Seaborg, G.T. (Eds.), *Encyclopedia of Chemical Technology, Parts 1 → 24*, 3rd edition, Interscience Publishers, New York, NY, USA, 1978.

McNaught, A.D., Nomenclature of carbohydrates, *Carbohydr. Res.* 297(1) (1997), 1–92.

Morris, D.G., *Stereochemistry*, The Royal Society of Chemistry, Cambridge, UK, 2001.

Morrison, R.T. and Boyd, R.N., *Organic Chemistry*, 3rd edition, Allyn and Bacon, Boston, MA, USA, 1975.

Natta, G. and Farina, M., *Stereochemistry*, Longman, London, UK, 1972.

Nevell, T.P. and Zeronian, S.H. (Eds.), *Cellulose Chemistry and its Applications*, Ellis Horwood, Chichester, UK, 1985.

Orzechowska, A., Bicentennial of the Fourdrinier: A rough start, *Pulp Pap. Can.* 108(9) (2007), 3–6.

Osborn, H. and Khan, T., *Oligosaccharides — Their Synthesis and Biological Roles*, Oxford University Press, Oxford, UK, 2000.

Parker, S.P. (Ed.), *McGraw-Hill Encyclopedia of Chemistry*, McGraw-Hill, New York, NY, USA, 1982.

Pigman, W. and Horton, D., *The Carbohydrates — Chemistry and Biochemistry, Volume IA*, 2nd edition, Academic Press, New York, NY, USA, 1972.

Pigman, W., Horton, D., and Herp, A., *The Carbohydrates — Chemistry and Biochemistry, Volume IIA*, 2nd edition, Academic Press, New York, NY, USA, 1970.

Rao, V.S.R., Qasba, P.K., Balaji, P.V., and Chandrasekaran, R., *Conformation of Carbohydrates*, Harwood Academic Publishers, Amsterdam, The Netherlands, 1998.

Roberts, J.D., Stewart, R., and Caserio, M.C., *Organic Chemistry*, W.A. Benjamin, Menlo Park, CA, USA, 1971.

Robinson, M.J.T., *Organic Stereochemistry*, Oxford University Press, Oxford, UK, 2002.

Robyt, J.F., *Essentials of Carbohydrate Chemistry*, Springer-Verlag, Heidelberg, Germany, 1998.

Rowell. R.M. (Ed.), *Handbook of Wood Chemistry and Wood Composites*, Taylor & Francis, Boca Raton, FL, USA, 2005.

Rowell, R.M. and Young, R.A., *Modified Cellulosics*, Academic Press, New York, NY, USA, 1978.

Sax, N.I. and Lewis, R.J., Sr., (Eds.), *Hawley's Condensed Chemical Dictionary*, 11th edition, Van Nostrand Reinhold Company, New York, NY, USA, 1987.

Scott, W.E., Abbott, J.C., and Trosset, S., *Properties of Paper: An Introduction*, TAPPI Press, Atlanta, GA, USA, 1995, s. 2-4.

Simmonds, R.J., *Chemistry of Biomolecules — An Introduction*, The Royal Society of Chemistry, Cambridge, UK, 1992.

Sinsky, A.J., Organic chemicals from biomass: An overview, in: D.L. Wise (Ed.), *Chemicals from Biomass*, The Benjamin/Cummins Publishing Company, London, UK, 1983, pp. 1–67.

Sjöström, E., *Wood Chemistry — Fundamentals and Applications*, 2nd edition, Academic Press, San Diego, CA, New York, USA, 1993.

Sjöström, E. & Alén, R. (Eds.), *Analytical Methods in Wood Chemistry, Pulping, and Papermaking*, Springer-Verlag, Heidelberg, Germany, 1999.

Sundholm, J., History of mechanical pulping, in: J. Sundholm (Ed.), *Mechanical Pulping*, Fapet Oy, Helsinki, Finland, 1999, pp. 22–33.

Taiz, L. and Zeiger, E., *Plant Physiology*, The Benjamin/Cummins Publishing Company, Redwood City, CA, USA, 1991.

Thompson, N.S., Hemicellulose, in: *Kirk-Othmer — Encyclopedia of Chemical Technology, Volume 13*, 4th edition, John Wiley & Sons, New York, NY, USA, 1993, pp. 54–72.

Viikari, L. and Alén, R., Biochemical and chemical conversion of forest biomass, in: R. Alén (Ed.), *Biorefining of Forest Resources*, Paper Engineers' Association, Helsinki, Finland, 2011, pp. 225–261.

Weast, R.C. (Ed.), *CRC Handbook of Chemistry and Physics*, 58th edition, CRC Press, West Palm Beach, FL, USA, 1978.

Weissermel, K. and Arpe, H.-J., *Industrial Organic Chemistry*, 3rd edition, VCH, Weinheim, Germany, 1997.

Whistler, R.L., BeMiller, J.N., and Paschall, E.F. (Eds.), *Starch — Chemistry and Technology*, 2nd edition, Academic Press, New York, NY, USA, 1984.

Whistler, R.L. and Chen, C.-C., Hemicelluloses, in: M. Lewin and I.S. Goldstein (Eds.), *Wood Structure and Composition*, Marcel Dekker, New York, NY, USA, 1991, pp. 321–407.

Wise, D.L. (Ed.), *Organic Chemicals from Biomass*, The Benjamin/Cummings Publishing Company, London, UK, 1983.

Yalpani, M. (Ed.), *Industrial Polysaccharides — Genetic Engineering, Structure/Property Relations and Applications — Progress in Biotechnology, Volume 3*, Elsevier, New York, NY, USA, 1987.

Young, R.A. and Rowell, R.M. (Eds.), *Cellulose — Structure, Modification and Hydrolysis*, John Wiley & Sons, New York, NY, USA, 1986.

Zoebelein, H. (Ed.), *Dictionary of Renewable Resources*, VCH, Weinheim, Germany, 1997.

Indexes

Index of General Subjects

D

E

F

H

Index of Selected Compounds

"F" refers to "figure" and "T" to "table"

A

B

C

D

E

F

H

P

S

T

U

X

Z

Index of Names of Persons

Y

Z

Å

Printed in the United States
By Bookmasters